INTERNATIONAL SERIES OF MONOGRAPHS IN

NATURAL PHILOSOPHY

GENERAL EDITOR: D. TER HAAR

VOLUME 4

QUANTUM FIELD THEORETICAL METHODS IN STATISTICAL PHYSICS

OTHER TITLES IN THE SERIES IN NATURAL PHILOSOPHY

Vol. 1 DAVYDOV—*Quantum Mechanics*

Vol. 2 FOKKER—*Time and Space, Weight and Inertia*

Vol. 3 KAPLAN—*Interstellar Gas Dynamics*

Vol. 5 OKUN'—*Weak Interaction of Elementary Particles*

Vol. 6 SHKLOVSKII—*Physics of the Solar Corona*

A. A. ABRIKOSOV, L. P. GOR'KOV
AND I. YE. DZYALOSHINSKII

Quantum Field Theoretical Methods in Statistical Physics

SECOND EDITION

Translated from the Russian by

D. E. BROWN

English translation edited by

D. TER HAAR

PERGAMON PRESS

OXFORD · LONDON · EDINBURGH · NEW YORK

PARIS · FRANKFURT

Pergamon Press Ltd., Headington Hill Hall, Oxford
4 & 5 Fitzroy Square, London W. 1
Pergamon Press (Scotland) Ltd., 2 & 3 Teviot Place, Edinburgh 1
Pergamon Press Inc., 122 East 55th Street, New York 10022
Pergamon Press GmbH, Kaiserstrasse 75, Frankfurt-am -Main

Second edition 1965

Library of Congress Catalog Card No. 63-21130

This second, revised and enlarged edition is based on the translation of the original Russian volume entitled Методы квантовой теории поля в статистической физике (Metody kvantovoi teoriyi polya v statisticheskoi fizike), published in 1962 by Fizmatgiz, Moscow, and on additional material supplied in 1964 by the authors.

CONTENTS

FOREWORD

SUBSTANTIAL success has been achieved in statistical physics in the last few years as a result of the wide use of methods borrowed from quantum field theory. The fruitfulness of these methods is connected with a new formulation of perturbation theory and above all with the extensive use of so-called Feynman diagrams. The main advantage of the diagram technique lies in ease of visualisation based on the concepts of the single-particle problem, it enables the structure of any approximation to be established and the necessary expressions to be written down with the aid of correspondence rules. The new methods are the most powerful and effective ones in quantum statistics at the present time, and have made it possible to solve a large number of problems which were insoluble in the old formulation of the theory; moreover, many new general relationships have been obtained.

Numerous articles in many journals have recently been devoted to an exposition of field theory methods in quantum statistics and to their application to concrete problems. At the same time, workers in statistical physics are not always acquainted with these methods. It has therefore seemed to us that the time is ripe for a logical and reasonably complete treatment of the subject to be accessible to a wide readership.

A few words about the subject-matter of this volume. We have tried primarily to demonstrate the practical nature of the new methods. We have therefore considered various concrete problems of quantum statistics in addition to providing a detailed treatment of the mathematical side. Naturally, the questions touched upon do not exhaust the new discoveries made in this field in recent years. They have been chosen with a view to their general interest, and the possibility of using them to illustrate the general method.

We have confined ourselves to just one possible variant of the statement of quantum statistics in the language of field theory (for instance, we have not touched upon the so-called three-dimensional perturbation theory). The Green function method, upon which the present book is based, is the simplest and most convenient one from our point of view.

The reader is assumed to be familiar with the fundamentals of statistical physics and quantum mechanics. We describe the method of second quantisation, and give all the information needed for deriving the field theoretical technique. This derivation is the subject of the first chapter, which briefly describes some modern ideas regarding the nature of energy spectra and includes simple examples.

A system of units in which $\hbar = 1$ is used. Temperatures are expressed in energy units ($k = 1$).

The authors express their gratitude to Academician L. D. Landau and to L. P. Pitayevskii for valuable discussions of the problems dealt with in this book.

PREFACE TO THE SECOND EDITION

We are very pleased that our book, in translation, will be accessible to English readers. We are especially gratified that this publication was undertaken by Pergamon Press who publish all books by Landau and Lifshitz from which we learned our physics. Finally we much regret that we had to disappoint Dr. ter Haar and were not able — because of lack of time — to write a complete chapter on kinetics. It is, however, possible that this decision, although not a pleasant one to take, is all the same the correct one.

The study of kinetic problems by means of quantum field theoretical methods is usually much more complicated than the theory of equilibrium properties of substances and the proposed chapter would in its style appreciably differ from the remainder of the book. We restricted ourselves to adding a short chapter on the derivation of the kinetic equation for a Fermi liquid. From this exposition the reader will already see how complex and cumbersome these problems are.

We wish to express our thanks to Dr. ter Haar for his patience and for his efforts in the completion of this publication.

Moscow
2 April 1965

A. Abrikosov
L. Gor'kov
I. Dzyaloshinskii

TRANSLATION EDITOR'S NOTE

THE PRESENT monograph by three prominent Russian theoreticians from Landau's Institute is by now well known to Western readers, and the question may be asked why a second English version is being brought out. When the authors kindly sent me a copy of their book on publication in 1962, I immediately got in touch with Dr. Abrikosov, asking him whether he would be agreeable to an English edition being brought out, and whether he knew of any other Western publisher bringing out such a translation. He replied that he would be pleased to see an English edition published, and that he preferred it to be published by the publishers of the Landau and Lifshitz series of textbooks, that is, by Pergamon Press. He also told me that as far as he knew no other Western publisher was preparing a translation. I then got in touch with Pergamon Press who obtained approval from the Russians and started to prepare a translation. When this translation was about finished, we were informed by Dr. Abrikosov that an American publisher had translated his book. In order to improve the presentation Dr. Abrikosov sent both to Pergamon and to the American publisher some material (about 25 pages in the present translation, in sections 16.2, 17, 19.1, 21, and 22.3–4) and at the same time told me that the authors were preparing an extra chapter on transport properties of a Fermi liquid. Rather than rush the translation without the extra chapter Pergamon decided to wait for the extra chapter, but at the end of 1964 Dr. Abrikosov informed me that he now did not think that the chapter as he had envisaged it would be written. He sent me, however, a shortened form of the extra chapter (Chapter VIII) as well as some new material for sections 4, 5, 17, 39 and a complete new subsection (19.6). Compared with the original Russian edition, about one-sixth is new material, so that this translation can truly be called a second edition.

I should like to express my thanks to Dr. Abrikosov for his assistance in the preparation of this second edition.

D. T. H.

Oxford

March 1965

CHAPTER I

PROPERTIES OF MANY-PARTICLE SYSTEMS AT LOW TEMPERATURES

§ 1. ELEMENTARY EXCITATIONS. ENERGY SPECTRUM AND PROPERTIES OF LIQUID He^4 AT LOW TEMPERATURES

1. *Introduction. Quasi-particles*

Statistical physics studies the behaviour of systems composed of very large numbers of particles. The macroscopic properties of liquids, gases and solids are due to the microscopic interactions between the particles composing the system. A full solution of the problem, involving a determination of the behaviour of each individual particle, is obviously unthinkable. Besides, the over-all macroscopic characteristics define only certain average properties of the system as a whole.

To fix the ideas, let us consider the thermodynamic properties. The macroscopic state of a system is defined by specifying three independent thermodynamic variables, say the pressure P, the temperature T, and the average number N of particles in the system. From the point of view of quantum mechanics a closed system of N particles is characterised by its energy levels E_n. We split off from this system a subsystem such that it can again be regarded to be macroscopic. Since the number of particles in the subsystem is, as before, extremely large whilst the interaction forces between the particles act at distances of atomic order, the subsystem, neglecting boundary effects, can in turn be regarded as closed and characterised by the energy levels for the given number of particles in the subsystem. Since the subsystem in fact interacts with the remaining particles of the closed system, it does not have a strictly fixed energy and number of particles, and can find itself at any level with a finite probability.

It is well known from statistical physics (see [1]), that the microscopic deduction of the thermodynamical formulae is based on the so-called Gibbs distribution, which gives the following probability of the subsystem finding itself in a state with energy E_{nN} and number of particles N:

$$w_{nN} = Z^{-1} \exp[-(E_{nN} - \mu N)/T]. \qquad (1.1)$$

Here T denotes the absolute temperature, μ the chemical potential, and Z a normalising factor which is defined by the condition

$$\sum_{nN} w_{nN} = 1. \tag{1.2}$$

We have from (1.1):

$$Z = \sum_{nN} \exp[-(E_{nN} - \mu N)/T]. \tag{1.3}$$

The quantity Z is called the grand partition function. If the energy levels E_{nN} are known, the partition function can be calculated. This in turn determines the thermodynamic functions, since Z is connected with the thermodynamic potential Ω (the potential, if the independent variables are V, T, and μ) by the relationship

$$\Omega = -T \ln Z. \tag{1.4}$$

Obviously, we can very easily compute the thermodynamic functions of ideal gases from these formulae, since their energy is the sum of the energies of the individual particles. A determination of the energy levels is impossible in the general case of a system with a large number of interacting particles. It has therefore only been possible so far in quantum statistics to take interactions between the particles into account when these are sufficiently weak. In practice, one can only obtain the first one or two approximations when one uses perturbation theory to evaluate the thermodynamic functions. For the majority of physical problems, in which the interaction is far from small, an approach based on the direct use of formulae (1.1)—(1.4) is out of the question.

The case of very low temperatures is a rather special one. As $T \to 0$ the energy levels of importance in the partition function are not far above the ground state energy (weakly excited states). The nature of the energy spectrum of a system in this energy region can be established in fair detail on the basis of extremely general considerations, which hold independently of the size and particular features of the interactions between the particles.

To make these last remarks clearer, let us take as an example the vibrational excitations of a crystal lattice. Provided the oscillations are small, the lattice can be regarded as a set of coupled harmonic oscillators. Introducing normal coordinates, we get a set of $3N$ (N is the number of atoms) linear oscillators with eigenfrequencies ω_i. In accordance with quantum mechanics, the energy spectrum of such a system is defined by the formula $E = \sum_{i=1}^{3N} (n_i + 1/2)\, \omega_i$, where the n_i are positive integers or zero. Different sets of n_i yield different energy levels of the system.

The lattice vibrations can be described as a superposition of monochromatic plane waves propagated in the crystal. Each wave is characterised by a wave vector, a frequency, and a number s, defining the type of wave. The possibility of different types of waves being propagated is equivalent to the frequency ω not being a single-valued function of the wave vector $\boldsymbol{k}$; instead, it is composed of several branches $\omega_s(\boldsymbol{k})$, the total number of which is equal to $3r$, where r is the number of atoms in an elementary cell of the crystal. In the case of small momenta, three of the branches (the so-called acoustic branches) are characterised by a linear dependence of the frequency on the wave vector: $\omega_s(\boldsymbol{k}) = u_s(\theta, \varphi)\,|\boldsymbol{k}|$. For the other branches, the curve $\omega_s(\boldsymbol{k})$ starts at a finite value for $\boldsymbol{k} = 0$ and depends weakly on $\boldsymbol{k}$ for small wave vectors (†).

Whilst a knowledge of the frequency spectrum, the energy levels, and the matrix elements of the displacements in the lattice (the oscillator coordinates) always enables us in principle to work out completely both the thermodynamic and kinetic characteristics of the oscillating lattice, it proves very convenient in practice to use a picture obtained with the aid of the quantum mechanical correspondence principle rather that the coupled oscillator picture. According to the correspondence principle, every plane wave can be associated with a set of moving "particles". The wave vector $\boldsymbol{k}$ will define the momentum of these particles (‡), and the frequency $\omega_s(\boldsymbol{k})$ their energy (††). The excited state of the lattice can be pictured as a collection of such "particles" (called phonons), moving freely in the volume of the solid. This is in complete correspondence with an expression for the energy levels of the system which is similar to the energy of an ideal gas.

The number n_i can be interpreted as the number of phonons in the state i ($i = (\boldsymbol{k}, s)$). The n_i can take on any values. It follows that the phonons are subject to Bose statistics even in the case when atoms composing the system have half-odd-integer spin.

At very low temperatures the phonons with small energies will play the most important role. It follows from what we have said earlier about the branches of the frequency spectrum that the smallest energies are those of the acoustic phonons with small momenta. The function $\omega(\boldsymbol{k})$ is a linear one in this case, and we can deduce from this fact alone a whole series of qualitative conclusions, such as the T^3 law for the specific heat of the lattice.

The so-called Debye isotropic model is often used instead of the real lattice for quantitative calculations. Instead of the three acoustic branches, the low frequency part of the spectrum is assumed to be the same as in an isotropic solid, i.e. composed of longitudinal phonons with energies $\omega_l(\boldsymbol{k}) = u_l k$ and transverse phonons with two possible polarisations and

(†) For more detailed information on the lattice vibration spectra see, e.g. R. E. Peierls [2].

(‡) In reality $\boldsymbol{k}$ is a "quasi-momentum" rather than a momentum (see [2]), through the distinction is of no importance here.

(††) Remember that $\hbar = 1$ in the system of units used here. This means that energy has dimensions $\sec^{-1}$, and momentum cm^{-1}. To pass to ordinary units, all the energies and momenta must be multiplied by $\hbar$.

the same dependence of the energy on the momentum $\omega_t(\boldsymbol{k}) = u_t k$. It is also assumed in this model that the phonon momenta do not exceed some limiting value k_D, determined by the normalisation to the correct number of degrees of freedom. Obviously $k_D \sim 1/a$, where a is the interatomic distance. This model leads to the familiar Debye interpolation formula for the specific heat of solids. We shall use this model later for investigating the electron-phonon interactions in a metal.

If the anharmonic terms are taken into account in the potential energy of the vibrating lattice, the expression given above for the energy ceases to be accurate. There are then non-vanishing transition-probabilities between states with different values of the n_i. This can be interpreted in phonon language as different interaction processes between phonons, leading to mutual scattering and to the creation of phonons. In other words, in a rigorous treatment the phonons can only be approximately regarded as independent particles.

The role of the anharmonic terms will increase with the amplitude of the vibrations, i.e. with an increase in temperature. In the phonon picture, the number of phonons increases with increase of temperature, which means that the role of the interactions between phonons increases in importance. The very concept of phonons as independent particles is thus only applicable in the fairly low temperature region (well below the melting point).

We now turn to the general case. By analogy with the example discussed, the fundamental picture of the energy spectrum for weakly excited states of a system is provided by the assumption that the energy levels can be constructed, to a first approximation, in accordance with the same principle as the energy levels of ideal gases.

In other words, it is assumed that any energy level can be written as the sum of the energies of a certain number of "quasi-particles", or elementary excitations, moving in the volume of the solid and possessing momentum p and energy $\varepsilon(p)$. (The dispersion law of the perturbations $\varepsilon(p)$ is in general not the same as the expression for the energy of the free particles, $\varepsilon_0(p) = p^2/2m$.) It must be emphasised right away that the elementary excitations arise as a result of collective interactions between the particles of the system, so that they relate to the system as a whole, and not to individual particles. In particular, their number is by no means the same as the total number of particles in the system.

All energy spectra can be divided into two major groups — Bose type and Fermi type spectra. In the first case the excitations possess integral angular momentum eigenvalues (spin) and obey Bose statistics; in the second, they have half-odd-integral spin and obey Fermi statistics. According to quantum mechanics, the angular momentum of any system can only change by an integer. It follows that Bose excitations can appear and vanish singly, but Fermi excitations only in pairs.

As already mentioned in our example involving lattice vibrations, the statistics of the elementary excitations are not necessarily the same as the statistics of the particles composing the system. The only obvious fact is that a Bose system cannot have excitations with half-odd-integral spin.

Elementary excitations do not correspond to strictly stationary states of a system, but represent a superposition of a large number of strictly stationary states with a narrow energy spread (packets). In view of this, there is a finite probability of a transition from one such state to another, which leads to the diffusing of a packet, i.e. to the damping of the excitation. Hence a description of a system by means of elementary excitations is valid only in as far as the energy width of the packet, which defines its damping, is small compared with its energy.

The diffusing of a packet and the related damping of the elementary excitations can be regarded as the result of interaction of "quasi-particles" with one another. The laws of conservation of energy and momentum are fulfilled here. Obviously, all these transitions can be split into processes of "decomposition" of one excitation into several others and processes of "scattering" of excitations by one another.

As we shall see below, decomposition of excitations can occur only for fairly large energies. Scattering processes become important only when the number of excitations is fairly large. Thus at low temperatures, where excitations with small energies are important and the number of them is small, neither type of process leading to the damping of excitations will be important. The weakness of the interactions between excitations at low temperatures enables them to be regarded as an ideal gas of "quasi-particles".

It is now possible to assume, on the basis of experimental data and direct theoretical calculations, that the ideas we have just outlined regarding the structure of the spectra are solidly established facts. The energy spectra of different physical materials (e.g. the liquid isotopes He^3 and He^4, metals, dielectrics, etc.) are of course quite different. Thus, liquid He^4 has a Bose type spectrum, whereas the spectra of liquid He^3 and the electron spectra of metals (†) are of the Fermi type.

2. *Spectrum of a Bose liquid* (‡)

An example of a system possessing a Bose type spectrum is provided by a so-called Bose liquid, i.e. a liquid consisting of atoms with integral spins. Only one such liquid is found in nature, namely liquid helium

(†) To avoid confusion we shall stipulate that the isotropic model of a metal (which is naturally a long way from the true picture) is considered throughout what follows. The electron spectra are sharply anisotropic in real metals, so that many of the results described in this book are merely qualitative as applied to metals.

(‡) The ideas outlined here regarding the spectrum of a Bose liquid were first proposed by L. D. Landau [3, 4].

(more precisely the isotope He^4), which does not solidify at the absolute zero of temperature under its own vapour pressure. Since He^4 atoms possess zero spin, we can essentially confine ourselves to this case.

The dependence of the excitation energy of a Bose liquid on the momentum, at limitingly small values of the latter, is determined from extremely general considerations. The domain of small momenta p corresponds to long-wave oscillations of the liquid. An oscillation of this type is in fact ordinary sound. We at once conclude from this that the elementary excitations at small p are identical to sound — phonons, for which the connection between energy and momentum is well known. Indeed, noticing that the frequency of sound ω is connected with the wave vector by the relationship $\omega = uk$, where u is the sound velocity, we immediately obtain the relationship of interest:

$$\varepsilon = up. \tag{1.5}$$

Thus, at small momenta the excitation energy in a Bose liquid is linearly dependent on its momentum, the coefficient of proportionality being the sound velocity.

As the momentum increases the function $\varepsilon(p)$ ceases to be linear and the further course of the $\varepsilon(p)$ curve cannot be determined by so simple a method. The following discussion is of interest in this connection, since it enables a number of deductions to be made regarding the function $\varepsilon(p)$ for arbitrary momenta (†).

The energy of a liquid is a functional of its density $\varrho(\boldsymbol{r})$ and of the hydrodynamic velocity $\boldsymbol{v}(\boldsymbol{r})$:

$$E(\varrho, \boldsymbol{v}) = \frac{1}{2}\int \varrho \boldsymbol{v}^2 \mathrm{d}^3\boldsymbol{r} + E^{(1)}(\varrho), \tag{1.6}$$

where $E^{(1)}$ is that part of the energy which is independent of the velocity. We shall consider small vibrations. In that case, $\varrho(\boldsymbol{r}) = \bar{\varrho} + \delta\varrho(\boldsymbol{r})$, where $\bar{\varrho}$ is the equilibrium density, independent of the coordinates, and $\delta\varrho(\boldsymbol{r})$ and $\boldsymbol{v}(\boldsymbol{r})$ are small quantities describing the vibrations. Notice that, by definition,

$$\bar{\varrho} = \frac{1}{V}\int \varrho(\boldsymbol{r})\mathrm{d}^3\boldsymbol{r}, \qquad \int \delta\varrho \, \mathrm{d}^3\boldsymbol{r} = 0.$$

Neglecting second order terms in $\delta\varrho$ and $\boldsymbol{v}$, the function $\varrho(\boldsymbol{r})$ in the first term on the right-hand side of (1.6) can be replaced by its mean value $\bar{\varrho}$. The expression for $E^{(1)}$ may be written to the same accuracy as

$$E^{(1)}(\varrho) = E^{(1)}(\bar{\varrho}) + \int \psi(\boldsymbol{r})\delta\varrho(\boldsymbol{r})\mathrm{d}^3\boldsymbol{r} + \frac{1}{2}\int\int \varphi(\boldsymbol{r}, \boldsymbol{r}')\delta\varrho(\boldsymbol{r})\delta\varrho(\boldsymbol{r}')\mathrm{d}^3\boldsymbol{r}\mathrm{d}^3\boldsymbol{r}'.$$

The functions $\psi(\boldsymbol{r})$ and $\varphi(\boldsymbol{r}, \boldsymbol{r}')$ are defined solely by properties of the liquid when not excited by vibrations, i.e. when homogeneous and

(†) The derivation given here is due to L. P. PITAEVSKII [5].

isotropic; in view of this $\psi(\boldsymbol{r})$ must be constant: $\psi(\boldsymbol{r}) = \text{const} = \psi$, whilst $\varphi(\boldsymbol{r}, \boldsymbol{r}')$ depends only on $|\boldsymbol{r} - \boldsymbol{r}'| : \varphi(\boldsymbol{r}, \boldsymbol{r}') = \varphi(|\boldsymbol{r} - \boldsymbol{r}'|)$. The first order term in the expansion of $E^{(1)}$ is therefore simply proportional to $\int \delta\varrho(\boldsymbol{r}) \mathrm{d}^3\boldsymbol{r} \equiv 0$. Finally,

$$E = E^{(1)}(\bar{\varrho}) + \frac{\bar{\varrho}}{2} \int \boldsymbol{v}^2 \mathrm{d}^3\boldsymbol{r} + \frac{1}{2} \int \int \varphi(|\boldsymbol{r} - \boldsymbol{r}'|) \delta\varrho(\boldsymbol{r}) \delta\varrho(\boldsymbol{r}') \mathrm{d}^3\boldsymbol{r} \mathrm{d}^3\boldsymbol{r}'.$$

The velocity $\boldsymbol{v}$ is connected with the density oscillations by the equation of continuity

$$\dot{\varrho} + \operatorname{div}(\varrho \boldsymbol{v}) = 0,$$

which, up to first order terms in $\delta\varrho$ and $\boldsymbol{v}$, can be written as

$$\dot{\varrho} + \bar{\varrho} \operatorname{div} \boldsymbol{v} = 0. \tag{1.7}$$

We now change to Fourier components:

$$\delta\varrho(\boldsymbol{r}) = \frac{1}{V} \sum_{\boldsymbol{p}} \varrho_{\boldsymbol{p}} e^{i(\boldsymbol{p}\cdot\boldsymbol{r})}, \quad \boldsymbol{v}(\boldsymbol{r}) = \frac{1}{V} \sum_{\boldsymbol{p}} \boldsymbol{v}_{\boldsymbol{p}} e^{i(\boldsymbol{p}\cdot\boldsymbol{r})}, \quad \varphi(\boldsymbol{r}) = \frac{1}{V} \sum_{\boldsymbol{p}} \varphi_{\boldsymbol{p}} e^{i(\boldsymbol{p}\cdot\boldsymbol{r})},$$

and taking into account that small vibrations of a liquid are always longitudinal, i.e. the velocity $\boldsymbol{v}_{\boldsymbol{p}}$ in a wave with wave vector $\boldsymbol{p}$ is always directed along $\boldsymbol{p}$:

$$\boldsymbol{v}_{\boldsymbol{p}} = a_{\boldsymbol{p}} \boldsymbol{p}.$$

We now easily find from (1.7) that

$$\boldsymbol{v}_{\boldsymbol{p}} = i \frac{\dot{\varrho}_{\boldsymbol{p}}}{\bar{\varrho}} \frac{\boldsymbol{p}}{p^2}$$

and

$$E = E^{(1)}(\bar{\varrho}) + \frac{1}{V} \sum_{\boldsymbol{p}} \left(\frac{|\dot{\varrho}_{\boldsymbol{p}}|^2}{2\bar{\varrho} p^2} + \frac{1}{2} \varphi_{\boldsymbol{p}} |\varrho_{\boldsymbol{p}}|^2 \right). \tag{1.8}$$

The first term in (1.8) represents the energy of the undisturbed liquid, the second splits into a sum of terms, each of which is the energy of a harmonic oscillation of frequency $\omega_{\boldsymbol{p}}$, where

$$\omega_{\boldsymbol{p}}^2 = \bar{\varrho} p^2 \varphi_{\boldsymbol{p}}. \tag{1.9}$$

We see thus that any small vibration of a liquid can be split into elementary oscillations, i.e. elementary excitations described by the equations for a harmonic oscillator.

In the quantum case the energy of each such oscillator is expressed by

$$\varepsilon_{\boldsymbol{p}} = \omega_{\boldsymbol{p}} \left(n + \frac{1}{2} \right); \qquad n = 0, 1, 2, \ldots$$

The resulting structure of the spectrum is in complete agreement with the picture outlined above of elementary excitations. The spectrum is the sum of the energies of different numbers of elementary excitations and the dependence of the energy of an elementary excitation $\varepsilon(\boldsymbol{p})$ on its momentum is determined by (1.9) and the obvious relationship

$$\varepsilon(\boldsymbol{p}) = \omega_{\boldsymbol{p}}.$$

To complete our solution, we must express φ_p in terms of the characteristics of the system. We have to note that in the quantum case the ground state energy of the system is not $E^{(1)}(\bar{\varrho})$, as in the classical case, since we have to take into account the so-called zero-point energy of the oscillators, which we know to be equal to $^1/_2\,\omega_p$ for each oscillator. Thus the ground state energy of a Bose liquid is equal to

$$E_0 = E^{(1)}(\bar{\varrho}) + \sum_p \frac{\omega_p}{2},$$

where (compare (1.8))

$$V\frac{\omega_p}{2} = \frac{1}{2\bar{\varrho}p^2}\overline{|\dot{\varrho}_p|^2} + \frac{1}{2}\varphi_p\overline{|\varrho_p|^2} = \varphi_p\overline{|\varrho_p|^2}. \tag{1.10}$$

We immediately obtain (†) from (1.9) and (1.10):

$$\varepsilon(p) = \omega_p = \frac{p^2}{2mS(p)}, \tag{1.11}$$

where $S(\boldsymbol{p}) = \overline{|\varrho_p|^2}/Vm\bar{\varrho}$ is the Fourier component of the so-called density correlation function

$$S(\boldsymbol{r}-\boldsymbol{r}') = \frac{\overline{[n(\boldsymbol{r})-\bar{n}]\,[n(\boldsymbol{r}')-\bar{n}]}}{\bar{n}}. \tag{1.12}$$

Here $n(\boldsymbol{r}) = \varrho(\boldsymbol{r})/m$ is the number of particles per unit volume.

Although it is impossible to evaluate $S(\boldsymbol{p})$, (1.11) enables us to draw a number of extremely important conclusions about the form of $\varepsilon(\boldsymbol{p})$. Alternatively, if we know certain general properties of the spectrum $\varepsilon(\boldsymbol{p})$, we can draw conclusions about the behaviour of $S(\boldsymbol{p})$, which determines the interaction processes between the liquid and various particles (such as neutrons; see below, Chap. III, § 17).

As already mentioned, in the small momenta region the excitation energy is linearly dependent on the momentum: $\varepsilon \approx up$. It follows that $S(\boldsymbol{p})$ is also linearly dependent on the momentum: $S \approx p/2mu$.

In the region of small distances, or what amounts to the same thing, of large momenta, the function $S(\boldsymbol{r})$ has the familiar form (see [1], § 114):

$$S(\boldsymbol{r}) = \delta(\boldsymbol{r}) + \nu(\boldsymbol{r}), \tag{1.13}$$

where $\nu(\boldsymbol{r})$ has no singularities as $\boldsymbol{r} \to 0$. We have then for the Fourier components:

$$S(\boldsymbol{p}) = 1 + \nu(\boldsymbol{p}),$$

$\nu(\boldsymbol{p}) \to 0$ as $p \to \infty$. Hence $S(\boldsymbol{p})$ tends to unity at large momenta and

$$\varepsilon(\boldsymbol{p}) \approx \frac{p^2}{2m},$$

(†) Formula (1.11) was first obtained by R. P. Feynman [6] by another method. His derivation is a good deal more complicated and does not seem to us more general than the method described above.

i.e. the energy of an elementary excitation is the same as the energy of a free atom of the liquid (He^4 atom).

For intermediate momenta, $S(\mathbf{p})$ may either increase monotonically from zero to unity as p increases, or may have a maximum at $p \sim 1/a$, where a is the interatomic distance (this follows from dimensional considerations, since there is only one length parameter, i.e. the interatomic distance, in the liquid problem). In the latter case, the spectrum of the elementary excitations may be of the form shown in Fig. 1. The suggestion that the excitation spectrum of liquid He^4 may have a minimum at $p \sim 1/a$ was first made by L. D. Landau [3, 4].

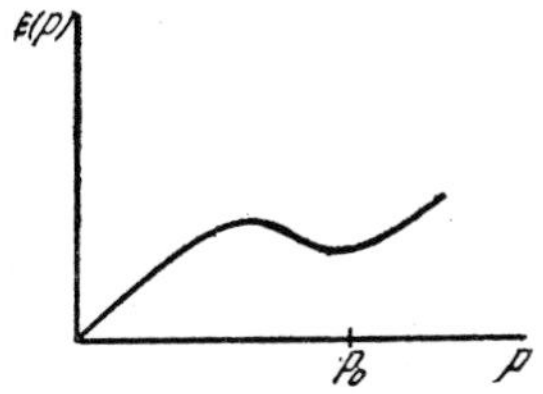

Fig. 1

It should be noted that the above derivation of (1.11) is based on a hydrodynamic approximation, in which the liquid is regarded as a continuous medium. This approximation loses its validity in cases where distances of interatomic order, or momenta of order $1/a$, are important. Hence (1.11), correct for small momenta, must be looked on as an interpolation between the small momenta region and that of very large momenta, for which the particles are, in fact, free and the elementary excitations are the same as the particles, i.e. have energies $p^2/2m$.

The elementary excitation spectrum in liquid He^4 can naturally not be evaluated in full detail. The most accurate $\varepsilon(p)$ curves were obtained quite recently from experiments on neutron scattering in He^4 [7].

A knowledge of the energy spectrum enables us to work out the thermodynamic functions of liquid He^4 (more precisely, the differences between their values at a given temperature and at $T = 0$). The part of the spectrum that plays the leading role depends on the value of T (see Fig. 1).

At the very lowest temperatures the part corresponding to small p, i.e. the phonons, will be the most important ones. At higher temperatures the excitations in the neighbourhood of the minimum of $\varepsilon(p)$ (at $p = p_0$) become the most important. On expanding the energy ε in powers of $p - p_0$, we get(†):

$$\varepsilon(p) = \Delta + \frac{1}{2m^*}(p - p_0)^2. \tag{1.14}$$

The elementary excitations are called "rotons" in this part of the spectrum(‡).

All the thermodynamic quantities will be sums of "phonon" and "roton" parts. To find the thermodynamic potential we only need to

(†) The numerical values of the constants appearing in this formula are, for He^4 [7]:

$$\Delta = 11{\cdot}4 \times 10^{11}\ \text{sec}^{-1},\quad p_0 = 1{\cdot}92 \times 10^8\ \text{cm}^{-1},\quad m^* = 0{\cdot}16\, m_{He^4}.$$

(‡) The term "rotons" is due to Tamm.

substitute (1.5), (1.14) in the formula (see [1], § 53):

$$\Omega = VT \int \ln\left(1 - e^{[\mu - \varepsilon(p)/T]}\right) . \frac{d^3 \boldsymbol{p}}{(2\pi)^3} . \tag{1.15}$$

The following facts should be borne in mind here. Firstly, the number of excitations is not fixed, but is itself determined from the equilibrium condition — that the free energy be a minimum with respect to any variation in the number of particles; this gives:

$$\left(\frac{\partial F}{\partial N}\right)_{V,T} = \mu = 0 \qquad (F = \Omega + \mu N). \tag{1.16}$$

When $\mu = 0$, the potential Ω is the same as the free energy F. Secondly, in view of the fact that the roton energy is always large compared to the temperatures under discussion, the Bose distribution for the rotons may be replaced by a Boltzmann distribution. This is connected with the fact that in the case $T \ll \varepsilon_{\text{rot}}$ we can simply confine ourselves to the first term in the expansion of $\ln(1 - \exp[-\varepsilon_{\text{rot}}/T])$ in powers of the small quantity $\exp[-\varepsilon_{\text{rot}}/T]$ when evaluating the integral in (1.15), whence follows the Boltzmann formula

$$F_{\text{rot}} = -VT \int e^{-\varepsilon_{\text{rot}}/T} \frac{d^3 \boldsymbol{p}}{(2\pi)^3} .$$

We find with the aid of these substitutions:

$$\left.\begin{aligned} F_{\text{ph}} &= -V \frac{\pi^2 T^4}{90 u^3}, \\ F_{\text{rot}} &= -V \frac{2 m^{*1/2} T^{3/2} p_0^2}{(2\pi)^{3/2}} e^{-\Delta/T} . \end{aligned}\right\} \tag{1.17}$$

All the remaining thermodynamic functions are now easily obtained.

3. *Superfluidity*

The most interesting property of a Bose liquid is its "superfluidity", i.e. its ability to flow through capillary tubes without friction. Landau showed [3] that this property follows from the form of the excitation spectrum that he proposed.

Let us take a Bose liquid at absolute zero, flowing along a capillary with velocity $\boldsymbol{v}$. In a coordinate system fixed in the liquid the latter is at rest, whilst the capillary moves with velocity $-\boldsymbol{v}$. The presence of friction between the liquid and the wall means that the liquid starts to be carried along by the walls. This implies that non-zero momentum and energy are present in the liquid. This is only possible when elementary excitations appear in the liquid. If one such excitation occurs, the liquid acquires momentum $\boldsymbol{p}$ and energy $\varepsilon(\boldsymbol{p})$. Let us now change to a coordinate system fixed in the capillary. The energy of the liquid in this system is equal to

$$\varepsilon + (\boldsymbol{p} \cdot \boldsymbol{v}) + \frac{Mv^2}{2}.$$

Hence the appearance of an excitation changes the energy of the liquid by an amount $\varepsilon + (\boldsymbol{p} \cdot \boldsymbol{v})$. This change must be negative if such an excitation is to appear, i.e.

$$\varepsilon + (\boldsymbol{p} \cdot \boldsymbol{v}) < 0.$$

This quantity is a minimum when $\boldsymbol{p}$ and $\boldsymbol{v}$ have opposite directions. It is therefore always necessary that $\varepsilon - pv < 0$, i.e. $v > \varepsilon/p$. Finally, in order that excitations appear in the liquid as a whole, the velocity must satisfy

$$v > \left(\frac{\varepsilon}{p}\right)_{\min}. \tag{1.18}$$

The minimum value of ε/p corresponds to the point of the $\varepsilon(p)$ curve at which

$$\frac{d\varepsilon}{dp} = \frac{\varepsilon}{p}, \tag{1.19}$$

i.e. the point where a straight line from the origin touches the $\varepsilon(p)$ curve.

Superfluid flow can therefore only occur when the liquid velocity is less than the elementary excitation velocity at points satisfying condition (1.19). (Remember that $d\varepsilon/dp$ is the elementary excitation velocity.)

There always exists for any Bose liquid at least one point at which condition (1.19) is fulfilled. This point is the origin $p = 0$. Since the excitations move with the velocity of sound for p close to zero, the superfluidity condition is clearly violated at flow velocities exceeding the velocity of sound u.

There is still one danger point in the excitation spectrum of liquid He^4. It is clear from the curve in Fig. 1 that it lies to the right of the roton minimum. We easily find by using (1.14) that the superfluid flow velocity must be

$$v < \frac{1}{m^*}\left(\sqrt{p_0^2 + 2m^*\Delta} - p_0\right)$$

or, if we make use of the numerical values of the constants (from which it is clear that $p_0^2 \gg 2m^*\Delta$),

$$v < \frac{\Delta}{p_0}.$$

We arrive at the final conclusion that the motion in He^4 is certainly not superfluid at velocities exceeding Δ/p_0.

Excitations occur in a Bose liquid at non-zero temperatures. It may easily be seen that this does not alter the above discussion regarding the possibility of the appearance of new excitations during the flow.

However, it is interesting to consider what effect excitations already present may have on the liquid motion.

To do this, we imagine that a "gas of elementary excitations" moves in the liquid with some macroscopic velocity $\boldsymbol{v}$. The distribution function is obtained in this case from the distribution function of the gas at rest by replacing ε by $\varepsilon - (\boldsymbol{p} \cdot \boldsymbol{v})$. The momentum of the gas per unit volume is obtained from the integral

$$\boldsymbol{P} = \int \boldsymbol{p} n[\varepsilon - (\boldsymbol{p} \cdot \boldsymbol{v})] \frac{d^3\boldsymbol{p}}{(2\pi)^3}. \tag{1.20}$$

We can expand $n[\varepsilon - (\boldsymbol{p} \cdot \boldsymbol{v})]$ at small velocities in powers of $(\boldsymbol{p} \cdot \boldsymbol{v})$. We obtain

$$\boldsymbol{P} = -\int \boldsymbol{p}(\boldsymbol{p} \cdot \boldsymbol{v}) \frac{\partial n}{\partial \varepsilon} \frac{d^3\boldsymbol{p}}{(2\pi)^3} = -\frac{v}{3} \int p^2 \frac{\partial n}{\partial \varepsilon} \frac{d^3\boldsymbol{p}}{(2\pi)^3}. \tag{1.21}$$

It follows from (1.21) that the momentum $\boldsymbol{P}$ of the moving gas of excitations is proportional to the velocity $\boldsymbol{v}$ of the motion. The coefficient of proportionality between $\boldsymbol{P}$ and $\boldsymbol{v}$ is obviously the mass of the moving body. Hence it may be seen that the gas motion relative to the liquid is accompanied by a mass transfer. The individual excitations may, of course, interact with the walls and be scattered from them. In the case of scattering, a momentum exchange takes place between the gas and the walls. This means that the gas motion will be viscous. Since the gas motion is accompanied by mass transfer, as we have just seen, we arrive at the conclusion that viscous flow can occur in a Bose liquid in which excitations are already present, with velocities at which the superfluidity condition (1.18) is certainly not violated. It is essential, however, that the viscous motion be accompanied by transfer of a mass which is by no means the same as the mass of all the liquid; this mass is determined by (1.21) and depends on the number of excitations (in particular, $\boldsymbol{P} = 0$ at $T = 0$).

We now consider the general picture of the motion of a Bose liquid when the velocity is such that the superfluidity condition is not violated.

We start with the absolute zero. If the liquid is in the ground state initially, i.e. there are no excitations in it, none can appear in it later and the motion will be superfluid.

The picture is quite different when $T \neq 0$. Excitations are now present in the liquid, the number of them being determined by the relevant statistical formulae. Although no new excitations can arise, there is nothing to prevent the excitations already present from colliding with the walls and exchanging momentum with them, as already described. Only a part of the liquid mass, as described by (1.21), will participate in this viscous motion. The remaining part of the mass will move as before, without friction either with the walls or with the part of the liquid participating in the viscous motion. A Bose liquid at $T \neq 0$ is therefore so to speak a mixture of two fluids, "superfluid" and "normal", moving without relative friction.

There is no such division in reality, of course; there are simply two motions in the liquid, each corresponding to its own effective mass or density. The "normal" density is the coefficient of proportionality between the momentum per unit volume of the moving excitation gas and its velocity. On substituting in (1.21) first the Bose distribution formula with $\varepsilon = up$, and then the Boltzmann distribution formula with ε from (1.14), we can find the phonon and roton parts of the normal density:

$$\varrho_{n\mathrm{ph}} = \frac{2\pi^2 T^4}{45 u^5}, \qquad \varrho_{n\mathrm{rot}} = \frac{2 m^{*1/2} p_0^4 e^{-\Delta/T}}{3(2\pi)^{3/2} T^{1/2}}. \tag{1.22}$$

The remaining part ϱ_s of the liquid density corresponds to the superfluid motion. Hence

$$\varrho = \varrho_n + \varrho_s. \tag{1.23}$$

Let $\boldsymbol{v}_n$ denote the macroscopic velocity of the gas of excitations, and $\boldsymbol{v}_s$ the superfluid liquid velocity. This latter has an important property. If a Bose liquid is placed in a cylinder and the latter rotated about its axis, the normal part is carried along by the cylinder wall and starts to rotate with it. On the other hand, the superfluid part remains at rest. It is thus impossible to introduce a rotation into the superfluid part. In other words, the motion of this part is always potential. This is expressed mathematically as

$$\operatorname{curl} \boldsymbol{v}_s = 0. \tag{1.24}$$

The motion of the superfluid part plays the role of boundary conditions for the excitations. It should be noticed that in the system of reference fixed in the superfluid part the function $\varepsilon(p)$ has the form described above. We obviously obtain in the system at rest:

$$\varepsilon' = \varepsilon(p) + (\boldsymbol{p} \cdot \boldsymbol{v}_s), \tag{1.25}$$

where $\boldsymbol{p}$ is the momentum in the system of reference fixed in the superfluid liquid.

This must be taken into account when writing out the transport equation for the excitations, which thus has the form

$$\frac{\partial n}{\partial t} + \left(\frac{\partial n}{\partial \boldsymbol{r}} \cdot \frac{\partial \varepsilon'}{\partial \boldsymbol{p}}\right) - \left(\frac{\partial n}{\partial \boldsymbol{p}} \cdot \frac{\partial \varepsilon'}{\partial \boldsymbol{r}}\right) = I(n), \tag{1.26}$$

where $I(n)$ is the collision integral.

The presence in a Bose liquid of two types of motion with two distinct velocities leads to a very particular form of hydrodynamics. The equation of hydrodynamics can be obtained from the transport equation (1.26).

The derivation (which we shall omit here) was carried out by I. M. Khalatnikov and is described in his review article [8](†).

The two-velocity hydrodynamics of a Bose liquid differs a good deal from ordinary hydrodynamics. In particular, oscillations of two different types can occur in such a liquid, with different velocities of propagation.

Oscillations of the first type are ordinary sound, or what is called first sound. The liquid moves as a whole in a wave of this sound; the normal and superfluid parts are not separated. The velocity of propagation of first sound is equal to u. Oscillations of the second type — so-called second sound — are propagated with a velocity

$$u_2 = \sqrt{\frac{\varrho_s T S^2}{\varrho_n \varrho C}}, \tag{1.27}$$

where C and S are respectively the specific heat and entropy per unit volume. In a wave of this type the normal and superfluid parts oscillate in counterphase, in such a way that the total flux of the liquid is $\boldsymbol{j} = \varrho_n \boldsymbol{v}_n + \varrho_s \boldsymbol{v}_s \approx 0$.

We shall not discuss in more detail problems connected with the hydrodynamics of a superfluid liquid. A discussion of sound in liquid He^4, and also excitation interaction processes leading to various dissipative phenomena (viscosity, heat conductivity, and so on), may be found in numerous specialist works and are described in detail in the reviews by E. M. Lifshitz [9] and I. M. Khalatnikov [8], to which the reader is referred.

We now consider what can be said regarding the behaviour of a Bose liquid at higher temperatures, when the number of excitations in it becomes large. In this case it is no longer possible to neglect interactions between excitations and our picture of the excitations as a gas of free particles ceases to correspond to reality. Formulae (1.17) for the thermodynamic functions, being based on the gas model, now lose their validity. This equally applies to formulae (1.22) for the normal density. On the other hand, the idea of two types of motion in a Bose liquid, occurring with corresponding effective densities, is not directly connected with the picture accepted above of the state of excitations, and we can suppose that this idea will still hold good at relatively high temperatures. The same applies to the hydrodynamical equations, which are no more than consequences of the laws of conservation, from which they can be deduced (see [8]). As the temperature rises, the normal density ϱ_n increases up to the point where it reaches a value equal to ϱ. A phase transition takes place in helium at this point (the so-called λ-point). Superfluid flow is possible below but not above this point; above it, the Bose liquid is subject to ordinary hydrodynamics.

(†) Hydrodynamical equations of superfluid He^4, valid at reasonably low volocities, were first obtained by L. D. Landau [3].

In principle, the transition from $\varrho_n \neq \varrho$ to $\varrho_n = \varrho$ could occur continuously or discontinuously. Experiment shows that the phase transition in helium is of the second kind and is not accompanied by emission or absorption of any latent heat (see [10], § 130). It follows from this that the normal density ϱ_n increases continuously as the temperature rises until it reaches the value ϱ at the transition point.

Well above the transition point, helium has no special features as compared with an ordinary liquid. As regards the neighbourhood of the λ-point, we have reason to expect a whole range of striking new properties here. The behaviour of the characteristics (in particular the thermodynamic functions) of a system close to a transition point of the second kind represents an as yet unsolved problem which is at the same time one of the most interesting of the physics of condensed media.

§ 2. FERMI LIQUIDS

1. *Excitations in Fermi liquids*

Let us consider a system of interacting particles obeying Fermi statistics. We shall confine ourselves to the case when the particle spin is $^1/_2$, since the discussion is necessarily limited to liquid He^3, electrons in metals or nuclear material. We shall refer to a system of interacting particles with spin $^1/_2$ as a Fermi liquid.

A theory of the weakly excited states of a Fermi liquid was expounded by L. D. Landau [11, 12]. His theory is based on the assumption that the excitation spectrum of a Fermi liquid has a similar type of structure to that of an ideal Fermi gas. Hence, before turning to the Fermi liquid, it will be useful to link up the familiar picture of the excited states of a Fermi gas with the idea of elementary excitations.

It is well known, that in the ground state of an ideal Fermi gas at $T = 0$, the particles fill all the quantum states with momenta less than some limiting value p_0, whilst no states with momenta greater than p_0 are filled. The filled states form in momentum space a sphere of radius p_0, called the Fermi sphere. The value of p_0 is determined from the condition that the number of states with $p < p_0$ be equal to the number of particles:

$$p_0 = \left(\frac{3\pi^2 N}{V}\right)^{1/3}, \tag{2.1}$$

where N/V is the particle density.

The particles will have a different momentum distribution in an excited state. It is easily seen that every such state can be formed from the ground state by successive translations of particles from inside to outside the Fermi sphere. The state obtained with each such elementary action differs from the initial state by the presence of particles with $p > p_0$

and "holes" with $p < p_0$. These particles with $p > p_0$ and holes with $p < p_0$ obviously play the role of elementary excitations in an ideal Fermi gas. They possess spin $^1/_2$, they can appear and vanish only in pairs, and they possess momenta in the region of p_0 for weakly excited states. The energy of such elementary excitations is conveniently measured from the Fermi level (i.e. from $p_0^2/2m$). The particle-type excitation energy is measured upwards from the Fermi boundary, and the hole energy downwards (i.e. the particle energy is $\xi = (p^2/2m) - (p_0^2/2m) \approx v(p - p_0)$, $v = p_0/m$, whilst the hole energy is $-\xi = (p_0^2/2m) - (p^2/2m) \approx v(p_0 - p)$).

It is assumed in Landau's theory that the weakly excited state of a Fermi liquid is very similar to the corresponding state of a Fermi gas. It can be described with the aid of a set of elementary excitations with spin $^1/_2$ and momenta in the neighbourhood of p_0. An essential hypothesis of the theory is that p_0 is connected with the liquid particle density by the same formula (2.1) as in the ideal gas case (a proof of this assertion will be given in Chap. IV). As in a gas, the excitations in the Fermi liquid are of two types — "particles" with momentum greater than p_0, and "holes" with momentum less than p_0, which can appear and vanish only in pairs. It follows from this that the number of "particles" is necessarily equal to the number of "holes".

In spite of the great similarity between excitations in a Fermi liquid and in an ideal Fermi gas, there are important differences, due to the fact that the excitations in the liquid mutually interact. The clearest outward sign of such interaction is the existence of superfluid (or superconducting, if we are speaking of electrons in a metal) Fermi liquids. The excitation spectrum described above for a Fermi gas does not lead to superfluidity, for an arbitrarily small energy is sufficient for excitation of a Fermi gas, i.e. for the formation of "particles" with $p > p_0$ and "holes" with $p < p_0$. At the same time, the total momentum of this pair can reach the value $2p_0$. In view of this $(\varepsilon/p)_{\min} = 0$, and hence it follows, by (1.18), that the critical velocity is zero, in other words, superfluidity is absent. The appearance of superfluidity is bound up with the fact that a definite type of quasi-particle interaction leads to a radical change in the spectrum. In particular, the excitation of such a Fermi liquid requires an expenditure of energy which cannot be made less than a certain definite value. The excitation spectrum is said to have a gap in such cases.

We defer further discussion of superfluid Fermi liquids to Chapter VII, and turn now to the properties of normal Fermi system excitations.

The interaction between excitations means that the very idea of elementary excitations only has a meaning close to the Fermi momentum p_0. As mentioned earlier, we can only speak of elementary excitations in the case when their damping is small compared with their energy.

The magnitude of the damping is determined either by processes of decomposition of one excitation into several others, or by collisions of excitations with each other. If the excitation energy is large compared with the temperature of the liquid, the chief role is played by the decomposition processes, and the damping is proportional to the probability of these processes. On taking into account the laws of conservation of energy and momentum, and the condition that the number of "particles" be equal to the number of "holes", the probability of decomposition may easily be seen to the proportional to $(p - p_0)^2$ (†). On the other hand, the excitation energy is proportional to $p - p_0$. Hence it is clear that the damping will be relatively small for excitations with momenta close to p_0.

If we are talking of an equilibrium Fermi liquid at finite temperatures, the mean energy of the "particles" and "holes" will be of order T. In view of the fact that the excitations are subject to Fermi statistics, the number of them is also proportional to T. It is easily seen that the probability of decomposition and scattering will be of the same order for such excitations, i.e. they are both proportional to T^2 (‡). Hence it follows that a description of a Fermi liquid with the aid of elementary excitations only applies at fairly low temperatures.

The properties of the energy spectrum of a Fermi liquid can be visualised more easily with the aid of a model based on the analogy with a Fermi gas. Suppose that the ground state of the liquid corresponds to a set of quasi-particles filling the Fermi sphere with limiting momentum

(†) The probability is best estimated by using the analogy with a Fermi gas. Let us consider the following process: a particle with momentum $\mathbf{p}_1 (p_1 > p_0)$ interacts with a particle inside the Fermi sphere with momentum $\mathbf{p}_2 (p_2 < p_0)$. As a result two particles are obtained, with momenta $\mathbf{p}_3$ and $\mathbf{p}_4 = \mathbf{p}_1 + \mathbf{p}_2 - \mathbf{p}_3$, where $p_3, p_4 > p_0$. The particle with momentum $\mathbf{p}_1$ has thus "decomposed" into particles with momenta $\mathbf{p}_3$ and $\mathbf{p}_4$ and a hole with momentum $\mathbf{p}_2$. The total probability of such a process is proportional to

$$\int \delta(\varepsilon_1 + \varepsilon_2 - \varepsilon_3 - \varepsilon_4)\, \mathrm{d}^3\mathbf{p}_2\, \mathrm{d}^3\mathbf{p}_3,$$

$$p_2 < p_0,\ p_3 > p_0,\ p_4 = |\mathbf{p}_1 + \mathbf{p}_2 - \mathbf{p}_3| > p_0.$$

It is easily seen that, for $p_1 - p_0 \ll p_0$, the permissible domains of variation of the moduli of the vectors p_2 and p_3 are

$$p_0 < p_3 < p_1 + p_2 - p_0,\quad 2p_0 - p_1 < p_2 < p_0.$$

The angle between $\mathbf{p}_1$ and $\mathbf{p}_2$ can be arbitrary. The angle between $\mathbf{p}_3$ and $\mathbf{p}_1 + \mathbf{p}_2$ is defined from the conservation of energy condition, whilst the integral over this angle makes the δ-function vanish. The remaining integral over $\mathrm{d}^3\mathbf{p}_2\, \mathrm{d}^3\mathbf{p}_3$ is taken close to $p_2 \approx p_3 \approx p_0$ and gives the factor $(p_1 - p_0)^2$.

(‡) These processes are in essence aspects of the same phenomenon for an almost ideal Fermi gas, and the relevant probability is proportional to

$$\int \delta(\varepsilon_1 + \varepsilon_2 - \varepsilon_3 - \varepsilon_4)\, n(\varepsilon_2)\, (1 - n(\varepsilon_3))\, (1 - n(\varepsilon_4))\, \mathrm{d}^3\mathbf{p}_2\, \mathrm{d}^3\mathbf{p}_3.$$

It can be assumed formally that it is a question of scattering when $|\mathbf{p}_2| > p_0$, and of decomposition when $|\mathbf{p}_2| < p_0$. The integral is proportional to T^2 in both cases when $\varepsilon_1 - \mu \sim T$.

p_0. The relation (2.1) can be interpreted as an equality between the number of quasi-particles and the number of liquid particles. Excitations in such a model are in complete accord with the idea of "particles" and "holes". In particular, the equality between the number of "particles" and the number of "holes" is expressed as the conservation of the number of quasi-particles in the model. If we introduce a quasi-particle distribution function $n(\boldsymbol{p})$, its variations will be restricted by the condition

$$\int \delta n \, \mathrm{d}^3 \boldsymbol{p} = 0. \tag{2.2}$$

The gas model is convenient for further investigations into the properties of a Fermi liquid. It must be borne in mind, however, that the concept of quasi-particles only has a meaning in the neighbourhood of the Fermi surface. Hence, it follows that the properties of the gas model, in which quasi-particles far from the Fermi surface play an essential role, by no means correspond to an actual Fermi liquid.

2. *Quasi-particle energy*

Apart from the assumptions made about the nature of the elementary excitations, Landau's theory is based on a hypothesis concerning the quasi-particle interaction. He assumes that this interaction can be described through a self-consistent field, acting on a quasi-particle and produced by the surrounding quasi-particles.

The energy of the system will now be no longer equal to the sum of the energies of the separate quasi-particles; instead, it is a functional of their distribution function. The energy of an individual quasi-particle is defined in a natural manner as the variational derivative of the total energy with respect to the distribution function:

$$\delta E = 2 \int \varepsilon \, \delta n \frac{\mathrm{d}^3 \boldsymbol{p}}{(2\pi)^3} V \tag{2.3}$$

(the factor 2 comes from summation over the spin).

In fact, it is clear from this formula that ε is precisely the variation in the system energy due to adding one quasi-particle of momentum $\boldsymbol{p}$ (†).

It is assumed in (2.2) and (2.3) that the quasi-particles are distributed uniformly in space. This restriction means in practice that non-uniformity in space may occur only at distances substantially exceeding the quasi-particle wavelength. Since we are only considering excitations in the neighbourhood of the Fermi boundary, i.e. with momenta close to p_0, it follows from (2.1) that the wavelength is of the order of the interatomic spacing. The requirement of spatial uniformity thus implies no restrictions in practice.

(†) Remember that $n(\boldsymbol{p})$ is the quasi-particle momentum distribution, i.e. $2 \int n(\boldsymbol{p}) \, \mathrm{d}^3\boldsymbol{p}/(2\pi)^3$ is the number of quasi-particles per unit volume.

In the presence of a magnetic field, and also in the case of a ferromagnetic system, the distribution function must be regarded as an operator acting on the spin indices (density matrix): $n_{\alpha\beta}$. The quasi-particle energy $\varepsilon_{\alpha\beta}$ is also an operator. When there is no magnetic field and the system is not ferromagnetic, the operators $n_{\alpha\beta}$ and $\varepsilon_{\alpha\beta}$ are proportional to the unit matrix. Formula (2.3) must therefore be written in the general case as

$$\delta\left(\frac{E}{V}\right) = \sum_{\alpha,\beta} \int \varepsilon_{\alpha\beta}\delta n_{\beta\alpha} \frac{d^3\boldsymbol{p}}{(2\pi)^3}.$$

This last expression can be conveniently written in the abbreviated form

$$\delta\left(\frac{E}{V}\right) = \mathrm{Tr}_\sigma \int \varepsilon\,\delta n \frac{d^3\boldsymbol{p}}{(2\pi)^3}, \tag{2.4}$$

provided we remember that ε and n stand for the respective matrices; the sign Tr_σ denotes as usual the sum of the diagonal elements of the product of matrices ε and δn.

The definition of the quasi-particle energy in accordance with (2.4) means that their equilibrium distribution function is in fact a Fermi function. This is most conveniently proved by using the familiar expression for the entropy (†)

$$\frac{S}{V} = -\mathrm{Tr}_\sigma \int [n \ln n + (1-n)\ln(1-n)] \frac{d^3\boldsymbol{p}}{(2\pi)^3}. \tag{2.5}$$

This formula has a purely combinatorial origin, and its applicability to a Fermi liquid is determined by the fact that the classification of the quasi-particle levels corresponds by hypothesis to the classification of the particle levels in an ideal gas.

Given that the number of particles and the energy are constant,

$$\delta N = 0, \ \delta E = 0,$$

the distribution function

$$n(\varepsilon) = n_F(\varepsilon) = \frac{1}{\exp[(\varepsilon-\mu)/T]+1} \tag{2.6}$$

can be found from the condition that the entropy be a maximum, by taking the variation with respect to δn. The energy ε is here a functional of n, so that (2.6) is in fact an extremely complicated implicit expression for $n(\varepsilon)$.

Being a functional of n, ε depends on the temperature. This dependence can be represented as follows. If we denote by $\varepsilon^{(0)}(\boldsymbol{p})$ the quasi-particle equilibrium energy at $T = 0$, it will be expressed at small deviations from equilibrium or at fairly low temperatures by the formula

$$\begin{aligned}\varepsilon(\boldsymbol{p},\sigma) &= \varepsilon^{(0)}(\boldsymbol{p},\sigma) + \delta\varepsilon(\boldsymbol{p},\sigma)\\ &= \varepsilon^{(0)}(\boldsymbol{p},\sigma) + \mathrm{Tr}_{\sigma'} \int f(\boldsymbol{p},\sigma;\boldsymbol{p}',\sigma')\,\delta n(\boldsymbol{p}',\sigma') \frac{d^3\boldsymbol{p}}{(2\pi)^3}.\end{aligned} \tag{2.7}$$

(†) As usual, we understand by $\mathrm{Tr}_\sigma \ln n$ the sum of the logarithms of the diagonal elements $n_{\alpha\alpha}$.

Here $\delta n = n - n_F(T = 0)$, whilst f is an operator depending on the momenta and spin operators of two quasi-particles. Formula (2.7) uses a notation indicating the matrix nature of the quantities appearing in it. As already said, ε and n are matrices in the spin variables. To emphasise this fact, we have written them as $\varepsilon(\boldsymbol{p}, \boldsymbol{\sigma})$, $n(\boldsymbol{p}, \boldsymbol{\sigma})$, where σ_x, σ_y, σ_z are the familiar Pauli matrices, connected with the quasi-particle spin operator $\boldsymbol{s}$ by the relationship $\boldsymbol{s} = {}^1/_2\boldsymbol{\sigma}$. The operator f is a matrix both with respect to the spin variables appearing on the left-hand side of (2.7) and with respect to the spin variables of the operator δn, under the integral sign on the right-hand side of (2.7). We can write (2.7) in a more detailed form (here, and in the following, summation over repeated indices is implied):

$$\varepsilon_{\alpha\beta}(\boldsymbol{p}) = \varepsilon^{(0)}_{\alpha\beta}(\boldsymbol{p}) + \int f_{\alpha\beta;\,\gamma\delta}(\boldsymbol{p}, \boldsymbol{p}')\,\delta n_{\delta\gamma}(\boldsymbol{p}')\,\frac{d^3\boldsymbol{p}}{(2\pi)^3}.$$

This shows the significance of the notation $f(\boldsymbol{p}, \boldsymbol{\sigma}; \boldsymbol{p}', \boldsymbol{\sigma}')$.

The function f thus defined is the second variational derivative of the energy per unit volume with respect to δn (cf. (2.7) and (2.4)), and is therefore symmetric with respect to an interchange of $\boldsymbol{p}, \boldsymbol{\sigma}$ with $\boldsymbol{p}', \boldsymbol{\sigma}'$. The function f is a very important characteristic of a Fermi liquid. As we shall see below (see Chap. IV), it is connected with the amplitude of scattering of two quasi-particles at zero angle.

The dependence of f on the spin variables can be written in the general form

$$f(\boldsymbol{p}, \boldsymbol{\sigma}; \boldsymbol{p}', \boldsymbol{\sigma}') = \varphi(\boldsymbol{p}, \boldsymbol{p}') + \sigma_i\sigma'_k\zeta_{ik}(\boldsymbol{p}, \boldsymbol{p}'). \tag{2.8}$$

If the interaction of the spins has an exchange origin (†), the second term in (2.8) has the form $(\boldsymbol{\sigma}\cdot\boldsymbol{\sigma}')\,\zeta(\boldsymbol{p}, \boldsymbol{p}')$.

The quasi-particle energy ε is independent of the spin in the absence of a magnetic field. The function $\varepsilon^{(0)}$ in (2.7) depends only on p, and can be expanded into a series in $p - p_0$:

$$\xi(p) = \varepsilon^{(0)}(p) - \mu(0) = v(p - p_0), \tag{2.9}$$

where $\mu(0)$ is the chemical potential at $T = 0$, and v is a constant. We can write v, the excitation velocity at the Fermi surface, as

$$v = \frac{p_0}{m^*}, \tag{2.10}$$

(†) Several types of particle interaction are usually distinguished, depending on the spins: exchange interaction, connected with the possibility of exchange of identical particles; spin-orbit interaction, originating from the relativistic interaction of the moving magnetic moment with electric fields; direct magnetic interaction of the angular momenta. Exchange interactions are usually much larger than the other types, and are distinguished by its invariance with respect to a rotation of the total angular momentum of the particle system in space. The scalar product $(\boldsymbol{\sigma}\cdot\boldsymbol{\sigma}')$ has this property.

where m^* is the effective mass. Landau has shown [11] that there is a definite connection between m^* and f.

Let us write a relationship expressing the simple fact that the momentum per unit volume of liquid is the same as the mass flow. The momentum per unit volume is obviously the same as the quasi-particle momentum, i.e. equal to

$$2\int \boldsymbol{p} n \frac{\mathrm{d}^3\boldsymbol{p}}{(2\pi)^3}.$$

On the other hand, by virtue of the assumption that the number of Fermi liquid particles is equal to the number of quasi-particles, the flux of liquid particles is the same as the flux of quasi-particles and is equal to

$$2\int \boldsymbol{v} n \frac{\mathrm{d}^3\boldsymbol{p}}{(2\pi)^3},$$

where $\boldsymbol{v}$ is the quasi-particle velocity. We get the mass flux density from this expression simply by multiplying it by the atomic mass m of the particles in the liquid. Bearing in mind that by definition $\boldsymbol{v}$ is equal to $\partial\varepsilon/\partial\boldsymbol{p}$, we can write the condition that the momentum and mass flux be equal as

$$\int \boldsymbol{p} n \frac{\mathrm{d}^3\boldsymbol{p}}{(2\pi)^3} = m\int \frac{\partial\varepsilon}{\partial\boldsymbol{p}} n \frac{\mathrm{d}^3\boldsymbol{p}}{(2\pi)^3}. \tag{2.11}$$

We vary (2.11) with respect to n, noting that the change in the energy ε due to this is connected with δn by (2.7), which when there is no magnetic field (i.e. when n and ε are independent of spin), can be written as

$$\delta\varepsilon = \frac{1}{2}\mathrm{Tr}_{\sigma}\mathrm{Tr}_{\sigma'}\int f(\boldsymbol{p},\boldsymbol{\sigma};\boldsymbol{p}',\boldsymbol{\sigma}')\,\delta n' \frac{\mathrm{d}^3\boldsymbol{p}}{(2\pi)^3}.$$

It follows that:

$$\int \frac{\boldsymbol{p}}{m}\delta n \frac{\mathrm{d}^3\boldsymbol{p}}{(2\pi)^3} = \int \frac{\partial\varepsilon}{\partial\boldsymbol{p}}\delta n \frac{\mathrm{d}^3\boldsymbol{p}}{(2\pi)^3} + \frac{1}{2}\mathrm{Tr}_{\sigma}\mathrm{Tr}_{\sigma'}\int n\,\delta n' \frac{\partial}{\partial\boldsymbol{p}} f(\boldsymbol{p},\boldsymbol{\sigma};\boldsymbol{p}',\boldsymbol{\sigma}')\,\mathrm{d}^3\boldsymbol{p}\frac{\mathrm{d}^3\boldsymbol{p}'}{(2\pi)^6}.$$

We integrate by parts with respect to $\boldsymbol{p}$ in the second integral and redesignating the variables $\boldsymbol{p}\boldsymbol{\sigma} \rightleftarrows \boldsymbol{p}'\boldsymbol{\sigma}'$:

$$\int \frac{\boldsymbol{p}}{m}\delta n \frac{\mathrm{d}^3\boldsymbol{p}}{(2\pi)^3} = \int \frac{\partial\varepsilon}{\partial\boldsymbol{p}}\delta n \frac{\mathrm{d}^3\boldsymbol{p}}{(2\pi)^3} - \frac{1}{2}\mathrm{Tr}_{\sigma}\mathrm{Tr}_{\sigma'}\int\int \delta n f(\boldsymbol{p},\boldsymbol{\sigma};\boldsymbol{p}',\boldsymbol{\sigma}')\frac{\partial n'}{\partial\boldsymbol{p}'}\frac{\mathrm{d}^3\boldsymbol{p}\,\mathrm{d}^3\boldsymbol{p}'}{(2\pi)^6}.$$

As the δn are arbitrary, this leads at once to

$$\frac{\boldsymbol{p}}{m} = \frac{\partial\varepsilon}{\partial\boldsymbol{p}} - \frac{1}{2}\mathrm{Tr}_{\sigma}\mathrm{Tr}_{\sigma'}\int f(\boldsymbol{p},\boldsymbol{\sigma};\boldsymbol{p}',\boldsymbol{\sigma}')\frac{\partial n'}{\partial\boldsymbol{p}'}\frac{\mathrm{d}^3\boldsymbol{p}}{(2\pi)^3}.$$

When $T = 0$ the energy ε close to the Fermi surface has the form (2.9), whilst $\partial n'/\partial \boldsymbol{p}' \approx -(\boldsymbol{p}'/p')\,\delta(p'-p_0)$. We note that f depends only on the angle χ between $\boldsymbol{p}$ and $\boldsymbol{p}'$, inasmuch as the liquid is isotropic, and then get

$$\frac{1}{m^*} = \frac{1}{m} - \frac{p_0}{2(2\pi)^3}\mathrm{Tr}_\sigma \mathrm{Tr}_{\sigma'} \int f(\chi) \cos\chi \, \mathrm{d}\Omega, \qquad (2.12)$$

where $f(\chi)$ is the value of f for $|\boldsymbol{p}| = |\boldsymbol{p}'| = p_0$. The integration in (2.12) is performed over the direction of the vector $\boldsymbol{p}'$. This relationship connects the atomic mass of the liquid with the effective mass of the quasiparticles. It remains valid to a fair approximation as long as the temperature is sufficiently low.

The specific heat of a Fermi liquid is expressed in terms of m^* in accordance with the usual formula for a Fermi gas. Indeed, we have, from (2.3), for the specific heat per unit volume:

$$C_V = \left(\frac{\partial(E/V)}{\partial T}\right)_{N,V} = 2\int \varepsilon \left(\frac{\partial n}{\partial T}\right)_N \frac{\mathrm{d}^3\boldsymbol{p}}{(2\pi)^3}. \qquad (2.13)$$

It may easily be shown that replacing ε by $\varepsilon^{(0)}$ in the integrand gives a relative error of the order $(T/\mu(0))^3$. We therefore obtain, in a linear approximation in T, the usual gas formula

$$C_V = \frac{1}{3} m^* p_0 T. \qquad (2.14)$$

The entropy at low temperatures is given by the same formula(†).

3. *Sound*

The propagation of sound has a number of special features in a Fermi, just as in a Bose, liquid (though the features are different in the two cases). If we consider sound of a given frequency, it will be propagated in accordance with the laws of ordinary hydrodynamics provided the temperature is not too low. The damping will be proportional to the time τ between collisions of the excitations. As the temperature falls the probability of collisions will diminish as the square of the smearing-out of the Fermi distribution, i.e. the time of collisions will increase in accordance with a T^{-2}-law. In general sound ceases to be propagated at temperatures for which τ is of the order $1/\omega$.

The propagation of sound nevertheless becomes possible again as the temperature is further lowered. Its velocity is different, generally speaking, and it is no longer simply a wave of compression and rarefaction. The phenomenon was predicted by L. D. Landau [12] and called by

(†) Formula (2.13) can be used to determine m^* from experimental data on specific heats. By (2.1), the momentum p_0 is determined from the density of the liquid. Thus we find for liquid He^3 (see [13, 14]):

$$p_0 = 0{\cdot}76 \times 10^8\ \mathrm{cm}^{-1}, \qquad m^* = 2 m_{\mathrm{He}^3}.$$

him "zero sound". In view of the fact that the nature of sound is determined solely by the relationship between ω and τ, the two types can be characterised as low-frequency sound ($\omega\tau \ll 1$) and high-frequency sound ($\omega\tau \gg 1$).

At moderate temperatures, where the condition $\omega\tau \ll 1$ is satisfied, the sound velocity is determined by the compressibility in the usual way. It proves to be substantially dependent on the function f [11].

The compressibility may be conveniently expressed in terms of the derivative of the chemical potential with respect to the number of particles $\partial\mu/\partial N$. Using the fact that the chemical potential depends solely on N/V, we find:

$$\frac{\partial\mu}{\partial N} = -\frac{V^2}{N^2}\frac{\partial P}{\partial V} = \frac{1}{N}\frac{\partial P}{\partial\left(\frac{N}{V}\right)} \tag{2.15}$$

(P is the pressure). A relationship between $\partial\mu/\partial N$ and u^2 follows immediately from this:

$$u^2 = \frac{\partial P}{\partial\varrho} = \frac{\partial P}{\partial\left(\frac{mN}{V}\right)} = \frac{1}{m}N\frac{\partial\mu}{\partial N}. \tag{2.16}$$

We evaluate $\partial\mu/\partial N$ as follows. Since $\mu \approx \varepsilon(p_0)$, the variation of μ will be the combined result of the variation of p_0 and the variation of the form of the function $\varepsilon(p)$:

$$\delta\mu = \frac{1}{2}\mathrm{Tr}_\sigma \mathrm{Tr}_{\sigma'} \int f\,\delta n' \frac{d^3\boldsymbol{p}'}{(2\pi)^3} + \frac{\partial\varepsilon^{(0)}(p_0)}{\partial p_0}\delta p_0. \tag{2.17}$$

(We are assuming that no magnetic field is present.) By (2.1), the variations δN and δp_0 are connected by

$$\delta N = \frac{1}{\pi^2}p_0^2\,\delta p_0\,V.$$

Since only variations δn close to the Fermi surface are important in the integral of (2.17), the integration can be performed over the absolute value of the momentum. This gives:

$$\int f\,\delta n' \frac{d^3\boldsymbol{p}}{(2\pi)^3} = \frac{\delta N}{8\pi V}\int f\,d\Omega.$$

Hence

$$\frac{\partial\mu}{\partial N} = \frac{1}{16\pi V}\mathrm{Tr}_\sigma \mathrm{Tr}_{\sigma'} \int f\,d\Omega + \frac{\pi^2}{p_0 m^* V}. \tag{2.18}$$

Using expression (2.12) for the effective mass and the relation (2.1), we get

$$u^2 = \frac{p_0^2}{3m^2} + \frac{1}{6m}\left(\frac{p_0}{2\pi}\right)^3 \mathrm{Tr}_\sigma \mathrm{Tr}_{\sigma'} \int f(\chi)\,(1-\cos\chi)\,d\Omega. \tag{2.19}$$

The sound velocity is thus determined by (2.19) in the frequency region $\omega\tau \ll 1$. It differs from the sound velocity $u^2 = p_0^2/3m^2$ when interactions are absent.

To investigate the propagation in the frequency region $\omega\tau \gg 1$, we use the ordinary transport equation

$$\frac{\partial n}{\partial t} + \left(\frac{\partial n}{\partial \boldsymbol{r}} \cdot \frac{\partial \varepsilon}{\partial \boldsymbol{p}}\right) - \left(\frac{\partial n}{\partial \boldsymbol{p}} \cdot \frac{\partial \varepsilon}{\partial \boldsymbol{r}}\right) = I(n), \tag{2.20}$$

where $I(n)$ is the collision integral. For small deviations from equilibrium we can write the distribution function as

$$n = n_F + \delta n,$$

where n_F is the equilibrium function and δn a small additional term which is a periodic function of time:

$$\delta n \sim e^{i[(\boldsymbol{k}\cdot\boldsymbol{r})-\omega t]}.$$

The collision integral is of the order

$$I(n) \sim \frac{\delta n}{\tau},$$

and it can be neglected compared with $\partial n/\partial t$. It must be borne in mind when linearising equation (2.20) that ε is a functional of n, so that $\partial\varepsilon/\partial\boldsymbol{r}$ does not vanish. By (2.7):

$$\frac{\partial \varepsilon}{\partial \boldsymbol{r}} = \mathrm{Tr}_{\sigma'} \int f \frac{\partial \delta n'}{\partial \boldsymbol{r}} \frac{\mathrm{d}^3 \boldsymbol{p}'}{(2\pi)^3}.$$

Taking account of our remark, we get

$$\big((\boldsymbol{k}\cdot\boldsymbol{v}) - \omega\big)\delta n - (\boldsymbol{k}\cdot\boldsymbol{v})\frac{\partial n_F}{\partial \varepsilon} \mathrm{Tr}_{\sigma'} \int f \delta n' \frac{\mathrm{d}^3 \boldsymbol{p}'}{(2\pi)^3} = 0. \tag{2.21}$$

It follows from the form of this equation that δn is proportional to $\partial n_F/\partial\varepsilon \approx -\delta(\varepsilon - \mu)$. Denoting $\delta n = (\partial n_F/\partial \varepsilon)\,\nu$, we get:

$$((\boldsymbol{k}\cdot\boldsymbol{v}) - \omega)\nu + (\boldsymbol{k}\cdot\boldsymbol{v})\frac{1}{2}\mathrm{Tr}_{\sigma'} \int F\nu' \frac{\mathrm{d}\Omega'}{4\pi} = 0, \tag{2.22}$$

where

$$F(\chi) = f(\chi)\frac{p_0 m^*}{\pi^2}. \tag{2.23}$$

If we take $\boldsymbol{k}$ as the polar axis and introduce the notations $\tilde{u} = \omega/k$ for the wave propagation velocity and $s = \tilde{u}/v$, equation (2.22) becomes

$$(s - \cos\theta)\nu(\theta, \varphi, \boldsymbol{\sigma}) = \cos\theta \frac{1}{2}\mathrm{Tr}_{\sigma'} \int F(\chi)\nu(\theta', \varphi', \boldsymbol{\sigma}')\frac{\mathrm{d}\Omega}{4\pi}. \tag{2.24}$$

Equation (2.24) reveals the fundamental difference between ordinary sound and the sound propagated in a Fermi liquid when $\omega\tau \gg 1$. In the former case the distribution function remains isotropic in a system of reference in which the liquid as a whole is at rest. This implies a variation of the radius of the Fermi sphere and in addition, an oscillation of its centre relative to the point $\boldsymbol{p} = 0$. In the latter case the distribution function varies in a more complicated way, and the Fermi surface does not remain spherical. The variation of the Fermi surface is determined by the function ν.

Let us first of all consider the solution of (2.24) that is independent of spin. The only part that remains of the function $F(\chi)$ is now $\Phi(\chi)$, which is connected with the function φ in (2.8). We start with the simplest possible case, namely $\Phi = \Phi_0 = \text{const}$. We obtain from (2.24):

$$\nu = \frac{\text{const} \cdot \cos\theta}{s - \cos\theta} e^{i[(\boldsymbol{k}\cdot\boldsymbol{r}) - \omega t]}. \tag{2.25}$$

As we shall soon see, s must be greater than unity. This means that the Fermi surface is stretched out in the direction of motion.

On substituting (2.25) in (2.24) with $F = \Phi_0$, we obtain an equation for s. This gives after integration:

$$\frac{s}{2} \ln \frac{s+1}{s-1} - 1 = \frac{1}{\Phi_0}. \tag{2.26}$$

It is clear from this that, if s is real (which corresponds to undamped waves), it must be greater than unity, i.e.

$$\tilde{u} > v. \tag{2.27}$$

From (2.24) it follows that this condition remains valid for any function Φ. Further, in view of the fact that the left-hand side of (2.26) is always positive, the condition for the existence of zero-point sound is evidently $\Phi_0 > 0$.

If Φ_0 is large, s will also be large. We obtain from (2.26): $s \to \sqrt{\Phi_0/3}$ as $\Phi_0 \to \infty$. On the other hand, $s \to 1$ as $\Phi_0 \to 0$, i.e. $\tilde{u} \to v$. This is the case of an almost ideal Fermi gas.

The conclusion that $s \to 1$ as $\Phi \to 0$ may easily be seen to be independent of the form of Φ. For it follows from (2.24) that $s \to 1$ as $\Phi \to 0$, whilst ν differs from zero only for small θ. By (2.19), $u^2 = p_0^2/3m^2$ in a weakly non-ideal Fermi gas, i.e. $u \approx v/\sqrt{3} \approx \tilde{u}/\sqrt{3}$. The zero-sound velocity is therefore $\sqrt{3}$ times that of ordinary sound. It must be mentioned that, in the limit of an almost ideal Fermi gas, τ is substantially increased, as a result of which the frequency range corresponding to zero-point sound is increased, whereas ordinary sound only exists in the very low frequency region.

Equation (2.24) ceases to admit such a simple solution in the general case of an arbitrary function $\Phi(\chi)$. If we expand $\nu(\theta, \varphi)$ and $\Phi(\chi)$ into a series in spherical harmonics, separate equations can be written for the amplitudes corresponding to spherical functions with different azimuthal numbers m (i.e. factors $e^{im\varphi}$). The number m does not exceed the maximum index l in the expansion of $\Phi(\chi)$ in Legendre polynomials $\Phi(\chi) = \sum_l \Phi_l P_l(\cos\chi)$. We can therefore conclude that several kinds of "zero sound" can be produced in the general case, for which the variations of the distribution functions are non-isotropic in a plane perpendicular to the direction of propagation $\boldsymbol{k}$. As in the elementary case, the possibility of such vibrations being generated is determined by the form of the function Φ. For example, if $\Phi = \Phi_0 + \Phi_1 \cos\chi$, the condition for the generation of vibrations with $\nu \sim e^{i\varphi}$ is $\Phi_1 > 6$.

Attention should be drawn to the absence of compression and rarefaction of the liquid in such waves.

When the function f depends on the particle spins, peculiar waves, which we can call spin waves, may be propagated in the liquid. Indeed, suppose $F(\chi)$ has the form (exchange interaction of the spins)

$$F(\chi) = \Phi(\chi) + Z(\chi)\,(\boldsymbol{\sigma}\cdot\boldsymbol{\sigma}'). \tag{2.28}$$

In this case, apart from spin-independent solutions, (2.24) is satisfied by a function ν of the form

$$\nu = (\boldsymbol{\nu}\cdot\boldsymbol{\sigma}), \tag{2.29}$$

where $\boldsymbol{\nu}$ is an unknown vector. We obtain the equation

$$(s - \cos\theta)\,\boldsymbol{\nu} = \cos\theta \int Z\,\boldsymbol{\nu}' \frac{d\Omega}{4\pi} \tag{2.30}$$

for $\boldsymbol{\nu}$. The equation for the component of the vector $\boldsymbol{\nu}$ differs from the equation for a spin-independent $\boldsymbol{\nu}$ only in the substitution of Z for Φ. Hence, all our future arguments will still hold for spin waves. It can be shown [11] that the zero-order term in the expansion of Z in spherical harmonics determines the expression for the magnetic susceptibility of a Fermi liquid. It proves to be negative for liquid He^3, which in all probability indicates that spin waves cannot be propagated in this liquid.

The case of electrons in a metal is rather special. Obviously, oscillations accompanied by variations of the electron density alone without any vibrations of the crystal lattice cannot be propagated in a metal. Such oscillations would lead to the appearance of an uncompensated electric charge, i.e. their excitation requires a very large energy. This implies in all probability that the function f contains in the case of Coulomb forces, an infinite constant, independent of angle (see also § 22). From (2.26) it then follows that $s = \infty$. This argument only refers to density oscillations; however, under certain conditions, higher order "sounds" with $\nu \sim e^{im\varphi}$ (where $m \neq 0$), and spin waves unconnected with density variations, may be propagated in an electron liquid.

The possibility of the propagation of acoustic waves at $T = 0$ means that the excitation spectrum of the liquid contains Bose type phonon branches with energy linearly dependent on the momentum: $\varepsilon_i = u_i p$.

However, the corrections to the thermodynamic functions due to the phonons contain higher degrees of T (the specific heat $\sim T^3$), which are not to be taken into account in the above approximation.

We shall show later (Chap. IV) how the basic propositions of the theory may be obtained from microscopic considerations of a system of fermions with arbitrary short-range interaction forces.

Landau's theory in the form described refers primarily to the low temperature properties of liquid He^3. The existence of Coulomb interactions between the particles leads to a number of special features. Some of them will be demonstrated in terms of a simple model in § 22. Superfluid (superconducting) Fermi systems differ in an even more essential way from an ordinary Fermi liquid. The properties of superconductors will be discussed in Chap. VII. Finally, mention must also be made of ferromagnetic Fermi systems, which also differ from the model considered. The properties of such Fermi liquids were investigated by A. A. Abrikosov and I. E. Dzyaloshinskii [15], to whose work the reader is referred.

§ 3. SECOND QUANTISATION

The theory of Bose and Fermi liquids described above has been to some extent phenomenological. It has been based on definite assumptions regarding the temperature dependent excitation spectrum. We shall be concerned in the following with providing a microscopic basis for this theory. The present section will be devoted to an auxiliary mathematical device known as the method of second quantisation (†).

Suppose we have a system of N non-interacting particles, which may find themselves in states with wave functions $\varphi_1(\xi), \varphi_2(\xi), \ldots$, which form a complete orthonormal system. Here, ξ denotes any set of variables characterising the states of the particles — usually the coordinates and spin components. The system can evidently be described by specifying the number of particles in states $\varphi_1, \varphi_2, \ldots$, rather than by specifying the complete wave function. This implies passing to a new representation, called the second quantisation representation. The numbers $N_1, N_2, \ldots$, play the role of variables in it. Let us start with the case of particles obeying Bose statistics. We know that the complete wave function of a system of bosons is symmetric with respect to any permutation of the variables corresponding to different particles. It may easily be shown that the wave function corresponding to the occupation numbers $N_1, N_2, \ldots$, has the form

$$\Phi_{N_1 N_2 \ldots} = \left(\frac{N_1! N_2! \ldots}{N!}\right)^{1/2} \sum_P \varphi_{p_1}(\xi_1)\varphi_{p_2}(\xi_2) \cdots \varphi_{p_N}(\xi_N); \tag{3.1}$$

(†) It seems useful to give a brief description here of the method of second quantisation (see e.g. [80]), inasmuch as this method is basic for the methods developed later.

here, the p_i denote the states, whilst the summation is over all possible permutations of different p_i. The factor in front of the sum is for normalisation $\left(\int |\Phi|^2 \prod_i d\xi_i = 1\right)$. We shall consider $\Phi_{N_1,N_2,\ldots}$ as a function of the variables N_i.

Let $F^{(1)}$ be an operator symmetrical with respect to all the particles, of the form

$$F^{(1)} = \sum_a f_a^{(1)}, \tag{3.2}$$

where f_a is an operator acting only on functions of ξ_a. It is easily seen that when such an operator acts upon the function $\Phi_{N_1,N_2,\ldots}$, it either maps it on to the same function, or on to some other, corresponding to a change in the state of one of the particles. In view of this, the matrix elements of $F^{(1)}$ with respect to the functions (3.1) have the form:

diagonal elements:

$$\overline{F^{(1)}} = \sum_i f_{ii}^{(1)} N_i,$$

off-diagonal elements:

$$(F^{(1)})_{N_i-1 N_k}^{N_i N_k-1} = f_{ik}^{(1)} \sqrt{N_i N_k}, \tag{3.3}$$

where

$$f_{ik}^{(1)} = \int \varphi_i^*(\xi) f^{(1)} \varphi_k(\xi)\, d\xi.$$

The operator $F^{(1)}$ can be pictured as acting on the numbers N_i if we introduce operators a_i, which decrease by one the number of particles in the ith state and possess the matrix elements

$$(a_i)_{N_i}^{N_i-1} = \sqrt{N_i}. \tag{3.4}$$

The Hermitian conjugate operators a_i^+ obviously have the matrix elements

$$(a_i^+)_{N_i-1}^{N_i} = (a_{iN_i}^{N_i-1})^* = \sqrt{N_i}, \tag{3.5}$$

i.e. they increase the number of particles by one. It is easily shown that the operator $F^{(1)}$ can be written as

$$F^{(1)} = \sum f_{ik}^{(1)} a_i^+ a_k. \tag{3.6}$$

Indeed, the matrix elements of this operator are the same as those of (3.3). This is in fact the expression for $F^{(1)}$ in the second quantisation form.

By (3.4) and (3.5), the products of the operators a_i^+ and a_i are the diagonal operators

$$\begin{aligned} a_i^+ a_i &= N_i, \\ a_i a_i^+ &= N_i + 1. \end{aligned} \tag{3.7}$$

The following commutation relations for the operators a_i are a consequence of (3.4), (3.5) and (3.7):

$$[a_i, a_k^+]_- = a_i a_k^+ - a_k^+ a_i = \delta_{ik}, \qquad (3.8)$$
$$[a_i, a_k]_- = [a_i^+, a_k^+]_- = 0.$$

The symmetrised operator

$$F^{(2)} = \sum_{a,b} f_{ab}^{(2)}, \qquad (3.9)$$

where $f_{ab}^{(2)}$ acts on functions of ξ_a and ξ_b, can be similarly written. In the second quantisation form $F^{(2)}$ becomes

$$F^{(2)} = \sum_{iklm} f^{(2)\,ik}_{\;\;lm} a_i^+ a_k^+ a_l a_m, \qquad (3.10)$$

where

$$f^{(2)\,ik}_{\;\;lm} = \int \varphi_i^*(\xi_1)\, \varphi_k^*(\xi_2) f^{(2)} \varphi_l(\xi_1) \varphi_m(\xi_2) \mathrm{d}\xi_1 \mathrm{d}\xi_2.$$

The same applies to more complicated operators.

Let us take the Hamiltonian of a system of interacting particles situated in an external field,

$$H = \sum H_a^{(1)} + \sum_{a,b} U^{(2)}(\boldsymbol{r}_a, \boldsymbol{r}_b) + \sum_{a,b,c} U^{(3)}(\boldsymbol{r}_a, \boldsymbol{r}_b, \boldsymbol{r}_c) + \cdots, \qquad (3.11)$$

where $H_a^{(1)} = (-\nabla_a^2/2m) + U(\boldsymbol{r}_a)$. It becomes in the second quantisation form:

$$H = \sum H_{ik}^{(1)} a_i^+ a_k + \sum_{iklm} U^{(2)\,ik}_{\;\;lm} a_i^+ a_k^+ a_l a_m + \cdots. \qquad (3.12)$$

If we take for the φ_i the eigenfunctions of the Hamiltonian $H_a^{(1)}$ the first term in (3.12) becomes equal to

$$H^{(1)} = \sum \varepsilon_i a_i^+ a_i = \sum \varepsilon_i N_i. \qquad (3.13)$$

In the case of Fermi statistics, the complete wave function of a system must be anti-symmetric with respect to all the variables. This means that the occupation numbers can only be 0 or 1 in the case of non-interacting particles, and the wave function has the form

$$\Phi_{N_1 N_2 \cdots} = \frac{1}{\sqrt{N!}} \sum_P (-1)^P \varphi_{p_1}(\xi_1) \varphi_{p_2}(\xi_2) \cdots \varphi_{p_N}(\xi_N), \qquad (3.14)$$

where all the numbers $p_1, p_2, \ldots, p_N$ are different. The symbol $(-1)^P$ indicates that odd permutations appear with the "minus" sign in the sum (3.14). For definiteness, we shall take with the "plus" sign that term in the sum in which

$$p_1 < p_2 < p_3 < \cdots < p_N. \qquad (3.15)$$

The matrix elements of an operator $F^{(1)}$ of the type (3.2) are in the present case:
diagonal elements:

$$\overline{F^{(1)}} = \sum_i f_{ii} N_i, \qquad (3.16)$$

off-diagonal elements:

$$(F^{(1)})^{1_i\, 0_k}_{0_i\, 1_k} = \pm f_{ik}^{(1)},$$

where the "plus" or "minus" sign is taken, depending on whether the total number of particles in states between the ith and kth is even or odd. We introduce operators a_i with matrix elements

$$(a_i)_1^0 = (a_i^+)_0^1 = (-1)^{\sum_{l=1}^{i-1} N_l}. \tag{3.17}$$

The operator $F^{(1)}$ can be written with the aid of these operators in the form (3.6).

The products of operators a_i and a_i^+ are equal to

$$\begin{aligned} a_i^+ a_i &= N_i, \\ a_i a_i^+ &= 1 - N_i. \end{aligned} \tag{3.18}$$

Hence

$$\{a_i a_i^+\} = a_i a_i^+ + a_i^+ a_i = 1.$$

All the remaining anti-commutators are equal to zero. Thus

$$\begin{aligned} \{a_i a_k^+\} &= \delta_{ik}, \\ \{a_i a_k\} &= \{a_i^+ a_k^+\} = 0. \end{aligned} \tag{3.19}$$

The more complicated operators, and in particular the Hamiltonian, can be written in terms of operators a_i, a_i^+ precisely as in the case of bosons.

§ 4. DILUTE BOSE GAS

A weakly non-ideal gas, i.e. a gas in which the role of the particle interactions is relatively small, provides a simple example of a quantum liquid. As we shall see, a necessary condition for such a gas is that the particle scattering amplitude be small compared with the mean wavelength $\lambda = 1/\bar{p}$, which, for a degenerate gas, is of the same order of magnitude as the mean distance between the particles.

In these circumstances, due to the smallness of the momentum of the colliding particles, it is sufficient to a first approximation to take only the s-scattering into account. If we denote the s-scattering amplitude by a, the p-scattering will be of the order $a(a/\lambda)^2$(†). Triple collisions also

(†) If r_0 characterises the range of the forces, quantum mechanics shows (see [16]) that, for $\lambda \gg r_0$, the scattering amplitudes with different momenta l will be of the order $r_0(r_0/\lambda)^{2l}$. It will be clear from what follows that the wavelengths of importance are different for boson and for fermion gases. Since p-scattering can occur only for particles with non-vanishing momentum, the order of magnitude the most important momenta in the case of a boson gas can be obtained by considering the integrals (4.20) and (4.15) over the non-condensed state. The essential momenta in these integrals are of order $(aN/V)^{1/2}$, or $\lambda \sim (V/aN)^{1/2}$: hence it follows that the contribution to the energy from p-scattering will be $(a/\lambda)^2 \sim a^3 N/V$ times the contribution from s-scattering. In the case of a fermion gas, the essential momenta are those near the Fermi level, or $\lambda \sim (V/N)^{1/3}$. The correction from p-scattering will thus be about $a^2(N/V)^{2/3}$ times the contribution from s-scattering.

contribute little. A very elementary estimate shows that taking such collisions into account must lead to the appearance of extra powers of the volume in the denominator; this means that the correction is of the order a^3N/V. However, this is in actual fact only correct for a boson gas while for a fermion gas the contribution from triple collisions is even less $(\sim (a^3N/V)^{5/3})$ (†). We shall assume that the interaction between the particles is repulsive, i.e. that the scattering amplitude has the "plus" sign. This is connected with the fact that, no matter how weak the attraction, a Bose gas can never remain dilute at low temperatures. In a Fermi gas, the attraction between the particles leads to superfluidity. We shall not discuss this case here.

We shall evaluate in this section the ground state energy and energy spectrum of a dilute boson gas $(T = 0)$ (‡). A dilute Fermi gas will be discussed in the next section. We shall assume for simplicity that the Bose gas particles have zero spin. The energy of interaction may be written in this case as

$$H_{int} = \frac{U}{2V} \sum_{p_1+p_2=p_3+p_4} a^+_{p_4} a^+_{p_3} a_{p_2} a_{p_1}. \tag{4.1}$$

Taking U outside the summation sign corresponds to assuming the interaction to be the same between any pairs of particles, the scattering amplitude being independent of the angle (s-scattering). To a first approximation, U is connected with the scattering amplitude by the relationship

$$U = \frac{4\pi}{m} a. \tag{4.2}$$

This equation is easily obtained by the following argument. By definition (see [16]), the s-scattering amplitude is connected with the effective cross-section for the scattering of two identical particles by the relation (in the centre of mass system)

$$d\sigma = (2a)^2 d\Omega.$$

On the other hand, $d\sigma$ can be determined with the aid of the Hamiltonian (4.1). We obtain in the Born approximation (see [16]):

$$d\sigma = \left(\frac{m}{4\pi}\right)^2 (2U)^2 d\Omega,$$

whence follows (4.2).

(†) This last statement follows from the fact that the wave function of three colliding fermions must be antisymmetric. This requires that the third particle possesses odd orbital angular momentum relative to that particle of the first two that has the same spin z-component. As a result, at least one extra factor $(a/\lambda)^2$ appears.

(‡) The energy spectrum of a dilute boson gas was first obtained by N. N. Bogolyubov [17]; the ground state energy was found by Huang and Yang [18] and by Brueckner and Sawada [19]. We shall in essence follow references [17] and [19] in the present section.

N. N. Bogolyubov has shown [17] that, when we are concerned with the ground state for the weakly excited states of a dilute boson gas, the interaction energy operator (4.1) can be considerably simplified, and it becomes possible to carry out the diagonalisation of the Hamiltonian and hence obtain the energy spectrum. The simplification is based on the following idea. In the ground state the particles of an ideal Bose gas are in the lowest level with zero energy, or, "in the condensate". In view of the weakness of the interaction in a dilute gas, its ground state will only differ slightly from the state of an ideal gas, i.e. the number of particles in the condensate will still substantially exceed the number in other energy levels: $N - N_0 \ll N_0$. The same applies to weakly excited states. Since the matrix elements of the Bose operators a_i are equal to $\sqrt{N_i}$, we can clearly neglect the interactions of particles above the zero energy level with each other, and simply take into account only the interactions of condensate particles with one another and with the excited particles. This means that we need only retain the following terms in the sum of (4.1):

$$H_{int} = \frac{U}{2V}\Big[a_0^+ a_0^+ a_0 a_0 + \sum_{p\neq 0} (2 a_p^+ a_0^+ a_p a_0 + 2 a_{-p}^+ a_0^+ a_{-p} a_0 + a_p^+ a_{-p}^+ a_0 a_0 + a_0^+ a_0^+ a_p a_{-p})\Big]. \tag{4.3}$$

In view of the fact that N_0 is an extremely large number, we are justified in regarding the operators a_0^+ and a_0 simply as c-numbers, and in replacing them by $\sqrt{N_0}$. Indeed, the commutators of these operators with one another or with any other operators a_i, a_i^+ give 1 or 0, i.e. are always small by comparision with the matrix elements of the operators a_0 and a_0^+. We thus obtain:

$$H_{int} = \frac{U}{2V}\Big[N_0^2 + 2N_0 \sum_{p\neq 0} (a_{-p}^+ a_{-p} + a_p^+ a_p) + N_0 \sum_{p\neq 0} (a_p^+ a_{-p}^+ + a_p a_{-p})\Big]. \tag{4.4}$$

The total number of particles in the system can be written as

$$N = N_0 + \frac{1}{2} \sum_{p\neq 0} (a_p^+ a_p + a_{-p}^+ a_{-p}). \tag{4.5}$$

This enables us to express the total number N_0 in (4.4) in terms of N. On confining ourselves to terms in H_{int} of not less than the first degree in N, and adding the kinetic energy operator, we get the following Hamiltonian:

$$H = \frac{UN^2}{2V} + \frac{1}{2}\sum_{p\neq 0}\Big[\Big(\frac{p^2}{2m} + \frac{UN}{V}\Big)(a_p^+ a_p + a_{-p}^+ a_{-p}) + \frac{UN}{V}(a_p^+ a_{-p}^+ + a_p a_{-p})\Big]. \tag{4.6}$$

The last term in the Hamiltonian is non-diagonal. To diagonalise, we carry out a linear transformation of the operators a_p and a_p^+:

$$a_p = \frac{1}{\sqrt{1-A_p^2}}(\alpha_p + A_p \alpha_{-p}^+),$$
$$a_p^+ = \frac{1}{\sqrt{1-A_p^2}}(\alpha_p^+ + A_p \alpha_{-p}). \tag{4.7}$$

The new operators α_p and α_p^+ satisfy the same permutation relations as the old ones. On expressing the operators a_p and a_p^+ in (4.6) in terms of α_p and α_p^+, we get

$$H = \frac{UN^2}{2V} + \sum_{p\neq 0} \frac{1}{1-A_p^2}\left[\left(\frac{p^2}{2m}+\frac{NU}{V}\right)A_p^2 + \frac{NU}{V}A_p\right]$$
$$+\frac{1}{2}\sum_{p\neq 0}\frac{1}{1-A_p^2}\left[\left(\frac{p^2}{2m}+\frac{NU}{V}\right)(1+A_p^2) + 2\frac{UN}{V}A_p\right](\alpha_p^+\alpha_p + \alpha_{-p}^+\alpha_{-p})$$
$$+\frac{1}{2}\sum_{p\neq 0}\frac{1}{1-A_p^2}\left[\left(\frac{p^2}{2m}+\frac{NU}{V}\right)2A_p + \frac{NU}{V}(1+A_p^2)\right](\alpha_p^+\alpha_{-p}^+ + \alpha_p\alpha_{-p}). \tag{4.8}$$

For the non-diagonal terms to vanish, the coefficient A_p must satisfy the relationship

$$\left(\frac{p^2}{2m}+\frac{NU}{V}\right)2A_p + \frac{NU}{V}(1+A_p^2) = 0.$$

This gives us

$$A_p = \frac{V}{UN}\left[-\frac{p^2}{2m} - \frac{NU}{V} + \sqrt{\left(\frac{p^2}{2m}+\frac{UN}{V}\right)^2 - \left(\frac{UN}{V}\right)^2}\right]. \tag{4.9}$$

The "plus" sign is required in front of the radical in order for the excited states to possess positive energy. On writing the coefficients A_p in (4.8) in accordance with (4.9), we get

$$H = \frac{UN^2}{2V} - \frac{1}{2}\sum_{p\neq 0}\left[\left(\frac{p^2}{2m}+\frac{UN}{V}\right) - \sqrt{\left(\frac{p^2}{2m}+\frac{UN}{V}\right)^2 - \left(\frac{UN}{V}\right)^2}\right]$$
$$+\frac{1}{2}\sum_{p\neq 0}\sqrt{\left(\frac{p^2}{2m}+\frac{UN}{V}\right)^2 - \left(\frac{UN}{V}\right)^2}\,(\alpha_p^+\alpha_p + \alpha_{-p}^+\alpha_{-p}). \tag{4.10}$$

The expression obtained consists of two terms. The first is a constant, whilst the second is a diagonal operator which can be written in the form

$$\sum_{p\neq 0} n_p \varepsilon(p),$$

where the n_p are the occupation numbers corresponding to operators α_p. The least energy is obtained when all the n_p are zero, i.e. $\sum_p n_p \varepsilon(p)$ is the excitation energy. This expression has the same form as (3.13) for the energy of a system of non-interacting particles. It follows from this that the weakly excited state of a dilute Bose gas can be described with the aid of an elementary excitation model with the energy spectrum(†).

$$\varepsilon(\mathbf{p}) = \sqrt{\left(\frac{p^2}{2m} + \frac{UN}{V}\right)^2 - \left(\frac{UN}{V}\right)^2}. \tag{4.11}$$

The expression becomes in the limit of small momenta,

$$\varepsilon(\mathbf{p}) \approx \frac{\sqrt{4\pi a N/V}}{m} p, \tag{4.12}$$

i.e. it corresponds to the phonon part of the spectrum of a Bose liquid. In the case of large momenta the energy $\varepsilon(\mathbf{p})$ becomes the energy of a free particle:

$$\varepsilon(\mathbf{p}) \approx \frac{p^2}{2m}, \tag{4.13}$$

which is also in accordance with the results of § 2.

The first terms in (4.10) obviously represent the ground state energy of a Bose liquid. It is easily seen that the sum over $\mathbf{p}$ in this expression diverges as $\sum_{\mathbf{p}} 1/p^2$ for large momenta. This is connected with the fact that the energy cannot actually be expanded in powers of U. The presence of the constant U leads to an infinity in the energy, as may be seen at once from (4.10). It is essential in the present case that the scattering amplitude a be finite and small, thus enabling the energy to be expanded in powers of a.

Equation (4.2) between U and a is not exact; it only holds up to first order terms. Since we are in fact interested in higher order terms in the energy, (4.2) needs to be corrected. On considering in second order perturbation theory the scattering of two particles in the condensate accompanied by a transition to the states $\mathbf{p}$, $-\mathbf{p}$, we get

$$U - \frac{U^2}{V} \sum_{\mathbf{p} \neq 0} \frac{1}{p^2/m} = \frac{4\pi a}{m}. \tag{4.14}$$

On now expressing U in terms of a and substituting the result in (4.10), we find for the ground state energy:

$$E = \frac{2\pi a}{m} \frac{N^2}{V} + \frac{8\pi^2 a^2}{m^2} \left(\frac{N}{V}\right)^2 \sum_{\mathbf{p} \neq 0} \frac{1}{p^2/m}$$
$$- \frac{1}{2} \sum_{\mathbf{p} \neq 0} \left(\frac{p^2}{2m} + \frac{4\pi a N}{mV}\right) \left[1 - \sqrt{1 - \left(\frac{4\pi a N/mV}{p^2/2m + 4\pi a N/mV}\right)^2}\right]. \tag{4.15}$$

Expression (4.15) is convergent for large p. On integrating over the momenta, we get

$$\frac{E}{V} = \frac{2\pi a}{m} \left(\frac{N}{V}\right)^2 \left[1 + \frac{128}{15\sqrt{\pi}} a^{3/2} \left(\frac{N}{V}\right)^{1/2}\right]. \tag{4.16}$$

(†) Notice that the Born approximation has been used in our derivation. Actually, (4.11), when expressed in terms of the scattering amplitude a with the aid of equation (4.2), holds whenever the condition $a/\lambda \ll 1$ is fulfilled, and not just in the Born approximation. We shall prove this in Chap. V. The same applies as regards (4.16), (5.20) and (5.21).

Notice that the expansion is in $[a(N/V)^{1/3}]^{3/2}$.

The sound velocity can be found from this formula:

$$u = \sqrt{\frac{V^2}{mN}\frac{\partial^2 E}{\partial V^2}} = \frac{\sqrt{4\pi a N/V}}{m}. \tag{4.17}$$

This expression is the same, as it must be, as the coefficient of p in expression (4.12) for the phonon part of the spectrum.

We mentioned at the start of this section that the amplitude a must be positive in a Bose gas. This is also clear from (4.17), since the sound velocity would be imaginary (unstable state) if $a < 0$.

The momentum distribution of the excitations is given by the usual Bose formula

$$\bar{n}_{\boldsymbol{p}} = \overline{\alpha^+_{\boldsymbol{p}}\alpha_{\boldsymbol{p}}} = \frac{1}{e^{\varepsilon(p)/T} - 1}. \tag{4.18}$$

As regards the momentum distribution of the particles themselves in a Bose liquid, this can be found by evaluating

$$\bar{N}_{\boldsymbol{p}} = \overline{a^+_{\boldsymbol{p}} a_{\boldsymbol{p}}}.$$

On using (4.7), we get

$$\bar{N}_{\boldsymbol{p}} = \frac{\bar{n}_{\boldsymbol{p}} + A^2_{\boldsymbol{p}}(\bar{n}_{\boldsymbol{p}} + 1)}{1 - A^2_{\boldsymbol{p}}}. \tag{4.19}$$

This expression naturally refers only to $\boldsymbol{p} \neq 0$. The number of particles with zero energy is obtained from the formula

$$N_0 = N - \sum_{\boldsymbol{p}\neq 0} \bar{N}_{\boldsymbol{p}}.$$

At absolute zero $\bar{n}_{\boldsymbol{p}} = 0$, so that we have from (4.19):

$$\bar{N}_{\boldsymbol{p}} = \frac{\dfrac{8\pi^2 a^2}{m^2}\left(\dfrac{N}{V}\right)^2}{\varepsilon(\boldsymbol{p})\left[\varepsilon(\boldsymbol{p}) + \dfrac{p^2}{2m} + \dfrac{4\pi a N}{m V}\right]}, \tag{4.20}$$

$$\frac{N_0}{N} = 1 - \frac{8}{3\sqrt{\pi}} a^{3/2}\left(\frac{N}{V}\right)^{1/2}. \tag{4.21}$$

It is clear from this that, even in the ground state, in a non-ideal Bose gas, not all particles have zero momentum.

§ 5. DILUTE FERMI GAS

We now turn to a Fermi gas. We shall find the ground state energy, the effective mass of the excitations and the f-function(†) up to terms of order $(a/\lambda)^2$, where a is the s-scattering amplitude.

(†) The ground state energy was obtained by HUANG and YANG [20], and by LEE and YANG [21], and the effective mass of the excitations by A. A. ABRIKOSOV and I. M. KHALATNIKOV [22], and by V. M. GALITSKII [23]. The f-function was obtained in [22].

The excitation energy operator cannot be described with the aid of (4.1) as in the case of bosons. For, if we take i, k, l, m in (4.1) to indicate not only the momenta but also the spin, the sum vanishes by virtue of the anti-commutativity of the Fermi operators. This is connected with the fact that the Hamiltonian (4.1) takes no account of the specific nature of fermion scattering. According to quantum mechanics (see [12], § 114), s-scattering can only occur for identical particles with spin $^1/_2$ when the spins are anti-parallel. The amplitude is then twice what it is when the particles are different. On taking this into account, we can write the interaction energy as

$$H_{int} = \frac{U}{V} \sum_{p_1+p_2=p_3+p_4} a^+_{p_3,1/2} a^+_{p_4,-1/2} a_{p_2,-1/2} a_{p_1,1/2}, \tag{5.1}$$

or equivalently

$$H_{int} = \frac{U}{2V} \sum_{p_1+p_2=p_3+p_4} a^+_{p_3\alpha} a^+_{p_4\beta} a_{p_2\beta} a_{p_1\alpha}. \tag{5.1'}$$

As before, U is connected to a first approximation with the s-scattering amplitude by

$$U = \frac{4\pi a}{m}. \tag{5.2}$$

Let us apply perturbation theory with respect to H_{int}. The first order correction to the ground state energy is equal to the diagonal matrix element of H_{int}:

$$E^{(1)} = \frac{U}{V} \sum n_i n_k Q_{ik}, \tag{5.3}$$

where the subscripts i, k correspond to given momenta and spin, n_i is the occupation number at $T = 0$(†), equal to 1 for $p < p_0$ and 0 for $p > p_0$ ($p_0 = (3\pi^2 N/V)^{1/3}$), whilst the factor Q_{ik} takes into account the fact that the spins of particles in states i and k are anti-parallel. This factor may conveniently be written in the form

$$Q_{ik} = \frac{1}{4}\big(1 - (\boldsymbol{\sigma}_i \cdot \boldsymbol{\sigma}_k)\big), \tag{5.4}$$

where $^1/_2\boldsymbol{\sigma}_i$ is the spin operator of a particle in the state i. On substituting (5.2) and (5.4) in (5.3), we get

$$E^{(1)} = \frac{\pi a}{m} \frac{N^2}{V}. \tag{5.5}$$

To find the second order correction we use the perturbation theory expression:

$$E_n^{(2)} = \sum_{m \neq n} \frac{|(H_{int})_{nm}|^2}{E_n - E_m}. \tag{5.6}$$

(†) The n_i here denote the occupation numbers for non-interacting particles. It may easily be realised that, at $T = 0$, they are the same as the occupation numbers of the quasi-particles and differ from the occupation numbers N_i for a system of interacting particles.

On substituting (5.1) into this, we get the sum

$$\frac{U^2}{V^2}\sum_{iklm}\frac{n_i\, n_k(1-n_l)\,(1-n_m)\,Q_{ik}\,Q_{lm}}{(p_i^2+p_k^2-p_l^2-p_m^2)/2m}. \tag{5.7'}$$

Since our aim is to obtain an expansion of the energy in powers of a, we must recall as in § 4 that the relation (5.2) between U and the scattering amplitude is not exact, and only holds to first order in U. If second order terms are taken into account, we get instead of (5.2):

$$U+\frac{2U^2}{V}\sum_{l,m}\frac{Q_{lm}}{(p_i^2+p_k^2-p_l^2-p_m^2)/2m}=\frac{4\pi a}{m}. \tag{5.2'}$$

If we use this to express U in terms of a and substitute the result in (5.3), terms proportional to a^2 are obtained in the expression for $E^{(1)}$, and these naturally belong to the second order correction. Taking this into account, we get the following second approximation to the energy:

$$E^{(2)}=\frac{16a^2\pi^2}{m^2V^2}\sum_{iklm}\left[\frac{n_i n_k(1-n_l)\,(1-n_m)\,Q_{ik}\,Q_{lm}}{(p_i^2+p_k^2-p_l^2-p_m^2)/2m}-\frac{n_i n_k\,Q_{ik}\,Q_{lm}}{(p_i^2+p_k^2-p_l^2-p_m^2)/2m}\right]. \tag{5.7}$$

In contrast to (5.7′), this expression is not divergent for large $\boldsymbol{p}$. Consequently, as in a Bose gas, a renormalization of U leads to the elimination of a divergence in the energy.

Notice that (5.2′) appears to be contradictory at first glance, because the left-hand side depends on the angle between $\boldsymbol{p}_i$ and $\boldsymbol{p}_k$, whilst the right-hand side does not. This might suggest that the form we took for the interaction energy operator is incorrect. In reality, however, this fact should not be detrimental. If we split the integral over the momenta in (5.2′) into its principal value and a circuit around the pole, we see easily that only the last term depends on the angle between $\boldsymbol{p}_i$ and $\boldsymbol{p}_k$. The principal value is a divergent integral over $\boldsymbol{p}_l$ and we can assume in it that $\boldsymbol{p}_i$ and $\boldsymbol{p}_k \approx 0$. The circuit around the pole leads to a purely imaginary expression while the correction to the energy must be real. It follows therefore that this term does not contribute to the energy.

The term containing four n_i that appears in the first part of (5.7) vanishes, because the denominator is anti-symmetric with respect to the substitutions $i, k \leftrightarrows l, m$, whereas the numerator is symmetric and all the domains of summation are the same. The remaining two terms with products of three n_i are equal to one another. Hence we finally obtain

$$E^{(2)}=-\frac{32\,a^2\pi^2}{m^2V^2}\sum_{iklm}\frac{n_i n_k n_l\,Q_{ik}}{(p_i^2+p_k^2-p_l^2-p_m^2)/2m}. \tag{5.8}$$

On passing from summations to integrations, this expression can be written as

$$\frac{E^{(2)}}{V} = -\frac{32a^2\pi^2}{m^2(2\pi)^9} \int\limits_{|p_1|<p_0} d^3p_1 \int\limits_{|p_2|<p_0} d^3p_2 \int\limits_{|p_3|<p_0} d^3p_3 \int d^3p_4 \frac{\delta(p_1+p_2-p_3-p_4)}{(p_1^2+p_2^2-p_3^2-p_4^2)/2m}. \tag{5.9}$$

In accordance with § 2, the excitation energy is given by(†)

$$\varepsilon_i = \frac{\delta E}{\delta n_i}. \tag{5.10}$$

Variation of (5.3) and (5.8) with respect to n_i gives

$$\varepsilon(p) = \frac{p^2}{2m} + \frac{2\pi a N}{mV} + \frac{16\pi^2 a^2}{m^2(2\pi)^9} \int\limits_{|p_1|<p_0} d^3p_1 \int\limits_{|p_2|<p_0} d^3p_2 \int d^3p_3$$
$$\times \left[\frac{\delta(p_1+p_2-p-p_3)}{(p^2+p_3^2-p_1^2-p_2^2)/2m} - 2\frac{\delta(p_1+p-p_2-p_3)}{(p^2+p_1^2-p_2^2-p_3^2)/2m}\right]. \tag{5.11}$$

The evaluation of the ground state energy and effective mass of the excitations thus requires the evaluation of the integrals (5.9) and (5.11). The integration is fairly laborious, due to the multiplicity of the integrals and the awkward domain of integration.

An alternative, simpler method can be used, based on using the function f. If we introduce

$$f_{ik} = \frac{\delta^2 E}{\delta n_i\, \delta n_k}, \tag{5.12}$$

we shall be able to find the effective mass and low frequency sound velocity according to (2.12) and (2.19) of § 2. The ground state energy can be found from the sound velocity after suitable integration.

The problem therefore amounts to finding f. On varying (5.3) and (5.8) first with respect to n_i, then with respect to n_k, we find the following expression for f:

$$f = \frac{8a\pi}{m} Q_{\sigma\sigma'} - \frac{64\pi^2 a^2}{m^2(2\pi)^3} \int\limits_{|p_1|<p_0} d^3p_1 \int d^3p_2 \left[Q_{\sigma\sigma'} \frac{\delta(p+p'-p_1-p_2)}{(p^2+p'^2-p_1^2-p_2^2)/2m}\right.$$
$$\left. + \frac{1}{4}\frac{\delta(p+p_1-p'-p_2)}{(p^2+p_1^2-p'^2-p_2^2)/2m} + \frac{1}{4}\frac{\delta(p'+p_1-p-p_2)}{(p'^2+p_1^2-p^2-p_2^2)/2m}\right]. \tag{5.13}$$

(†) This formula may appear to be incorrect at first sight, since ε is the variational derivative of E with respect to the distribution function of the quasi-particles, and not with respect to the particle distribution. But the derivative in (5.10) is not with respect to the true particle distribution, but with respect to the distribution of the non-interacting particles, which, as remarked earlier (footnote on p. 36), is the same at $T = 0$ as the distribution of the quasi-particles of an interacting system.

If we put $|\mathbf{p}| = |\mathbf{p}'| = p_0$ right away, the integration in (5.13) is much simpler than in (5.9) and (5.11). We get

$$f(\chi) = \frac{2\pi a}{m}\left[1 + 2a\left(\frac{3N}{\pi V}\right)^{1/3}\left(2 + \frac{\cos\chi}{2\sin\frac{\chi}{2}}\ln\frac{1+\sin\frac{\chi}{2}}{1-\sin\frac{\chi}{2}}\right)\right]$$
$$-\frac{2\pi a}{m}(\boldsymbol{\sigma}\cdot\boldsymbol{\sigma}')\left[1 + 2a\left(\frac{3N}{\pi V}\right)^{1/3}\left(1 - \frac{\sin\frac{\chi}{2}}{2}\ln\frac{1+\sin\frac{\chi}{2}}{1-\sin\frac{\chi}{2}}\right)\right]. \quad (5.14)$$

A special feature of (5.14) deserves attention. The function f for particles with opposite spins has a logarithmic singularity for angles χ close to π:

$$f(\chi) \sim [1 - (\boldsymbol{\sigma}\cdot\boldsymbol{\sigma}')]\ln\frac{1}{\pi - \chi}. \quad (5.15)$$

The approximation that we have used is evidently not strictly applicable here. The singularity in f for $\chi = \pi$ is a reflection of the singularity in the scattering amplitude of excitations colliding at an angle π (see Chap. IV). The correct expression can be obtained for this case by summing the main terms of the perturbation theory series, i.e. the terms in which the logarithm appears to a maximum power (one less than the power of a). If we regard χ as exactly equal to π, but $\lambda = p^2 + p'^2 - 2p_0^2 \neq 0$, summation leads to the appearance in f of the factor

$$\frac{1}{1 + a\left(\frac{3N}{\pi V}\right)^{1/3}\left(\ln\frac{p_0^2}{\lambda} + \frac{i\pi}{2}\right)} \quad (5.16)$$

(the real part is written down neglecting terms of higher order than the logarithm). Since a is positive by our hypothesis, this expression tends to zero as $\lambda \to 0$.

The case $a < 0$ is also possible in principle in the case of a Fermi gas. As distinct from a Bose gas, the Fermi gas will remain dilute by virtue of the Pauli principle, and at first sight all the formulae will retain their validity. If we look at (5.16), however, it becomes clear that the scattering amplitude will have a pole for some small imaginary value of λ. This is connected with the instability of the ground state relative to the formation of bound pairs of quasi-particles with opposite momenta and spins (Cooper effect), which is the main reason for the superconductivity of metals (see Chap. VII). We shall confine ourselves here to the case $a > 0$.

All in all, the expression that we have obtained for f does not hold at angles close to π. But in view of the fact that the singularity is logarithmic, it is only important in the immediate neighbourhood of the singularity. Moreover, only integrals of f with regular functions appear in the quantities of interest to us, so that the logarithmic singularity of f is of no importance.

On substituting (5.14) into (2.12), we find for the effective mass

$$\frac{m}{m^*} = 1 - \frac{8}{15}(7\ln 2 - 1)a^2\left(\frac{3N}{\pi V}\right)^{2/3}. \tag{5.17}$$

Similarly, we get for the sound velocity, from (2.19):

$$u^2 = \frac{\pi^{4/3}}{3^{1/3}}\left(\frac{N}{V}\right)^{2/3}\frac{1}{m^2} + 2\frac{\pi a}{m^2}\frac{N}{V}\left[1 + \frac{4}{15}a\left(\frac{3N}{\pi V}\right)^{1/3}(11 - 2\ln 2)\right]. \tag{5.18}$$

The ground state energy of a Fermi liquid is readily obtained from this formula. We make use of (2.16): $u^2 = (N/m)(\partial\mu/\partial N)$, and obtain after integrating (5.18) twice:

$$E = \int \mu \, dN = E^{(0)} + \frac{\pi a}{m}\frac{N^2}{V}\left[1 + \frac{6}{35}a\left(\frac{3N}{\pi V}\right)^{1/3}(11 - 2\ln 2)\right]. \tag{5.19}$$

The results (5.17) and (5.19) can also be obtained directly, by integration of (5.9) and (5.11). This demonstrates the validity of the main propositions of the theory of a Fermi liquid using our present model. A general derivation of these ideas will be offered in Chap. IV.

As in the case of a Bose gas, it will be interesting to conclude by finding the momentum distribution of the particles. To do this, we have to evaluate the matrix element

$$\bar{N}_{p,1/2} = \bar{N}_{p,-1/2} = \langle \Psi^* a^+_{p,1/2} a_{p,1/2} \Psi \rangle, \tag{5.20}$$

where Ψ is the true wave function of the interacting particles. We substitute the function Ψ, obtained from perturbation theory as far as second order terms (see [16]):

$$\Psi_n = \Psi_n^{(0)} + \sum_m{}' \frac{(H_{int})_{mn}\Psi_m^{(0)}}{E_n - E_m} + \sum_m{}' \sum_k{}' \frac{(H_{int})_{mk}(H_{int})_{kn}\Psi_m^{(0)}}{(E_n - E_k)(E_n - E_m)}$$
$$- (H_{int})_{nn}\sum_m{}' \frac{(H_{int})_{mn}\Psi_m^{(0)}}{(E_n - E_m)^2} - \frac{\Psi_n^{(0)}}{2}\sum_m{}' \frac{|(H_{int})_{mn}|^2}{(E_n - E_m)^2}. \tag{5.21}$$

On noting that the operator $a^+_{p1/2}\, a_{p1/2}$ is diagonal in the representation in terms of the functions $\Psi_n^{(0)}$, we get

$$\bar{N}_{p1/2} - n_{p1/2} = \sum_m{}' \frac{|(H_{int})_{m0}|^2 \left(n^{(m)}_{p1/2} - n_{p1/2}\right)}{(E_0 - E_m)^2}, \tag{5.22}$$

where $n^{(m)}_{p1/2}$ is the number of particles with momentum $\boldsymbol{p}$ and spin up in the state $\Psi_m^{(0)}$ of the non-interacting system and $n_{p1/2}$ is the number in the ground state. As already mentioned, the distribution of the non-interacting particles $n_{p1/2}$ is the same as the distribution of excitations at $T = 0$.

On substituting here H_{int} from (5.1), we get

$$\overline{N}_{\mathbf{p}1/2} - n_{\mathbf{p}1/2}$$

$$= \begin{cases} -\dfrac{16\pi^2 a^2}{(2\pi)^6 m^2} \displaystyle\int\limits_{|\mathbf{p}_1|<p_0} d^3\mathbf{p}_1 \int\limits_{|\mathbf{p}_2|>p_0} d^3\mathbf{p}_2 \int\limits_{|\mathbf{p}_3|>p_0} d^3\mathbf{p}_3 \frac{\delta(\mathbf{p}+\mathbf{p}_1-\mathbf{p}_2-\mathbf{p}_3)}{[(p^2+p_1^2-p_2^2-p_3^2)/2m]^2}, \\ \qquad \text{for } |\mathbf{p}| < p_0, \\ \dfrac{16\pi^2 a^2}{(2\pi)^6 m^2} \displaystyle\int\limits_{|\mathbf{p}_1|>p_0} d^3\mathbf{p}_1 \int\limits_{|\mathbf{p}_2|<p_0} d^3\mathbf{p}_2 \int\limits_{|\mathbf{p}_3|<p_0} d^3\mathbf{p}_3 \frac{\delta(\mathbf{p}+\mathbf{p}_1-\mathbf{p}_2-\mathbf{p}_3)}{[(p^2+p_1^2-p_2^2-p_3^2)/2m]^2}, \\ \qquad \text{for } |\mathbf{p}| > p_0. \end{cases} \tag{5.23}$$

Thus it turns out that there is only a second order difference in a between the momentum distribution of the particles and the distribution of the quasi-particles $n_{\mathbf{p}1/2}$. The integral of (5.23) over all $\mathbf{p}$ obviously vanishes, in agreement with the fact that the number of particles in the liquid is equal to the number of quasi-particles. It is of interest that the function $N_{\mathbf{p}1/2}$ has a discontinuity at $|\mathbf{p}| = p_0$.

It will be shown in Chap. III that this is a general property of Fermi liquids.

Evaluation of the integrals in (5.23)(†) leads to fairly unwieldy expressions, which we shall not write down here in full. We shall just give some limiting values:

$$\begin{aligned} N_{0,1/2} &= 1 - 2a^2\left(\frac{3}{\pi}\frac{N}{V}\right)^{2/3}\left(1 - \frac{1}{2}\ln 2\right), \\ N_{p_0-0,1/2} &= 1 - 2a^2\left(\frac{3}{\pi}\frac{N}{V}\right)^{2/3}\left(\frac{1}{3} + \ln 2\right), \\ N_{p_0+0,1/2} &= 2a^2\left(\frac{3}{\pi}\frac{N}{V}\right)^{2/3}\left(\ln 2 - \frac{1}{3}\right), \\ N_{p_0-0,1/2} - N_{p_0+0,1/2} &= 1 - 4a^2\left(\frac{3}{\pi}\frac{N}{V}\right)^{2/3}\ln 2, \\ N_{p\gg p_0,1/2} &= \frac{16a^2}{9}\left(\frac{3}{\pi}\frac{N}{V}\right)^{2/3}\left(\frac{p_0}{p}\right)^4. \end{aligned} \tag{5.24}$$

Thus N_p is close to 1 for $p < p_0$, decreasing slightly as p increases from 0 to p_0; N_p then drops discontinuously to a value of the order $a^2(N/V)^{2/3}$ and, for $p \gg p_0$, decreases proportional to $a^2(N/V)^{2/3}(p_0/p)^4$.

(†) This has been done by V. A. Belyakov [24].

CHAPTER II

QUANTUM FIELD THEORETICAL METHODS AT $T=0$

§ 6. THE INTERACTION REPRESENTATION

IN THE form described in the previous chapter, the method of second quantisation is unsuitable for solving a wide range of problems. It can actually only be applied in the case of weak interactions between the particles. Either perturbation theory is applicable in this case, or the Hamiltonian is so far simplified that it can easily be diagonalised. But we often find ourselves in a position where it is impossible to confine ourselves to the first few terms of the perturbation theory series. In these cases we need a method which will give reasonably simple and translucent rules for describing any term of this series.

Quite often, by virtue of the physical situation, it is possible to extract from the perturbation theory series a sequence (as a rule infinite) of the so-called "main" terms, of a higher order of magnitude than the remainder. The problem then reduces to the summation of this sequence.

In the general case, however, when all the terms of the perturbation theory series are of the same order, the problem consists in obtaining various general relationships (e.g. (2.1), connecting the Fermi boundary momentum p_0 and the number of particles of the liquid; this formula is at the basis of Landau's theory of a Fermi liquid). The most convenient approach for these purposes is the diagram technique developed in the present chapter, and borrowed from quantum field theory (†).

We shall start our exposition of the methods of quantum field theory by putting the method of second quantisation in a rather different form. We introduce the operators of a "field of particles":

$$\begin{aligned} \psi(\xi) &= \sum_i \varphi_i(\xi) a_i, \\ \psi^+(\xi) &= \sum_i \varphi_i^*(\xi) a_i^+, \end{aligned} \tag{6.1}$$

where a_i, a_i^+ are the second quantisation operators introduced in the previous chapter, $\varphi_i(\xi)$ is the wave function of the particle in state i. We can interpret $\psi(\xi)$ and $\psi^+(\xi)$ as operators producing the annihilation or creation of particles at a given point in ξ-space. The commutation

(†) See e.g. [25].

relations follow for these operators from § 3:

$$\begin{aligned} \psi(\xi)\psi^+(\xi') \mp \psi^+(\xi')\psi(\xi) &= \delta(\xi - \xi'), \\ \psi(\xi)\psi(\xi') \mp \psi(\xi')\psi(\xi) &= 0, \\ \psi^+(\xi)\psi^+(\xi') \mp \psi^+(\xi')\psi^+(\xi) &= 0, \end{aligned} \tag{6.2}$$

where the upper sign corresponds to Bose, and the lower to Fermi statistics. The one-particle operator $F^{(1)}$ may be written in the new representation in the form

$$F^{(1)} = \int \psi^+(\xi) f^{(1)} \psi(\xi) \, d\xi. \tag{6.3}$$

The two-particle and more complicated operators may be expressed similarly.

The Hamiltonian is readily expressed in terms of the operators ψ and ψ^+. For instance, the Hamiltonian for a system of spin $^1/_2$ particles in the absence of a magnetic field is

$$H = \int \left[\frac{1}{2m} \nabla \psi_\alpha^+(\boldsymbol{r}) \nabla \psi_\alpha(\boldsymbol{r}) + U(\boldsymbol{r}) \psi_\alpha^+(\boldsymbol{r}) \psi_\alpha(\boldsymbol{r}) \right] d^3\boldsymbol{r} + \frac{1}{2} \int\int \psi_\alpha^+(\boldsymbol{r}) \psi_\beta^+(\boldsymbol{r}') U^{(2)}(\boldsymbol{r}, \boldsymbol{r}') \psi_\beta(\boldsymbol{r}') \psi_\alpha(\boldsymbol{r}) \, d^3\boldsymbol{r} \, d^3\boldsymbol{r}' + \cdots. \tag{6.4}$$

We are assuming here that the interaction between the particles is independent of their spin. The indices α and β denote the z-component of the spin, summation being understood over pairs of repeated indices. The Hamiltonian for a system of bosons with zero spin only differs in having no indices for the operators ψ. The extension to more complicated cases presents no difficulty.

The form of (6.4) is the same as the expression for the mean energy of a system of N particles in identical states $\psi_\alpha(\boldsymbol{r})$, normalised by the relationship $\int |\psi_\alpha|^2 d\boldsymbol{r} = N$. The Hamiltonian in the second quantisation representation can always be easily found on the basis of this similarity.

The operator of the particle density at a given point is of importance as well as the Hamiltonian. Since it is $n(\boldsymbol{r}) = \sum_a \delta(\boldsymbol{r} - \boldsymbol{r}_a)$ in the usual representation, we get here:

$$n(\boldsymbol{r}) = \int \psi_\alpha^+(\boldsymbol{r}_a) \delta(\boldsymbol{r} - \boldsymbol{r}_a) \psi_\alpha(\boldsymbol{r}_a) \, d^3\boldsymbol{r}_a = \psi_\alpha^+(\boldsymbol{r}) \psi_\alpha(\boldsymbol{r}). \tag{6.5}$$

The operator of the number of particles is correspondingly

$$N = \int n(\boldsymbol{r}) \, d^3\boldsymbol{r} = \int \psi_\alpha^+(\boldsymbol{r}) \psi_\alpha(\boldsymbol{r}) \, d^3\boldsymbol{r}.$$

Now suppose that we have a system of particles with Hamiltonian H. Let us determine how the states of the system vary in time. To do this, we have to solve the Schrödinger equation

$$i \frac{\partial \Phi}{\partial t} = H \Phi \tag{6.6}$$

(Φ is the wave function of the system). The solution of (6.6) can be written in the symbolic form (†):

$$\Phi(t) = e^{-iHt}\Phi_H, \tag{6.7}$$

where Φ_H is a function independent of time.

The time variation of the matrix element of any operator F can be found from (6.7):

$$F_{nm}(t) = \langle \Phi_n^*(t) F \Phi_m(t) \rangle = \langle \Phi_{Hn}^* e^{iHt} F e^{-iHt} \Phi_{Hm} \rangle. \tag{6.8}$$

The last expression can be interpreted as a matrix element in the functions Φ_H of the operator

$$\tilde{F}(t) = e^{iHt} F e^{-iHt}. \tag{6.9}$$

This means the transformation to a new, the so-called Heisenberg representation. The representation considered earlier, in which the operators F are independent of time (for instance, $\psi(\boldsymbol{r})$ and $\psi^+(\boldsymbol{r})$), is known as the Schrödinger representation. The most important property of the Heisenberg representation is that the wave functions Φ_H are independent of time. The time dependence is carried over to the operators; we find from (6.9):

$$\frac{\partial \tilde{F}}{\partial t} = i(H\tilde{F} - \tilde{F}H) \equiv i[H, \tilde{F}]_-. \tag{6.10}$$

The situation is precisely the opposite in the Schrödinger representation. The operators are time independent (provided we are not speaking of a variable external field), whilst the wave function depends on time. It is clear from (6.9), that both representations are the same for the Hamiltonian itself.

If we consider the stationary state of the system, the wave function Φ_{Hn} satisfies the equation

$$H\Phi_{Hn} = E_n \Phi_{Hn}. \tag{6.11}$$

We have in this case, from (6.8):

$$F_{nm}(t) = \langle \Phi_{Hn}^* F \Phi_{Hm} \rangle e^{i(E_n - E_m)t}. \tag{6.12}$$

Let us take, for instance, a system of non-interacting particles without spin. We choose as the $\varphi_i(\xi)$ the free-particles eigenfunctions $(1/\sqrt{V})e^{i(\boldsymbol{p}\cdot\boldsymbol{r})}$ (V is the volume). In the Schrödinger representation, the operator ψ will be

$$\psi(\boldsymbol{r}) = \frac{1}{\sqrt{V}} \sum_{\boldsymbol{p}} a_{\boldsymbol{p}} e^{i(\boldsymbol{p}\cdot\boldsymbol{r})}. \tag{6.13}$$

We find by using (6.4) that the Hamiltonian in the Schrödinger representation has the form (3.13), i.e. $H = \sum \varepsilon_0(\boldsymbol{p}) n_{\boldsymbol{p}}$, where $\varepsilon_0(\boldsymbol{p})$ is the energy of the free particles. Hence, by (6.9), the operator $\tilde{\psi}(\boldsymbol{r}, t)$ in the

(†) The operator e^{-iHt} represents a symbolic way of writing the series $1 - iHt + \cdots + (1/n!)(-iHt)^n + \cdots$.

Heisenberg representation turns out to be

$$\tilde{\psi}(\mathbf{r}, t) = \frac{1}{\sqrt{V}} \sum_{\mathbf{p}} e^{i \sum_{\mathbf{p}'} \varepsilon_0(\mathbf{p}') n_{\mathbf{p}'} t} a_{\mathbf{p}} e^{-i \sum_{\mathbf{p}''} \varepsilon_0(\mathbf{p}'') n_{\mathbf{p}''} t} e^{i(\mathbf{p}\cdot\mathbf{r})}$$

$$= \frac{1}{\sqrt{V}} \sum_{\mathbf{p}} a_{\mathbf{p}} e^{i[(\mathbf{p}\cdot\mathbf{r}) - \varepsilon_0(\mathbf{p})t]}. \tag{6.14}$$

It must be remarked that the Heisenberg operators $\tilde{\psi}(\mathbf{r}, t)$ do not in general satisfy the commutation rules (6.2) for the corresponding Schrödinger operators. However, when the operators ψ are taken at a single instant, it follows from (6.9) and (6.2) that the commutation rules for these are the same as the rules for the Schrödinger operators $\psi(\mathbf{r})$.

In addition to the two representations mentioned, there is another one which is extremely important for what follows; it is of an intermediate type, and is known as the interaction representation. The properties of this representation are fundamental to the methods of quantum field theory.

We split off from the Hamiltonian the part H_{int}, corresponding to the particle interaction:

$$H = H_0 + H_{int}, \tag{6.15}$$

and carry out the following transformation of the Schrödinger wave function of the system:

$$\Phi_i = e^{iH_0 t} \Phi. \tag{6.16}$$

If we differentiate the function Φ_i with respect to time, we get

$$i \frac{\partial \Phi_i}{\partial t} = - H_0 \Phi_i + e^{iH_0 t}(H_0 + H_{int}) \Phi = e^{iH_0 t} H_{int} e^{-iH_0 t} \Phi_i. \tag{6.17}$$

Hence

$$i \frac{\partial \Phi_i}{\partial t} = \mathbf{H}_{int}(t) \Phi_i,$$

$$\mathbf{H}_{int}(t) = e^{iH_0 t} H_{int} e^{-iH_0 t}. \tag{6.18}$$

The interaction representation is achieved in terms of the functions Φ_i. Any operator is obtained in this representation from the Schrödinger representation in accordance with the same formula (6.18) as $\mathbf{H}_{int}(t)$. It follows that any operator $\mathbf{F}(t)$ in this representation satisfies the equation

$$\frac{\partial \mathbf{F}(t)}{\partial t} = i [H_0, \mathbf{F}(t)]_-, \tag{6.19}$$

i.e. the same equation as the Heisenberg operator for a system of non-interacting particles. We therefore arrive at the conclusion that all the operators in the interaction representation have the same form as the Heisenberg operators for the corresponding non-interacting system, whilst the wave function satisfies the Schrödinger equation with Hamiltonian $H_{int}(t)$. A great advantage of this representation is that it is possible to use "free" operators.

We now determine the time-dependence of the function $\Phi_i(t)$ in the interaction representation. Since the operators $H_{int}(t)$ do not commute with one another at different instants, we cannot simply write down the solution of equation (6.17) as

$$\Phi_i(t) = \text{const} \exp\left[-i \int^{t} H_{int}(t')\,dt'\right].$$

We therefore proceed as follows. Suppose we know the value of Φ_i at the instant t_0. We transform the differential equation (6.17) into an integral equation by integrating both sides with respect to t from t_0 to $t\,(t > t_0)$. We have:

$$\Phi_i(t) = \Phi_i(t_0) - i \int_{t_0}^{t} H_{int}(t')\,\Phi_i(t')\,dt'.$$

We look for the solution of this equation in the form of a series in H_{int}:

$$\Phi_i(t) = \Phi_i^{(0)}(t) + \Phi_i^{(1)}(t) + \cdots.$$

In the zero-th approximation, $\Phi_i^{(0)}(t) = \Phi_i(t_0)$. To a first approximation,

$$\Phi_i^{(1)}(t) = -i \int_{t_0}^{t} H_{int}(t_1)\,dt_1\,\Phi_i(t_0);$$

to a second approximation,

$$\Phi_i^{(2)}(t) = -\int_{t_0}^{t} H_{int}(t_1)\,dt_1 \int_{t_0}^{t_1} H_{int}(t_2)\,dt_2\,\Phi_i(t_0);$$

to the nth approximation,

$$\Phi_i^{(n)}(t) = (-i)^n \int_{t_0}^{t} H_{int}(t_1)\,dt_1 \int_{t_0}^{t_1} H_{int}(t_2)\,dt_2 \cdots \int_{t_0}^{t_{n-1}} H_{int}(t_n)\,dt_n\,\Phi_i(t_0).$$

It follows from the structure of the series for $\Phi_i(t)$ that the complete result can be written as

$$\Phi_i(t) = S(t, t_0)\,\Phi_i(t_0), \tag{6.20}$$

where the matrix $S(t, t_0)$ is given by the series

$$S(t, t_0) = 1 - i \int_{t_0}^{t} H_{int}(t_1)\,dt_1 + \cdots$$
$$\cdots + (-i)^n \int_{t_0}^{t} H_{int}(t_1)\,dt_1 \cdots \int_{t_0}^{t_{n-1}} H_{int}(t_n)\,dt_n + \cdots. \tag{6.21}$$

The characteristic feature of series (6.21) is that the operators H_{int}, taken at later instants, always appear to the left of operators at earlier instants, since we always have

$$t > t_1 > t_2 > \cdots > t_n > t_0.$$

We can make (6.21) more symmetrical. Let us take the nth term

$$(-i)^n \int \cdots \int_{t > t_1 \cdots > t_0} H_{int}(t_1)\,H_{int}(t_2) \cdots H_{int}(t_n)\,dt_1\,dt_2 \cdots dt_n$$

and arbitrarily transform the variables of integration $t_1, \ldots, t_n \rightarrow t_{p_1}, t_{p_2}, \ldots, t_{p_n}$, which naturally leaves the term unchanged. On carrying out all the possible permutations of the variables $t_1, \ldots, t_n$, adding all the expressions and dividing by the number of permutations $n!$, the domain of integration will be extended for each variable to the full interval from t_0 to t. At the same time, it is essential that all the operators $\boldsymbol{H}_{int}$ be arranged in decreasing order of time from left to right. On denoting by T the operator of this ordering, the time-ordering operator, we can write the nth term of the series as

$$S^{(n)}(t, t_0) = \frac{(-i)^n}{n!} \int_{t_0}^{t} \cdots \int_{t_0}^{t} T\{\boldsymbol{H}_{int}(t_1) \cdots \boldsymbol{H}_{int}(t_n)\} \mathrm{d}t_1 \cdots \mathrm{d}t_n. \quad (6.22)$$

It is now easily verified that (6.21) can be written as

$$S(t, t_0) = T \exp\left\{-i \int_{t_0}^{t} \boldsymbol{H}_{int}(t') \mathrm{d}t'\right\}, \quad (6.23)$$

as can be seen by expanding the exponent into a series and using the definition of T. The operator $S(t, t_0)$ has the obvious property

$$S(t_2, t_1) S(t_1, t_0) = S(t_2, t_0), \quad t_2 > t_1 > t_0. \quad (6.24)$$

Equations (6.16) and (6.18) establish the connection between the Schrödinger and the interaction representation. The connection between the interaction and Heisenberg representations may be found from (6.20). Suppose that the transformation of the wave functions is given by

$$\Phi_i(t) = Q(t) \Phi_H,$$

where Q is a unitary operator. We have from (6.20):

$$Q(t) = S(t, t_0) Q(t_0),$$

whence it follows, by (6.24), that

$$Q(t) = S(t, \alpha) P,$$

where α is a certain instant, and P is a time-independent operator. To find P, we substitute into the relation $\Phi_i(t) = Q(t) \Phi_H$ expressions (6.16) and (6.7) for Φ_i and Φ_H in terms of the Schrödinger function Φ. This gives

$$e^{iH_0 t} = S(t, \alpha) P e^{iHt}.$$

Observing that $S(\alpha, \alpha) = 1$, we have

$$P = e^{iH_0\alpha} e^{-iH\alpha}.$$

It is convenient at this stage to bring in an assumption regarding the so-called "adiabatic switching on of the interaction" (†). Suppose that, at the instant $t = -\infty$, there is no interaction between the particles,

(†) It must be emphasised at once that our use here of "adiabatic switching on of the interaction" is purely formal. It enables the correct result to be obtained in the shortest way, but is by no means necessary (see e.g. [26]).

but that interaction is then "brought in" infinitely slowly. If we now let $\alpha \to -\infty$, then $P \to 1$, i.e.

$$\Phi_i(t) = S(t)\Phi_H, \tag{6.25}$$

where

$$S(t) = S(t, -\infty). \tag{6.26}$$

Using (6.24), we get

$$S(t_2, t_1) = S(t_2)S^{-1}(t_1), \quad t_1 < t_2. \tag{6.27}$$

The relationship between the operators in the interaction representation and the Heisenberg operators becomes, from (6.25):

$$\tilde{F}(t) = S^{-1}(t)\,\boldsymbol{F}(t)\,S(t). \tag{6.28}$$

We shall often encounter the following time-ordered products of several Heisenberg operators, averaged over the ground state of the system Φ_H^0:

$$\langle \Phi_H^{0*}\, T\,[\tilde{A}(t)\,\tilde{B}(t')\,\tilde{C}(t'')\cdots]\Phi_H^0\rangle. \tag{6.29}$$

In the case of Fermi operators we shall somewhat extend the definition of T-ordering as compared with that given in the derivation of (6.23); whereas we retain this definition for Bose operators. We shall now understand, by the T-product of operators $A(t_1)\,B(t_2)\,C(t_3)\ldots$ their product from left to right in order of decreasing time, multiplied by $(-1)^P$, where P is the number of permutations of the Fermi operators with one another, required to obtain the time-ordered product from $A(t_1)\,B(t_2)\,C(t_3)\ldots$ Thus if $F_1(t_1)$, $F_2(t_2)$ are Fermi, and $B_1(t_3)$, $B_2(t_4)$ are Bose operators, we have

$$T\{F_1(t_1)\,F_2(t_2)\} = \begin{cases} F_1(t_1)\,F_2(t_2), & t_1 > t_2, \\ -F_2(t_2)\,F_1(t_1), & t_1 < t_2, \end{cases}$$

$$T\{B_1(t_3)\,F_1(t_1)\} = \begin{cases} B(t_3)\,F_1(t_1), & t_3 > t_1, \\ F_1(t_1)\,B_1(t_3), & t_3 < t_1, \end{cases}$$

$$T\{B_1(t_3)\,B_2(t_4)\} = \begin{cases} B_1(t_3)\,B_2(t_4), & t_3 > t_4, \\ B_2(t_4)\,B_1(t_3), & t_3 < t_4. \end{cases}$$

The new definition of T-ordering is the same as the old one in the case of $H_{int}(t)$, since Fermi operators always appear in pairs in H_{int}. Of course all the rules for T-ordering are the same for operators in the Heisenberg and in the interaction representation.

Let the time order in (6.29) be such that

$$t > t' > t'' > \cdots.$$

We use (6.28) to change to the operators in the interaction representation. This gives:

$$\begin{aligned} &\langle \Phi_H^{0*} S^{-1}(t)\,\boldsymbol{A}(t)\,S(t)\,S^{-1}(t')\,\boldsymbol{B}(t')\,S(t')\cdots\Phi_H^0\rangle \\ &= \langle \Phi_H^{0*} S^{-1}(\infty)\,S(\infty, t)\,\boldsymbol{A}(t)\,S(t, t')\,\boldsymbol{B}(t')\cdots\Phi_H^0\rangle \\ &= \langle \Phi_H^{0*} S^{-1}(\infty)\,T\,[\boldsymbol{A}(t)\boldsymbol{B}(t')\boldsymbol{C}(t'')\cdots S(\infty)]\Phi_H^0\rangle. \end{aligned} \tag{6.30}$$

The transformation from (6.29) to (6.30) is obviously independent of the order of the times t, t', t'', i.e. it holds always.

It now remains for us to determine $\Phi_H^{0*} S^{-1}(\infty) = [S(\infty)\Phi_H^0]^*$, i.e. the result of the action of the operator $S(\infty)$ on the ground state wave function. It follows from (6.20) and (6.25) that

$$\Phi_H^0 = \Phi_i(-\infty), \quad S(\infty)\Phi_H^0 = \Phi_i(\infty).$$

Hence $S(\infty)\Phi_H^0$ is the function $\Phi_i(\infty)$ which was obtained from the ground state wave function $\Phi_i(-\infty)$ as a result of adiabatically switching on the interaction between the particles. We know that the ground state of the system, i.e. the state in which the energy is minimal, must be non-degenerate. But by the general principles of quantum mechanics (see [16]), a system which is in a non-degenerate stationary state cannot be carried to another state under the action of infinitely slow excitation. We can therefore conclude that the function $\Phi_i(\infty) = S(\infty)\Phi_H^0$ can only differ from Φ_H^0 by a phase factor:

$$S(\infty)\Phi_H^0 = e^{iL}\Phi_H^0. \tag{6.31}$$

Hence follows finally the relation

$$\langle\Phi_H^{0*}\, T[\tilde{A}(t)\tilde{B}(t')\tilde{C}(t'')\cdots]\Phi_H^0\rangle = \frac{\langle\Phi_H^{0*}\, T[\boldsymbol{A}(t)\boldsymbol{B}(t')\boldsymbol{C}(t'')\cdots S(\infty)]\Phi_H^0\rangle}{\langle\Phi_H^{0*}S(\infty)\Phi_H^0\rangle}. \tag{6.32}$$

Let us emphasise that this conclusion only holds for averaging over the ground state of the system, since any other energy level of the system is multiply degenerate, and the system in general passes to another state as a result of collisions between the particles. Thus (6.30) holds for averaging over an excited state, but not (6.32).

We shall consider in this chapter systems at $T = 0$, i.e. in the ground state. For simplicity, we shall denote the corresponding averages simply by $\langle\cdots\rangle$ and write operators in the interaction representation in ordinary type. In cases where Schrödinger operators are required, we shall emphasize that they depend only on the coordinates (e.g. $\psi(\boldsymbol{r})$), whilst at the same time specifically indicating such cases.

§ 7. THE GREEN FUNCTION(†)

1. *Definition. Free particle Green functions*

One of the most important characteristics of the microscopic properties of a system in quantum field theory is the one-particle Green

(†) This section is largely based on a paper by V. M. Galitskii and A. B. Migdal [27].

function (†). It is defined as

$$G_{\alpha\beta}(x, x') = -i\langle T(\tilde{\psi}_\alpha(x)\tilde{\psi}_\beta^+(x'))\rangle. \quad (7.1)$$

We understand by x (or x') a set of four variables — the coordinates $\boldsymbol{r}$ and the time t; α and β are the spin indices.

A knowledge of the Green function enables us to find the average over the ground state of any one-particle operator of the type (3.2). Indeed, we have from (6.3):

$$\overline{F^{(1)}} = \pm i \int \left[\lim_{\substack{t' \to t+0 \\ \boldsymbol{r}' \to \boldsymbol{r}}} f_{\alpha\beta}^{(1)}(x) G_{\alpha\beta}(x, x') \right] \mathrm{d}^3\boldsymbol{r}$$

(the plus (minus) sign for Bose (Fermi) statistics). For instance, the density of the number of particles and the particle flux density are, respectively, equal to

$$n(x) = \pm i \lim_{\substack{t' \to t+0 \\ \boldsymbol{r}' \to \boldsymbol{r}}} G_{\alpha\alpha}(x, x'),$$

$$j(x) = \pm \frac{1}{2m} \lim_{\substack{t' \to t+0 \\ \boldsymbol{r}' \to \boldsymbol{r}}} (\nabla_{\boldsymbol{r}} - \nabla_{\boldsymbol{r}'}) G_{\alpha\alpha}(x, x').$$

We show below that the Green function can be used to find the energy as a function of the volume, and hence the equations of state of the system (the dependence of the pressure on the density) at $T = 0$. It will further be shown that the poles of the Fourier transform of the Green function (7.1) determine the excitation spectrum. This enables the thermodynamic functions of the system to be found at temperatures different from zero (though reasonably low, of course).

A fact of extreme importance is that the Green function can be evaluated by means of so-called diagram techniques (see §§ 8 and 9), which have considerable advantages over perturbation theory in the ordinary form.

We shall be concerned in this section with an analysis of the general properties of the Green function. For typographical simplicity we shall omit the indices α, β. This cannot lead to misunderstandings, since in the absence of ferromagnetism and external magnetic field $G_{\alpha\beta}$ must be of the form $G_{\alpha\beta} = G\delta_{\alpha\beta}$. We shall confine ourselves to these cases.

We shall consider in this chapter the properties of systems of Fermi particles; we know that a Bose system has a number of special features at absolute zero, connected with the existence of the condensate; these

(†) The term "Green function" has not the same meaning in field theory as in the theory of linear equations. Although the Green function satisfies an equation with a δ-function on the right-hand side, this equation is generally speaking non-linear (see § 10). The free particle Green functions are exceptions: they are the Green functions of the linear equations for the Heisenberg field operators $\tilde{\psi}(\boldsymbol{r}, t)$. The term Green function was originally only applied to this case, but was later extended to expression (7.1) for any interacting system.

latter systems will be discussed in Chap. V. The phonons (lattice-vibration quanta) provide an exception. In view of the fact that their number is not given, no condensation can occur in the phonon gas and its properties can be considered by ordinary methods.

The Green functions of homogeneous, spatially infinite systems depend, in the absence of external fields, only on the differences $\boldsymbol{r}-\boldsymbol{r}'$ and $t-t'$. We expand G into a Fourier integral:

$$G(x-x') = \int \frac{d^4p}{(2\pi)^4} G(\boldsymbol{p}, \omega)\, e^{i[(\boldsymbol{p}\cdot\boldsymbol{r}-\boldsymbol{r}')-\omega(t-t')]} \quad (d^4p = d^3\boldsymbol{p}\, d\omega). \tag{7.2}$$

We can find $G(\boldsymbol{p}, \omega)$ very simply for a system of non-interacting particles. In the case of a system of fermions, on substituting in (7.1) expression (6.14) for the Heisenberg operators of the free field and taking into account that all the levels with $|\boldsymbol{p}| < p_0$ are occupied, whilst those with $|\boldsymbol{p}| > p_0$ are empty, we have:

$$G^{(0)}(x) = -\frac{i}{V} \sum_{\boldsymbol{p}} e^{i[(\boldsymbol{p}\cdot\boldsymbol{r})-\varepsilon_0(\boldsymbol{p})t]} \begin{cases} 1-n_{\boldsymbol{p}}, & t>0, \\ -n_{\boldsymbol{p}}, & t<0, \end{cases} \tag{7.3}$$

where

$$n_{\boldsymbol{p}} = a_{\boldsymbol{p}}^{+} a_{\boldsymbol{p}} = \begin{cases} 1, & |\boldsymbol{p}| < p_0, \\ 0, & |\boldsymbol{p}| > p_0. \end{cases}$$

We go over to the momentum representation. We have by (7.3):

$$G^{(0)}(\boldsymbol{p}, \omega) = -i \left\{ \theta(|\boldsymbol{p}| - p_0) \int_0^\infty e^{i[\omega-\varepsilon_0(\boldsymbol{p})]t} dt \right.$$
$$\left. - \theta(p_0 - |\boldsymbol{p}|) \int_0^\infty e^{-i[\omega-\varepsilon_0(\boldsymbol{p})]t} dt \right\}, \tag{7.4}$$

where

$$\theta(z) = \begin{cases} 1, & z>0, \\ 0, & z<0. \end{cases}$$

The expression for $G(\boldsymbol{p}, \omega)$ contains two integrals of the type

$$\int_0^\infty e^{ist} dt.$$

An integral of this type may be found as the limit

$$\lim_{\delta\to+0} \int_0^\infty e^{ist-\delta t} dt = i \lim_{\delta\to+0} \frac{1}{s+i\delta}. \tag{7.5}$$

The $i\delta$ in the denominator indicates the method of going round the pole $s=0$ when integrating this function, namely,

$$\int F(s) \frac{ds}{s+i\delta} = \fint \frac{F(s)}{s} ds - i\pi F(0),$$

where $⨍$ denotes the principal value of the integral. We can therefore write:

$$\frac{1}{s+i\delta} = \frac{1}{s} - i\pi\delta(s).$$

The symbol $\delta_+(s)$ is sometimes used for denoting $(1/\pi)\,(i/(s+i\delta))$:

$$\delta_+(s) = \delta(s) - \frac{1}{i\pi s}. \tag{7.6}$$

We obtain from (7.4) and (7.5):

$$G^{(0)}(\boldsymbol{p}, \omega) = \frac{\theta(|\boldsymbol{p}| - p_0)}{\omega - \varepsilon_0(\boldsymbol{p}) + i\delta} + \frac{\theta(p_0 - |\boldsymbol{p}|)}{\omega - \varepsilon_0(\boldsymbol{p}) - i\delta}.$$

Observing that the only difference between the formula for G when $|\boldsymbol{p}| < p_0$ and the formula when $|\boldsymbol{p}| > p_0$ is in the change of sign in front of δ, we can finally write

$$G^{(0)}(\boldsymbol{p}, \omega) = \frac{1}{\omega - \varepsilon_0(\boldsymbol{p}) + i\delta \operatorname{sign}(|\boldsymbol{p}| - p_0)}. \tag{7.7}$$

Let us now consider a system of phonons, and confine ourselves to the elementary case of longitudinal vibrations in a continuous isotropic medium.

Let us first of all determine what is to be understood by the phonon field operators.

Let $\boldsymbol{q}(\boldsymbol{r}, t)$ denote the displacement of a point of the medium. The momentum per unit volume will be equal to $\varrho\,\dot{\boldsymbol{q}}(\boldsymbol{r}, t)$, where ϱ is the density. In accordance with quantum mechanics, the $\boldsymbol{q}$ and $\dot{\boldsymbol{q}}$ are replaced by operators with commutation relations

$$\varrho\,[\dot{q}_i(\boldsymbol{r}, t), q_k(\boldsymbol{r}', t)] = -i\delta(\boldsymbol{r} - \boldsymbol{r}')\,\delta_{ik}. \tag{7.8}$$

The integral of (7.8) over a small volume $d^3\boldsymbol{r}$ yields the usual commutation rule for a coordinate and momentum.

We expand the operator $\boldsymbol{q}$ in plane waves. Here, a specification of the wave vector $\boldsymbol{k}$ uniquely determines the frequency, which we denote by $\omega_0(\boldsymbol{k})$. We therefore have

$$\boldsymbol{q}(\boldsymbol{r}, t) = \frac{1}{\sqrt{V}} \sum_{\boldsymbol{k}} \frac{\boldsymbol{k}}{|\boldsymbol{k}|} \{q_{\boldsymbol{k}} e^{i[(\boldsymbol{k}\cdot\boldsymbol{r}) - \omega_0(\boldsymbol{k})t]} + q_{\boldsymbol{k}}^+ e^{-i[(\boldsymbol{k}\cdot\boldsymbol{r}) - \omega_0(\boldsymbol{k})t]}\}. \tag{7.9}$$

We are considering longitudinal waves, so that the Fourier components of the vector $\boldsymbol{q}$ will be directed along the wave vector $\boldsymbol{k}$. In view of this we shall in future use the projection $\boldsymbol{q}_{\boldsymbol{k}}$ on the direction $\boldsymbol{k}$, which we denote by $q_{\boldsymbol{k}}$.

We now introduce operators $b_{\boldsymbol{k}}$, connected with the $q_{\boldsymbol{k}}$ by

$$q_{\boldsymbol{k}} = \frac{b_{\boldsymbol{k}}}{\sqrt{2\varrho\,\omega_0(\boldsymbol{k})}}. \tag{7.10}$$

It then follows from (7.8) that the operators b_k satisfy the usual commutation relations for Bose creation and annihilation operators.

The operator of the kinetic energy of the vibrations is equal to

$$K = \frac{\varrho}{2} \int [\dot{q}(\boldsymbol{r}, t)]^2 \mathrm{d}^3\boldsymbol{r}. \tag{7.11}$$

Using the fact that the mean kinetic energy of the vibrations is equal to the mean potential energy, we arrive at

$$\overline{H} = 2\overline{K} = \sum_{\boldsymbol{k}} \omega_0(\boldsymbol{k}) \left(n_{\boldsymbol{k}} + \frac{1}{2}\right), \tag{7.12}$$

where $n_{\boldsymbol{k}} = b_{\boldsymbol{k}}^+ b_{\boldsymbol{k}}$.

The displacement operators $\boldsymbol{q}$ could be taken as the free phonon field operators. However, it is more convenient to use a rather different definition, in view of the investigation of the interaction of the phonons with the electrons in a metal (see § 8), namely

$$\tilde{\varphi}(x) = \frac{i}{\sqrt{V}} \sum_{\boldsymbol{k}} \sqrt{\frac{\omega_0(\boldsymbol{k})}{2}} \{b_{\boldsymbol{k}} e^{i[(\boldsymbol{k}\cdot\boldsymbol{r}) - \omega_0(\boldsymbol{k})t]} - b_{\boldsymbol{k}}^+ e^{-i[(\boldsymbol{k}\cdot\boldsymbol{r}) - \omega_0(\boldsymbol{k})t]}\}. \tag{7.13}$$

This formula refers to the longitudinal phonons in the Debye model (see § 1), if we restrict the summation over $\boldsymbol{k}$ by the condition $|\boldsymbol{k}| < k_D$.

Let us emphasise once more that the phonon field operators are real, since they correspond to real displacements of the atoms of the lattice. This property is obviously retained when discussing the interaction of phonons with one another and with other particles.

The Green function of the phonons is usually denoted by D. The definition of this function is similar to (7.1):

$$D(x, x') = -i\langle T(\tilde{\varphi}(x)\tilde{\varphi}(x'))\rangle. \tag{7.14}$$

On substituting the free operators (7.13) as the $\tilde{\varphi}(x)$ in this, and recalling that there are no phonons in the ground state, we get

$$D^{(0)}(x) = -\frac{i}{V} \sum_{\boldsymbol{k}} \frac{\omega_0(\boldsymbol{k})}{2} \begin{cases} e^{i[(\boldsymbol{k}\cdot\boldsymbol{r}) - \omega_0(\boldsymbol{k})t]}, & t > 0, \\ e^{-i[(\boldsymbol{k}\cdot\boldsymbol{r}) - \omega_0(\boldsymbol{k})t]}, & t < 0. \end{cases} \tag{7.15}$$

If we take the Fourier components of this expression with respect to $\boldsymbol{r}$ and t, we get

$$\begin{aligned} D^{(0)}(\boldsymbol{k}, \omega) &= \frac{\omega_0(\boldsymbol{k})}{2} \left[\frac{1}{\omega - \omega_0(\boldsymbol{k}) + i\delta} - \frac{1}{\omega + \omega_0(\boldsymbol{k}) - i\delta}\right] \\ &= \frac{\omega_0^2(\boldsymbol{k})}{\omega^2 - \omega_0^2(\boldsymbol{k}) + i\delta}. \end{aligned} \tag{7.16}$$

2. *Analytic properties*

Let us now consider the general properties of the Green functions of systems of interacting particles. We shall start with a Fermi system.

In terms of Schrödinger operators, we get

$$G(\mathbf{r}-\mathbf{r}', t-t') = -i\langle e^{iHt}\psi(\mathbf{r})e^{-iH(t-t')}\psi^{+}(\mathbf{r}')e^{-iHt'}\rangle$$

$$= -i\sum_{s}\langle \Phi_H^{0*}e^{iHt}\psi(\mathbf{r})e^{-iHt}\Phi_s\rangle\langle\Phi_s^{*}e^{iHt'}\psi(\mathbf{r}')e^{-iHt'}\Phi_H^0\rangle$$

$$= -i\sum_{s}\psi_{0s}(\mathbf{r})\psi_{s0}^{+}(\mathbf{r}')e^{-i(E_s-E_0)(t-t')}, \qquad t>t',$$

$$G(\mathbf{r}-\mathbf{r}', t-t') = i\sum_{s'}\psi_{0s'}^{+}(\mathbf{r}')\psi_{s'0}(\mathbf{r})e^{i(E_{s'}-E_0)(t-t')}, \qquad t<t'.$$

The dependence of the matrix elements $\psi_{nm}(\mathbf{r})$ and $\psi_{nm}(\mathbf{r})$ on the coordinates is, for a homogeneous system:

$$\psi_{nm}(\mathbf{r}) = \psi_{nm}(0)e^{-i(\mathbf{p}_{nm}\cdot\mathbf{r})}, \qquad \psi_{nm}^{+}(\mathbf{r}) = \psi_{nm}^{+}(0)e^{i(\mathbf{p}_{nm}\cdot\mathbf{r})},$$

where $\mathbf{p}_{nm} = \mathbf{p}_n - \mathbf{p}_m$, and $\mathbf{p}_n, \mathbf{p}_m$ are the momenta of the system in states n and m(†). We have, taking $\mathbf{p}_0 = 0$:

$$G(\mathbf{r}-\mathbf{r}', t-t') = -i\sum_{s}|\psi_{0s}(0)|^2 e^{i(\mathbf{p}_s\cdot\mathbf{r}-\mathbf{r}')}e^{-i(E_s-E_0)(t-t')}, \quad t>t',$$

$$G(\mathbf{r}-\mathbf{r}', t-t') = i\sum_{s'}|\psi_{s'0}(0)|^2 e^{-i(\mathbf{p}_{s'}\cdot\mathbf{r}-\mathbf{r}')}e^{i(E_{s'}-E_0)(t-t')}. \tag{7.17}$$

The operator $\psi^{+}(r)$ increases the number of particles by one. In view of this, the summation over s for $t > t'$ is performed over the states where the number of particles is $N+1$. Conversely, the summation over s' for $t < t'$ is performed over the states where the number of particles is $N-1$. We introduce the notation

$$E_s - E_0(N) = \varepsilon_s + \mu, \tag{7.18}$$

where

$$\varepsilon_s = E_s - E_0(N+1) \tag{7.19}$$

is the excitation energy of the system, which is positive by definition, whilst $\mu = E_0(N+1) - E_0(N)$ is the chemical potential at $T = 0$. Similarly,

$$E_{s'} - E_0(N) = E_{s'} - E_0(N-1) - [E_0(N) - E_0(N-1)] = \varepsilon_{s'} - \mu'. \tag{7.18'}$$

The $\varepsilon_{s'}$ and μ' in the last formula refer to a system of $N-1$ particles. It is possible, however, to assume $\varepsilon_s = \varepsilon_{s'}$, $\mu = \mu'$. This only introduces an error of order $1/N$.

Further, we introduce the functions

$$\begin{aligned} A(\mathbf{p}, E)\,\mathrm{d}E &= (2\pi)^3\sum_{s}|\psi_{0s}(0)|^2\delta(\mathbf{p}-\mathbf{p}_s), \qquad E<\varepsilon_s<E+\mathrm{d}E,\\ B(\mathbf{p}, E)\,\mathrm{d}E &= (2\pi)^3\sum_{s'}|\psi_{s'0}(0)|^2\delta(\mathbf{p}+\mathbf{p}_s), \qquad E<\varepsilon_{s'}<E+\mathrm{d}E. \end{aligned} \tag{7.20}$$

(†) This follows from the fact that, in accordance with quantum mechanics (see [16]), the operator corresponding to a translation is $e^{i(\hat{\mathbf{p}}\cdot\mathbf{r})}$ ($\hat{\mathbf{p}}$ is the momentum operator). Consequently, $\psi(\mathbf{r}) = e^{-i(\hat{\mathbf{p}}\cdot\mathbf{r})}\,\psi(0)\,e^{i(\hat{\mathbf{p}}\cdot\mathbf{r})}$. Notice, by the way, that if we write $\psi(\mathbf{r})$ as $\psi(\mathbf{r}) = (1/\sqrt{V})\,\Sigma\, a_{\mathbf{p}}\, e^{i(\mathbf{p}\cdot\mathbf{r})}$, we evidently get $\psi_{nm}(0) = (1/\sqrt{V})\,(a_{-\mathbf{p}_{nm}})_{nm}$.

We now expand G as a Fourier integral(†):

$$G(\mathbf{p}, \omega) = \int_0^{\infty} \mathrm{d}E \left\{\frac{A(\mathbf{p}, E)}{\omega - E - \mu + i\delta} + \frac{B(\mathbf{p}, E)}{\omega + E - \mu - i\delta}\right\}. \tag{7.21}$$

The coefficients A, B in this expression are real and positive. The analytic properties of the function $G(\mathbf{p}, \omega)$ can be investigated using (7.21).

On separating the real and imaginary parts of G, we find that

$$\mathrm{Re}\, G(\mathbf{p}, \omega) = \fint_0^{\infty} \mathrm{d}E \left\{\frac{A(\mathbf{p}, E)}{\omega - E - \mu} + \frac{B(\mathbf{p}, E)}{\omega + E - \mu}\right\}, \tag{7.22}$$

$$\mathrm{Im}\, G(\mathbf{p}, \omega) = \begin{cases} -\pi A(\mathbf{p}, \omega - \mu), & \omega > \mu, \\ \pi B(\mathbf{p}, \mu - \omega), & \omega < \mu \end{cases} \tag{7.23}$$

($\fint$ denotes the principal part of the integral). The imaginary part of the Green function thus changes sign at $\omega = \mu$. A comparison of (7.23) and (7.22) leads to the following relation between the real and imaginary parts of G:

$$\mathrm{Re}\, G(\mathbf{p}, \omega) = \frac{1}{\pi} \int_{-\infty}^{\infty} \frac{\mathrm{Im}\, G(\mathbf{p}, \omega') \,\mathrm{sign}\,(\omega' - \mu)}{\omega' - \omega} \mathrm{d}\omega'. \tag{7.24}$$

An asymptotic formula for G as $\omega \to \infty$ can be got from (7.21) and (7.20):

$$G(\mathbf{p}, \omega) \to \frac{1}{\omega} \int_0^{\infty} \mathrm{d}E \,[A(\mathbf{p}, E) + B(\mathbf{p}, E)]$$

$$= \frac{1}{\omega}\left\{(2\pi)^3 \sum_s |\psi_{0s}(0)|^2 \delta(\mathbf{p} - \mathbf{p}_s) + (2\pi)^3 \sum_{s'} |\psi_{s'0}(0)|^2 \delta(\mathbf{p} + \mathbf{p}_{s'})\right\}.$$

It is easily seen that the coefficient of $1/\omega$ is equal to the Fourier component of the anti-commutator

$$\psi(\mathbf{r})\psi^+(\mathbf{r}') + \psi^+(\mathbf{r}')\psi(\mathbf{r}) = \delta(\mathbf{r} - \mathbf{r}'),$$

i.e. unity. This is proved simply by averaging the anticommutator over the ground state (which does not change its value), transforming the resulting average as in (7.17) and taking the Fourier component with respect to $\mathbf{r} - \mathbf{r}'$. We obtain in this way

$$G(\mathbf{p}, \omega) \to \frac{1}{\omega} \quad \text{as} \quad \omega \to \infty. \tag{7.21'}$$

As regards the properties of G as a function of the complex variable ω, it follows from (7.24) that it is not analytic. The connection between the real and imaginary parts of a function analytic in the upper half-plane is an expression that differs from (7.24) in replacing sign $(\omega' - \mu)$ by unity. We have -1 instead of sign $(\omega' - \mu)$ in the case of a function analytic in the lower half-plane.

(†) A formula of this type was first obtained by LEHMANN [28] in quantum field theory.

Let us consider alongside G the two functions G_R, G_A, analytic in the upper and lower half-planes respectively and defined by the relations (for real ω):

$$\begin{aligned} \operatorname{Re} G &= \operatorname{Re} G_R = \operatorname{Re} G_A\,, \\ \operatorname{Im} G_R &= \operatorname{Im} G \operatorname{sign}(\omega - \mu)\,, \\ \operatorname{Im} G_A &= -\operatorname{Im} G \operatorname{sign}(\omega - \mu)\,. \end{aligned} \tag{7.25}$$

It follows from (7.25) that G_R is the same as G^* on the real semi-axis $\omega - \mu < 0$; similarly, G_A is the same as G^* for $\omega - \mu > 0$. We can thus write:

$$\begin{aligned} G_R(\mathbf{p}, \omega) &= \begin{cases} G(\mathbf{p}, \omega), & \omega > \mu, \\ G^*(\mathbf{p}, \omega), & \omega < \mu, \end{cases} \\ G_A(\mathbf{p}, \omega) &= \begin{cases} G^*(\mathbf{p}, \omega), & \omega > \mu, \\ G(\mathbf{p}, \omega), & \omega < \mu. \end{cases} \end{aligned} \tag{7.25'}$$

It follows from (7.25′) that G_R is the analytic continuation of G from the semi-axis $\omega > \mu$, and G_A the continuation of G from the semi-axis $\omega < \mu$.

The functions G_R and G_A are defined as follows in the coordinate representation:

$$\left.\begin{aligned} G_R(x - x') &= \begin{cases} -i\langle \tilde{\psi}(x)\tilde{\psi}^+(x') + \tilde{\psi}^+(x')\tilde{\psi}(x)\rangle & \text{as } t > t', \\ 0 & \text{as } t < t', \end{cases} \\ G_A(x - x') &= \begin{cases} 0 & \text{as } t > t', \\ i\langle \tilde{\psi}^+(x')\tilde{\psi}(x) + \tilde{\psi}(x)\tilde{\psi}^+(x')\rangle & \text{as } t < t'. \end{cases} \end{aligned}\right\} \tag{7.26}$$

Indeed, on carrying out the same operations as were performed when deriving (7.21), we get

$$G_R(\mathbf{p}, \omega) = \int_0^\infty \mathrm{d}E \left\{ \frac{A(\mathbf{p}, E)}{\omega - E - \mu + i\delta} + \frac{B(\mathbf{p}, E)}{\omega + E - \mu + i\delta} \right\}, \tag{7.27}$$

$$G_A(\mathbf{p}, \omega) = G_R^*(\mathbf{p}, \omega).$$

On comparing the real and imaginary parts of G_R and G_A with (7.22) and (7.23), these functions are easily seen to satisfy (7.25). G_R, G_A are known as the retarded and advanced Green functions.

We now turn to the phonons. The phonon field operator is real, i.e. $\tilde{\varphi}(x) = \tilde{\chi}(x) + \tilde{\chi}^+(x)$. It must be remembered in addition that the chemical potential $\mu = 0$ (see § 1) and that there are no particles present in the ground state. We find in the same way as above that

$$D(\mathbf{r} - \mathbf{r}', t - t') = \begin{cases} -i \sum_s |\chi_{0s}(0)|^2 e^{-i(E_s - E_0)(t-t')} e^{i(\mathbf{k}_s \cdot \mathbf{r} - \mathbf{r}')}, & t > t', \\ -i \sum_s |\chi_{0s}(0)|^2 e^{i(E_s - E_0)(t-t')} e^{-i(\mathbf{k}_s \cdot \mathbf{r} - \mathbf{r}')}, & t < t'. \end{cases} \tag{7.28}$$

We introduce the function

$$P(\mathbf{k}, E)\mathrm{d}E = (2\pi)^3 \sum_s |\chi_{0s}(0)|^2 \delta(\mathbf{k} - \mathbf{k}_s) = (2\pi)^3 \sum_s |\chi_{0s}(0)|^2 \delta(\mathbf{k} + \mathbf{k}_s),$$

where the summation over s refers to those states whose energies E_s are between the limits $E < E_s - E_0 < E + \mathrm{d}E$. Expanding (7.28) into a Fourier integral we have

$$D(\boldsymbol{k}, \omega) = \int_0^\infty P(\boldsymbol{k}, E)\left\{\frac{1}{\omega - E + i\delta} - \frac{1}{\omega + E - i\delta}\right\}\mathrm{d}E. \quad (7.29)$$

The imaginary part of this function is always negative:

$$\operatorname{Im} D(\boldsymbol{k}, \omega) = -\pi P(\boldsymbol{k}, |\omega|). \quad (7.30)$$

The real and imaginary parts are connected by the same expression as for $G(\boldsymbol{p}, \varepsilon)$. It follows from this that the phonon Green function has the same analytic properties as the Green function of a system of fermions with $\mu = 0$.

We can therefore construct two analytic functions D_R and D_A, satisfying (7.25) with $\mu = 0$. These functions are, in the coordinate representation:

$$\begin{aligned} D_R(x - x') &= \begin{cases} -i\langle\tilde{\varphi}(x)\tilde{\varphi}(x') - \tilde{\varphi}(x')\tilde{\varphi}(x)\rangle, & t > t', \\ 0, & t < t', \end{cases} \\ D_A(x - x') &= \begin{cases} 0, & t > t', \\ -i\langle\tilde{\varphi}(x')\tilde{\varphi}(x) - \tilde{\varphi}(x)\tilde{\varphi}(x')\rangle, & t < t'. \end{cases} \end{aligned} \quad (7.31)$$

3. *Physical meaning of the poles*

We have already mentioned, that a knowledge of the Green function enables us to find a whole series of physical characteristics of a system. In particular, the elementary excitation spectrum can be determined from it.

Let us take a Fermi system, which is described at the initial instant t' by the wave function

$$\Psi_0(t') = \psi_{\boldsymbol{p}}^+(t')\Phi_i(t'), \quad (7.32)$$

where $\psi_{\boldsymbol{p}}^+(t')$ is the operator of the creation of a particle with momentum $\boldsymbol{p}$ in the interaction representation, i.e. $a_{\boldsymbol{p}}^+ e^{i\varepsilon_0(\boldsymbol{p})t'}$, and $\Phi_i(t')$ is the wave function of the ground state of the particle system in the interaction representation. At an instant $t > t'$ the wave function of the system will have the form

$$\Psi(t) = S(t, t')\psi_{\boldsymbol{p}}^+(t')\Phi_i(t').$$

Let us find the probability amplitude of the state $\Psi_0(t)$. It is equal to

$$\begin{aligned} \langle\Psi_0^*(t)\Psi(t)\rangle &= \langle\Phi_i^*(t)\psi_{\boldsymbol{p}}(t)S(t, t')\psi_{\boldsymbol{p}}^+(t')\Phi_i(t')\rangle \\ &= \langle\Phi_H^{0*}S^{-1}(t)\psi_{\boldsymbol{p}}(t)S(t, t')\psi_{\boldsymbol{p}}^+(t')S(t')\Phi_H^0\rangle \\ &= \langle\tilde{\psi}_{\boldsymbol{p}}(t)\tilde{\psi}_{\boldsymbol{p}}^+(t')\rangle = iG(\boldsymbol{p}, t - t'), \quad t - t' > 0. \end{aligned} \quad (7.33)$$

We have passed here from the interaction to the Heisenberg representation.

To find $G(\mathbf{p},t)$, we have to evaluate the integral

$$G(\mathbf{p}, t) = \int_{-\infty}^{\infty} \frac{d\omega}{2\pi} G(\mathbf{p}, \omega) e^{-i\omega t}. \tag{7.34}$$

In view of the fact that $G(\mathbf{p}, \omega)$ is not analytic, we split the integral into two parts: from $-\infty$ to μ and from μ to ∞. In the first interval G is the same as the analytic function G_A, and in the second the same as G_R. As already remarked above, the function G_A has no singularities in the lower half-plane. We can therefore deform the contour of integration in the first integral (Fig. 2). If the horizontal piece of the contour is removed a sufficient distance into the lower half-plane, the integral over

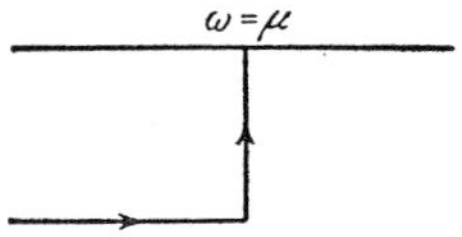

Fig. 2

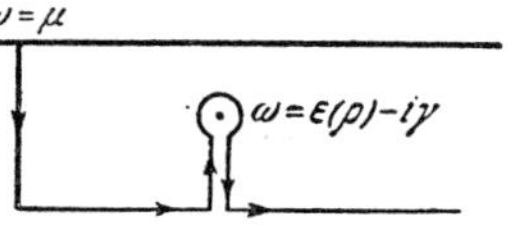

Fig. 3

this piece will be extremely small due to the factor $e^{-i\omega t}$ in (7.34), and we are left only with $\int_{\mu-i\infty}^{\mu} (d\omega/2\pi)\, G_A\, e^{-i\omega t}$. We now turn to the second integral. The function G_R generally speaking has singularities in the lower half-plane. Let us suppose that, in the fourth quadrant of the complex variable $\omega - \mu$ the singularity closest to the real axis is a simple pole at the point $\omega = \varepsilon(\mathbf{p}) - i\gamma$, where $\gamma \ll \varepsilon(\mathbf{p}) - \mu$. We deform the contour of integration as illustrated in Fig. 3. The horizontal piece of this contour must evidently lie above the next singularity and cannot be displaced to $-i\infty$. However, the integral can be made small by taking a sufficiently large time t. There now remains the integral over the vertical part and the circuit round the pole:

$$\int_{\mu}^{\mu-i\infty} \frac{d\omega}{2\pi} G_R e^{-i\omega t} - i a e^{-i\varepsilon(\mathbf{p})t-\gamma t},$$

where a is the residue of G_R at the pole. We shall show below that, when $t \gg [\varepsilon(\mathbf{p}) - \mu]^{-1}$, the contribution from both integrals over the vertical parts of the contours of Figs. 2 and 3 is small. We therefore obtain in the limit, for large t:

$$iG(\mathbf{p}, t) \approx a e^{-i\varepsilon(\mathbf{p})t-\gamma t}. \tag{7.35}$$

If we had one free particle in the initial state with momentum $\mathbf{p}$ and energy $\varepsilon_0(\mathbf{p})$, the quantity analogous to (7.33) would be $\exp[-i\varepsilon_0(\mathbf{p})(t - t')]$. It follows from this that there is a wave packet in state (7.32) that behaves like a quasi-particle with energy $\varepsilon(\mathbf{p})$, and is damped in time according to the law $\exp[-\gamma(t - t')]$. Hence the energy and damping of the quasi-particles are determined by the real and imaginary parts of the pole of G_R in the lower half-plane. The wave packet amplitude is connected with the residue of G_R at the pole.

We now show that the parts of the integrals that we have neglected can be regarded as small in the momentum region where $\varepsilon(\mathbf{p}) \approx \mu$. Recalling that $G_A = G_R^*$, we find for the sum of the integrals over the vertical pieces of the contours of Figs. 2 and 3:

$$\int_{\mu}^{\mu-i\infty} (G_R - G_R^*) e^{-i\omega t} \frac{d\omega}{2\pi} = 2i \int_{\mu}^{\mu-i\infty} \operatorname{Im} G_R e^{-i\omega t} \frac{d\omega}{2\pi}.$$

In accordance with the phenomenological considerations described in § 2, the condition $\gamma \ll \varepsilon(\mathbf{p}) - \mu$ only holds in the neighbourhood $\varepsilon(\mathbf{p}) \approx \mu$ (†). Assuming that $t \gg 1/(\varepsilon(\mathbf{p}) - \mu)$, we can therefore replace G_R by $a/(\omega - \varepsilon(\mathbf{p}) + i\gamma)$. On introducing the new variable $i(\omega - \mu) = u$, we get

$$-\frac{2\gamma a e^{-i\mu t}}{2\pi} \int_0^{\infty} \frac{du\, e^{-ut}}{\gamma^2 + [\varepsilon(\mathbf{p}) - \mu + iu]^2}.$$

Since $t \gg [\varepsilon(\mathbf{p}) - \mu]^{-1}$, the integral is equal to

$$-\frac{\gamma a e^{-i\mu t}}{\pi t [\varepsilon(\mathbf{p}) - \mu]^2}.$$

If t is regarded as not too large compared with $1/\gamma$, this quantity is much less that the result of the circuit round the pole in Fig. 3.

Similar arguments can be developed for the state with wave function

$$\Psi_0'(t) = \psi_{\mathbf{p}}(t) \Phi_i(t). \tag{7.36}$$

On considering this state at a later instant t', we get

$$\langle \Psi_0'^*(t') \Psi'(t') \rangle = -iG(\mathbf{p}, t - t') \text{ for } t - t' < 0.$$

When evaluating $G(\mathbf{p}, t)$ in accordance with (7.34), the poles of $G_A(\mathbf{p}, \omega)$ in the upper half-plane are important, since $t < 0$. When $|t| \gg |\varepsilon(\mathbf{p}) - \mu|^{-1}$, we find as above:

$$-iG(\mathbf{p}, t) \approx a e^{-i\varepsilon(\mathbf{p})t - \gamma t},$$

where $\varepsilon(\mathbf{p}) < \mu$, $\gamma < 0$. We obtain a wave packet corresponding to a hole with $\varepsilon(\mathbf{p}) < \mu$. Consequently the energy and damping of the holes are given by the poles of $G_A(\mathbf{p}, \omega)$ in the upper half-plane. Notice that γ has opposite signs for "particles" and for "holes".

The same results hold in regard to the phonons. It may easily be seen from (7.29) that, in this case, for each pole of the function $D_R(\mathbf{k}, \omega)$ in the lower half-plane, there is a corresponding pole, located symmetrically with respect to the point $\omega = 0$, of the function $D_A(\omega, \mathbf{k})$ in the upper

(†) The validity of the assumption that $\gamma \ll \varepsilon(\mathbf{p}) - \mu$ as $\varepsilon(\mathbf{p}) - \mu \to 0$ has been strictly proved for numerous concrete examples (see e.g. §§ 21, 22).

half-plane. Thus both methods of determination of the excitation spectrum yield the same result.

Apart from the energy spectrum, the Green function can be used to find the connection between the chemical potential and the number of particles per unit volume, and also the ground state energy and the momentum distribution of the particles (naturally, given our restrictions, all this refers only to Fermi systems).

It follows from the definition (7.1) of the Green function that

$$\frac{N}{V} = \langle \tilde{\psi}_\alpha^+(x) \tilde{\psi}_\alpha(x) \rangle = -i \lim_{\substack{r=r' \\ t' \to t+0}} G_{\alpha\alpha}(x-x') = -2i \lim_{\substack{r=r' \\ t' \to t+0}} G(x-x'),$$

where $G_{\alpha\beta} = \delta_{\alpha\beta} G$.

On changing to the momentum representation for G, we get

$$\frac{N}{V} = -2i \lim_{t \to +0} \int \frac{d^3 p \, d\omega}{(2\pi)^4} G(\mathbf{p}, \omega) e^{i\omega t}. \tag{7.37}$$

Since the integral in (7.37) depends only on μ, we arrive at the function $N(\mu)$. If we find the inverse $\mu(N)$ and use the equation $\mu = (\partial E_0/\partial N)_V$, the ground state energy can be found. Actually, this approach is not the most convenient one in practice. We shall return to the ground state energy in § 9.

To find the momentum distribution of the particles we only need to evaluate the expression

$$N_{1/2}(\mathbf{p}) = N_{-1/2}(\mathbf{p}) = \langle \Phi_0^* a_{\mathbf{p}\,1/2}^+ a_{\mathbf{p}\,1/2} \Phi_0 \rangle = \langle a_{\mathbf{p}\,1/2}^+ a_{\mathbf{p}\,1/2} \rangle,$$

where $\Phi_0 = e^{-iE_0 t} \Phi_H^0$ is the Schrödinger wave function of the ground state of the system. On comparing this expression with (7.17) (see also the remark on p. 54), we find that

$$N_{1/2}(\mathbf{p}) = N_{-1/2}(\mathbf{p}) = -2i \lim_{t \to +0} \int_{-\infty}^{\infty} G(\mathbf{p}, \omega) e^{i\omega t} \frac{d\omega}{2\pi}. \tag{7.38}$$

An interesting property of the momentum distribution can be obtained from (7.38) (A. B. Migdal [29]). We define the limiting Fermi momentum p_0 for the excitations using the equation $\varepsilon(p_0) = \mu$. We consider $N(\mathbf{p})$ close to $|\mathbf{p}| = p_0$. In accordance with the hypothesis of § 2, the excitations of a Fermi liquid are "particles" and "holes" with momenta close to p_0. The damping of these quasi-particles is small compared with $|\varepsilon(\mathbf{p}) - \mu|$. These data can be used for finding the poles of the functions G_A and G_R. When $|\mathbf{p}| < p_0$, G_A has a pole in the upper half-plane close to the real axis; this pole disappears when $|\mathbf{p}| > p_0$, and a pole of G_R makes its appearance in the lower half-plane. Now suppose that integral (7.38) is written as two contour integrals of G_A and G_R, as was done with integral (7.34). We displace the horizontal pieces of the contours in Figs. 2 and 3 into the lower half-plane through a distance from the real axis much greater that $\varepsilon(\mathbf{p}) - \mu$. The integrals over these pieces will now be insensitive to small changes in the momentum $\boldsymbol{p}$. As regards the integrals over the vertical pieces of the contours, they can be combined into

$$2\int_{\mu}^{\mu-iL} \operatorname{Im} G_R(\boldsymbol{p}, \omega)\frac{d\omega}{2\pi}.$$

This integral can be split into a part over a region remote from the point $\varepsilon = \mu$ by a distance greater than $\varepsilon(\boldsymbol{p}) - \mu$, and an integral over a near-by region. The integral over the remote region will only depend slightly on variations in $|\boldsymbol{p}|$, whilst we can substitute $G_R \approx a/(\omega - \varepsilon(\boldsymbol{p}) + i\gamma)$ in the integral over the near-by region and verify that it is negligibly small ($\sim \gamma/(\varepsilon(\boldsymbol{p}) - \mu)$). Hence it follows that the only difference between expressions (7.38) for $N(\boldsymbol{p})$ when $|\boldsymbol{p}| < p_0$ and $|\boldsymbol{p}| > p_0$ lies in the fact that, in the first case, there is no circuit round the pole in Fig. 3, whereas there is in the second case. It follows that

$$N_{1/2}(p_0 - 0) - N_{1/2}(p_0 + 0) = a. \tag{7.39}$$

By (7.21), the constant a must be positive. We therefore arrive at the conclusion that the momentum distribution of the particles has a jump at the same point $|\boldsymbol{p}| = p_0$ as the excitation distribution. By the fundamental assumption of the theory of a Fermi liquid, the Fermi limiting momentum p_0 of the excitations is connected with the density of the number of particles by relationship (2.1) (the validity of this assumption will be proved in Chap. VI). The jump of the momentum distribution of interacting particles therefore occurs at the same point as for non-interacting particles. Since $0 \le N_{1/2}(\boldsymbol{p}) \le 1$, we have for the size of the jump:

$$0 \le a \le 1. \tag{7.40}$$

An example is provided by the momentum distribution of the particles in a dilute Fermi gas, which we found in § 5.

4. *Green function of a system in an external field*

We now turn to a system in an external field, which is independent of time. The Green function will depend in this case on the variables $t - t'$, $\boldsymbol{r}$ and $\boldsymbol{r}'$. We now get, instead of (7.17):

$$\begin{aligned} G(\boldsymbol{r}, \boldsymbol{r}', t-t') &= -i \sum_s \left(\psi(\boldsymbol{r})_{0s}\psi^*_{0s}(\boldsymbol{r}')\right) e^{-i(E_s-E_0)(t-t')}, \quad t > t', \\ G(\boldsymbol{r}, \boldsymbol{r}', t-t') &= i \sum_s \left(\psi^*(\boldsymbol{r}')_{s0}\psi_{s0}(\boldsymbol{r})\right) e^{i(E_s-E_0)(t-t')}, \quad t < t'. \end{aligned} \tag{7.41}$$

where $\psi(\boldsymbol{r})$ and $\psi(\boldsymbol{r}')$ are Schrödinger operators. If we proceed as in the case when there is no external field, we get a formula such as (7.21) with complex functions A and B. We can get round this difficulty by taking the symmetrised combination

$$\frac{1}{2}\left[G(\boldsymbol{r}, \boldsymbol{r}', t-t') + G(\boldsymbol{r}', \boldsymbol{r}, t-t')\right]. \tag{7.42}$$

As far as the dependence on ω is concerned, the Fourier component of this function has all the properties of the function G in the absence of external fields. Formulae (7.21)—(7.27) will all hold for it, the only difference being that all the quantities will depend on the parameters $\boldsymbol{r}, \boldsymbol{r}'$ instead of $\boldsymbol{p}$.

If we consider fermions, not interacting with one another, in an external field, it is convenient to take the operators $\psi(\boldsymbol{r})$ in the form $\psi(\boldsymbol{r}) = \sum a_s \varphi_s(\boldsymbol{r})$, where $\varphi_s(\boldsymbol{r})$ are the eigenfunctions of the particle in the field. We find in this case, instead of (7.17):

$$G(\boldsymbol{r}, \boldsymbol{r}', t-t') = -i \sum_s \varphi_s^*(\boldsymbol{r}')\varphi_s(\boldsymbol{r}) e^{-i\varepsilon_s(t-t')} \begin{cases} 1 - n_s, & t > 0, \\ -n_s, & t < 0, \end{cases}$$

where

$$n_s = \begin{cases} 1, & \varepsilon_s < \mu, \\ 0, & \varepsilon_s > \mu, \end{cases} \tag{7.43}$$

and ε_s denotes the energy of a particle in the state φ_s. On taking the Fourier component with respect to time, we get

$$G(\boldsymbol{r}, \boldsymbol{r}', \omega) = \sum_{\varepsilon_s > \mu} \frac{\varphi_s^*(\boldsymbol{r}')\varphi_s(\boldsymbol{r})}{\omega - \varepsilon_s + i\delta} + \sum_{\varepsilon_s < \mu} \frac{\varphi_s^*(\boldsymbol{r}')\varphi_s(\boldsymbol{r})}{\omega - \varepsilon_s - i\delta}. \tag{7.44}$$

We now introduce a quantity similar to A and B in (7.20):

$$A(\boldsymbol{r}, \boldsymbol{r}', E)\,\mathrm{d}E = \sum_s \varphi_s^*(\boldsymbol{r}')\varphi_s(\boldsymbol{r}), \quad E < \varepsilon_s < E + \mathrm{d}E.$$

We put $\boldsymbol{r} = \boldsymbol{r}'$ and integrate this relation over $\mathrm{d}^3\boldsymbol{r}$. In view of the normalisation of the functions $\varphi_s(\boldsymbol{r})$, we simply obtain on the right-hand side the number of levels $\mathrm{d}N$ in the interval $\mathrm{d}E$. Hence

$$\int \mathrm{d}^3\boldsymbol{r}\, A(\boldsymbol{r}, \boldsymbol{r}, E) = \frac{\mathrm{d}N(E)}{\mathrm{d}E}.$$

Equation (7.44) may be written in terms of the function A as

$$G(\boldsymbol{r}, \boldsymbol{r}', \omega) = \int \mathrm{d}E \frac{A(\boldsymbol{r}, \boldsymbol{r}', E)}{\omega - E + i\delta \operatorname{sign}(E - \mu)}.$$

Hence it follows that the imaginary part of $G(\boldsymbol{r}, \boldsymbol{r}, \omega)$ (A is real and positive when $\boldsymbol{r} = \boldsymbol{r}'$) is equal to

$$\operatorname{Im} G(\boldsymbol{r}, \boldsymbol{r}, \omega) = \begin{cases} -\pi A(\boldsymbol{r}, \boldsymbol{r}, \omega), & \omega > \mu, \\ \pi A(\boldsymbol{r}, \boldsymbol{r}, \omega), & \omega < \mu. \end{cases}$$

We therefore obtain

$$\frac{\mathrm{d}N(E)}{\mathrm{d}E} = -\frac{1}{\pi} \operatorname{sign}(E - \mu) \int \operatorname{Im} G(\boldsymbol{r}, \boldsymbol{r}, E)\,\mathrm{d}^3\boldsymbol{r}. \tag{7.45}$$

§ 8. BASIC PRINCIPLES OF DIAGRAM TECHNIQUES

1. *Transformation from N to μ as independent variable*

Before starting to evaluate the Green function, we shall transform to new variables. We have so far considered a system with a given number of particles. It will be convenient for us in the following to regard this number as variable and to specify the chemical potential. In essence, we have already made use of variables of this sort for phonons, where we had $\mu = 0$ and the number of particles in the system was not specified. In the case of a Fermi system, however, it was precisely the number of particles that was specified, whilst the chemical potential μ appearing in the formulae had to be regarded as a function of this number. In practical calculations it is more convenient to regard μ as the independent variable, then pass to a specified number of particles in the final result.

The transformation from one independent variable to another can be performed as follows. We know that the wave functions and energy levels of the system can be obtained from the variational principle

$$\langle \Psi^* \hat{H} \Psi \rangle = \min \tag{8.1}$$

under the condition that

$$\langle \Psi^* \hat{N} \Psi \rangle = \text{const}, \tag{8.2}$$

where $\hat{H}$ and $\hat{N}$ are the Hamiltonian and the operator of the number of particles. Instead of this, we can use the method of Lagrangian multipliers and find the absolute minimum of the expression

$$\langle \Psi^* (\hat{H} - \mu \hat{N}) \Psi \rangle,$$

where μ is a constant, determined by using condition (8.2). Changing from a given N to a given μ therefore amounts to replacing the Hamiltonian by the operator $\hat{H} - \mu \hat{N}$. In view of the fact that the operator $\hat{N}$ commutes with the Hamiltonian, formulae are easily found for transforming the operators $\tilde{\psi}(x)$:

$$\tilde{\psi}_\mu(x) = e^{-i\mu\hat{N}t} \tilde{\psi}_N(x) e^{i\mu\hat{N}t} = e^{i\mu t} \tilde{\psi}_N(x), \tag{8.3}$$

inasmuch as the operator $\tilde{\psi}_N$ decreases the number of particles by one. Similarly, we have for the operator $\tilde{\psi}_\mu^+(x)$:

$$\tilde{\psi}_\mu^+(x) = e^{-i\mu t} \tilde{\psi}_N^+(x). \tag{8.4}$$

The Green function is defined as

$$G_\mu(x, x') = G_N(x, x') e^{i\mu(t-t')}. \tag{8.5}$$

It follows from this that all the results of the previous section still hold for G_μ, provided we perform the substitution

$$\omega_{(N)} \to \omega_{(\mu)} + \mu. \tag{8.6}$$

Since we never require more than the values of G at $t = t'$ when evaluating the number of particles and their momentum distribution, the relevant formulae (7.37) and (7.38) evidently remain unchanged. The poles of the new Green function give the excitation energy, measured from the level of the chemical potential.

As already mentioned, it is more convenient in practice to make use of the functions G_μ. In future, therefore, we shall as a rule have in mind this definition of the Green function, and denote it simply by G. A special proviso will be made when an analysis of the general properties of the function G requires that the number of particles be assumed to be given (as in the previous section).

2. *Wick's theorem*

We now turn to the evaluation of the Green function. Formula (6.32) of § 6 for changing to the interaction representation enables us to write the perturbation theory series in a simple and compact form. As applied to the Green function, (6.32) has the form

$$G(x, x') = \frac{-i\langle T\psi(x)\psi^+(x')S(\infty)\rangle}{\langle S(\infty)\rangle}, \tag{8.7}$$

where

$$S(\infty) = T\left[\exp\left(-i\int_{-\infty}^{\infty} H_{int}\,\mathrm{d}t\right)\right]. \tag{8.8}$$

Let us again emphasise that the operators ψ appearing in (8.7) (and also in H_{int}) are subject to the equations for non-interacting particles.

We expand $S(\infty)$ in the numerator of (8.7) in powers of H_{int}. We get

$$S(\infty) = 1 - i\int_{-\infty}^{\infty} H_{int}\,\mathrm{d}t + \frac{(-i)^2}{2}\iint_{-\infty}^{\infty} T[H_{int}(t_1)H_{int}(t_2)]\,\mathrm{d}t_1\mathrm{d}t_2 + \cdots,$$

$$G(x, x') = -\frac{i}{\langle S(\infty)\rangle}\sum_{n=0}^{\infty}\frac{(-i)^n}{n!}\int_{-\infty}^{\infty}\cdots\int_{-\infty}^{\infty}\mathrm{d}t_1\cdots\mathrm{d}t_n$$
$$\times\langle T[\psi(x)\psi^+(x')H_{int}(t_1)\cdots H_{int}(t_n)]\rangle. \tag{8.9}$$

We shall not as yet expand the $\langle S(\infty)\rangle$ in the denominator. The interaction Hamiltonian H_{int} is as a rule the integral over the spatial variables (and sometimes also over the time) of the product of a number of operators ψ (actual examples will be discussed below). Hence each term of the series (8.9) contains the average of the time-ordered product of several operators of the particle field in the interaction representation.

In view of this, we must first of all consider an expression of the form

$$\langle T(ABCD\cdots XYZ)\rangle,$$

where $A, B, \ldots, X, Y, Z$ are field operators in the interaction representation (remember that these operators are the same as the corresponding operators for non-interacting particles).

Each of the field operators can be split into two terms. One of them gives zero when it acts upon the ground state wave function. This part may be called the "annihilation operator". In the phonon operator (7.13), it is the sum containing the b_k whilst it is the part of the sum with $|\boldsymbol{p}| > p_0$ in the Fermion operator (6.14). The other part, which may be called the "creation operator", has the property that its Hermitian conjugate gives zero when it acts upon the ground state. We shall define the normal product of several operators

$$N(AB \cdots XYZ)$$

as the product in which all the "creation operators" are arranged to the left of all the "annihilation operators", the sign being dependent on whether the permutation of the Fermi operators is even or odd. Further, we define the "contraction" of two operators as the difference

$$A^c B^c = T(AB) - N(AB).$$

We now show that the T-product can always be expressed in terms of all possible N-products with all possible "contractions":

$$\begin{aligned} T(ABCD \cdots XYZ) &= N(ABCD \cdots XYZ) \\ &+ N(A^c B^c CD \cdots XYZ) + N(A^c BC^c D \cdots XYZ) + \cdots \\ &\cdots + N(A^c B^a C^a \cdots X^c Y^b Z^b). \end{aligned} \tag{8.10}$$

This relationship is known as Wick's theorem (see [25]).

We note first of all that simultaneous permutation of the operators on both sides of (8.10) does not destroy the relationship. We can thus assume without loss of generality that the time-ordering of the operators is in accordance with their arrangement in (8.10). In order to obtain the N-product from the T-product, we have to take all the creation operators and commute them in turn with all the annihilation operators to their left. This gives a sum of N-products of the type written down in (8.10). But it will only contain contractions of those operators whose order in the T-product is different from that in the N-product. But since the contractions of operators for which both orders are equivalent vanish, we can assume that the right-hand side of (8.10) contains normal products with all possible contractions. Hence (8.10) is proved.

It may easily be verified by means of (6.14) and (7.13) that contractions of the Fermi operators $\psi^+(x')$ and $\psi(x)$, and also of two phonon operators $\varphi(x)$ and $\varphi(x')$, are simply c-numbers, whilst all the rest vanish.

For example,

$$\psi^{+c}(x')\psi^{c}(x) = \frac{1}{V}\sum_{p,p'}\Big\{a^{+}_{p'}a_{p} - \underset{|p|>p_0}{a^{+}_{p'}a_{p}} + \underset{|p|<p_0}{a_{p}a^{+}_{p'}}\Big\} e^{i[(p\cdot r)-(p'\cdot r')-\varepsilon_0(p)t+\varepsilon_0(p')t']}$$

$$= \frac{1}{V}\sum_{|p|<p_0} e^{i(p\cdot r-r')-i\varepsilon_0(p)(t-t')} \quad \text{for } t' > t,$$

$$\psi^{+c}(x')\,\psi^{c}(x) = \frac{1}{V}\sum_{p,p'}\Big\{-a_{p}a^{+}_{p'} - \underset{|p|>p_0}{a^{+}_{p'}a_{p}} + \underset{|p|<p_0}{a_{p}a^{+}_{p'}}\Big\} e^{i[(p\cdot r)-(p'\cdot r')-\varepsilon_0(p)t+\varepsilon_0(p')t']}$$

$$= -\frac{1}{V}\sum_{|p|>p_0} e^{i(p\cdot r-r')-i\varepsilon_0(p)(t-t')} \quad \text{for } t' < t.$$

By definition, the average of a normal product over the ground state vanishes. Consequently,

$$A^c B^c = \langle T(AB)\rangle.$$

We obtain after this, on taking the average over the ground state of (8.10):

$$\langle T(ABCD\cdots XYZ)\rangle = \langle T(AB)\rangle\,\langle T(CD)\rangle\cdots\langle T(YZ)\rangle$$
$$\pm\langle T(AC)\rangle\,\langle T(BD)\rangle\cdots\langle T(YZ)\rangle \pm\cdots. \tag{8.11}$$

Our average therefore splits up into the sum of all possible products of averages over the ground state of individual pairs of operators. It follows in particular from (8.11) that the operators $A, B, C, \ldots$ necessarily include an even number of operators of each field. If we take into account the definition (7.1) of the Green function, we arrive at the conclusion that the average of the T-product of any number of field operators is given by the sum of the products of the free Green functions.

3. *Feynman diagrams*

We now return to the initial expression (8.9). Since H_{int} is the integral of a product of operators ψ, each term of the sum in (8.9) can be transformed in accordance with (8.11). The result can be expressed in a translucent form using Feynman diagrams. This may best be illustrated by an actual example. Suppose our system consists of identical fermions with binary spin-independent interaction forces. In accordance with § 6, H_{int} has the form

$$H_{int} = \frac{1}{2}\int \psi^{+}_{\alpha}(r_1)\,\psi^{+}_{\beta}(r_2)\,U(r_1 - r_2)\,\psi_{\beta}(r_2)\psi_{\alpha}(r_1)\,d^3r_1 d^3r_2. \tag{8.12}$$

If we write $V(x_1 - x_2) = U(r_1 - r_2)\,\delta(t_1 - t_2)$, the operator $\int H_{int}\,dt$ will contain two four-dimensional integrals.

Let us now consider the terms of sum (8.9). The first term is the Green function of non-interacting particles. The next one is

$$\delta G^{(1)} = -\frac{1}{2\langle S(\infty)\rangle}\int d^4x_1 d^4x_2$$
$$\times\langle T\big(\psi_{\alpha}(x)\psi^{+}_{\beta}(x')\psi^{+}_{\gamma}(x_1)\psi^{+}_{\delta}(x_2)\psi_{\delta}(x_2)\psi_{\gamma}(x_1)\big)\rangle V(x_1 - x_2).$$

By (8.11), the matrix element under the integral is

$$\langle T(\psi_\alpha(x)\psi_\gamma^+(x_1))\rangle \langle\psi_\delta^+(x_2)\psi_\delta(x_2)\rangle \langle T(\psi_\gamma(x_1)\psi_\beta^+(x'))\rangle$$
$$-\langle T(\psi_\alpha(x)\psi_\gamma^+(x_1))\rangle \langle\psi_\delta^+(x_2)\psi_\gamma(x_1)\rangle \langle T(\psi_\delta(x_2)\psi_\beta^+(x'))\rangle$$
$$+\langle T(\psi_\alpha(x)\psi_\delta^+(x_2))\rangle \langle\psi_\gamma^+(x_1)\psi_\gamma(x_1)\rangle \langle T(\psi_\delta(x_2)\psi_\beta^+(x'))\rangle$$
$$-\langle T(\psi_\alpha(x)\psi_\delta^+(x_2))\rangle \langle\psi_\gamma^+(x_1)\psi_\delta(x_2)\rangle \langle T(\psi_\gamma(x_1)\psi_\beta^+(x'))\rangle$$
$$+\langle T(\psi_\alpha(x)\psi_\beta^+(x'))\rangle \langle\psi_\gamma^+(x_1)\psi_\gamma(x_1)\rangle \langle\psi_\delta^+(x_2)\psi_\delta(x_2)\rangle$$
$$-\langle T(\psi_\alpha(x)\psi_\beta^+(x'))\rangle \langle\psi_\gamma^+(x_1)\psi_\delta(x_2)\rangle \langle\psi_\delta^+(x_2)\psi_\gamma(x_1)\rangle.$$

In accordance with definition (7.1) of the Green function, this expression can be written as

$$\begin{aligned} &iG^{(0)}_{\alpha\gamma}(x, x_1)G^{(0)}_{\delta\delta}(x_2, x_2)G^{(0)}_{\gamma\beta}(x_1, x') - iG^{(0)}_{\alpha\gamma}(x, x_1)G^{(0)}_{\gamma\delta}(x_1, x_2)G^{(0)}_{\delta\beta}(x_2, x') \\ &+ iG^{(0)}_{\alpha\delta}(x, x_2)G^{(0)}_{\gamma\gamma}(x_1, x_1)G^{(0)}_{\delta\beta}(x_2, x') - iG^{(0)}_{\alpha\delta}(x, x_2)G^{(0)}_{\delta\gamma}(x_2, x_1)G^{(0)}_{\gamma\beta}(x_1, x') \\ &- iG^{(0)}_{\alpha\beta}(x, x')G^{(0)}_{\gamma\gamma}(x_1, x_1)G^{(0)}_{\delta\delta}(x_2, x_2) + iG^{(0)}_{\alpha\beta}(x, x')G^{(0)}_{\delta\gamma}(x_2, x_1)G^{(0)}_{\gamma\delta}(x_1, x_2). \end{aligned} \tag{8.13}$$

Our expression thus splits up into a sum of terms, each of which contains three Green functions of non-interacting particles.

Feynman showed that each such term can be associated with a special diagram, drawn in accordance with the following principle. The set of space-time coordinates and the spin, on which the operators ψ appearing in our expression depend, may be pictured as points in a plane. We then join by continuous lines the points which appear as arguments in one function $G^{(0)}$, and by a wavy line the points x_1, x_2, appearing in the function $V(x_1 - x_2)$. The quantity $\delta G^{(1)}$ will now correspond to six such diagrams, as illustrated in Fig. 4(†). Each of them has two external

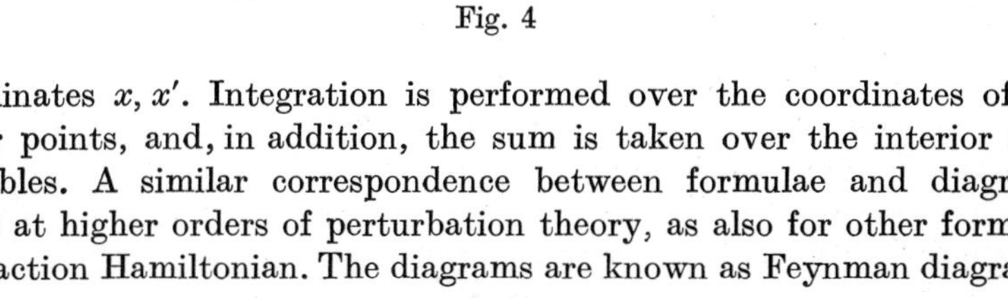

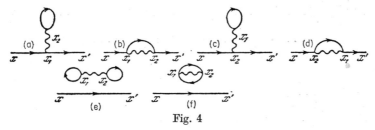

Fig. 4

coordinates x, x'. Integration is performed over the coordinates of interior points, and, in addition, the sum is taken over the interior spin variables. A similar correspondence between formulae and diagrams holds at higher orders of perturbation theory, as also for other forms of interaction Hamiltonian. The diagrams are known as Feynman diagrams.

(†) For simplicity, the spin variables are omitted in the figure.

A definite analytic expression corresponds to each Feynman diagram. Evaluation of the perturbation theory series amounts to drawing all the possible Feynman diagrams and evaluating the corresponding integrals. The rules for drawing the diagrams and writing the corresponding formulae depend on the actual form of the interaction. However, a general pattern may always be observed independently of this, and as a result the calculations are greatly simplified.

All the Feynman diagrams for G can be split into two groups — connected and unconnected diagrams. We shall describe as connected diagrams those in which all the points are linked somehow with the lines with endpoints x and x'. For instance, diagrams a, b, c, d of Fig. 4 are connected, whilst e and f are unconnected. In the general case, when we have a term of the perturbation theory series (8.9), the connected diagrams will be those in which $\psi(x)$ is paired with ψ^+ of $H_{int}(t_{p_1})$, ψ of $H_{int}(t_{p_1})$ with ψ^+ of $H_{int}(t_{p_2})$ and so on, the net result being that we arrive at $\psi^+(x')$ without leaving out any H_{int} (Fig. 5a). The remaining diagrams, in which one or more of the operators H_{int} are not connected by any pairings with $\psi(x)$ and $\psi^+(x')$, are called unconnected (Fig. 5b).

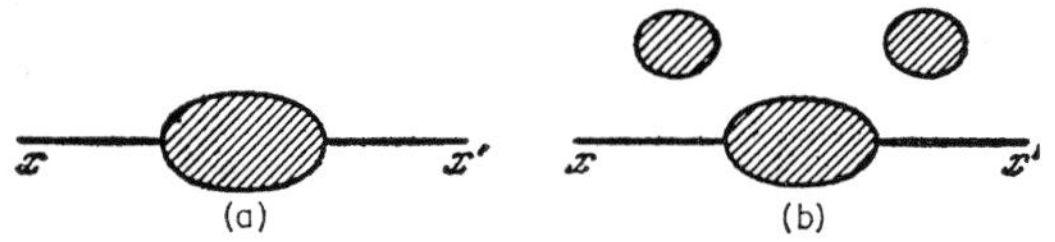

Fig. 5

We now consider the correction to the Green function corresponding to an unconnected diagram. It is obviously made up of two factors. The first includes all the H_{int} connected with $\psi(x)$ and $\psi^+(x')$; in other words, it includes the expression corresponding to the connected block of Fig. 5b, which contains the end-points. The second factor describes the remaining part of the diagram. The expression of the correction is therefore

$$-i\frac{(-i)^n}{n!}\int\cdots\int dt_1\cdots dt_m\langle T\left(\psi(x)\psi^+(x')H_{int}(t_1)\cdots H_{int}(t_m)\right)\rangle_c$$
$$\times\int\cdots\int dt_{m+1}\cdots dt_n\langle T\left(H_{int}(t_{m+1})\cdots H_{int}(t_n)\right)\rangle.$$

We understand here by $\langle\cdots\rangle_c$ and $\langle\cdots\rangle$ a well-defined method of splitting operators ψ, ψ^+ into pairs in accordance with Wick's theorem. The symbol $\langle\cdots\rangle_c$ emphasises that the pairing is carried out so that a connected diagram is obtained.

It is easily seen that the diagrams will include ones that give precisely the same contribution. Indeed, if we change the pairing in such a way that it is simply a question of rearranging the different H_{int} between the brackets $\langle\cdots\rangle_c$ and $\langle\cdots\rangle$, this will correspond simply to a transformation of the variables of integration and does not alter the value of the approximation to G. The number of such diagrams is evidently equal to the number of ways of splitting the n operators H_{int} into groups of m and $n-m$ operators, i.e. $(n!)/[m!(n-m)!]$.

The total contribution of all these diagrams will be

$$(-i)\frac{(-i)^m}{m!}\int\cdots\int \mathrm{d}t_1\cdots \mathrm{d}t_m\langle T\left(\psi(x)\psi^+(x')H_{int}(t_1)\cdots H_{int}(t_m)\right)\rangle_c$$
$$\times\frac{(-i)^{n-m}}{(n-m)!}\int\cdots\int \mathrm{d}t_{m+1}\cdots \mathrm{d}t_n\langle T\left(H_{int}(t_{m+1})\cdots H(t_n)\right)\rangle.$$

We sum the contributions of all the diagrams, of any orders, containing a definite connected part and arbitrary unconnected parts. We obviously get

$$(-i)\frac{(-i)^m}{m!}\int\cdots\int \mathrm{d}t_1\cdots \mathrm{d}t_m\langle T\left(\psi(x)\psi^+(x')H_{int}(t_1)\cdots H_{int}(t_m)\right)\rangle_c$$
$$\times\left\{1-i\int_{-\infty}^{\infty}\mathrm{d}t_{m+1}\langle H_{int}(t_{m+1})\rangle-\frac{1}{2}\int\int \mathrm{d}t_{m+1}\mathrm{d}t_{m+2}\langle T\left(H_{int}(t_1)H_{int}(t_{m+2})\right)\rangle\right.$$
$$\left.\cdots+\frac{(-i)^k}{k!}\int\cdots\int \mathrm{d}t_{m+1}\cdots \mathrm{d}t_{m+k}\langle T\left(H_{int}(t_{m+1})\cdots H_{int}(t_{m+k})\right)\rangle+\cdots\right\}.$$

We return to the initial formula (8.7). If we expand the denominator $\langle S(\infty)\rangle$ in powers of H_{int}, precisely the same expression is obtained as in the curly brackets in the last formula. Hence

$$\langle T\left(\psi(x)\psi^+(x')S(\infty)\right)\rangle=\langle T\left(\psi(x)\psi^+(x')S(\infty)\right)\rangle_c\langle S(\infty)\rangle$$

and, by (8.7),

$$G(x,x')=-i\langle T\left(\psi(x)\psi^+(x')S(\infty)\right)\rangle_c. \tag{8.14}$$

The rule obtained holds not only in regard to the Green function, but also when evaluating any expression of the type (6.32) with any number of field operators. This conclusion will be of importance later. The rule enables us in practice to omit the factor $\langle S(\infty)\rangle$ in the denominator of (8.9) and at the same time to ignore the contribution of the unconnected diagrams.

A further simplification arises from the fact that all the types of pairing in the expression

$$-i\frac{(-i)^m}{m!}\int\cdots\int \mathrm{d}t_1\cdots \mathrm{d}t_m\langle T\left(\psi(x)\psi^+(x')H_{int}(t_1)\cdots H_{int}(t_m)\right)\rangle_c,$$

differing only in commutations of the H_{int}, give the same contribution. In view of this, we can omit the factor $1/m!$ and only take into account those pairings that lead to topologically non-equivalent diagrams, i.e. those that cannot be obtained from one another by a commutation of the operators H_{int}. The contribution from each such diagram no longer contains a factor which is strongly dependent on the order m of the diagram. Due to this, each diagram can be split into elements which can be regarded individually as corrections to the appropriate Green functions. A factor λ^m, where λ is a constant, can evidently be regarded as a trivial dependence on m. This factor does not hinder the splitting of the diagram into elements. On the contrary, the appearance of a factor of the type $1/m$ prevents this splitting and the summation of the parts of the diagram individually.

§ 9. RULES FOR DRAWING DIAGRAMS FOR DIFFERENT TYPES OF INTERACTION

1. The diagram technique in coordinate space. Examples

We now turn to a detailed consideration of the rules for drawing Feynman diagrams in different cases. The basis of each diagram is a line representing the Green function of the fermion or phonon. We shall represent the former by a full-drawn, and the latter by a dotted, line. We draw an arrow on the line to indicate its direction: the line departs from the point with coordinates x and spin α and arrives at the point with coordinates x' and spin β. Thus the line in Fig. 6*a* denotes the Green function

$$G^{(0)}_{\alpha\beta}(x, x') \equiv G^{(0)}_{\alpha\beta}(x - x'),$$

and in Fig. 6b the Green function

$$G^{(0)}_{\beta\alpha}(x', x) \equiv G^{(0)}_{\beta\alpha}(x' - x).$$

We can omit the arrow on a phonon line (Fig. 7), since, as we saw in § 7, $D^{(0)}$ is an even function of $x - x'$. Integration is carried out with

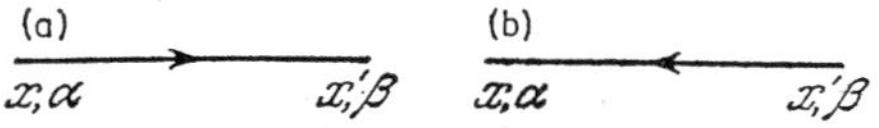

Fig. 6

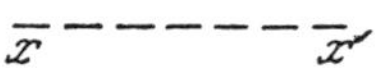

Fig. 7

respect to the coordinates of the vertices, i.e. the points joining lines (over all space and over t from $-\infty$ to ∞). In addition, summation is carried out over the spin variables of these vertices.

The following is an analysis of actual cases.

A. Two-particle interaction. We have already considered the simplest Feynman diagrams for this interaction (Fig. 4), where our aim was to explain the connection between the diagrams and formulae. As already said, unconnected diagrams must be discarded along with the factor $\langle S(\infty)\rangle^{-1}$. Thus there only remain, to first order, diagrams 4 *a*, 4 *b*, 4 *c* and 4 *d*. But in view of the fact that integration is carried out over the coordinates x_1 and x_2 (and summation over the corresponding spin variables), it turns out that diagram *c* is equal to diagram *a*, and diagram *d* to diagram *b*. This leads to a compensation of the factor $^1/_2$ in H_{int}. A similar situation holds for higher approximations. Thus the rule is to leave out this factor and to consider only topologically non-equivalent diagrams (for instance, *a* and *b*).

Attention must also be paid to the following. As already mentioned earlier, the sign attaching to each diagram depends on whether the permutation of the Fermi operators ψ is even or not. It is easily seen that a change of sign is connected with the formation of a closed loop in the diagram. The sign of the diagram is therefore determined by the factor $(-1)^F$, where F is the number of closed loops.

Another point deserving attention is the case when the times are the same in both arguments of one of the functions $G^{(0)}$. This only happens when two operators of one Hamiltonian pair off. In view of the fact that the order of the operators in H_{int} is given (all the ψ^+ are to the left of all the ψ), such $G^{(0)}$ have to be understood as $\lim_{\delta\to+0} G(t, t+\delta) \equiv \lim_{\delta\to 0} G(-\delta) = i\langle\psi^+(\boldsymbol{r}_1)\,\psi(\boldsymbol{r}_2)\rangle$. We now state the rules by which the approximations of any order are evaluated.

(1) We draw all the topologically non-equivalent diagrams with $2n$ vertices and two endpoints. Two full-drawn and one wavy line join at each vertex.

(2) Each full-drawn line is associated with a Green function $G^{(0)}_{\alpha\beta}(x, x')$ (x, α are the coordinates at the beginning of the line, x', β the coordinates of the end).

(3) Each wavy line is associated with a potential $V(x - x') = U(\boldsymbol{r} - \boldsymbol{r}')\,\delta(t - t')$.

(4) Integration is carried out with respect to the coordinates of all the vertices ($d^4x = d^3\boldsymbol{r}\,dt$), summation with respect to all the interior spin variables α.

(5) The expression obtained is multiplied by $i^n(-1)^F$, where F is the number of closed loops.

(6) If there are any $G^{(0)}$ with time argument zero in the expression, they are to be understood as the limit $G^{(0)}(-0)$.

Let us take the second order correction as an example. The relevant topologically non-equivalent connected diagrams are shown in Fig. 8.

In accordance with the rules stated, the corresponding analytic expressions are

$$-\int d^4x_1 d^4x_2 d^4x_3 d^4x_4 G^{(0)}_{\alpha\gamma_1}(x-x_1) G^{(0)}_{\gamma_1\gamma_2}(x_1-x_2) \times G^{(0)}_{\gamma_2\beta}(x_2-x') G^{(0)}_{\gamma_3\gamma_3}(0) G^{(0)}_{\gamma_4\gamma_4}(0)\, V(x_1-x_3)\, V(x_2-x_4), \quad \text{(a)}$$

$$-\int d^4x_1 \cdots d^4x_4 G^{(0)}_{\alpha\gamma_1}(x-x_1) G^{(0)}_{\gamma_1\gamma_2}(x_1-x_2) G^{(0)}_{\gamma_2\gamma_3}(x_2-x_3) \times G^{(0)}_{\gamma_3\gamma_4}(x_3-x_4) G^{(0)}_{\gamma_4\beta}(x_4-x')\, V(x_1-x_2)\, V(x_3-x_4), \quad \text{(b)}$$

$$+\int d^4x_1 \cdots d^4x_4 G^{(0)}_{\alpha\gamma_1}(x-x_1) G^{(0)}_{\gamma_1\gamma_2}(x_1-x_2) G^{(0)}_{\gamma_2\gamma_3}(x_2-x_3) \times G^{(0)}_{\gamma_3\beta}(x_3-x') G^{(0)}_{\gamma_4\gamma_4}(0)\, V(x_1-x_4)\, V(x_2-x_3), \quad \text{(c)}$$

$$+\int d^4x_1 \cdots d^4x_4 G^{(0)}_{\alpha\gamma_1}(x-x_1) G^{(0)}_{\gamma_1\gamma_2}(x_1-x_2) G^{(0)}_{\gamma_2\gamma_3}(x_2-x_3) \times G^{(0)}_{\gamma_3\beta}(x_3-x') G^{(0)}_{\gamma_4\gamma_4}(0)\, V(x_1-x_2)\, V(x_3-x_4), \quad \text{(d)}$$

$$-\int d^4x_1 \cdots d^4x_4 G^{(0)}_{\alpha\gamma_1}(x-x_1) G^{(0)}_{\gamma_1\beta}(x_1-x') G^{(0)}_{\gamma_2\gamma_3}(x_2-x_3) \times G^{(0)}_{\gamma_3\gamma_2}(x_3-x_2) G^{(0)}_{\gamma_4\gamma_4}(0)\, V(x_1-x_2)\, V(x_3-x_4), \quad \text{(e)}$$

$$+\int d^4x_1 \cdots d^4x_4 G^{(0)}_{\alpha\gamma_1}(x-x_1) G^{(0)}_{\gamma_1\beta}(x_1-x') G^{(0)}_{\gamma_2\gamma_3}(x_2-x_3) \times G^{(0)}_{\gamma_3\gamma_4}(x_3-x_4) G^{(0)}_{\gamma_4\gamma_2}(x_4-x_2)\, V(x_1-x_2)\, V(x_3-x_4), \quad \text{(f)}$$

$$+\int d^4x_1 \cdots d^4x_4 G^{(0)}_{\alpha\gamma_1}(x-x_1) G^{(0)}_{\gamma_1\gamma_2}(x_1-x_2) G^{(0)}_{\gamma_2\gamma_3}(x_2-x_3) \times G^{(0)}_{\gamma_3\beta}(x_3-x') G^{(0)}_{\gamma_4\gamma_4}(0)\, V(x_1-x_3)\, V(x_2-x_4), \quad \text{(g)}$$

$$-\int d^4x_1 \cdots d^4x_4 G^{(0)}_{\alpha\gamma_1}(x-x_1) G^{(0)}_{\gamma_1\gamma_2}(x_1-x_2) G^{(0)}_{\gamma_2\gamma_3}(x_2-x_3) \times G^{(0)}_{\gamma_3\gamma_4}(x_3-x_4) G^{(0)}_{\gamma_4\beta}(x_4-x')\, V(x_1-x_4)\, V(x_2-x_3), \quad \text{(h)}$$

$$-\int d^4x_1 \cdots d^4x_4 G^{(0)}_{\alpha\gamma_1}(x-x_1) G^{(0)}_{\gamma_1\gamma_2}(x_1-x_2) G^{(0)}_{\gamma_2\gamma_3}(x_2-x_3) \times G^{(0)}_{\gamma_3\gamma_4}(x_3-x_4) G^{(0)}_{\gamma_4\beta}(x_4-x')\, V(x_1-x_3)\, V(x_2-x_4), \quad \text{(i)}$$

$$+\int d^4x_1 \cdots d^4x_4 G^{(0)}_{\alpha\gamma_1}(x-x_1) G^{(0)}_{\gamma_1\gamma_2}(x_1-x_2) G^{(0)}_{\gamma_2\beta}(x_2-\beta) \times G^{(0)}_{\gamma_3\gamma_4}(x_3-x_4) G^{(0)}_{\gamma_4\gamma_3}(x_4-x_3)\, V(x_1-x_3)\, V(x_2-x_4). \quad \text{(j)}$$

Perturbation theory can be presented in a rather different and more symmetrical form for the case of two-particle interactions. This proves convenient, when the interaction depends on the spins. The Hamiltonian of such an interaction is

$$H_{int} = \frac{1}{2}\int \psi^+_\alpha(\mathbf{r}_1)\psi^+_\beta(\mathbf{r}_2)\, U_{\alpha\beta\gamma\delta}(\mathbf{r}_1-\mathbf{r}_2)\,\psi_\delta(\mathbf{r}_2)\psi_\gamma(\mathbf{r}_1)\, d^3\mathbf{r}_1 d^3\mathbf{r}_2. \quad (9.1)$$

The integral $\int H_{int} dt$ which appears in the operator S is written in a form symmetrical with respect to all the variables:

$$\int H_{int} dt = \frac{1}{4}\int \cdots \int d^4x_1 \cdots d^4x_4 \psi^+_{\gamma_1}(x_1)\psi^+_{\gamma_2}(x_2)\, \Gamma^{(0)}_{\gamma_1\gamma_2,\gamma_3\gamma_4}(x_1x_2, x_3x_4) \times \psi_{\gamma_4}(x_4)\psi_{\gamma_3}(x_3). \quad (9.2)$$

In view of the anti-commutativity of the ψ-operators, $\Gamma^{(0)}$ can be regarded as an anti-symmetric function with respect to the interchanges $x_1\gamma_1 \rightleftarrows x_2\gamma_2$ or $x_3\gamma_3 \rightleftarrows x_4\gamma_4$. This function can be obtained from

$$U_{\gamma_1\gamma_2,\gamma_3\gamma_4}(\mathbf{r}_1 - \mathbf{r}_2)\,\delta(t_1 - t_2)\,\delta(x_1 - x_3)\,\delta(x_2 - x_4)$$

by subtracting the analogous expression with arguments 3 and 4 interchanged(†). The first order correction to the Green function is

$$\delta G^{(1)} = -\frac{1}{4}\int d^4x_1 \cdots d^4x_4 \Gamma^{(0)}_{\gamma_1\gamma_2,\gamma_3\gamma_4}(x_1x_2, x_3x_4)$$
$$\times \langle T\left(\psi_\alpha(x)\psi^+_\beta(x')\psi^+_{\gamma_1}(x_1)\psi^+_{\gamma_2}(x_2)\psi_{\gamma_4}(x_4)\psi_{\gamma_3}(x_3)\right)\rangle$$

(we shall everywhere omit the index "c" on the averaging symbol $\langle\cdots\rangle$). Since $\Gamma^{(0)}$ is anti-symmetric in its arguments, we obtain from this the single term

$$i\int d^4x_1 d^4x_2 d^4x_3 d^4x_4 G^{(0)}_{\alpha\gamma_1}(x - x_1)G^{(0)}_{\gamma_3\gamma_2}(x_3 - x_2)$$
$$\times G^{(0)}_{\gamma_4\beta}(x_4 - x')\Gamma^{(0)}_{\gamma_1\gamma_2;\gamma_3\gamma_4}(x_1x_2; x_3x_4).$$

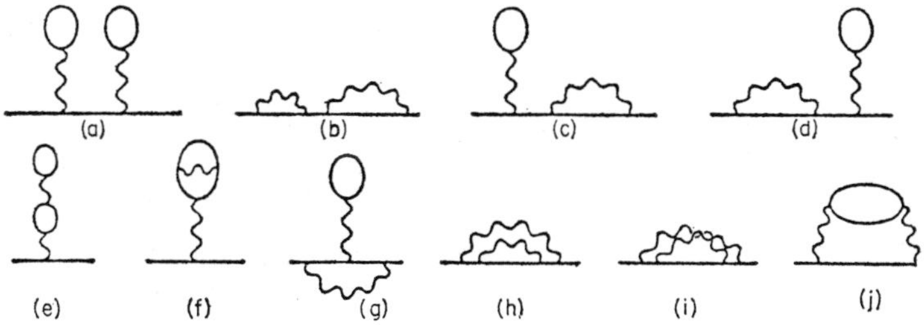

Fig. 8

We shall denote $\Gamma^{(0)}$ by an open square on the diagrams. The first order diagram is of the form illustrated in Fig. 9.

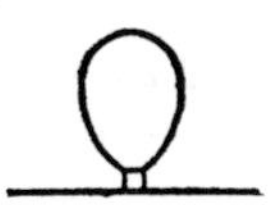

Fig. 9

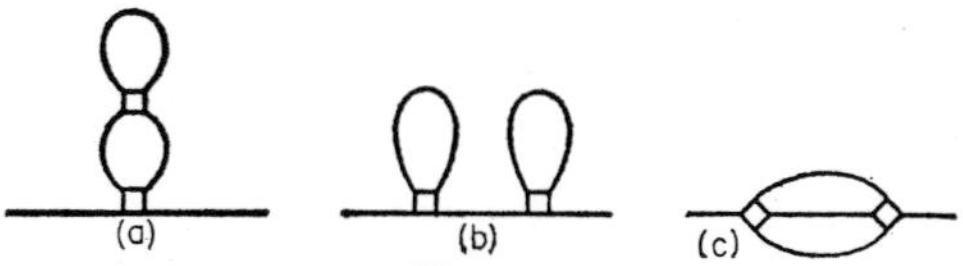

Fig. 10

There are altogether three connected and topologically distinct diagrams for the second approximation (Fig. 10). The corresponding expres-

(†) With this method of writing, the "plus" sign precedes the term corresponding to the "transition" $x_1\gamma_1 \to x_3\gamma_3$, $x_2\gamma_2 \to x_4\gamma_4$ (cf. (9.1)). This has to be borne in mind when determining the sign of the diagram (see below).

sions are

$$-\int \mathrm{d}^4x_1 \cdots \mathrm{d}^4x_8\, G^{(0)}_{\alpha\gamma_1}(x-x_1)\, G^{(0)}_{\gamma_3\beta}(x_3-x')\, G^{(0)}_{\gamma_4\gamma_5}(x_4-x_5)$$
$$\times G^{(0)}_{\gamma_7\gamma_2}(x_7-x_2)\, G^{(0)}_{\gamma_8\gamma_6}(x_8-x_6)\, \Gamma^{(0)}_{\gamma_1\gamma_2,\gamma_3\gamma_4}(x_1x_2; x_3x_4)\, \Gamma^{(0)}_{\gamma_5\gamma_6,\gamma_7\gamma_8}(x_5x_6, x_7x_8), \quad \text{(a)}$$

$$-\int \mathrm{d}^4x_1 \cdots \mathrm{d}^4x_8\, G^{(0)}_{\alpha\gamma_1}(x-x_1)\, G^{(0)}_{\gamma_3\gamma_5}(x_3-x_5)\, G^{(0)}_{\gamma_7\beta}(x_7-x')$$
$$\times G^{(0)}_{\gamma_4\gamma_2}(x_4-x_2)\, G^{(0)}_{\gamma_8\gamma_6}(x_8-x_6)\, \Gamma^{(0)}_{\gamma_1\gamma_2,\gamma_3\gamma_4}(x_1x_2, x_3x_4)\, \Gamma^{(0)}_{\gamma_5\gamma_6,\gamma_7\gamma_8}(x_5x_6, x_7x_8), \quad \text{(b)}$$

$$-\frac{1}{2}\int \mathrm{d}^4x_1 \cdots \mathrm{d}^4x_8\, G^{(0)}_{\alpha\gamma_1}(x-x_1)\, G^{(0)}_{\gamma_3\gamma_5}(x_3-x_5)\, G^{(0)}_{\gamma_7\gamma_2}(x_7-x_2)$$
$$\times G^{(0)}_{\gamma_8\beta}(x_8-x')\, G^{(0)}_{\gamma_4\gamma_6}(x_4-x_6)\, \Gamma^{(0)}_{\gamma_1\gamma_2,\gamma_3\gamma_4}(x_1x_2, x_3x_4)\, \Gamma^{(0)}_{\gamma_5\gamma_6,\gamma_7\gamma_8}(x_5x_6, x_7x_8). \quad \text{(c)}$$

Notice that the last term contains the factor $^1/_2$.

Evaluation of the nth order approximation proceeds as follows:

(1) All the topologically non-equivalent diagrams are drawn (in the present case, all the diagrams obtained by commutation of the vertices of the square are topologically equivalent).

(2) With each line we associate a Green function $G^{(0)}_{\alpha\beta}(x-x')$.

(3) With each square we associate the function $\Gamma^{(0)}_{\gamma_1\gamma_2,\gamma_3\gamma_4}(x_1x_2, x_3x_4)$.

(4) We integrate over the coordinates of all the vertices of the quadrilaterals and sum over the spin variables.

(5) Each diagram is multiplied by $(m/2^n)\,(i)^n$, where m is the number of different diagrams which correspond to the given diagram in the non-symmetrised technique. The sign of the diagram is also determined from a comparison with the non-symmetrised technique.

The last statement may be illustrated as follows. Take diagram 10 a, for example; it corresponds in the non-symmetrised technique to diagrams 8 e, f, g, h. Thus $m = 4$. As regards diagram 10 c, there are only two corresponding diagrams 8 i, j, so that $m = 2$ and the corresponding expression appears with the coefficient $^1/_2$(†).

We shall again use diagram 10 a to illustrate the choice of sign in front of the diagram. The quantity $\Gamma^{(0)}$ is obtained by antisymmetrising an expression, in which the point 3 is the same as 1, and 2 as 4. If we now regard these coordinates as the same in the expression corresponding to Fig. 10 a, we immediately get the diagram of Fig. 8 e, which contains two loops and appears with the coefficient $(i)^2$. In practical calculations it is simplest to start by writing down the arguments in all the $\Gamma^{(0)}$, then fix the arguments in $G^{(0)}$, whilst bearing in mind the correspondence with a diagram in Fig. 8.

(†) This programme can prove difficult to carry out in complicated diagrams. It is easier to obtain the analytic expression directly from (8.14), and use the diagram merely as a guide to the different methods of pairing.

According to the rules laid down, the third order approximation diagram in Fig. 11 corresponds to the expression

$$-\frac{(i)^3}{2}\int d^4x_1 \cdots d^4x_{12} G^{(0)}_{\alpha\gamma_1}(x-x_1) G^{(0)}_{\gamma_3\gamma_5}(x_3-x_5) G^{(0)}_{\gamma_7\gamma_9}(x_7-x_9)$$
$$\times G^{(0)}_{\gamma_{11}\beta}(x_{11}-x') G^{(0)}_{\gamma_4\gamma_{10}}(x_4-x_{10}) G^{(0)}_{\gamma_{12}\gamma_6}(x_{12}-x_6) G^{(0)}_{\gamma_8\gamma_2}(x_8-x_2)$$
$$\times \Gamma^{(0)}_{\gamma_1\gamma_2,\gamma_3\gamma_4}(x_1x_2, x_3x_4) \Gamma^{(0)}_{\gamma_5\gamma_6,\gamma_7\gamma_8}(x_5x_6, x_7x_8) \Gamma^{(0)}_{\gamma_9\gamma_{10},\gamma_{11}\gamma_{12}}(x_9x_{10}, x_{11}x_{12}).$$

Obviously, if we substitute for $\Gamma^{(0)}$ its expression in terms of the potential $U_{\alpha\beta,\gamma\delta}(\mathbf{r}_1-\mathbf{r}_2)$, all the expressions concerned transform to the corresponding formulae of the non-symmetrised theory.

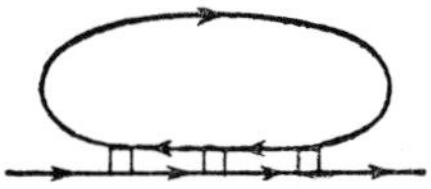

Fig. 11

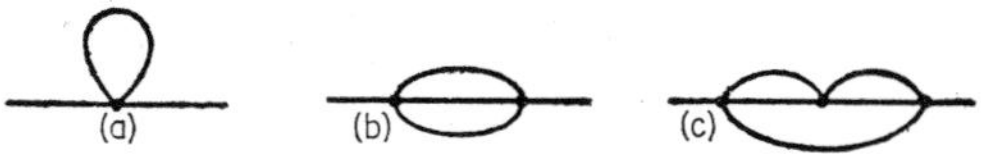

Fig. 12

These expressions take a particularly simple form when we are concerned with spin-independent point interactions, i.e. with a potential

$$U_{\alpha\beta,\gamma\delta}(\mathbf{r}_1-\mathbf{r}_2) = \lambda\delta_{\alpha\gamma}\delta_{\beta\delta}\delta(\mathbf{r}_1-\mathbf{r}_2).$$

In this case $\Gamma^{(0)}$ is

$$\Gamma^{(0)}_{\gamma_1\gamma_2,\gamma_3\gamma_4} = \lambda(\delta_{\gamma_1\gamma_3}\delta_{\gamma_2\gamma_4} - \delta_{\gamma_1\gamma_4}\delta_{\gamma_2\gamma_3})\delta(x_1-x_2)\delta(x_1-x_3)\delta(x_1-x_4)$$
$$= \lambda L_{\gamma_1\gamma_2,\gamma_3\gamma_4}\delta(x_1-x_2)\delta(x_1-x_3)\delta(x_1-x_4). \qquad (9.3)$$

There remains only one of the four integrations over the vertices of the squares in Figs. 9 and 10. These squares can therefore be replaced by points. For example, the diagrams in Figs. 9, 10*c* and 11 will have the form illustrated in Fig. 12, whilst the corresponding expressions will be

$$i\lambda L_{\gamma_1\gamma_2,\gamma_3\gamma_4} G^{(0)}_{\gamma_3\gamma_2}(0)\int d^4x_1 G^{(0)}_{\alpha\gamma_1}(x-x_1) G^{(0)}_{\gamma_4\beta}(x_1-x'), \qquad \text{(a)}$$

$$-\frac{\lambda^2}{2} L_{\gamma_1\gamma_2,\gamma_3\gamma_4} L_{\gamma_5\gamma_6,\gamma_7\gamma_8}\int d^4x_1 d^4x_2 G^{(0)}_{\alpha\gamma_1}(x-x_1) G^{(0)}_{\gamma_3\gamma_5}(x_1-x_2)$$
$$\times G^{(0)}_{\gamma_7\gamma_2}(x_2-x_1) G^{(0)}_{\gamma_8\beta}(x_2-x') G^{(0)}_{\gamma_4\gamma_6}(x_1-x_2), \qquad \text{(b)}$$

$$-\frac{i\lambda^3}{2} L_{\gamma_1\gamma_2,\gamma_3\gamma_4} L_{\gamma_5\gamma_6,\gamma_7\gamma_8} L_{\gamma_9\gamma_{10},\gamma_{11}\gamma_{12}}\int d^4x_1 d^4x_2 d^4x_3 G^{(0)}_{\alpha\gamma_1}(x-x_1) G^{(0)}_{\gamma_3\gamma_5}(x_1-x_2)$$
$$\times G^{(0)}_{\gamma_7\gamma_9}(x_2-x_3) G^{(0)}_{\gamma_{11}\beta}(x_3-x') G^{(0)}_{\gamma_4\gamma_{10}}(x_1-x_3) G^{(0)}_{\gamma_{12}\gamma_6}(x_3-x_2) G^{(0)}_{\gamma_8\gamma_2}(x_2-x_1). \text{ (c)}$$

B. Interaction of fermions with phonons. We shall assume, with a view to future applications, that we are discussing the isotropic model of a metal, in which the electrons interact with the phonons. The mechanism of the interaction is equivalent to the appearance of a polarisation as a result of lattice vibrations. The electron energy changes by an amount

$$-e \int n(\boldsymbol{r}) K(\boldsymbol{r} - \boldsymbol{r}') \operatorname{div} \boldsymbol{P}(\boldsymbol{r}') \mathrm{d}^3\boldsymbol{r} \mathrm{d}^3\boldsymbol{r}', \qquad (9.4)$$

where $n(\boldsymbol{r})$ is the electron density at the point $\boldsymbol{r}$, $\boldsymbol{P}$ is the polarisation, and $K(\boldsymbol{r} - \boldsymbol{r}')$ is the interaction function. When $|\boldsymbol{r} - \boldsymbol{r}'|$ is less than the lattice constant, $K(\boldsymbol{r} - \boldsymbol{r}') \approx 1/|\boldsymbol{r} - \boldsymbol{r}'|$. At great distances $K(\boldsymbol{r} - \boldsymbol{r}')$ rapidly drops to zero due to the screening of the polarisation charge by the electrons. This enables us to replace $K(\boldsymbol{r} - \boldsymbol{r}')$ by $a^2 \delta(\boldsymbol{r} - \boldsymbol{r}')$, where a is a constant of the order of the lattice period. The polarisation $\boldsymbol{P}$ is proportional to the displacement of the medium:

$$\boldsymbol{P}(\boldsymbol{r}) = C \boldsymbol{q}(\boldsymbol{r}),$$

where C is a constant of the order ZeN/V (N/V is the number of ions per unit volume, and Ze is their charge).

Since $\operatorname{div} \boldsymbol{P} = C \operatorname{div} \boldsymbol{q}$ enters into the energy of interaction of the electrons with the lattice vibrations, it follows that the electrons interact only with the longitudinal vibrations. By (9.4), the interaction energy operator can be written as

$$e a^2 C \int \psi^+(\boldsymbol{r}) \psi(\boldsymbol{r}) \operatorname{div} \boldsymbol{q}(\boldsymbol{r}) \mathrm{d}^3\boldsymbol{r}.$$

Since the operators $q_{\boldsymbol{k}}$ can themselves be expressed in terms of the creation and annihilation operators with certain coefficients of proportionality, we can include an additional factor in the definition of the field operators, such that a more convenient form of the operator H_{int} is obtained. It is easily seen that, given our choice (7.13) of operators $\varphi(x)$, the Hamiltonian of the electron-phonon interaction may be written as

$$H_{int} = g \int \psi_\alpha^+(\boldsymbol{r}) \psi_\alpha(\boldsymbol{r}) \varphi(\boldsymbol{r}) \mathrm{d}^3\boldsymbol{r}, \qquad (9.5)$$

where the constant of interaction g is equal to

$$g = \frac{e a^2 C}{u_0 \sqrt{\varrho}}$$

($u_0 = \omega_0(\boldsymbol{k})/|\boldsymbol{k}|$ is the sound velocity). If we substitute in this the orders of magnitude of all the constants, expressed in terms of the electron parameters, we get

$$g^2 = \frac{2\pi^2 \zeta}{p_0 m}, \qquad (9.6)$$

where m is the mass of an electron. The constant ζ is dimensionless in this definition and turns out to be of the order of unity from experimental data.

When finding the Green functions, it is only necessary to take into account the even terms of the expansion of $S(\infty)$ in powers of H_{int}. Since the averaging of the electron and phonon operators proceeds independently, the diagrams for the electron Green function turn out the same as in the case of two-particle interactions of fermions with one another. The only thing we need do is replace the wavy lines everywhere by dotted lines corresponding to the Green function of the phonons, and to carry out the substitution

$$V(x_1 - x_2) \rightarrow g^2 D^{(0)}(x_1 - x_2)$$

in the relevant expressions.

We now consider the Green function of the phonons. The first non-vanishing corrections to this function are found in the second approximation in H_{int} and are represented by the diagrams in Fig. 13. The corresponding expressions are

$$-g^2 i \int d^4x_1 d^4x_2 D^{(0)}(x - x_1) D^{(0)}(x_2 - x') G^{(0)}_{\alpha\beta}(x_1 - x_2) G^{(0)}_{\beta\alpha}(x_2 - x_1), \quad \text{(a)}$$

$$+g^2 i \int d^4x_1 d^4x_2 D^{(0)}(x - x_1) D^{(0)}(x_2 - x') G^{(0)}_{\alpha\alpha}(0) G^{(0)}_{\beta\beta}(0). \quad \text{(b)}$$

We show that the second term in this formula must be put equal to zero. In fact, by definition in the function $D^{(0)}$ appear the quantities φ that are proportional to div $\boldsymbol{q}$, where $\boldsymbol{q}$ is the displacement vector. It follows that $D^{(0)}(x - x_1)$ is proportional to

$$\langle T(\varphi(x), \operatorname{div} \boldsymbol{q}(x_1))\rangle = \operatorname{div}_{\boldsymbol{r}_1} \langle T(\varphi(x)\, \boldsymbol{q}(x_1))\rangle.$$

Since, in the expression for diagram 13 *b*, the coordinate $\boldsymbol{r}_1$ only appears in $D^{(0)}(x - x_1)$, and this function is a divergence, the integral over $d^3\boldsymbol{r}_1$ transforms into a surface integral and vanishes, independently of whether the displacement on the boundary is regarded as zero or is subject to periodic boundary conditions. By the same reasoning, all the diagrams for D-functions in which the end-points are unconnected vanish.

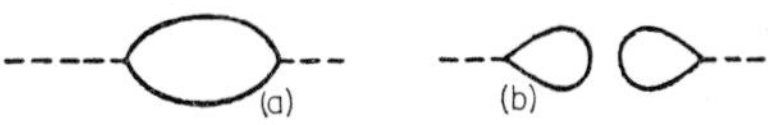

Fig. 13

General rules for evaluating the corrections to the Green functions of electrons and phonons may be stated as follows. To find the $2n$th order approximation, we must

(1) draw all the topologically non-equivalent connected diagrams with $2n$ vertices (connected in the sense of also excluding diagrams like 13 *b*);

(2) associate with each continuous line a function $G^{(0)}_{\alpha\beta}(x - x')$, and with each dotted line a function $D^{(0)}(x - x')$;

(3) carry out integration over the coordinates of all the vertices and summation over the spins;

(4) multiply the expression obtained by $g^{2n}(-1)^F (i)^n$, where F is the number of closed loops formed by the Fermi $G^{(0)}$-lines.

We give as an example the expression for the diagram of Fig. 14:

$$g^4 \int d^4x_1 \cdots d^4x_4 D^{(0)}(x - x_1) D^{(0)}(x_2 - x_3) D^{(0)}(x_4 - x')$$
$$\times G^{(0)}_{\gamma_1\gamma_2}(x_1 - x_2) G^{(0)}_{\gamma_2\gamma_4}(x_2 - x_4) G^{(0)}_{\gamma_4\gamma_3}(x_4 - x_3) G^{(0)}_{\gamma_3\gamma_1}(x_3 - x_1).$$

C. External field. Our last example is the interaction of particles with an external field. In accordance with § 6, the interaction Hamiltonian is

$$H_{int} = \int \psi^+_\alpha(\mathbf{r}) V_{\alpha\beta}(\mathbf{r}, t) \psi_\beta(\mathbf{r}) d^3\mathbf{r}. \tag{9.7}$$

The indices $\alpha\beta$ of the potential V refer to the case when the influence of an external magnetic field on the spins is considered. Then, $V_{\alpha\beta}(\mathbf{r}, t) = \mu_0 [\boldsymbol{\sigma}_{\alpha\beta} \cdot \mathbf{H}(\mathbf{r}, t)]$, where μ_0 is the magnetic moment of a particle and the components of $\boldsymbol{\sigma}$ are the Pauli matrices.

Fig. 14

Fig.15

It is easily seen that the diagrams have in this example the elementary shape shown in Fig. 15. A cross on the diagram corresponds to the potential $V_{\alpha\beta}(x)$. For instance, the expression for diagram 15 *b* is

$$\int d^4x_1 d^4x_2 G^{(0)}_{\alpha\gamma_1}(x - x_1) G^{(0)}_{\gamma_2\gamma_3}(x_1 - x_2) G^{(0)}_{\gamma_4\beta}(x_2 - x') V_{\gamma_1\gamma_2}(x_1) V_{\gamma_3\gamma_4}(x_2).$$

The rules for forming the diagrams and corresponding expressions are trivial. The diagrams of all orders have the same coefficient 1. The only thing that needs to be mentioned is the destruction of the uniformity of space and time. As a result, the Green function now depends on x and x' separately, and not only on the difference $x - x'$.

2. *The diagram technique in momentum space. Examples*

The technique described above enables us to write down without difficulty any term of the perturbation theory series in integral form. Evaluation of the integrals is extremely difficult, however, due to the fact that $G^{(0)}$ and $D^{(0)}$ are discontinuous functions in the time argument. To find the corrections to the Green functions by this method would mean integrating over the time over a set of domains whose number would increase at catastrophic speed with the order of the approximation. We can get out of this situation by expanding all the quantities into Fourier integrals.

Let us start with two-particle interactions.

A. Two-particle interaction of fermions. We take the expression corresponding to the diagram of Fig. 4 *b*:

$$i \int G^{(0)}_{\alpha\gamma_1}(x - x_1) G^{(0)}_{\gamma_1\gamma_2}(x_1 - x_2) G^{(0)}_{\gamma_2\beta}(x_2 - x') V(x_1 - x_2) d^4x_1 d^4x_2.$$

We expand all the quantities into Fourier integrals in accordance with the formulae

$$G^{(0)}_{\alpha\gamma_1}(x_1-x_2)=\int\frac{d^4p}{(2\pi)^4}G^{(0)}_{\alpha\gamma}(p)\,e^{ip(x_1-x_2)},$$

$$V(x_1-x_2)=\int\frac{d^4q}{(2\pi)^4}V(q)\,e^{iq(x_1-x_2)},$$

where p and q are the four-dimensional vectors $p=(\boldsymbol{p},\omega)$, $q=(\boldsymbol{q},\omega)$, and the product $p(x_1-x_2)$ is equal to $(\boldsymbol{p}\cdot\boldsymbol{r}_1-\boldsymbol{r}_2)-\omega(t_1-t_2)$. The expression for the free Fermi function $G^{(0)}_{\alpha\gamma}(p)$ has already been found in § 7 (formula (7.7) with the substitution $\omega\to\omega+\mu$). If we substitute these expansions in the first approximation term in the Green function written above, the latter becomes

$$i(2\pi)^{-16}\int G^{(0)}_{\alpha\gamma_1}(p_1)G^{(0)}_{\gamma_1\gamma_2}(p_2)G^{(0)}_{\gamma_2\beta}(p_3)\,V(q)\,e^{ip_1(x-x_1)+ip_2(x_1-x_2)}$$
$$\times e^{ip_3(x_2-x')+iq(x_1-x_2)}\,d^4p_1\,d^4p_2\,d^4p_3\,d^4q\,d^4x_1\,d^4x_2.$$

We integrate with respect to d^4x_1 and d^4x_2. We get

$$i(2\pi)^{-8}\int G^{(0)}_{\alpha\gamma_1}(p_1)G^{(0)}_{\gamma_1\gamma_2}(p_2)G^{(0)}_{\gamma_2\beta}(p_3)\,V(q)\,\delta(p_1-p_2-q)$$
$$\times\delta(p_2+q-p_3)\,e^{ip_1x-ip_3x'}\,d^4p_1\,d^4p_2\,d^4p_3\,d^4q.$$

We now take the Fourier components of this integral in x and x'. We obtain:

$$\delta G^{(1)}_{\alpha\beta}(p,p')=i\int G^{(0)}_{\alpha\gamma_1}(p)G^{(0)}_{\gamma_1\gamma_2}(p_2)G^{(0)}_{\gamma_2\beta}(p')\,V(q)$$
$$\times\delta(p-p_2-q)\,\delta(p_2+q-p')\,d^4p_2\,d^4q.$$

On comparing this expression with diagram 5 *b*, we see that each continuous line now corresponds to $G^{(0)}(\boldsymbol{p})$, the wavy line to $V(q)$, each vertex to the δ-function $\delta(\Sigma p)=\delta(\Sigma\boldsymbol{p})\,\delta(\Sigma\omega)$, expressing the laws of conservation of energy and momentum, the integral being taken over the momenta of the interior lines. On performing the integration with respect to d^4p_2 and noting that $G^{(0)}_{\alpha\beta}(p)=G^{(0)}(p)\,\delta_{\alpha\beta}$, we get

$$\delta G^{(1)}_{\alpha\beta}(p,p')=\delta G^{(1)}(p)\,\delta(p-p')\,(2\pi)^4\,\delta_{\alpha\beta},$$

$$\delta G^{(1)}(p)=iG^{(0)}(p)\int G^{(0)}(p-q)\,V(q)\,\frac{d^4q}{(2\pi)^4}\,G^{(0)}(p).$$

The resulting expression for $\delta G^{(1)}(p)$, which is the first approximation term in the Fourier component of the function $G(x-x')$ with the variable $x-x'$, enables us to interpret the diagram in a very visual way. We can imagine a particle with momentum p, which in the course of its motion emits an "interaction quantum" with momentum q and itself acquires momentum $p-q$. After a certain time the particle absorbs this quantum and remains with momentum p.

Similar transformations can be performed on the other diagrams. For instance, the first approximation term $\delta G^{(1')}(p)$, corresponding to diagram 4 *a*, is

$$\delta G^{(1')} = -2iG^{(0)}(p)\,V(0)\int \frac{d^4p_1}{(2\pi)^4}\,G^{(0)}(p_1)\,e^{i\omega t}G^{(0)}(p),$$

where $t \to +0$. The factor $e^{i\omega t}$ is introduced under the integral sign because, in the coordinate representation, a G-function with equal arguments is present here, and, as already mentioned, this is defined as $\lim G^{(0)}(-0)$. The factor 2 appears when taking the trace over the spins. The diagrams of $\delta G^{(1)}$ and $\delta G^{(1')}$ are illustrated in the momentum form in Fig. 16.

We now take the diagram of any order n, containing $2n$ vertices, $2n+1$ continuous lines and n wavy lines. If we substitute the Fourier transforms, for $G^{(0)}$ and V and integrate over the $2n$ coordinates of the vertices, we get $2n$ factors of the type $\delta(\Sigma p)$, expressing the laws of conservation. One of these laws leads to the equality of the external momenta, as a result of which all the terms of the expansion of $G(x, x')$ into a perturbation theory series are only dependent on the difference $x - x'$, this being an immediate consequence of the homogeneity in space. The remaining $2n - 1$ δ-functions imply that, of the $3n - 1$ integrations over the 4-momenta of the interior lines (both continuous and wavy), only n remain.

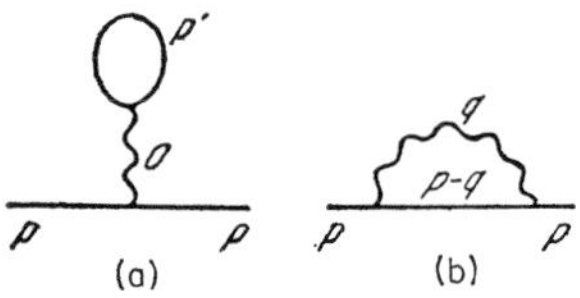

Fig.16

We now give the general rules for writing the expressions corresponding to different diagrams.

(1) Every line is associated with a definite 4-momentum. The two outer lines have an external momentum, whilst the momenta of the interior lines must satisfy the laws of conservation at each vertex.

(2) Every continuous line is associated with

$$G^{(0)}_{\alpha\beta}(p) = \frac{\delta_{\alpha\beta}}{\omega - \xi(\mathbf{p}) + i\delta\,\mathrm{sign}\,\xi(\mathbf{p})},$$

where

$$\xi(\mathbf{p}) = \varepsilon_0(\mathbf{p}) - \mu = \frac{\mathbf{p}^2}{2m} - \mu,\quad \delta \to +0.$$

(3) Every wavy line is associated with

$$V(q) \equiv U(\mathbf{q}).$$

(4) The integrations are performed over the n independent 4-momenta.

(5) A coefficient $(i)^n(2\pi)^{-4n}(-1)^F$, where F is the number of closed loops, is put in front of the expression obtained.

Any correction to a Green function is easily written down with the aid of these rules. For instance, the expression for Fig. 17 is

$$-i\delta_{\alpha\beta}G^{(0)^2}(p)(2\pi)^{-20}\int d^4q_1\cdots d^4q_4\,U(q_1)\,U(q_2)\,U(q_3)U(q_1+q_2+q_3)\,U(q_4)$$
$$\times G^{(0)}(p-q_1)G^{(0)}(p-q_1-q_2)G^{(0)}(p-q_1-q_2-q_4)G^{(0)}(p-q_1-q_2-q_3-q_4)$$
$$\times G^{(0)}(p-q_4)\int d^4p_1G^{(0)}(p_1)G^{(0)}(p_1+q_3)G^{(0)}(p_1+q_2+q_3)\,G^{(0)}(p_1+q_1+q_2+q_3).$$

We next consider the symmetrised version of the diagram technique for two-particle interactions. The symmetrised quantity $\Gamma^{(0)}_{\gamma_1\gamma_2,\gamma_3\gamma_4}(x_1, x_2; x_3, x_4)$ was introduced earlier. By definition, $\Gamma^{(0)}$ depends only on the coor-

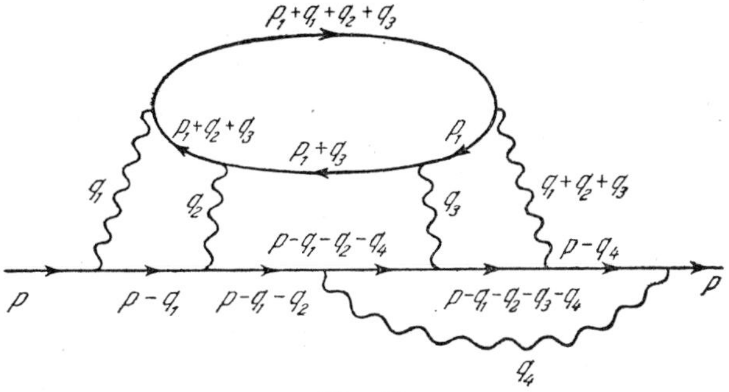

Fig. 17

dinate differences. The Fourier components of $\Gamma^{(0)}$ will therefore contain $\delta(p_1+p_2-p_3-p_4)$. In view of this it will be convenient for us to define the Fourier component of $\Gamma^{(0)}$ at once as

$$(2\pi)^4\delta(p_1+p_2-p_3-p_4)\Gamma^{(0)}(p_1, p_2; p_3, p_1+p_2-p_3)$$
$$=\int d^4x_1\cdots d^4x_4\Gamma^{(0)}(x_1x_2, x_3x_4)\,e^{-ip_1x_1-ip_2x_2+ip_3x_3+ip_4x_4}.$$

The Fourier transform of the first order term corresponding to the diagram of Fig. 9 is

$$-iG^{(0)^2}(p)\int\frac{d^4p_1}{(2\pi)^4}\Gamma_{\alpha\gamma,\beta\gamma}(pp_1; pp_1)G^{(0)}(p_1).$$

The diagram in momentum space is illustrated in Fig. 18. The general rules for drawing the diagrams are just the same as before. In particular,

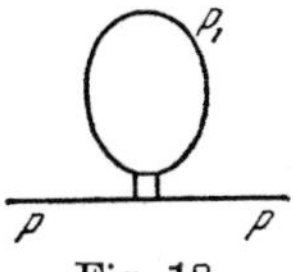

Fig. 18

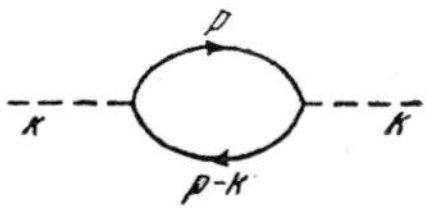

Fig. 19

the coefficient of the nth order diagram only differs from the coefficient in the coordinate form by the factor $(2\pi)^{-4n}$.

B. Electron-phonon interaction. The general rules for interpreting the diagram of order $2n$ for the Green function of an electron or a phonon are as follows:

(1) each full drawn line is associated with

$$G^{(0)}(\boldsymbol{p}) = \frac{1}{\omega - \xi(\boldsymbol{p}) + i\delta \operatorname{sign} \xi(\boldsymbol{p})},$$

where $\delta \to +0$;

(2) each phonon (dotted) line is associated with (see (7.16))

$$D_0(k) = \frac{\omega_0^2(\boldsymbol{k})}{\omega^2 - \omega_0^2(\boldsymbol{k}) + i\delta},$$

where $\delta \to +0$;

(3) integration is performed over the n independent momenta;

(4) the result of the integration is multiplied by

$$g^{2n}(2\pi)^{-4n}(i)^n(-1)^F,$$

where F is the number of closed loops.

For example, the second order diagram of Fig. 19 corresponds to the expression

$$-2[D^{(0)}(k)]^2 g^2 i \int \frac{d^4p}{(2\pi)^4} G^{(0)}(p) G^{(0)}(p-k).$$

C. External fields. As already mentioned, the space becomes inhomogeneous in the case of an external field and $G(x, x')$ ceases to be a function of $x - x'$ only. In view of this, we shall consider the Fourier transform $G_{\alpha\beta}(p, p')$ of $G_{\alpha\beta}(x, x')$ with respect to both variables:

$$G_{\alpha\beta}(x, x') = \int G_{\alpha\beta}(p, p') e^{ipx - ip'x'} \frac{d^4p}{(2\pi)^4} \frac{d^4p'}{(2\pi)^4}.$$

On Fourier transforming the expression corresponding to diagram 15 *a*, i.e.

$$\int d^4x_1 G^{(0)}_{\alpha\gamma_1}(x - x_1) G^{(0)}_{\gamma_2\beta}(x_1 - x') V_{\gamma_1\gamma_2}(x_1),$$

we get

$$G^{(0)}(p) V_{\alpha\beta}(p - p') G^{(0)}(p'),$$

where $V_{\alpha\beta}(p)$ is the Fourier component of $V_{\alpha\beta}(x)$:

$$V_{\alpha\beta}(x) = \int V_{\alpha\beta}(p) e^{ipx} \frac{d^4p}{(2\pi)^4}.$$

The corresponding diagram in momentum space is illustrated in Fig. 20 *a*. The next order diagram, illustrated in Fig. 20 *b*, corresponds to the expression

$$G^{(0)}(p) \int V_{\alpha\gamma}(p - p_1) G^{(0)}(p_1) V_{\gamma\beta}(p_1 - p') \frac{d^4p_1}{(2\pi)^4} G^{(0)}(p').$$

Thus, in the nth order diagram for $G(p, p')$:

(1) the left-hand end-point is associated with $G^{(0)}(p)$, and the right-hand one with $G^{(0)}(p')$;

(2) the cross denotes the Fourier component of the external potential with a momentum equal to the difference between the momenta of the $G^{(0)}$ lines to the left and the right of the cross;

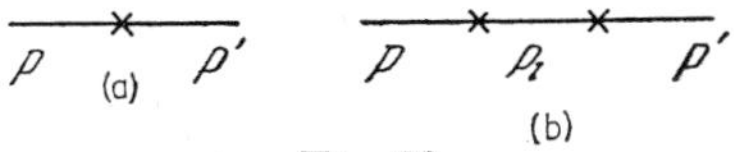

Fig. 20

(3) integration is carried out over all the momenta of the $G^{(0)}$-lines apart from the two external ones, and summation over the spin variables on which V depends (with the exception of the external ones);

(4) after integration and summation, the expression is multiplied by the coefficient $(2\pi)^{-4(n-1)}$.

§ 10. DYSON EQUATION. THE VERTEX PART. MANY-PARTICLE GREEN FUNCTIONS

1. Diagram summation. Dyson equation

In the majority of problems of quantum statistics it is as a rule impossible to confine ourselves to the first few terms of the perturbation theory series. Instead, we have to sum different infinite series of terms, corresponding to the so-called "main" diagrams, the contribution of which, by virtue of the conditions of the problem, is of the same order of magnitude. A remarkable property of the diagram technique for Green functions described above is that we can associate summation of an infinite (or finite) set of terms of the perturbation theory series with a special type of "graphical summation" of diagrams. The diagram representing the sum is composed of elements, each of which is in turn the result of a summation. For example, the lines of such a diagram may represent the sum of an infinite sequence ofterms of the perturbation theory series for the Green function (the "sum" of the diagrams). The association of the diagram with definite expressions is carried out in accordance with the same rules as those for evaluating the expressions in perturbation theory; each line of a diagram is associated with the corresponding sum of diagrams, and so on.

The possibility of a graphical summation is based upon the rules described above for evaluating the corrections to the Green function from the relevant diagrams. It may be seen at a glance that these corrections are formed as it were from separate bricks — the Green functions and vertex operators, the connecting element ("cement") being integration over the coordinates (or momenta). This enables us to form a diagram, not merely from basic elements — the zero Green functions $G^{(0)}$ and elementary vertices, but directly from whole blocks, composed of a large number of basic elements.

Let us take the diagram of Fig. 21 *a* as an example. We can first use the rules of the diagram technique to write down the expression corresponding to it. We now proceed as follows. We find first the contribution of the part of the initial diagram surrounded by the dotted line, then write down the expression corresponding to the diagram of 21 *b*, but taking instead of $G^{(0)}$ for the crossed line the more complicated expression. It may readily be seen by direct calculation that both methods yield the same result.

This conclusion is completely general. We can always distinguish from the diagram for G a part not containing external lines and joined to the rest by two $G^{(0)}$-lines, find its contribution, and write down the expression for the whole diagram by using the "abbreviated" diagram; here, we have to substitute the contribution of the extracted part for the corresponding line.

We shall describe as the self-energy part any part of a diagram joined with the rest by two $G^{(0)}$- (or $D^{(0)}$-) lines. A self-energy part will be called irreducible if it cannot be split into two parts joined by a single $G^{(0)}$-line. For example, the self-energy parts in Figs. 9, 10 *a* and 10 *c* are irreducible, whilst the part of Fig. 10 *b* is reducible. Any diagram for the functions G and D consists of a base line with irreducible self-energy parts threaded on it; these parts may be repeated an infinite number of times and may appear in any order.

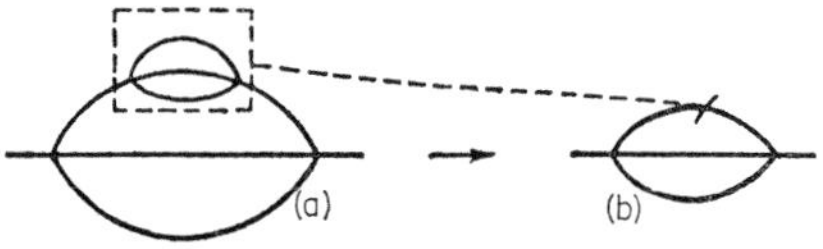

Fig. 21

It is impossible to sum all the diagrams for the Green functions in the general case. However, we can carry out a partial summation, in such a way that there only remains a sum over different irreducible self-energy parts.

Let us take any diagram for a G-function. It starts with a $G^{(0)}$-line. Then there is an irreducible self-energy part. It we chop off these two elements from the diagram, the remainder will again start with $G^{(0)}$ and may contain any number of arbitrary self-energy parts. Thus the remainder is again a complete G-function. This gives us the following equation for G:

$$G = G^{(0)} + G^{(0)}\Sigma G$$

or

$$G^{-1} = G^{(0)-1} - \Sigma, \tag{10.1}$$

where

$$\Sigma = \Sigma_1 + \Sigma_2 + \Sigma_3 + \cdots \tag{10.2}$$

is the sum of different irreducible self-energy parts. We shall call Σ the total irreducible self-energy part or the mass operator.

We can find Σ by using diagrams that differ from the diagrams for G in that the two end $G^{(0)}$-lines are absent. But in these cases where it is impossible to confine ourselves to evaluating the first diagrams and an infinite series has to be summed, it is as a rule more convenient to express Σ in terms of another set of diagrams, which we call the vertex part. This procedure depends on the actual type of interaction and will be illustrated by reference to the interactions considered in § 9.

A. Two-particle interactions. In this case it is most convenient to use the symmetrized form of the theory. The first order term in Σ corresponds to Fig. 9, without the external $G^{(0)}$-lines. From the terms of subsequent orders we first of all extract all the diagrams in which the self-energy part is linked with the basic G-line by means of a single shaded square $\Gamma^{(0)}$. An example is provided by Fig. 10 *a*. It is completely obvious that the set of all such diagrams for Σ can be obtained from the first order diagram by adding all possible self-energy parts to the interior $G^{(0)}$-line. The latter is now transformed into a complete G-line. Thus the set of all diagrams for Σ, linked with the basic G-line by a single square $\Gamma^{(0)}$, is equal to

$$\Sigma^{(1)}_{\alpha\beta}(p) = i\int \frac{d^4p_1}{(2\pi)^4}\Gamma^{(0)}_{\alpha\gamma;\delta\beta}(p, p_1; p_1, p)G_{\delta\gamma}(p_1). \tag{10.3}$$

Let us agree to use a heavy line for a complete G-function. We can now represent $\Sigma^{(1)}$ by the diagram of Fig. 22.

The simplest diagram not appearing in this sequence is the self-energy part in Fig. 10 *c*. Certain of the more complicated diagrams may be obtained by including self-energy parts in the interior of the $G^{(0)}$-line. It is not possible, however, to obtain the diagram of Fig. 11 by this method. Nevertheless, it can also be regarded as a more complex form of the diagram of Fig. 10 *c*. If we chop off the three interior $G^{(0)}$-lines issuing from the left-hand square $\Gamma^{(0)}$ from the rest of the diagram, it will be seen that diagram 11 follows from diagram 10 *c* by replacing the right-hand square by another diagram, illustrated in Fig. 23.

Fig. 22

Fig. 23

It is easily seen that, in general, all the diagrams for Σ not appearing in (10.3) follow from 10 *c*(†) by insertion of self-energy parts in the interior $G^{(0)}$-lines and by replacing the right-hand square by a set of all diagrams with four ends that do not decompose into unconnected parts. We call this set the vertex part, and denote it by $\Gamma_{\alpha\beta,\gamma\delta}(p_1, p_2; p_3, p_4)$; it will be represented on diagrams by a shaded square. Notice that, just as in $\Gamma^{(0)}$, the 4-momenta in Γ must satisfy the laws of conservation: $p_1 + p_2 = p_3 + p_4$.

(†) Since diagrams 9 and 10*c* provide the basis for obtaining more complicated diagrams they are sometimes known as skeletons.

The second part of Σ is therefore represented by the diagram of Fig. 24 and is equal to

$$\Sigma^{(2)}_{\alpha\beta} = -\frac{1}{2}\int \Gamma^{(0)}_{\alpha\xi,\eta\delta}(p, p_1; p_2, p+p_1-p_2)G_{\eta\mu}(p_2)G_{\nu\xi}(p_1) \times G_{\delta\gamma}(p+p_1-p_2)\Gamma_{\mu\gamma,\nu\beta}(p_2, p+p_1-p_2; p_1, p)\frac{d^4p_1 d^4p_2}{(2\pi)^8}. \quad (10.4)$$

On substituting $\Sigma = \Sigma^{(1)} + \Sigma^{(2)}$ in equation (10.1), we get

$$[\omega - \xi(\mathbf{p})]G_{\alpha\beta}(p) - i\int \frac{d^4p_1}{(2\pi)^4}\Gamma^{(0)}_{\alpha\xi,\eta\gamma}(pp_1; p_1p)G_{\eta\xi}(p_1)G_{\gamma\beta}(p) + \frac{1}{2}\int \Gamma^{(0)}_{\alpha\xi,\eta\delta}(p, p_1; p_2, p_1+p-p_2)G_{\eta\mu}(p_2)G_{\nu\xi}(p_1)G_{\delta\gamma}(p+p_1-p_2) \times \Gamma_{\mu\gamma,\nu\varrho}(p_2, p+p_1-p_2; p_1, p)\frac{d^4p_1 d^4p_2}{(2\pi)^8}G_{\varrho\beta}(p) = \delta_{\alpha\beta}. \quad (10.5)$$

This equation, connecting the G-function with the vertex part, is known as the Dyson equation. It has been obtained here by a summation of

Fig. 24

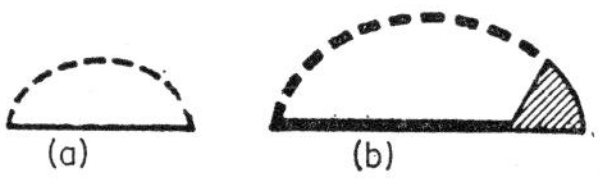

Fig. 25

diagrams. An analytic derivation of the Dyson equation and a more detailed consideration of the vertex part will be found below.

B. Electron-phonon interaction. The simplest diagram for Σ in the Green function of an electron is Fig. 25 *a*. It may readily be seen, in precisely the same way as above, that this diagram is the only skeleton, i.e. any more complicated diagram may be obtained by addition of self-energy parts to the interior $G^{(0)}$- and $D^{(0)}$-lines and by replacing the right-hand vertex by the set of all diagrams with one phonon and two electron ends. We shall call this quantity the vertex part $\Gamma(p, p-k; k)$ and represent it by a shaded triangle on a diagram.

In the case of the electron-phonon interaction, therefore, the total irreducible self-energy part for an electron Σ is represented by the diagram of Fig. 25*b* and is equal to

$$\Sigma = ig\int G(p-k)D(k)\Gamma(p-k, p; k)\frac{d^4k}{(2\pi)^4} \quad (10.6)$$

(we have put here $G_{\alpha\beta} = G\,\delta_{\alpha\beta}$).

Substitution of this expression in (10.1) gives the Dyson equation:

$$[\omega - \xi(p)]G(p) - ig\int G(p-k)D(k)\Gamma(p-k, p; k)\frac{d^4k}{(2\pi)^4}G(p) = 1. \quad (10.7)$$

The self-energy part for the phonons, which we denote by the symbol Π, can be similarly obtained from the skeleton diagram of Fig. 26 *a* by replacing the electron $G^{(0)}$-lines by complete G-lines and one of the constants g by the vertex part. Diagram 26 *a* now transforms to 26 *b*, equal to

$$\Pi(k) = -2ig \int G(p)G(p-k) \times \Gamma(p, p-k; k) \frac{d^4p}{(2\pi)^4}. \quad (10.8)$$

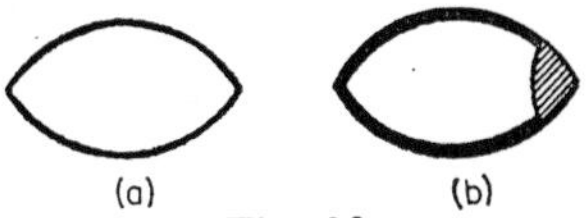

Fig. 26

The Dyson equation is in this case

$$[\omega_0^2(\boldsymbol{k})]^{-1} [\omega^2 - \omega_0^2(\boldsymbol{k})] D(k) + 2ig \int G(p)G(p-k)\Gamma(p, p-k; k) \frac{d^4p_1}{(2\pi)^4} D(k) = 1. \quad (10.9)$$

C. External fields. A Dyson type equation can also be written down for a system of fermions in an external field. Noting that all the diagrams for G are chains similar to those of Fig. 15, we arrive at the conclusion that the role of Σ is played by the Fourier component of the potential $V_{\alpha\beta}$. The Dyson equation is in this case

$$[\omega - \xi(\boldsymbol{p})] G_{\alpha\beta}(p, p') - \int V_{\alpha\gamma}(p - p_1) G_{\gamma\beta}(p_1, p') \frac{d^4p_1}{(2\pi)^4} = \delta_{\alpha\beta}. \quad (10.10)$$

2. *Vertex parts. Many-particle Green functions*

The Dyson equation also follows directly from the equations of motion for the Heisenberg operators

$$i \frac{\partial \tilde{\psi}_\alpha}{\partial t} = [\tilde{\psi}_\alpha(x), \hat{H} - \hat{N}\mu]_-, \qquad \hat{H} = -\int \psi_\alpha^+(\boldsymbol{r}) \frac{\nabla^2}{2m} \psi_\alpha(\boldsymbol{r}) d^3\boldsymbol{r} + H_{int}.$$

The operators $\hat{H}$ and $\hat{N}$ can be represented by expressions either in terms of the Schrödinger operators $\psi_\alpha(\boldsymbol{r})$, or in terms of the Heisenberg operators $\tilde{\psi}_\alpha(\boldsymbol{r}, t)$, since $\hat{H}$ and $\hat{N}$ are the same in both representations.

On separating $\hat{H}_{int}$ from $\hat{H} - \mu\hat{N}$ and using the commutation rules for the operators $\tilde{\psi}$ and $\tilde{\psi}^+$, taken at the same instant, we get

$$i \frac{\partial \tilde{\psi}_\alpha}{\partial t} = \left(-\frac{\nabla^2}{2m} - \mu\right) \tilde{\psi}_\alpha(x) + [\tilde{\psi}_\alpha(x), H_{int}].$$

We differentiate the G-function with respect to the first time argument:

$$i \frac{\partial}{\partial t} G_{\alpha\beta}(x, x') = \frac{\partial}{\partial t} \langle T(\tilde{\psi}_\alpha(x) \tilde{\psi}_\beta^+(x'))\rangle.$$

We write $T(\ldots)$ as

$$\theta(t-t')\tilde{\psi}_\alpha(x)\tilde{\psi}_\beta^+(x') - \theta(t'-t)\tilde{\psi}_\beta^+(x')\tilde{\psi}_\alpha(x),$$

where $\theta(t) = \begin{cases} 1, & t > 0, \\ 0, & t < 0. \end{cases}$

We now have:

$$i\frac{\partial}{\partial t}G_{\alpha\beta}(x, x') = \theta(t-t')\frac{\partial\tilde{\psi}_\alpha(x)}{\partial t}\tilde{\psi}^+_\beta(x') - \theta(t'-t)\tilde{\psi}^+_\beta(x')\frac{\partial\tilde{\psi}_\alpha(x)}{\partial t}$$

$$+ \delta(t-t')\left(\tilde{\psi}_\alpha(\boldsymbol{r}, t)\tilde{\psi}_\beta(\boldsymbol{r}', t) + \tilde{\psi}^+_\beta(\boldsymbol{r}', t)\tilde{\psi}_\alpha(\boldsymbol{r}, t)\right)$$

$$= \left\langle T\left(\frac{\partial\tilde{\psi}_\alpha(x)}{\partial t}\tilde{\psi}^+_\beta(x')\right)\right\rangle + \delta(x-x')\delta_{\alpha\beta}.$$

We have used the commutation rule here. We finally have:

$$\left(i\frac{\partial}{\partial t} + \frac{\nabla^2}{2m} + \mu\right)G_{\alpha\beta}(x, x')$$

$$= \delta(x-x')\delta_{\alpha\beta} - i\langle T\left([\tilde{\psi}_\alpha(x), H_{int}]_-, \tilde{\psi}^+_\beta(x')\right)\rangle. \qquad (10.11)$$

The form of the right-hand side depends on the actual interaction, so let us turn to particular cases.

A. Two-particle interactions. The operator H_{int} is defined by (9.2). On carrying out the detailed calculation and writing the result in a symmetric form (just as was done when deriving (9.3)), we get for the last term in (10.11):

$$-\frac{i}{2}\int d^4x_2 d^4x_3 d^4x_4 \Gamma^{(0)}_{\alpha\gamma_2;\gamma_3\gamma_4}(xx_2; x_3x_4)\langle T\left(\tilde{\psi}^+_{\gamma_2}(x_2)\tilde{\psi}_{\gamma_4}(x_4)\tilde{\psi}_{\gamma_3}(x_3)\tilde{\psi}^+_\beta(x')\right)\rangle.$$

The problem therefore amounts to finding the mean of the time-ordered product of four ψ-operators. We shall call this the two-particle Green function:

$$G^{\text{II}}_{\alpha\beta,\gamma\delta}(x_1x_2; x_3x_4) = \langle T\left(\tilde{\psi}_\alpha(x_1)\tilde{\psi}_\beta(x_2)\tilde{\psi}^+_\gamma(x_3)\tilde{\psi}^+_\delta(x_4)\right)\rangle. \qquad (10.12)$$

By (6.32), G^{II} can be expressed in terms of the operators ψ in the interaction representation:

$$G^{\text{II}}_{\alpha\beta,\gamma\delta}(x_1x_2; x_3x_4) = \frac{\langle T\left(\psi_\alpha(x_1)\psi_\beta(x_2)\psi^+_\gamma(x_3)\psi^+_\delta(x_4)\right)S(\infty)\rangle}{\langle S(\infty)\rangle}. \qquad (10.13)$$

This expression can be evaluated in a similar way to the Green function. The operator $S(\infty)$ in the numerator is expanded into a series in powers of H_{int}. On then applying Wick's theorem, we can write each term of the series as a sum of terms containing products of the functions $G^{(0)}$. A Feynman diagram can be drawn for each of these terms. In contrast to the diagrams for the Green function, all the present diagrams will have four end-points. It is easily seen that, as previously, we only need to take into account the connected diagrams, i.e. those in which there is no part not connected with one of the ends; at the same time the factor $\langle S(\infty)\rangle$ must be thrown away in the denominator of (10.13).

Another rule still holds, namely, all the expressions depend on the order of the diagram only via the factors λ^n. This enables us to work with parts of diagrams and to carry out partial summations.

All the connected diagrams for G^{II} fall into two groups. One group contains the diagrams in which the point x_1 is connected by successive pairings with the point x_3, and x_2 with x_4, whereas e.g. x_1 and x_4 are isolated from one another. Such diagrams fall into two separate parts, with no lines joining the parts. We put in the same group the diagrams in which x_1 is connected with x_4, and x_2 with x_3, whilst there is no link between x_1 and x_3.

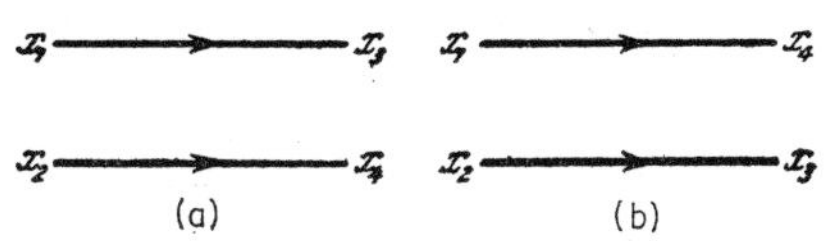

Fig. 27

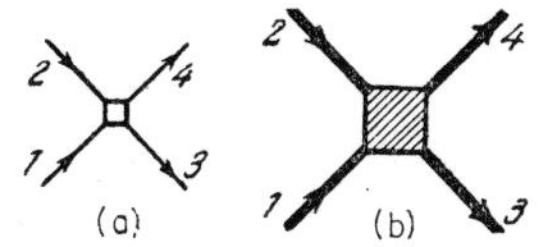

Fig. 28

The simplest diagrams of this type are to zero order in H_{int} and are illustrated in Fig. 27. The corresponding expressions are

$$G^{(0)}_{\alpha\gamma}(x_1 - x_3)G^{(0)}_{\beta\delta}(x_2 - x_4), \tag{a}$$

$$-G^{(0)}_{\alpha\delta}(x_1 - x_4)G^{(0)}_{\beta\gamma}(x_2 - x_3). \tag{b}$$

It is easily seen that any more complicated diagram of this group can be obtained by addition of self-energy parts to the $G^{(0)}$-lines, i.e. by replacing light $G^{(0)}$-lines by heavy G-lines.

The other group contains all the diagrams that do not decompose into separate parts. The simplest diagram of this type is to the first order in H_{int} and is as shown in Fig. 28 *a*. The corresponding expression is

$$i\int G^{(0)}_{\alpha\gamma_1}(x_1 - x_1')G^{(0)}_{\beta\gamma_2}(x_2 - x_2')G^{(0)}_{\gamma_3\gamma}(x_3' - x_3)G^{(0)}_{\gamma_4\delta}(x_4' - x_4)$$
$$\times \Gamma^{(0)}_{\gamma_1\gamma_2,\gamma_3\gamma_4}(x_1'x_2';x_3'x_4')\,d^4x_1'\,d^4x_2'\,d^4x_3'\,d^4x_4'.$$

More complicated diagrams are obtained from 28*a* by complicating the external $G^{(0)}$-lines and replacing the square by more complex arrangements with four vertices, as in Fig. 23, for instance. At the same time, the $G^{(0)}$ in the above formula are replaced by G, and $\Gamma^{(0)}$ by Γ, corresponding to the set of all possible diagrams with four outer ends; in other words, Fig. 28*a* becomes Fig. 28*b*.

It follows from our arguments that $G^{\mathrm{II}}_{\alpha\beta,\gamma\delta}(x_1\,x_2;\,x_3\,x_4)$ is conveniently written as

$$G^{\mathrm{II}}_{\alpha\beta,\gamma\delta}(x_1x_2;x_3x_4) = G_{\alpha\gamma}(x_1 - x_3)G_{\beta\delta}(x_2 - x_4) - G_{\alpha\delta}(x_1 - x_4)G_{\beta\gamma}(x_2 - x_3)$$
$$+\, i\int d^4x_1' \cdots d^4x_4' G_{\alpha\gamma_1}(x_1 - x_1')G_{\beta\gamma_2}(x_2 - x_2')$$
$$\times G_{\gamma_3\gamma}(x' - x_3)G_{\gamma_4\delta}(x_4' - x_4)\,\Gamma_{\gamma_1\gamma_2,\gamma_3\gamma_4}(x_1'x_2';x_3'x_4'), \tag{10.14}$$

where Γ corresponds to the vertex part introduced earlier.

The last term in (10.11) is equal to

$$\frac{i}{2}\int d^4x_2 d^4x_3 d^4x_4 \Gamma^{(0)}_{\alpha\gamma_2;\gamma_3\gamma_4}(xx_2;\,x_3x_4)\,G^{\mathrm{II}}_{\gamma_3\gamma_4;\gamma_2\beta}(x_3x_4,\,x_2x').$$

On expanding $G_{\gamma_3\gamma_4;\gamma_2\beta}$ in accordance with (10.14) and taking into account the anti-symmetry of $\Gamma^{(0)}$ with respect to the arguments with indices 3 and 4, we get from (10.11):

$$\left(i\frac{\partial}{\partial t} - H_0 + \mu\right)G_{\alpha\beta}(x-x') - i\int d^4x_2 d^4x_3 d^4x_4 \Gamma^{(0)}_{\alpha\gamma_2;\gamma_3\gamma_4}(xx_2;\,x_3x_4)$$

$$\times G_{\gamma_3\gamma_2}(x_3-x_2)\,G_{\gamma_4\beta}(x_4-x') + \frac{1}{2}\int d^4x_2\cdots d^4x_8 \Gamma^{(0)}_{\alpha\gamma_2;\gamma_3\gamma_4}(xx_2;\,x_3x_4)$$

$$\times G_{\gamma_4\gamma_6}(x_4-x_6)\,G_{\gamma_3\gamma_5}(x_3-x_5)\,G_{\gamma_7\gamma_2}(x_7-x_2)\,G_{\gamma_8\beta}(x_8-x')\,\Gamma_{\gamma_5\gamma_6,\gamma_7\gamma_8}(x_5x_6;\,x_7x_8)$$

$$= \delta(x-x')\,\delta_{\alpha\beta}. \qquad (10.15)$$

In view of the spatial homogeneity, Γ and G^{II} only depend on three coordinate differences. Their Fourier components may therefore be conveniently defined as in the case of $\Gamma^{(0)}$. For example,

$$\Gamma_{\alpha\beta,\gamma\delta}(p_1,\,p_2;\,p_3,\,p_1+p_2-p_3)\,(2\pi)^4\,\delta(p_1+p_2-p_3-p_4)$$

$$= \int \Gamma_{\alpha\beta,\gamma\delta}(x_1\,x_2;\,x_3x_4)\,e^{-ip_1x_1-ip_2x_2+ip_3x_3+ip_4x_4}d^4x_1\cdots d^4x_4. \qquad (10.16)$$

By (10.14), the connexion between the Fourier components of G^{II} and Γ is given by

$$G^{\mathrm{II}}_{\alpha\beta,\gamma\delta}(p_1,\,p_2;\,p_3,\,p_1+p_2-p_3) = G_{\alpha\gamma}(p_1)\,G_{\beta\delta}(p_2)\,\delta(p_1-p_3)\,(2\pi)^4$$

$$- G_{\alpha\delta}(p_1)\,G_{\beta\gamma}(p_2)\,\delta(p_2-p_3)\,(2\pi)^4 + iG_{\alpha\gamma_1}(p_1)\,G_{\beta\gamma_2}(p_2)\,G_{\gamma_3\gamma}(p_3)$$

$$\times G_{\gamma_4\delta}(p_1+p_2-p_3)\,\Gamma_{\gamma_1\gamma_2;\gamma_3\gamma_4}(p_1,\,p_2;\,p_3,\,p_1+p_2-p_3). \qquad (10.17)$$

Fourier transformation of equation (10.15) leads to equation (10.5). The Dyson equation has thus been derived analytically, the quantity Γ appearing in it being defined by (10.12), (10.17) and (10.16).

We can evaluate Γ by a summation of diagrams. Examples of such diagrams are given in Fig. 23, and also in Fig. 29 *a*, *b*, *c*. It follows from the fact that the diagrams for Γ can be regarded as a part of the diagrams for the G-functions that the rules for associating each diagram with the corresponding expressions remain the same as when evaluating G. This may easily be verified directly, if we use the analytic definition of Γ and proceed in precisely the same way as in the previous section.

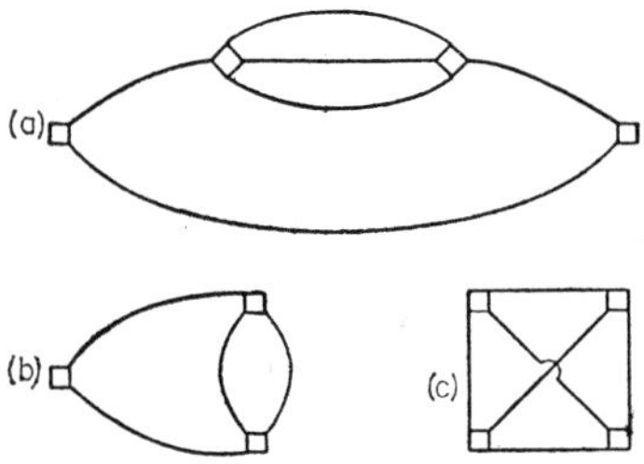

Fig. 29

When evaluating Γ it usually proves convenient to carry out a preliminary partial summation of the individual parts. We introduce for this purpose the concept of a compact diagram. This is a diagram that does not contain self-energy parts. For example, the diagrams of Figs. 23 and 29 *b*, *c* are compact, whereas the diagram of Fig. 29 *a* is not. All the diagrams for Γ follow from compact diagrams by the addition of self-energy parts to the interior $G^{(0)}$-lines, i.e. by replacing the $G^{(0)}$-lines by complete G-lines. It is therefore sufficient, when finding Γ, to confine ourselves to drawing compact diagrams and to associate complete G-functions with each continuous line on the diagram.

B. Electron-phonon interactions. On taking H_{int} in the form (9.5), we find for the last term in (10.11):

$$-ig\langle T\left(\tilde{\psi}_\alpha(x)\tilde{\psi}^+_\beta(x')\tilde{\varphi}(x)\right)\rangle.$$

We can also associate

$$P_{\alpha\beta}(x_1x_2;x_3)=\langle T\left(\tilde{\psi}_\alpha(x_1)\tilde{\psi}^+_\beta(x_2)\tilde{\varphi}(x_3)\right)\rangle \tag{10.18}$$

with a set of Feynman diagrams with one phonon and two electron ends. The simplest of these diagrams occurs in first order perturbation theory (Fig. 30 *a*) and is equal to

$$-g\delta_{\alpha\beta}\int d^4y G^{(0)}(x_1-y)G^{(0)}(y-x_2)D^{(0)}(y-x).$$

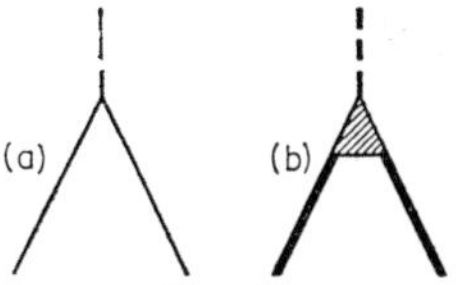

Fig. 30

On arguing as above, we can associate $P_{\alpha\beta}$ with diagram 30 *b*, and it is equal to

$$P_{\alpha\beta}(x_1x_2;x_3)=\delta_{\alpha\beta}P(x_1x_2;x_3)$$
$$=-\delta_{\alpha\beta}\int d^4x_1'd^4x_2'd^4x_3'G(x_1-x_1')G(x_2'-x_2)D(x_3'-x_3)\Gamma(x_1'x_2';x_3'). \tag{10.19}$$

The function Γ corresponds to the set of all diagrams with three (one phonon and two electron) end-points. Thus Γ is the vertex part for the electron-phonon interaction. In view of the spatial homogeneity, Γ and P depend only on two coordinate differences. We can therefore represent e.g. Γ as a Fourier integral:

$$\Gamma(p,p-k;k)\,(2\pi)^4\delta(p-p'-k)$$
$$=\int d^4x_1d^4x_2d^4x_3\Gamma(x_1x_2;x_3)e^{-ipx_1+ip'x_2+ikx_3}. \tag{10.20}$$

The relationship between the Fourier components of Γ and P is

$$P(p,p-k;k)=-G(p)G(p-k)D(k)\Gamma(p,p-k;k). \tag{10.21}$$

On using (10.18) and (10.19) to write the last term of equation (10.11) for the electron-phonon interaction, we get an equation for G in coordinate space. On Fourier transforming this equation with the aid of (10.20), we get the Dyson equation (10.7).

All that has been said regarding the evaluation of the vertex part for two-particle interactions still holds in the present case. To find Γ, we have to draw all the compact diagrams and associate them with analytic formulae in accordance with the same rules as when finding G. Each full-drawn line will now denote a complete G-function, and each dotted a complete D-function. Examples are shown in Fig. 31.

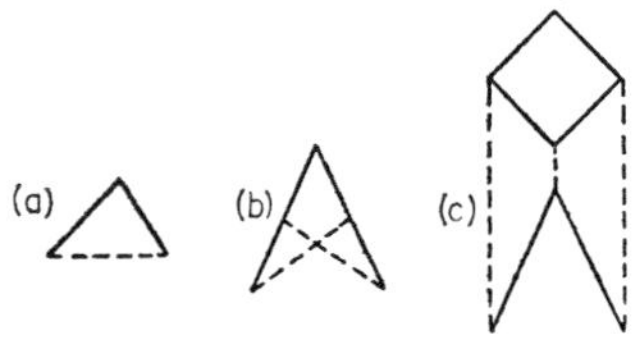

Fig. 31

Let us dwell on the meaning of the functions G^{II} and P, which we introduced when deriving the Dyson equations. These functions, and also other averages of time-ordered products of a larger number of field operators, are known as many-particle Green functions. The function G and D themselves are therefore called one-particle Green functions. Many-particle, like one-particle, Green functions determine the macroscopic properties of a system. In particular, the two-particle Green function G^{II} determines the behaviour of a system of electrons in an external electromagnetic field (see Chap. VI). In view of the fact that these functions depend on a large number of arguments, the analysis of their analytic properties presents considerable difficulties. The situation is simpler when several arguments can be regarded as equal. For instance, if we take $x_1 = x_3$, $x_2 = x_4$ in G^{II}, the analytic properties of the Fourier transform of this function with respect to the variable $x_1 - x_2$ are the same as those of the phonon Green function $D(\omega, \boldsymbol{k})$. Since it is precisely such particular cases that usually interest us, it is simpler to determine the analytic properties of particular Green functions, without attempting an investigation of the general case.

The poles of the Fourier components of the many-particle functions, like the poles of $G(p)$ and $D(k)$, determine the excitation spectrum of a system. These necessarily include all the poles of $G(p)$ and $D(k)$. Apart from these poles, there may be new ones, corresponding to other branches of the excitation spectrum. We shall not be concerned with a general analysis of this question. A concrete example is discussed in Chap. IV, § 19, where we find the equation for the poles of the two-particle Green function of a Fermi system and show that they determine the Bose branches of the excitation spectrum.

We could in principle evaluate many-particle Green functions by writing down equations analogous to the Dyson equations, connecting these functions with higher-order functions. In practice, however, this procedure gives no useful results and it is simpler to sum the diagrams directly. It often happens that a certain sequence of diagrams is the most important and summation of the diagrams does not usually present much difficulty in these cases.

3. *Ground state energy*

We shall conclude this section with some expressions that enable us to find the correction to the ground state energy resulting from interactions between the particles.

We subtract from equation (10.11) the corresponding equation for $G^{(0)}$. We get

$$\left(i\frac{\partial}{\partial t}+\frac{\nabla^2}{2m}+\mu\right)[G_{\alpha\beta}(x-x')-G^{(0)}_{\alpha\beta}(x-x')]$$
$$=-i\langle T([\tilde{\psi}_\alpha(x), H_{int}], \tilde{\psi}^+_\beta(x'))\rangle.$$

We let $\boldsymbol{r}\to\boldsymbol{r}'$ and $t'\to t+0$. We then integrate both sides over $\boldsymbol{r}$, and get

$$\nu\langle H_{int}\rangle=-i\int \mathrm{d}^3\boldsymbol{r}\lim_{\substack{\boldsymbol{r}'\to\boldsymbol{r}\\ t'\to t+0}}\left(i\frac{\partial}{\partial t}+\frac{\nabla^2}{2m}+\mu\right)[G_{\alpha\alpha}(x-x')-G^{(0)}_{\alpha\alpha}(x-x')],$$

where ν is the number of ψ^+ operators appearing in H_{int}. Let the interaction Hamiltonian be proportional to some constant g (such a constant can always be brought in). The ground state energy (more precisely, the potential $\Omega=E-\mu N$), considered as a function of μ, is equal to $\Omega=\langle\hat{H}-\mu\hat{N}\rangle$.

By virtue of a familiar statistical formula (see Landau and Lifshitz [1]), we have

$$\frac{\partial\Omega}{\partial g}=\left\langle\frac{\partial}{\partial g}(\hat{H}-\mu\hat{N})\right\rangle=\frac{1}{g}\langle H_{int}\rangle.$$

Integration of this with respect to $\mathrm{d}g$ from 0 to g gives

$$\Omega-\Omega_0=\int_0^g\frac{\mathrm{d}g_1}{g_1}\langle H_{int}\rangle, \tag{10.22}$$

where Ω_0 is the potential for non-interacting particles. On substituting in this the expression found above for $\langle H_{int}\rangle$ in terms of the Green function, we get

$$\Omega-\Omega_0=-\frac{i}{\nu}\int_0^g\frac{\mathrm{d}g_1}{g_1}\int\mathrm{d}^3\boldsymbol{r}\lim_{\substack{\boldsymbol{r}'\to\boldsymbol{r}\\ t'\to t+0}}\left(i\frac{\partial}{\partial t}+\frac{\nabla^2}{2m}+\mu\right)$$
$$\times[G_{\alpha\alpha}(x-x')-G^{(0)}_{\alpha\alpha}(x-x')].$$

Putting $G_{\alpha\beta}(x-x')=\delta_{\alpha\beta}G(x-x')$, changing to momentum space and using the equation for $G^{(0)}$, we finally get

$$\Omega-\Omega_0=-\frac{2i}{\nu}V\int_0^g\frac{\mathrm{d}g_1}{g_1}\int\frac{\mathrm{d}^4p}{(2\pi)^4}G^{(0)-1}(p)[G(p)-G^{(0)}(p)]e^{i\omega t}, \tag{10.23}$$

where $t\to+0$, and where V is the volume of the system.

We can get another useful formula from the following relation (see [1]):

$$\left(\frac{\partial\Omega}{\partial m}\right)_{T,V,\mu}=\left\langle\frac{\partial\hat{H}}{\partial m}\right\rangle.$$

Since

$$\frac{\partial \hat{H}}{\partial m} = \frac{1}{2m^2} \int \psi_\alpha^+(\boldsymbol{r}) \nabla^2 \psi_\alpha(\boldsymbol{r}) \, \mathrm{d}^3\boldsymbol{r},$$

we have

$$\frac{\partial \Omega}{\partial m} = -\frac{i}{2m^2} \int \left[\nabla_{\boldsymbol{x}}^2 \underset{\substack{\boldsymbol{x}' \to \boldsymbol{x} \\ t' \to t+0}}{G_{\alpha\alpha}}(x - x') \right] \mathrm{d}^3\boldsymbol{r}.$$

On changing to Fourier components, we get

$$\frac{\partial \Omega}{\partial m} = \frac{iV}{m^2} \int p^2 G(p) e^{i\omega t} \frac{\mathrm{d}^4 p}{(2\pi)^4}, \tag{10.24}$$

where $t \to +0$.

Finally, it is pertinent to recall here a formula mentioned in § 7:

$$\frac{\partial \Omega}{\partial \mu} = -N = iV \int G(p) e^{i\omega t} \frac{\mathrm{d}^4 p}{(2\pi)^4}. \tag{10.25}$$

CHAPTER III

THE DIAGRAM TECHNIQUE AT FINITE TEMPERATURES(†)

§ 11. TEMPERATURE-DEPENDENT GREEN FUNCTIONS

1. General properties

We have so far studied the properties of a large number particles at absolute zero. The subject becomes much more complex at finite temperatures.

The ordinary "classical" method of statistical physics consists in a direct evaluation of the thermodynamic functions of a system as functions of its temperature and density. Since no problem of this type can in fact be solved exactly, the answer is expressed as an expansion in powers of some small parameter. If we use ordinary thermodynamic perturbation theory (see Landau and Lifshitz [1]), we can easily write down the first two terms of the perturbation theory series for the free energy F:

$$F = F_0 + \sum_n V_{nn} \exp\{(F_0 - E_n^{(0)})/T\}$$
$$+ \frac{1}{2} \sum_{n,m} \frac{|V_{nm}|^2}{E_n^{(0)} - E_m^{(0)}} [\exp\{(F_0 - E_n^{(0)})/T\} - \exp\{(F_0 - E_m^{(0)})/T\}]$$
$$+ \frac{1}{2T}\Big(\sum_n V_{nn} \exp\{(F_0 - E_n^{(0)})/T\}\Big)^2 + \cdots.$$

But it is no easy matter to write down the next terms, not to mention to evaluate them directly. The summation of an infinite sequence of terms is a quite hopeless task. It is for these reasons that the diagram technique of quantum field theory becomes so attractive in statistics at finite temperatures; this technique, based on Green functions, enables a clear visual representation of the structure and nature of any approximation to be made.

(†) The arguments and calculations of this chapter are largely a repeat of corresponding passages of Chap. II. It seemed useful to write the chapters in parallel because of their importance for what follows. Actually, the reader already acquainted with the methods of quantum field theory, and only interested in the temperature diagram technique, can start his reading here.

The description of the diagram technique in the previous chapter does not permit us directly to generalise to the case of finite temperatures. A diagram technique can only be developed at finite temperatures for special quantities — the temperature-dependent Green functions, which, instead of depending on time like the Green functions considered earlier, depend on a fictitious imaginary "time" $i\tau$, varying from $-i/T$ to zero (Matsubara [30]).

Matsubara's method, like the technique at absolute zero, does not evaluate the thermodynamic functions themselves; instead it finds the above-mentioned temperature-dependent Green functions $\mathfrak{G}(\boldsymbol{r}, \tau)$. Any term of the perturbation theory series for the latter is described by the appropriate Feynman diagram, and is evaluated in accordance with the Feynman rules: each line of the diagram is associated with the temperature Green function for a free particle $\mathfrak{G}^{(0)}(\boldsymbol{r}, \tau)$, each vertex with an interaction operator, and so on. The only difference as compared with the $T = 0$ case is that, instead of integrating over a time t from $-\infty$ to ∞, we integrate at each vertex over τ from 0 to $1/T$.

The temperature-dependent Green function occurring in the diagram technique at non-zero temperature is defined as

$$\mathfrak{G}_{\alpha\beta}(\boldsymbol{r}_1, \tau_1; \boldsymbol{r}_2, \tau_2)$$
$$= \begin{cases} -\operatorname{Tr}[e^{(\Omega+\mu\hat{N}-\hat{H})/T} e^{(\hat{H}-\mu\hat{N})(\tau_1-\tau_2)}\psi_\alpha(\boldsymbol{r}_1) e^{-(\hat{H}-\mu\hat{N})(\tau_1-\tau_2)}\psi_\beta^+(\boldsymbol{r}_2)], & \tau_1 > \tau_2; \\ \pm\operatorname{Tr}[e^{(\Omega+\mu\hat{N}-\hat{H})/T} e^{-(\hat{H}-\mu\hat{N})(\tau_1-\tau_2)}\psi_\beta^+(\boldsymbol{r}_2) e^{(\hat{H}-\mu\hat{N})(\tau_1-\tau_2)}\psi_\alpha(\boldsymbol{r}_1)], & \tau_1 < \tau_2. \end{cases} \tag{11.1}$$

Here $\psi_\alpha(\boldsymbol{r})$, $\psi_\alpha(\boldsymbol{r})$ are the Schrödinger operators of the system, and the plus (minus) sign refers to the case of fermions (bosons). The operation Tr denotes taking the sum of all the diagonal matrix elements. The summation is carried out both over the number of particles in the system and over all the possible states of the system for a given number of particles. Hence $\mathfrak{G}$ is by definition a function of the temperature T and the chemical potential μ. The Ω occurring in (11.1) is the thermodynamic potential in the variables T, V, μ $(\mathrm{d}\Omega = -S\,\mathrm{d}T - P\,\mathrm{d}V - N\,\mathrm{d}\mu)$. Remember that the operation $\operatorname{Tr}[\exp(\Omega + \mu N - H)/T \ldots]$ is the usual grand ensemble average; we shall often denote this by $\langle\cdots\rangle$.

The temperature-dependent phonon Green function $\mathfrak{D}$ is similarly defined:

$$\mathfrak{D}(\boldsymbol{r}_1, \tau_1; \boldsymbol{r}_2, \tau_2)$$
$$= \begin{cases} -\operatorname{Tr}[e^{(\Omega-\hat{H})/T} e^{\hat{H}(\tau_1-\tau_2)}\varphi(\boldsymbol{r}_1) e^{-\hat{H}(\tau_1-\tau_2)}\varphi(\boldsymbol{r}_2)], & \tau_1 > \tau_2, \\ -\operatorname{Tr}[e^{(\Omega-\hat{H})/T} e^{-\hat{H}(\tau_1-\tau_2)}\varphi(\boldsymbol{r}_2) e^{\hat{H}(\tau_1-\tau_2)}\varphi(\boldsymbol{r}_1)], & \tau_1 < \tau_2, \end{cases} \tag{11.2}$$

where $\varphi(\boldsymbol{r})$ is the Schrödinger operator of a phonon field.

It follows at once from the definitions (11.1) and (11.2) that the temperature-dependent Green functions depend only on the "time" difference $\tau_1 - \tau_2$. If the system is isolated and homogeneous, they are of course only dependent on the differences between the spatial coordinates: $\mathfrak{G} = \mathfrak{G}(\boldsymbol{r}_1 - \boldsymbol{r}_2, \tau_1 - \tau_2)$. $\mathfrak{G}(\tau)$ is a discontinuous function of τ, with a discontinuity at the point $\tau = 0$. The size of this discontinuity can be found directly from the definition. For the case of fermions, we have

$$\mathfrak{G}(\tau) - \mathfrak{G}(-\tau)|_{\tau\to+0} = -\mathrm{Tr}\{e^{(\Omega+\mu\hat{N}-\hat{H})/T}[\psi_\alpha(\boldsymbol{r}_1)\psi_\beta^+(\boldsymbol{r}_2) + \psi_\beta^+(\boldsymbol{r}_2)\psi_\alpha(\boldsymbol{r}_1)]\}$$

or, using the commutation rules for ψ and ψ^+,

$$\Delta\mathfrak{G} = -\delta_{\alpha\beta}\delta(\boldsymbol{r}_1 - \boldsymbol{r}_2).$$

The discontinuity of the boson $\mathfrak{G}$-function is equal to that for fermions.

Expressions (11.1) and (11.2) can be written in a form analogous to the definition of the Green function at absolute zero. This is done by introducing the "Heisenberg" operators of the particles, dependent on the "time" τ, through the equations (†)

$$\begin{aligned} \tilde{\psi}_\alpha(\boldsymbol{r}, \tau) &= e^{\tau(\hat{H}-\mu\hat{N})}\psi_\alpha(\boldsymbol{r})e^{-\tau(\hat{H}-\mu\hat{N})}, \\ \bar{\tilde{\psi}}_\alpha(\boldsymbol{r}, \tau) &= e^{\tau(\hat{H}-\mu\hat{N})}\psi_\alpha^+(\boldsymbol{r})e^{-\tau(\hat{H}-\mu\hat{N})}, \\ \tilde{\varphi}(\boldsymbol{r}, \tau) &= e^{\tau\hat{H}}\varphi(\boldsymbol{r})e^{-\tau\hat{H}}. \end{aligned} \tag{11.3}$$

With the aid of these, the unwieldy expressions (11.1) may be written as (cf. (7.1) and (7.14)):

$$\begin{aligned} \mathfrak{G}_{\alpha\beta}(\boldsymbol{r}_1, \tau_1; \boldsymbol{r}_2, \tau_2) &= -\mathrm{Tr}\{e^{(\Omega+\mu\hat{N}-\hat{H})/T}T_\tau(\tilde{\psi}_\alpha(\boldsymbol{r}_1, \tau_1)\bar{\tilde{\psi}}_\beta(\boldsymbol{r}_2, \tau_2))\} \\ &\equiv -\langle T_\tau(\tilde{\psi}_\alpha(\boldsymbol{r}_1, \tau_1)\bar{\tilde{\psi}}_\beta(\boldsymbol{r}_2, \tau_2))\rangle. \end{aligned} \tag{11.4}$$

The symbol T_τ in (11.4) denotes the operation of T-ordering already familiar from the previous chapter. The operators under the sign of the T_τ-product are arranged from left to right in order of decreasing "time" τ (we give the T-product symbol an index τ in order to distinguish it from the temperature T). Recall that, in the fermion case

$$T_\tau(\psi_1\psi_2\cdots) = \delta_P\psi_{i_1}\psi_{i_2}\cdots,$$

where the operators ψ are time-ordered from the right, and δ_P is equal to $+1$ or -1, depending on whether the permutation

$$1, 2, \ldots \to i_1, i_2, \ldots$$

is even or odd. In particular,

$$\begin{aligned} T_\tau(\tilde{\psi}(1)\bar{\tilde{\psi}}(2)) &= \tilde{\psi}(1)\bar{\tilde{\psi}}(2), \qquad \tau_1 > \tau_2, \\ T_\tau(\tilde{\psi}(1)\bar{\tilde{\psi}}(2)) &= -\bar{\tilde{\psi}}(2)\tilde{\psi}(1), \qquad \tau_1 < \tau_2. \end{aligned}$$

(†) Notice right away that ψ and $\tilde{\psi}$ are no longer Hermitian conjugates of each other.

Many-particle Green functions may be defined in Matsubara's method by similar relationships. For instance, the two-particle temperature-dependent Green function is

$$\mathfrak{G}^{\text{II}}_{\alpha\beta;\gamma\delta}(1, 2; 3, 4) = -\langle T_\tau(\tilde{\psi}_\alpha(1)\tilde{\psi}_\beta(2)\tilde{\bar{\psi}}_\gamma(3)\tilde{\bar{\psi}}_\delta(4))\rangle. \tag{11.5}$$

The extension to the case of Green functions depending on a large number of variables is obvious.

In principle, the $\mathfrak{G}$-functions determine all the thermodynamic properties of the system. If we use e.g. the formula

$$N = \pm \int \mathfrak{G}_{\alpha\alpha}(\mathbf{r}, \tau; \mathbf{r}, \tau + 0)\mathrm{d}^3\mathbf{r}, \tag{11.6}$$

which follows at once from the definition of $\mathfrak{G}$ and the relationship $N = \int \psi_\alpha^+(\mathbf{r})\, \psi_\alpha(\mathbf{r})\, \mathrm{d}^3\mathbf{r}$, we can find the number of particles in the system as a function of its chemical potential μ or alternatively, by solving (11.6) for μ, the chemical potential as a function of the temperature and the density $n = N/V$. If we then integrate the familiar thermodynamic relationship $\partial f/\partial n = \mu(n,T)$, we can find the free energy $f(n,T)$ per unit volume.

If there are only binary interactions between particles in a system, described by the Hamiltonian

$$\hat{H} = -\frac{1}{2m}\int \psi_\alpha^+(\mathbf{r})\nabla^2\psi_\alpha(\mathbf{r})\mathrm{d}^3\mathbf{r} + \frac{1}{2}\int\int \mathrm{d}^3\mathbf{r}_1\mathrm{d}^3\mathbf{r}_2\psi_\alpha^+(\mathbf{r}_1)\psi_\beta^+(\mathbf{r}_2)\,U(\mathbf{r}_1 - \mathbf{r}_2)\psi_\beta(\mathbf{r}_2)\psi_\alpha(\mathbf{r}_1),$$

its energy may be expressed in terms of the two-particle temperature-dependent Green function:

$$E(\mu, T) = \langle\hat{H}\rangle = \mp\frac{1}{2m}\int \nabla^2_{\mathbf{r}_1}\mathfrak{G}_{\alpha\alpha}(1, 2)\Big|_{\substack{\mathbf{r}_1=\mathbf{r}_2\\ \tau_2=\tau_1+0}}\mathrm{d}^3\mathbf{r}_1 - \frac{1}{2}\int\int U(\mathbf{r}_1 - \mathbf{r}_2)\,\mathfrak{G}^{\text{II}}_{\alpha\beta;\beta\alpha}(1, 2; 3, 4)\Big|_{\substack{\mathbf{r}_3=\mathbf{r}_2,\mathbf{r}_4=\mathbf{r}_1\\ \tau_3=\tau_4+0,\tau_4=\tau_1+0\\ \tau_1=\tau_2+0}}\mathrm{d}^3\mathbf{r}_1\mathrm{d}^3\mathbf{r}_2.$$

We shall later give a further series of relationships between the temperature-dependent Green functions and the thermodynamic functions.

The range of problems that can be solved with the aid of the temperature-dependent Green functions is not confined to thermodynamics. The Green functions determine various correlation properties of a system, including, in particular, those that make their appearance in the interaction of condensed solids with neutrons, X-rays, and so on. For example, the two-particle Green function is connected by an obvious relation with the density correlation function

$$F(\mathbf{r}_1,\mathbf{r}_2) = \overline{n(\mathbf{r}_1)n(\mathbf{r}_2)} \equiv \langle\psi_\alpha^+(\mathbf{r}_1)\psi_\alpha(\mathbf{r}_1)\psi_\beta^+(\mathbf{r}_2)\psi_\beta(\mathbf{r}_2)\rangle,$$

which determines the elastic scattering of X-rays and neutrons. We shall also establish in the following a connection between the temperature-dependent Green functions and corresponding time-dependent quantities, which enables us to study various kinetic phenomena.

We now note an important property of the temperature-dependent Green function $\mathfrak{G}$. As already mentioned, it is a function of the "time" difference $\tau_1 - \tau_2 = \tau$, and as such, is given in the interval $-1/T$ to $1/T$. Let us carry out a cyclical permutation of the operators under the trace sign (†) in expression (11.1) for $\mathfrak{G}(\tau < 0)$:

$$\mathfrak{G}(\tau < 0) = \pm \operatorname{Tr}\{e^{\Omega/T} e^{(\hat{H}-\mu\hat{N})\tau} \psi(\boldsymbol{r}_1) e^{-(\hat{H}-\mu\hat{N})(\tau+1/T)} \psi^+(\boldsymbol{r}_2)\}$$

$$= \pm \operatorname{Tr}\{e^{(\Omega+\mu\hat{N}-\hat{H})/T} e^{(\hat{H}-\mu\hat{N})(\tau+1/T)} \psi(\boldsymbol{r}_1) e^{-(\hat{H}-\mu\hat{N})(\tau+1/T)} \psi^+(\boldsymbol{r}_2)\}. \quad (11.7)$$

On comparing (11.7) with the expression for $\mathfrak{G}$ when $\tau > 0$ and observing that $0 < \tau + 1/T < 1/T$ when $\tau < 0$, we get

$$\mathfrak{G}(\tau < 0) = \pm \mathfrak{G}\left(\tau + \frac{1}{T}\right), \quad (11.8)$$

relating $\mathfrak{G}$ at negative "times" with its values for $\tau > 0$. Of course,

$$\mathfrak{D}(\tau < 0) = \mathfrak{D}\left(\tau + \frac{1}{T}\right). \quad (11.8\text{a})$$

Another useful relation follows from the obvious fact that the $\mathfrak{D}$-function of a phonon is real (the operators $\varphi(\boldsymbol{r})$ are real!). Let us evaluate $\mathfrak{D}^*(\tau < 0)$ formally:

$$\mathfrak{D}(\tau < 0) = \mathfrak{D}^*(\tau < 0) = -\operatorname{Tr}\{e^{\Omega/T} \varphi(\boldsymbol{r}_1) e^{\hat{H}\tau} \varphi(\boldsymbol{r}_2) e^{-\hat{H}\tau} e^{-\hat{H}/T}\}.$$

On comparing the expression obtained with $\mathfrak{D}(\tau > 0)$, we arrive at the conclusion that the temperature-dependent phonon Green function is an even function of τ:

$$\mathfrak{D}(\tau) = \mathfrak{D}(-\tau). \quad (11.9)$$

This statement holds for the Green function for any real field.

2. *Temperature-dependent free particle Green functions*

In perturbation theory as based on the diagram technique, an important role is played by the Green functions for free particles. If there are no interactions, the statistical averaging in (11.1) is performed independently over the state of each individual particle. The energy levels E_n of the system (and with these the thermodynamic potential Ω) can be

(†) That such a permutation is possible follows at once from the definition of the trace of the matrix of a product of several operators:

$$\operatorname{Tr}(ABC \ldots DF) = \sum_{i,k,\ldots} A_{ik} B_{kl} C_{lm} \ldots D_{np} F_{pi}$$
$$= \sum_{i,k,\ldots} B_{kl} C_{lm} \ldots D_{np} F_{pi} A_{ik} = \operatorname{Tr}(BC \ldots DFA).$$

expressed as sums of the energies of the individual particles with given momenta $\boldsymbol{p}$ and spin α:

$$E_n^{(0)} = \sum_{\boldsymbol{p},\alpha} n_{\boldsymbol{p}\alpha}\,\varepsilon_0(\boldsymbol{p}), \quad \Omega_0 = \sum_{\boldsymbol{p},\alpha} \Omega_{\boldsymbol{p}\alpha}^{(0)}.$$

Because of Pauli's principle, the occupation numbers of the states can only take the values 0 and 1 in the case of Fermi statistics.

Definition (11.1) is the most convenient one for evaluating the Green functions of free particles. We substitute in this the Fourier expansions of the Schrödinger operators ψ:

$$\psi_\alpha(\boldsymbol{r}_1) = \frac{1}{\sqrt{V}} \sum_{\boldsymbol{p}_1} a_{\boldsymbol{p}_1\alpha} e^{i(\boldsymbol{p}_1\cdot\boldsymbol{r}_1)}, \qquad \psi_\beta^+(\boldsymbol{r}_2) = \frac{1}{\sqrt{V}} \sum_{\boldsymbol{p}_2} a_{\boldsymbol{p}_2\beta}^+ e^{-i(\boldsymbol{p}_2\cdot\boldsymbol{r}_2)}.$$

We have:

$$\mathfrak{G}_{\alpha\beta}^{(0)}(\tau > 0) = -\frac{1}{V} \sum_{\boldsymbol{p}_1\boldsymbol{p}_2} e^{i(\boldsymbol{p}_1\cdot\boldsymbol{r}_1) - i(\boldsymbol{p}_2\cdot\boldsymbol{r}_2)}$$

$$\times \operatorname{Tr}\{e^{(\Omega_0 + \mu\hat{N} - \hat{H}_0)/T}\, e^{\tau(\hat{H}_0 - \mu\hat{N})} a_{\boldsymbol{p}_1\alpha} e^{-\tau(\hat{H}_0 - \mu\hat{N})} a_{\boldsymbol{p}_2\beta}^+\}.$$

On observing further that the Hamiltonian $\hat{H}$ has the form

$$\hat{H}_0 = \sum_{\boldsymbol{p}\alpha} \hat{n}_{\boldsymbol{p}\alpha}\varepsilon_0(\boldsymbol{p}), \quad \hat{N} = \sum_{\boldsymbol{p}\alpha} \hat{n}_{\boldsymbol{p}\alpha}$$

in the momentum representation, we can easily prove the identities

$$\begin{aligned} e^{\tau(\hat{H}_0 - \mu\hat{N})} a_{\boldsymbol{p}\alpha} e^{-\tau(\hat{H}_0 - \mu\hat{N})} &= a_{\boldsymbol{p}\alpha} e^{-\tau(\varepsilon_0(\boldsymbol{p}) - \mu)}, \\ e^{\tau(\hat{N}_0 - \mu\hat{N})} a_{\boldsymbol{p}\alpha}^+ e^{-\tau(\hat{H}_0 - \mu\hat{N})} &= a_{\boldsymbol{p}\alpha}^+ e^{\tau(\varepsilon_0(\boldsymbol{p}) - \mu)}, \end{aligned} \tag{11.10}$$

simply by evaluating the only non-zero matrix element on the right and left-hand sides. Hence

$$\mathfrak{G}^{(0)}(\tau > 0) = -\frac{1}{V} \sum_{\boldsymbol{p}_1\boldsymbol{p}_2} e^{i(\boldsymbol{p}_1\cdot\boldsymbol{r}_1) - i(\boldsymbol{p}_2\cdot\boldsymbol{r}_2) - \tau(\varepsilon_0(\boldsymbol{p}_1) - \mu)}$$

$$\times \operatorname{Tr}\{\exp(\Omega_0 + \mu\hat{N} - \hat{H}_0)/T)\, a_{\boldsymbol{p}_1\alpha} a_{\boldsymbol{p}_2\beta}^+\}.$$

The product $a_{\boldsymbol{p}_1\alpha} a_{\boldsymbol{p}_2\beta}^+$ has non-zero diagonal matrix elements only when $\boldsymbol{p}_1 = \boldsymbol{p}_2$, $\alpha = \beta$, so that

$$\mathfrak{G}_{\alpha\beta}^{(0)}(\boldsymbol{r}_1 - \boldsymbol{r}_2, \tau > 0) = -\delta_{\alpha\beta} \frac{1}{V} \sum_{\boldsymbol{p}} e^{i(\boldsymbol{p}\cdot\boldsymbol{r}_1 - \boldsymbol{r}_2) - \tau(\varepsilon_0(\boldsymbol{p}) - \mu)} \langle a_{\boldsymbol{p}\alpha} a_{\boldsymbol{p}\alpha}^+\rangle.$$

The quantity $\langle a_{\boldsymbol{p}\alpha} a_{\boldsymbol{p}\alpha}^+\rangle$ is expressible in terms of the equilibrium occupation numbers $n(\boldsymbol{p})$, which depend on the temperature and the chemical potential. For particles obeying Fermi statistics,

$$\langle a_{\boldsymbol{p}\alpha} a_{\boldsymbol{p}\alpha}^+\rangle = 1 - n(\boldsymbol{p}), \qquad n(\boldsymbol{p}) = [e^{(\varepsilon_0(\boldsymbol{p}) - \mu)/T} + 1]^{-1}; \tag{11.11}$$

for bosons,

$$\langle a_{\boldsymbol{p}\alpha} a_{\boldsymbol{p}\alpha}^+\rangle = 1 + n(\boldsymbol{p}), \qquad n(\boldsymbol{p}) = [\exp\{\varepsilon^{(\varepsilon_0(\boldsymbol{p}) - \mu)T}\} - 1]^{-1}. \tag{11.12}$$

We now let the volume V of the system tend to infinity and change from a summation over the momenta to an integration in the usual way. We finally get:

$$\mathfrak{G}^{(0)}_{\alpha\beta}(\boldsymbol{r}, \tau > 0) = -\delta_{\alpha\beta}\frac{1}{(2\pi)^3}\int d^3\boldsymbol{p}\, e^{i(\boldsymbol{p}\cdot\boldsymbol{r})-\tau(\varepsilon_0(\boldsymbol{p})-\mu)}\left(1 \mp n(\boldsymbol{p})\right), \qquad (11.13a)$$

where the upper (lower) sign corresponds to fermions (bosons). We can most easily find $\mathfrak{G}^{(0)}$ for $\tau < 0$ by means of equation (11.8):

$$\mathfrak{G}^{(0)}_{\alpha\beta}(\boldsymbol{r}, \tau < 0) = \mp\, \mathfrak{G}^{(0)}_{\alpha\beta}\left(\boldsymbol{r}, \tau + \frac{1}{T}\right)$$
$$= \pm\, \delta_{\alpha\beta}\frac{1}{(2\pi)^3}\int d^3\boldsymbol{p}\, e^{i(\boldsymbol{p}\cdot\boldsymbol{r})-\tau(\varepsilon_0(\boldsymbol{p})-\mu)}n(\boldsymbol{p}). \qquad (11.13b)$$

The Green function for free phonons is similarly evaluated. On substituting in (11.2) the Fourier expansion of the operator $\varphi(\boldsymbol{r})$:

$$\varphi(\boldsymbol{r}) = \frac{1}{\sqrt{V}}\sum_{\boldsymbol{k}}\sqrt{\frac{\omega_0(\boldsymbol{k})}{2}}\left(b_{\boldsymbol{k}}e^{i(\boldsymbol{k}\cdot\boldsymbol{r})} + b^+_{\boldsymbol{k}}e^{-i(\boldsymbol{k}\cdot\boldsymbol{r})}\right),$$

where $\omega_0(\boldsymbol{k})$ is the phonon energy, we find after the necessary calculations:

$$\mathfrak{D}^{(0)}(\boldsymbol{r}, \tau) = -\frac{1}{2(2\pi)^3}\int d^3\boldsymbol{k}\,\omega_0(\boldsymbol{k})\{(N(\boldsymbol{k})+1)e^{i(\boldsymbol{k}\cdot\boldsymbol{r})+\omega_0(\boldsymbol{k})|\tau|}$$
$$+ N(\boldsymbol{k})e^{i(\boldsymbol{k}\cdot\boldsymbol{r})-\omega_0(\boldsymbol{k})|\tau|}\}, \qquad (11.14)$$
$$N(\boldsymbol{k}) = [e^{\omega_0(\boldsymbol{k})/T} - 1]^{-1}.$$

$\mathfrak{D}^{(0)}$ is an even function of τ, in accordance with (11.9).

§ 12. PERTURBATION THEORY

1. *The interaction representation*

If the particles in a system are not free, we can transform in the expression for the temperature-dependent Green function to a special type of interaction representation, similar to the interaction representation of quantum field theory (Matsubara [30]). We introduce for this purpose the matrix $\mathfrak{S}(\tau)$ $(0 < \tau < 1/T)$, the analogue of the S-matrix of field theory; it is defined by

$$e^{-\tau(\hat{H}-\mu\hat{N})} = e^{-\tau(\hat{H}_0-\mu\hat{N})}\mathfrak{S}(\tau),$$
$$e^{\tau(\hat{H}-\mu\hat{N})} = \mathfrak{S}^{-1}(\tau)e^{\tau(\hat{H}_0-\mu\hat{N})}. \qquad (12.1)$$

We also introduce the operators of the particle field in the interaction representation:

$$\psi(\boldsymbol{r}, \tau) = e^{\tau(\hat{H}_0-\mu\hat{N})}\psi(\boldsymbol{r})e^{-\tau(\hat{H}_0-\mu\hat{N})},$$
$$\overline{\psi}(\boldsymbol{r}, \tau) = e^{\tau(\hat{H}_0-\mu\hat{N})}\psi^+(\boldsymbol{r})e^{-\tau(\hat{H}_0-\mu\hat{N})}, \qquad (12.2)$$

which are the same as the Heisenberg operators mentioned in § 11 when $\hat{H} = \hat{H}_0$.

The other operators in the interaction representation are defined in analogy with (12.2). In particular,

$$\hat{H}(\tau) = e^{\tau(\hat{H}_0 - \mu\hat{N})}\hat{H}e^{-\tau(\hat{H}_0-\mu\hat{N})},$$
$$\hat{H}_{int}(\tau) = e^{\tau(\hat{H}_0 - \mu\hat{N})}\hat{H}_{int}e^{-\tau(\hat{H}_0-\mu\hat{N})}.$$

It follows from this definition that the operators $\hat{H}(\tau)$, $\hat{H}_{int}(\tau)$ are obtained from $\hat{H}$, $\hat{H}_{int}$ when the $\psi(\boldsymbol{r})$, $\psi^+(\boldsymbol{r})$ are replaced in the latter by $\psi(\boldsymbol{r}, \tau)$, $\bar{\psi}(\boldsymbol{r}, \tau)$ respectively.

Notice also that $\hat{H}_0(\tau)$, $\hat{N}(\tau)$ are in fact independent of τ (the Hamiltonian of free particles commutes with the operator $\hat{N}$):

$$\hat{H}_0(\tau) = e^{\tau(\hat{H}_0 - \mu\hat{N})}\hat{H}_0e^{-\tau(\hat{H}_0-\mu\hat{N})} = \hat{H}_0,$$
$$\hat{N}(\tau) = e^{\tau(\hat{H}_0 - \mu\hat{N})}\hat{N}e^{-\tau(\hat{H}_0-\mu\hat{N})} = \hat{N}.$$

The matrix $\mathfrak{S}(\tau)$ satisfies a simple equation which only differs from the corresponding equation for the S-matrix (6.17) in that t is replaced by $-i\tau$. We shall derive the equation afresh, however, by differentiating the first of equations (12.1) with respect to τ:

$$-(\hat{H} - \mu\hat{N})e^{-\tau(\hat{H}-\mu\hat{N})} = e^{-\tau(\hat{H}_0-\mu\hat{N})}\frac{\partial\mathfrak{S}(\tau)}{\partial\tau} - (\hat{H}_0 - \mu\hat{N})e^{-\tau(\hat{H}_0-\mu\hat{N})}\mathfrak{S}(\tau).$$

On multiplying both sides of this by $\exp\tau(\hat{H}_0 - \mu\hat{N})$, we get

$$\frac{\partial\mathfrak{S}(\tau)}{\partial\tau} = -\hat{H}_{int}(\tau)\mathfrak{S}(\tau). \tag{12.3}$$

The solution of (12.3), satisfying the condition $\mathfrak{S}(0) = 1$ (which is a consequence of the definition of $\mathfrak{S}$), is

$$\mathfrak{S}(\tau) = T_\tau \exp\left\{-\int_0^\tau \hat{H}_{int}(\tau')\,d\tau'\right\}. \tag{12.4}$$

The symbol T_τ in (12.4) denotes, as already remarked, that all the operators must be arranged from left to right in order of decreasing τ. We can easily prove (12.4) by direct differentiation, if we bear in mind the meaning of the operation T_τ.

We shall consider, in addition to $\mathfrak{S}(\tau)$, the matrix $\mathfrak{S}(\tau_1, \tau_2)$ $(\tau_1 > \tau_2)$:

$$\mathfrak{S}(\tau_1, \tau_2) = T_\tau \exp\left\{-\int_{\tau_2}^{\tau_1} \hat{H}_{int}(\tau')\,d\tau'\right\},$$
$$\mathfrak{S}(\tau) = \mathfrak{S}(\tau, 0).$$

This matrix has a number of obvious properties:

$$\begin{aligned} \mathfrak{S}(\tau_1, \tau_3) &= \mathfrak{S}(\tau_1, \tau_2)\mathfrak{S}(\tau_2, \tau_3) \qquad (\tau_1 > \tau_2 > \tau_3), \\ \mathfrak{S}(\tau_1, \tau_2) &= \mathfrak{S}(\tau_1)\mathfrak{S}^{-1}(\tau_2) \qquad (\tau_1 > \tau_2). \end{aligned} \tag{12.5}$$

We now turn to the interaction representation in expression (11.1) for the Green function; we have, on expressing all the exponents containing $\hat{H}$ in terms of H_0 and $\mathfrak{S}$:

$$\mathfrak{G}(\tau > 0) = -e^{\Omega/T}\,\mathrm{Tr}\left\{\exp[-(\hat{H}_0-\mu\hat{N})/T]\,\mathfrak{S}\left(\frac{1}{T}\right)\mathfrak{S}^{-1}(\tau_1)e^{(\hat{H}_0-\mu\hat{N})\tau_1}\,\psi(\boldsymbol{r}_1)\right.$$
$$\left.\times\, e^{-(\hat{H}_0-\mu\hat{N})\tau_1}\mathfrak{S}(\tau_1)\mathfrak{S}^{-1}(\tau_2)e^{(\hat{H}_0-\mu\hat{N})\tau_2}\psi^+(\boldsymbol{r}_2)e^{-(\hat{H}_0-\mu\hat{N})\tau_2}\mathfrak{S}(\tau_2)\right\},$$

or, on taking (12.1) and (12.5) into account:

$$\mathfrak{G}(\tau > 0) = -e^{\Omega/T}\,\mathrm{Tr}\left\{\exp[-(\hat{H}_0-\mu\hat{N})/T]\,\mathfrak{S}\left(\frac{1}{T},\tau_1\right)\right.$$
$$\left.\times\,\psi(\boldsymbol{r}_1,\tau_1)\mathfrak{S}(\tau_1,\tau_2)\overline{\psi}(\boldsymbol{r}_2,\tau_2)\mathfrak{S}(\tau_2)\right\}. \quad (12.6\text{a})$$

We can write $\mathfrak{G}$ for $\tau < 0$ in a similar way:

$$\mathfrak{G}(\tau < 0) = \pm e^{\Omega/T}\,\mathrm{Tr}\left\{\exp[-(\hat{H}_0-\mu\hat{N})/T]\,\mathfrak{S}\left(\frac{1}{T},\tau_2\right)\right.$$
$$\left.\times\,\overline{\psi}(\boldsymbol{r}_2,\tau_2)\mathfrak{S}(\tau_2,\tau_1)\psi(\boldsymbol{r}_1,\tau_1)\mathfrak{S}(\tau_1)\right\}. \quad (12.6\text{b})$$

Expressions (12.6a) and (12.6b) can be combined into the single formula:

$$\mathfrak{G}(\tau) = -e^{\Omega/T}\,\mathrm{Tr}\left\{\exp[-(\hat{H}_0-\mu\hat{N})/T]\;T_\tau\left[\psi(\boldsymbol{r}_1,\tau_1)\overline{\psi}(\boldsymbol{r}_2,\tau_2)\mathfrak{S}\left(\frac{1}{T}\right)\right]\right\}, \quad (12.6\text{c})$$

which follows immediately from the definition of the T_τ-ordering and (12.5).

It now remains for us to transform $e^{\Omega/T}$. We first notice that, by definition,

$$e^{-\Omega/T} = \mathrm{Tr}\{e^{-(\hat{H}-\mu\hat{N})/T}\},$$

whence it follows immediately that

$$e^{-\Omega/T} = \mathrm{Tr}\left\{\exp[-(\hat{H}_0-\mu\hat{N})/T]\,\mathfrak{S}\left(\frac{1}{T}\right)\right\}.$$

The expression for $\mathfrak{G}$ in the interaction representation may finally be written as

$$\mathfrak{G}(\boldsymbol{r}_1,\tau_1;\boldsymbol{r}_2,\tau_2)$$
$$= -\frac{\mathrm{Tr}\left\{\exp[-(\hat{H}_0-\mu\hat{N})/T]\;T_\tau\left[\psi(\boldsymbol{r}_1,\tau_1)\overline{\psi}(\boldsymbol{r}_2,\tau_2)\mathfrak{S}\left(\frac{1}{T}\right)\right]\right\}}{\mathrm{Tr}\left\{\exp[-(\hat{H}_0-\mu\hat{N})/T]\,\mathfrak{S}\left(\frac{1}{T}\right)\right\}},$$

or, if we bring in the symbol of ensemble averaging over the states of the system of non-interacting particles,

$$\mathfrak{G}(\boldsymbol{r}_1, \tau_1; \boldsymbol{r}_2, \tau_2) = -\frac{\langle T_\tau(\psi(\boldsymbol{r}_1, \tau_1)\overline{\psi}(\boldsymbol{r}_2, \tau_2)\mathfrak{S})\rangle_0}{\langle\mathfrak{S}\rangle_0}, \tag{12.7}$$

$$\langle\cdots\rangle_0 = \mathrm{Tr}\,\{\exp[\Omega_0 + \mu\hat{N}_0 - \hat{H}_0)/T]\},\ \mathfrak{S} \equiv \mathfrak{S}\left(\frac{1}{T}\right). \tag{12.8}$$

By an exact repetition of the above manipulations, we can find expressions for the phonon Green function and the many-particle Green functions in the interaction representation. The phonon Green function becomes

$$\mathfrak{D}(1, 2) = -\frac{\langle T_\tau(\varphi(1)\varphi(2)\mathfrak{S})\rangle_0}{\langle\mathfrak{S}\rangle_0}, \tag{12.9}$$

whilst the two-particle Green function is

$$\mathfrak{G}^{\mathrm{II}}(1, 2; 3, 4) = -\frac{\langle T_\tau(\psi(1)\psi(2)\overline{\psi}(3)\overline{\psi}(4)\mathfrak{S})\rangle_0}{\langle\mathfrak{S}\rangle_0}. \tag{12.10}$$

The formulae for Green functions depending on a large number of variables only differ from (12.7), (12.9), and (12.10) in the number of ψ-operators under the T-product sign.

We conclude by mentioning the relation between the thermodynamic potential Ω and the $\mathfrak{S}$ matrix:

$$\Omega = \Omega_0 - T\ln\langle\mathfrak{S}\rangle_0. \tag{12.11}$$

Here Ω_0 is the potential Ω when interactions are absent:

$$\Omega_0 = -T\ln\mathrm{Tr}\{\exp[-(\hat{H}_0 - \mu\hat{N})/T]\}.$$

2. *Wick's theorem*

We now return to our fundamental problem, the evaluation of the Green functions of systems of interacting particles. If the interactions between the particles can be assumed to be weak, the expressions for the temperature-dependent Green functions in the interaction representation enable us to write the perturbation theory series in powers of $\hat{H}_{int}$ in a particularly compact form.

The interaction Hamiltonian appears in the Green function only via the $\mathfrak{S}$ matrix. On expanding the exponent on the right-hand side of (12.4) in powers of $\hat{H}_{int}(\tau)$, we get:

$$\begin{aligned}\mathfrak{S} &= 1 - \int_0^{1/T}\hat{H}_{int}(\tau')\,d\tau' + \frac{1}{2}\int_0^{1/T}\int_0^{1/T} d\tau'\,d\tau''\,T_\tau\big(\hat{H}_{int}(\tau')\hat{H}_{int}(\tau'')\big) - \cdots \\ &= \sum_{n=0}^{\infty}\frac{(-1)^n}{n!}\int_0^{1/T}\cdots\int_0^{1/T} d\tau_1\cdots d\tau_n\,T_\tau\big(\hat{H}_{int}(\tau_1)\cdots\hat{H}_{int}(\tau_n)\big).\end{aligned} \tag{12.12}$$

On substituting this expansion in the numerator of (12.7), we get the perturbation theory series for the Green function

$$\mathfrak{G}_{\alpha\beta}(\boldsymbol{r}_1, \tau_1; \boldsymbol{r}_2, \tau_2) = -\frac{1}{\langle\mathfrak{S}\rangle_0}\sum_{n=0}^{\infty}\frac{(-1)^n}{n!}\int_0^{1/T}\cdots\int_0^{1/T} d\tau_{(1)}\cdots d\tau_{(n)}$$
$$\times\langle T_\tau(\psi_\alpha(\boldsymbol{r}_1, \tau_1)\bar{\psi}_\beta(\boldsymbol{r}_2, \tau_2)\hat{H}_{int}(\tau_{(1)})\cdots\hat{H}_{int}(\tau_{(n)}))\rangle_0, \quad (12.13)$$

the first term of which is of course, the same as the free Green function $\mathfrak{G}^{(0)} = -\langle T_\tau(\psi(1)\,\bar{\psi}(2))\rangle_0$, evaluated in § 11.

We shall not expand the $\mathfrak{S}$ matrix in the expression $\langle\mathfrak{S}\rangle_0$ in the denominator of (12.13), since it cancels with the same factor in the numerator. Moreover, $\langle\mathfrak{S}\rangle_0$ is a constant, independent of $\boldsymbol{r}, \tau$, and cannot make any appearance in further calculations.

In all practical problems $H_{int}(\tau)$ is the product of a (usually quite small) number of operators $\psi(\boldsymbol{r}, \tau)$, $\bar{\psi}(\boldsymbol{r}, \tau)$ (and possibly $\varphi(\boldsymbol{r}, \tau)$), integrated over the spatial variables. The problem of evaluating the Green function from perturbation theory therefore reduces to evaluating the average of the T-product of a number of operators ψ, taken at different points of space and "time":

$$\langle T_\tau(\psi_\alpha(\boldsymbol{r}, \tau)\cdots\bar{\psi}_{\alpha'}(\boldsymbol{r}', \tau')\cdots)\rangle_0. \quad (12.14)$$

We have already encountered a similar type of problem in the previous chapter, when evaluating the ordinary Green functions at absolute zero. It was shown there that the average of any number of operators reduces to the sum of the products of all possible paired averages, these latter being equal by definition to the free-particle Green functions (Wick's theorem). As we shall see shortly, the situation is similar in our present case.

What we do is replace the operators ψ in (12.14) by their coordinate Fourier expansions (†)

$$\psi(\boldsymbol{r}, \tau) = \frac{1}{\sqrt{V}}\sum_{\boldsymbol{p}} a_{\boldsymbol{p}}(\tau)\,e^{i(\boldsymbol{p}\cdot\boldsymbol{r})-\tau(\varepsilon_0(\boldsymbol{p})-\mu)},$$
$$\bar{\psi}(\boldsymbol{r}, \tau) = \frac{1}{\sqrt{V}}\sum_{\boldsymbol{p}} a_{\boldsymbol{p}}^+(\tau)\,e^{-i(\boldsymbol{p}\cdot\boldsymbol{r})+\tau(\varepsilon_0(\boldsymbol{p})-\mu)}. \quad (12.15)$$

The $a_{\boldsymbol{p}}(\tau)$ and $a_{\boldsymbol{p}}^+(\tau)$ in (12.15) are ordinary annihilation and creation operators and are in fact independent of τ. We retain the τ, however, to denote the place that the operator must occupy in the T-ordering.

After substituting expansions (12.15) in (12.14) and getting rid of the exponents, we get

$$\frac{1}{\sqrt{V}}\sum_{\boldsymbol{p}_1}\frac{1}{\sqrt{V}}\sum_{\boldsymbol{p}_2}\cdots\frac{1}{\sqrt{V}}\sum_{\boldsymbol{p}_1'}\frac{1}{\sqrt{V}}\sum_{\boldsymbol{p}_2'}\cdots$$
$$\cdots\langle T_\tau(a_{\boldsymbol{p}_1}(\tau_1)a_{\boldsymbol{p}_2}(\tau_2)\cdots a_{\boldsymbol{p}_1'}^+(\tau_1')a_{\boldsymbol{p}_2'}^+(\tau_2')\cdots)\rangle_0. \quad (12.16)$$

(†) We can most easily verify (12.15) by using the definition of the operators in the interaction representation and the identities (11.10).

The only non-zero terms in the sum over $p_1, \ldots, p_1', \ldots$ are those containing an equal number of creation and annihilation operators referring to the same momentum. In particular, the non-zero terms include those that contain only one creation and one annihilation operator with the same momentum; thus the term

$$\frac{1}{V}\sum_{p_1}\frac{1}{V}\sum_{\substack{p_2\\ p_1\neq p_2\neq\cdots}}\cdots\langle T_\tau(a_{p_1}(\tau_1)a_{p_2}(\tau_2)\cdots a^+_{p_1}(\tau_1')a^+_{p_2}(\tau_2')\cdots)\rangle \tag{12.17a}$$

is non-zero, along with the other terms that differ from (12.17a) by a permutation of the momenta $p_1, p_2, \ldots$ in the operators a^+.

When there are several (e.g. two) creation operators with the same momentum, the corresponding non-zero terms in the sum are of the form

$$\frac{1}{V}\cdot\frac{1}{V}\sum_{p_1}\frac{1}{V}\sum_{\substack{p_2\\ p_1\neq p_2\neq\cdots}}\cdots\langle T_\tau(a_{p_1}(\tau_1)a_{p_1}(\tau_2)a_{p_2}(\tau_3)\cdots a^+_{p_1}(\tau_1')a^+_{p_1}(\tau_2')a^+_{p_2}(\tau_3')\cdots)\rangle_0. \tag{12.17b}$$

Expressions such as (12.17a) have a distinctive feature as compared with the rest, namely, the number of factors $1/V$ in (12.17a) is the same as the number of summations, whereas the former number is always greater in the other terms. Suppose now that we have already carried out the averaging $\langle\cdots\rangle_0$ and that we let the volume V of the system tend to infinity, whilst maintaining the density of the number of particles N/V constant. (The sums are now replaced by integrals according to $V^{-1}\Sigma\ldots\to(2\pi)^{-3}\int\ldots$) In the limit, as $V-\infty$, the sum (12.17a), when expressed in terms of integrals over the momenta of the different combinations of Fermi or Bose functions, remains finite. (We have already encountered an elementary example of this kind in § 11, where we found the free Green function $\mathfrak{G}^{(0)}=-\langle T_\tau\{\psi(1)\,\bar{\psi}(2)\}\rangle_0$.) On the other hand, in expressions such as (12.17b), apart from the integrals over the momenta, a certain number of extra factors $1/V$ remains, with the result that these expressions vanish as $V\to\infty$.

Therefore, the only remaining terms in the sum in (12.16) when $V\to\infty$ are those of the form (12.17a), in which all the creation (and annihilation) operators have different momenta. This means that, when evaluating $\langle T_\tau\,\{a_{p_1}(\tau_1)\,a_{p_2}(\tau_2)\ldots a^+_{p_1'}(\tau_1')\,a^+_{p_2'}(\tau_2')\ldots\}\rangle_0$ we can in practice average each pair of operators a_p, a^+_p, independently. The mean value of the T-product of a large number of operators may now be expressed as the sum of all possible pairwise averages. For instance,

$$\langle T_\tau\{a_{p_1}(\tau_1)a_{p_2}(\tau_2)a^+_{p_1'}(\tau_1')a^+_{p_2'}(\tau_2')\}\rangle_0\to\langle T_\tau\{a_{p_1}(\tau_1)a^+_{p_2'}(\tau_2')\}\rangle_0 \tag{12.18a}$$
$$\times\langle T_\tau\{a_{p_2}(\tau_2)a^+_{p_1'}(\tau_1')\}\rangle_0\mp\langle T_\tau\{a_{p_1}(\tau_1)a^+_{p_1'}(\tau_1')\}\rangle_0\langle T_\tau\{a_{p_2}(\tau_2)a_{p_2'}(\tau_2')\}\rangle_0$$

(the minus (plus) sign for Fermi (Bose) statistics) (†).

(†) It can be shown that this is the case also for averages of T_τ-products of several a_p and a_p^+ with the same momenta (as in (12.17b)) excluding the operators a_0 and a_0^+ for a boson system below the condensation temperature (p. 108). The proof will not be given here.

In the coordinate representation form, these results mean that the average of the T-product of a number of ψ-operators splits into the sum of products of all possible pairwise averaged operators $\psi, \bar{\psi}$. In particular, we have instead of (12.18a):

$$\langle T_\tau\{\psi(\mathbf{r}_1, \tau_1)\psi(\mathbf{r}_2, \tau_2)\bar{\psi}(\mathbf{r}_1', \tau_1')\bar{\psi}(\mathbf{r}_2', \tau_2')\}\rangle_0$$

$$= \frac{1}{V^2} \sum_{\mathbf{p}_1', \mathbf{p}_2', \mathbf{p}_1, \mathbf{p}_2} \exp i[(\mathbf{p}_1 \cdot \mathbf{r}_1) + (\mathbf{p}_2 \cdot \mathbf{r}_2) - (\mathbf{p}_1' \cdot \mathbf{r}_1') - (\mathbf{p}_2' \cdot \mathbf{r}_2')]$$

$$\times \exp[-\tau_1(\varepsilon_0(\mathbf{p}_1) - \mu) - \tau_2(\varepsilon_0(\mathbf{p}_2) - \mu)] \exp[\tau_1'(\varepsilon_0(\mathbf{p}_1') - \mu) + \tau_2'(\varepsilon_0(\mathbf{p}_2') - \mu)]$$

$$\times \langle T_\tau\{a_{\mathbf{p}_1}(\tau_1) a_{\mathbf{p}_2}(\tau_2) a^+_{\mathbf{p}_1'}(\tau_1') a^+_{\mathbf{p}_2'}(\tau_2')\}\rangle_0$$

$$= \frac{1}{V} \sum_{\mathbf{p}_1, \mathbf{p}_2'} \exp[i(\mathbf{p}_1 \cdot \mathbf{r}_1) - i(\mathbf{p}_2' \cdot \mathbf{r}_2') - \tau_1(\varepsilon_0(\mathbf{p}_1) - \mu) + \tau_2'(\varepsilon_0(\mathbf{p}_2') - \mu)]$$

$$\times \langle T_\tau\{a_{\mathbf{p}_1}(\tau_1) a^+_{\mathbf{p}_2'}(\tau_2')\}\rangle_0 \frac{1}{V} \sum_{\mathbf{p}_1', \mathbf{p}_2} \exp[i(\mathbf{p}_2 \cdot \mathbf{r}_2) - i(\mathbf{p}_1' \cdot \mathbf{r}_1') - \tau_2(\varepsilon_0(\mathbf{p}_2) - \mu)$$

$$+ \tau_1'(\varepsilon_0(\mathbf{p}_1') - \mu)] \langle T_\tau\{a_{\mathbf{p}_2}(\tau_2) a^+_{\mathbf{p}_1'}(\tau_1')\}\rangle_0$$

$$\mp \frac{1}{V} \sum_{\mathbf{p}_1, \mathbf{p}_1'} \exp[i(\mathbf{p}_1 \cdot \mathbf{r}_1) - i(\mathbf{p}_1' \cdot \mathbf{r}_1') - \tau_1(\varepsilon_0(\mathbf{p}_1) - \mu) + \tau_1'(\varepsilon_0(\mathbf{p}_1') - \mu)]$$

$$\times \langle T_\tau\{a_{\mathbf{p}_1}(\tau_1) a^+_{\mathbf{p}_1'}(\tau_1')\}\rangle_0 \frac{1}{V} \sum_{\mathbf{p}_2, \mathbf{p}_2'} \exp[i(\mathbf{p}_2 \cdot \mathbf{r}_2) - i(\mathbf{p}_2' \cdot \mathbf{r}_2') - \tau_2(\varepsilon_0(\mathbf{p}_2) - \mu)$$

$$+ \tau_2'(\varepsilon_0(\mathbf{p}_2') - \mu)] \langle T_\tau\{a_{\mathbf{p}_2}(\tau_2) a^+_{\mathbf{p}_2'}(\tau_2')\}\rangle_0$$

$$= \langle T_\tau\{\psi(\mathbf{r}_1, \tau_1)\bar{\psi}(\mathbf{r}_2', \tau_2')\}\rangle_0 \langle T_\tau\{\psi(\mathbf{r}_2, \tau_2)\bar{\psi}(\mathbf{r}_1', \tau_1')\}\rangle_0$$

$$\mp \langle T_\tau \psi\{(\mathbf{r}_1, \tau_1)\bar{\psi}(\mathbf{r}_1', \tau_1')\}\rangle_0 \langle T_\tau\{\psi(\mathbf{r}_2, \tau_2)\bar{\psi}(\mathbf{r}_2', \tau_2')\}\rangle. \quad (12.18\text{b})$$

Similar relations also hold for a larger number of operators.

Apart from the sign, the averages on the right-hand side of (12.18b) are the same as the temperature-dependent Green functions of free particles. When evaluating the temperature-dependent Green functions we encounter, therefore, the same situation as was found in the absolute zero case. As there, an expansion (12.13) holds for the Green function $\mathfrak{G}$, which has the same form (apart from the factor i^n and the limits of integration over τ) as the expansion (8.9) for the function G. As earlier, we can evaluate the averages $\langle T_\tau(\ldots)\rangle_0$ appearing in (12.13) by using Wick's theorem, according to which these averages can be expressed in terms of the averages of pairs of creation and annihilation operators.

Notice that there is no concept of a normal product in the technique described. Wick's theorem holds only for the averages, and not for the T-products themselves.

If we use Wick's theorem to write out any term of the series (12.13) and replace $\langle T_\tau(\psi_1 \tilde{\psi})\rangle_0$ by the free Green function

$$\mathfrak{G}^{(0)}_{\alpha\beta}(\boldsymbol{r}_1 - \boldsymbol{r}_2, \tau_1 - \tau_2) = -\langle T_\tau\{\psi_\alpha(\boldsymbol{r}_1, \tau_1)\bar{\psi}_\beta(\boldsymbol{r}_2, \tau_2)\}\rangle_0,$$

we arrive at an expression which has precisely the same structure as the corresponding series for $T = 0$. This enables us to describe the different approximations of the perturbation theory series with the aid of the same Feynman diagrams as we used in the previous chapter. The only rules affected are those for associating specific expressions with the different parts of a diagram. In our present case each line of the diagram has to be associated with the temperature-dependent free particle Green function $\mathfrak{G}^{(0)}$ instead of with $G^{(0)}$, whilst the integration over time from $-\infty$ to ∞ at each vertex of the diagram is replaced by an integration over the imaginary "time" τ from 0 to $1/T$.

We have so far tacitly assumed that, when the volume of the system tends to infinity (with a given density), all the free particle Green functions and their integrals remain finite. It was on this basis, in particular, that we neglected terms of the form (12.17b) as $V \to \infty$, and were thus able to formulate Wick's theorem. The situation changes substantially in the case of a system of Bose particles at temperatures below the condensation temperature T_c and in Fermi systems possessing superfluidity properties.

In the case of a Bose gas at $T < T_c$, the creation and annihilation operators of particles in the zero momentum state are proportional to the root of the volume:

$$a_0 \sim a_0^+ \sim \sqrt{N} \sim \sqrt{V},$$

so that the terms of type (12.17b) remain finite as $V \to \infty$ and moreover do not satisfy Wick's theorem (see footnote on p. 106). A similar situation arises for Fermi superfluids. In both cases it is a question of using special techniques, which will be described in separate chapters (Chapters V and VII).

We now turn to the case when the ordinary diagram technique is suitable. Just as in the previous chapter, the diagrams for the Green functions have two external lines; one of them starts at the point $\boldsymbol{r}_1, \tau_1$, corresponding to the coordinates of the operator $\psi_\alpha(\boldsymbol{r}_1, \tau_1)$, whilst the other ends at the point $\boldsymbol{r}_2, \tau_2$, corresponding to the coordinates of the operator $\bar{\psi}_\beta(\boldsymbol{r}_2, \tau_2)$. As before, the diagrams for $\mathfrak{G}$ can be split into two classes — connected and unconnected. It can be verified, by means of exactly similar arguments, that taking into account the unconnected diagrams leads to cancelling the denominator in (12.7). We have thus for the $\mathfrak{G}$-function:

$$\mathfrak{G}_{\alpha\beta}(\boldsymbol{r}_1, \tau_1; \boldsymbol{r}_2, \tau_2) = -\langle T_\tau\{\psi_\alpha(\boldsymbol{r}_1, \tau_1)\bar{\psi}_\beta(\boldsymbol{r}_2, \tau_2)\mathfrak{S}\}\rangle_c, \qquad (12.19)$$

where $\langle \cdots \rangle_c$ denotes taking into account all the connected diagrams.

Since our discussion has made no use whatever of the fact that the number of external lines on the diagram is equal to two, it will still hold for many-particle Green functions. We can omit the $\langle\mathfrak{S}\rangle$ in the denominator of the relevant formulae (of the type (12.10)) and consider only the contribution of the connected diagrams when evaluating the averages.

As in the previous chapter, each diagram enters into the series for $\mathfrak{G}$ with a coefficient of the type λ^m, not essentially dependent on the order of the diagram. This fact is extremely important when summing infinite sequences of diagrams.

§ 13. THE DIAGRAM TECHNIQUE IN COORDINATE SPACE. EXAMPLES

The main result of the last section was to establish the fact that the ordinary Feynman diagram technique can be used for evaluating the temperature-dependent Green functions. The chief element in every diagram is the line representing the Green function for a free particle or a phonon. As in the previous chapter, we represent the Green function for a free particle by a full-drawn line (Fig. 32); the arrow indicates its direction: the line "departs" from the point with coordinates $\boldsymbol{r}_1, \tau_1$ and spin α_1 (this point corresponds to the operator ψ in the definition of $\mathfrak{G}$-function) and "enters" at the point $\boldsymbol{r}_2, \tau_2, \alpha_2$ (it corresponds to the operator $\bar{\psi}$). The coordinates of the point of "departure" are written to

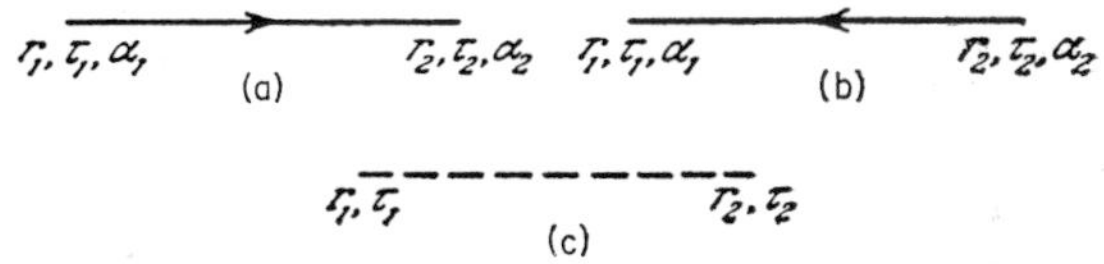

Fig. 32

the left in the argument of the Green function, the coordinates of the "entry" point to the right. For instance, the line in Fig. 32a represents the Green function

$$\mathfrak{G}^{(0)}_{\alpha_1\alpha_2}(\boldsymbol{r}_1, \tau_1; \boldsymbol{r}_2, \tau_2) \equiv \mathfrak{G}^{(0)}_{\alpha_1\alpha_2}(\boldsymbol{r}_1 - \boldsymbol{r}_2, \tau_1 - \tau_2),$$

and in Fig. 32b the function

$$\mathfrak{G}^{(0)}_{\alpha_2\alpha_1}(\boldsymbol{r}_2, \tau_2; \boldsymbol{r}_1, \tau_1) \equiv \mathfrak{G}^{(0)}_{\alpha_2\alpha_1}(\boldsymbol{r}_2 - \boldsymbol{r}_1, \tau_2 - \tau_1).$$

We shall represent the phonon Green function by a dotted line (Fig. 32c). We do not need to indicate the direction on the phonon line, since, as we saw in § 11, $\mathfrak{D}^{(0)}$ is an even function of $\boldsymbol{r}_1 - \boldsymbol{r}_2$ and $\tau_1 - \tau_2$.

The integration is over the coordinates of the points of intersection of the lines — the "vertices": over all space with respect to $\boldsymbol{r}$ and from 0 to $1/T$ with respect to τ. Summation over the spin variables is also carried out at the vertices.

The actual form of the diagrams depends on the type of interaction between the particles. The diagrams are drawn by using Wick's theorem, in accordance with which the average of the T-products of several operators appearing in the perturbation theory series for the Green function (12.13) is represented by the sum of the products of paired averages. These latter are connected with the free particle Green functions by the following relations:

$$\begin{aligned} \langle T_\tau\{\psi_\alpha(\mathbf{r}_1, \tau_1)\bar{\psi}_\beta(\mathbf{r}_2, \tau_2)\}\rangle &= -\mathfrak{G}^{(0)}_{\alpha\beta}(\mathbf{r}_1-\mathbf{r}_2, \tau_1-\tau_2), \\ \langle T_\tau\{\bar{\psi}_\beta(\mathbf{r}_2, \tau_2)\psi_\alpha(\mathbf{r}_1, \tau_1)\}\rangle &= \pm\,\mathfrak{G}^{(0)}_{\alpha\beta}(\mathbf{r}_1-\mathbf{r}_2; \tau_1-\tau_2) \end{aligned} \tag{13.1}$$

(the plus (minus) sign for fermions (bosons)).

The average of the product of two phonon operators can be expressed in terms of $\mathfrak{D}^{(0)}$:

$$\langle T_\tau\{\varphi(\mathbf{r}_{,1}\,\tau_1)\varphi(\mathbf{r}_2, \tau_2)\}\rangle = -\mathfrak{D}^{(0)}(\mathbf{r}_1-\mathbf{r}_2, \tau_1-\tau_2). \tag{13.2}$$

Let us take the different types of interaction.

A. Two-particle interaction. Let binary forces described by the potential $U(\mathbf{r}_1-\mathbf{r}_2)$ act between the particles. The Hamiltonian $\hat{H}_{int}$ is, in the interaction representation,

$$\hat{H}_{int}(\tau) = \frac{1}{2}\int\int \mathrm{d}^3\mathbf{r}_1\mathrm{d}^3\mathbf{r}_2\bar{\psi}_\alpha(\mathbf{r}_1, \tau)\bar{\psi}_\beta(\mathbf{r}_2, \tau)\,U(\mathbf{r}_1-\mathbf{r}_2)\,\psi_\beta(\mathbf{r}_2, \tau)\psi_\alpha(\mathbf{r}_1, \tau). \tag{13.3}$$

It is convenient to introduce instead of the potential $U(\mathbf{r}_1-\mathbf{r}_2)$ a potential $\mathfrak{V}(\mathbf{r}_1-\mathbf{r}_2,\ \tau_1-\tau_2)$, dependent on the "time" τ, through the formula

$$\mathfrak{V}(\mathbf{r}_1-\mathbf{r}_2, \tau_1-\tau_2) = U(\mathbf{r}_1-\mathbf{r}_2)\,\delta(\tau_1-\tau_2). \tag{13.4}$$

Using (13.4), we can write (12.4) for the $\mathfrak{S}$ matrix in the symmetric form:

$$\mathfrak{S} = T_\tau \exp\left\{-\frac{1}{2}\int\int \mathrm{d}^3\mathbf{r}_1\mathrm{d}^3\mathbf{r}_2\mathrm{d}\tau_1\mathrm{d}\tau_2\bar{\psi}_\beta(\mathbf{r}_2, \tau_2)\psi_\alpha(\mathbf{r}_1, \tau_1)\right.$$
$$\left.\times\,\mathfrak{V}(\mathbf{r}_1-\mathbf{r}_2, \tau_1-\tau_2)\psi_\beta(\mathbf{r}_2, \tau_2)\bar{\psi}_\alpha(\mathbf{r}_1, \tau_1)\right\}.$$

Let us find the first order approximation in U to the Green function. We have (†):

$$\mathfrak{G}^{(1)}_{\alpha\beta}(x-y) = \frac{1}{2}\int\int \mathrm{d}^4z_1\mathrm{d}^4z_2\langle T_\tau\{\psi_\alpha(x)\bar{\psi}_\beta(y)\,\mathfrak{V}(z_1-z_2)\,\bar{\psi}_{\gamma_1}(z_1)\bar{\psi}_{\gamma_2}(z_2)$$
$$\times\,\psi_{\gamma_2}(z_2)\psi_{\gamma_1}(z_1)\}\rangle. \tag{13.5}$$

(†) Throughout the rest of this section we shall use Latin italic type to denote the set of four variables $x = (\mathbf{r}, \tau)$. For instance, $\mathfrak{G}(x-y) = \mathfrak{G}(x-y, \tau_1-\tau_2)$, $\mathrm{d}^4x = \mathrm{d}^3\mathbf{r}\,\mathrm{d}\tau$.

It follows from Wick's theorem, that $\langle\cdots\rangle$ can be written as the sum of the following four terms:

$$\langle T_\tau\{\psi_\alpha(x)\bar{\psi}_\beta(y)\}\rangle\,\langle\{\bar{\psi}_{\gamma_1}(z_1)\psi_{\gamma_1}(z_1)\}\rangle\,\langle\{\bar{\psi}_{\gamma_2}(z_2)\psi_{\gamma_2}(z_2)\}\rangle, \tag{I}$$

$$\mp\langle T_\tau\{\psi_\alpha(x)\bar{\psi}_\beta(y)\}\rangle\,\langle\{\bar{\psi}_{\gamma_2}(z_2)\psi_{\gamma_1}(z_1)\}\rangle\,\langle\{\bar{\psi}_{\gamma_1}(z_1)\psi_{\gamma_2}(z_2)\}\rangle, \tag{II}$$

$$\langle T_\tau\{\psi_\alpha(x)\bar{\psi}_{\gamma_1}(z_1)\}\rangle\,\langle T_\tau\{\psi_{\gamma_1}(z_1)\bar{\psi}_\beta(y)\}\rangle\,\langle\{\bar{\psi}_{\gamma_2}(z_2)\psi_{\gamma_2}(z_2)\}\rangle, \tag{III}$$

$$\mp\langle T_\tau\{\psi_\alpha(x)\bar{\psi}_{\gamma_1}(z_1)\}\rangle\,\langle\{\bar{\psi}_{\gamma_2}(z_2)\psi_{\gamma_1}(z_1)\}\rangle\,\langle T_\tau\{\psi_{\gamma_2}(z_2)\bar{\psi}_\beta(y)\}\rangle \tag{IV}$$

together with the two terms that are obtained from (III) and (IV) by the substitutions $z_1\to z_2$, $\gamma_1\to\gamma_2$. The contribution of these latter to integral (13.5) is evidently the same as that of (I)—(IV), which simply leads to the disappearance of the $^1/_2$ in front of the integral.

On replacing $\langle T_\tau\{\cdots\}\rangle$ by the Green functions $\mathfrak{G}^{(0)}$ in accordance with (13.1), we find that the first order approximation is the sum of the following four expressions:

$$-\mathfrak{G}^{(0)}_{\alpha\beta}(x-y)\int\int \mathrm{d}^4z_1\mathrm{d}^4z_2\,\mathfrak{G}^{(0)}_{\gamma_1\gamma_1}(0)\mathfrak{V}(z_1-z_2)\,\mathfrak{G}^{(0)}_{\gamma_2\gamma_2}(0), \tag{I}$$

$$\pm\mathfrak{G}^{(0)}_{\alpha\beta}(x-y)\int\int \mathrm{d}^4z_1\mathrm{d}^4z_2\,\mathfrak{G}^{(0)}_{\gamma_1\gamma_2}(z_1-z_2)\,\mathfrak{G}^{(0)}_{\gamma_2\gamma_1}(z_2-z_1)\,\mathfrak{V}(z_1-z_2), \tag{II}$$

$$\pm\int\int \mathrm{d}^4z_1\mathrm{d}^4z_2\,\mathfrak{G}^{(0)}_{\alpha\gamma_1}(x-z_1)\,\mathfrak{G}^{(0)}_{\gamma_1\beta}(z_1-y)\,\mathfrak{G}^{(0)}_{\gamma_2\gamma_2}(0)\,\mathfrak{V}(z_1-z_2), \tag{III}$$

$$-\int\int \mathrm{d}^4z_1\mathrm{d}^4z_2\,\mathfrak{G}^{(0)}_{\alpha\gamma_1}(x-z_1)\,\mathfrak{G}^{(0)}_{\gamma_1\gamma_2}(z_1-z_2)\,\mathfrak{G}^{(0)}_{\gamma_2\beta}(z_2-y)\mathfrak{V}(z_1-z_2). \tag{IV}$$

Notice that we always take $\mathfrak{G}^{(0)}(\boldsymbol{r}_1-\boldsymbol{r}_2;\,0)$ as $\tau\to-0$.

We shall represent the function $\mathfrak{V}(z_1-z_2)$ by a wavy line on the diagrams. We can now associate the diagrams of Fig. 33 with expressions (I)—(IV). Diagrams I, II are of the unconnected type described in the previous section. We showed there that their contribution is to be neglected when evaluating the Green functions.

A contribution to the first order approximation is thus only supplied by diagrams III, VI, together with the diagrams that differ from these in a commutation of the coordinates of the vertices z_1, z_2. It will be recalled that such diagrams are termed topologically equivalent; all topologically equivalent diagrams give the same contribution.

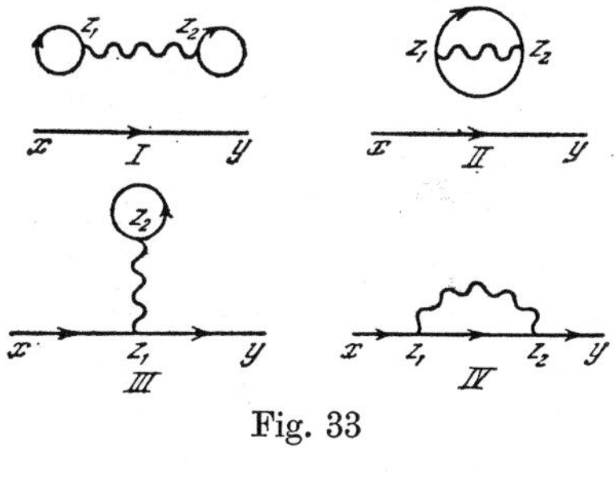

Fig. 33

It should be noted that the expressions corresponding to diagrams III and IV have opposite signs in the case of Fermi statistics. This is connected with the presence of the closed loop in diagram III. It can be shown for a diagram of any order that any closed fermion loop (not necessarily formed from a single line, as in the present case) enters into the relevant expression with the "minus" sign.

We now state the rules by which the correction of any order may be evaluated:

(1) First of all, all the connected topologically non-equivalent diagrams with $2n$ vertices and two end-points must be drawn; two full-drawn and one wavy line meet at each vertex.

(2) Every full-drawn line of the diagram is associated with the particle Green function $\mathfrak{G}^{(0)}_{\alpha\beta}(x-y)$ (x, α are the coordinates of the beginning, y, β the coordinates of the end of the line).

(3) Each wavy line is associated with the generalised potential $\mathfrak{B}(x-y)$.

(4) Integration is performed over the coordinates of each vertex z ($d^4z = d^3z\, d\tau$) and summation over the spin variable α.

(5) The expression obtained is multiplied by $(-1)^{n+F}$, where n is the order of the diagram, and F the number of closed fermion loops in it.

(6) If Green functions of the same time arguments $\mathfrak{G}^{(0)}(0)$ appear in an expression, they are to be regarded as $\lim_{\tau\to+0} \mathfrak{G}^{(0)}(r_1 - r_2, -\tau)$.

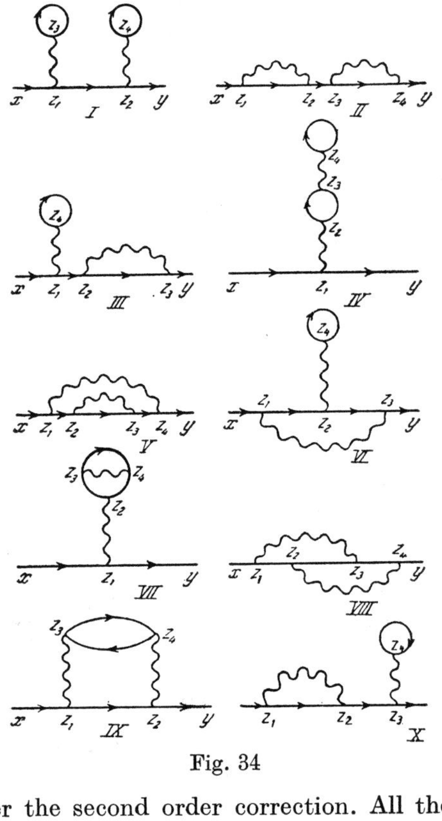

Fig. 34

Let us consider the second order correction. All the possible topologically non-equivalent diagrams with four vertices are illustrated in

Fig. 34. Using rules (1)—(6), we can easily write down the expressions corresponding to these diagrams:

$$\int \mathrm{d}^4z_1\mathrm{d}^4z_2\mathrm{d}^4z_3\mathrm{d}^4z_4\, \mathfrak{G}^{(0)}_{\alpha\gamma_1}(x-z_1)\,\mathfrak{G}^{(0)}_{\gamma_1\gamma_2}(z_1-z_2)\,\mathfrak{G}^{(0)}_{\gamma_2\beta}(z_2-y)$$
$$\times\,\mathfrak{V}(z_1-z_3)\,\mathfrak{V}(z_2-z_4)\,\mathfrak{G}^{(0)}_{\gamma_3\gamma_3}(0)\,\mathfrak{G}^{(0)}_{\gamma_4\gamma_4}(0), \qquad \text{(I)}$$

$$\int \mathrm{d}^4z_1\mathrm{d}^4z_2\mathrm{d}^4z_3\mathrm{d}^4z_4\, \mathfrak{G}^{(0)}_{\alpha\gamma_1}(x-z_1)\,\mathfrak{G}^{(0)}_{\gamma_1\gamma_2}(z_1-z_2)\,\mathfrak{G}^{(0)}_{\gamma_2\gamma_3}(z_2-z_3)$$
$$\times\,\mathfrak{G}^{(0)}_{\gamma_3\gamma_4}(z_3-z_4)\,\mathfrak{G}^{(0)}_{\gamma_4\beta}(z_4-y)\,\mathfrak{V}(z_1-z_2)\,\mathfrak{V}(z_3-z_4), \qquad \text{(II)}$$

$$\mp\int \mathrm{d}^4z_1\mathrm{d}^4z_2\mathrm{d}^4z_3\mathrm{d}^4z_4\, \mathfrak{G}^{(0)}_{\alpha\gamma_1}(x-z_1)\,\mathfrak{G}^{(0)}_{\gamma_1\gamma_2}(z_1-z_2)\,\mathfrak{G}^{(0)}_{\gamma_2\gamma_3}(z_2-z_3)$$
$$\times\,\mathfrak{G}^{(0)}_{\gamma_3\beta}(z_3-y)\,\mathfrak{G}^{(0)}_{\gamma_4\gamma_4}(0)\,\mathfrak{V}(z_1-z_4)\,\mathfrak{V}(z_2-z_3), \qquad \text{(III)}$$

$$\int \mathrm{d}^4z_1\mathrm{d}^4z_2\mathrm{d}^4z_3\mathrm{d}^4z_4\, \mathfrak{G}^{(0)}_{\alpha\gamma_1}(x-z_1)\,\mathfrak{G}^{(0)}_{\gamma_1\beta}(z_1-y)\,\mathfrak{G}^{(0)}_{\gamma_2\gamma_3}(z_2-z_3)$$
$$\times\,\mathfrak{G}^{(0)}_{\gamma_3\gamma_2}(z_3-z_2)\,\mathfrak{G}^{(0)}_{\gamma_4\gamma_4}(0)\,\mathfrak{V}(z_1-z_2)\,\mathfrak{V}(z_3-z_4), \qquad \text{(IV)}$$

$$\int \mathrm{d}^4z_1\mathrm{d}^4z_2\mathrm{d}^4z_3\mathrm{d}^4z_4\, \mathfrak{G}^{(0)}_{\alpha\gamma_1}(x-z_1)\,\mathfrak{G}^{(0)}_{\gamma_1\gamma_2}(z_1-z_2)\,\mathfrak{G}^{(0)}_{\gamma_2\gamma_3}(z_2-z_3)$$
$$\times\,\mathfrak{G}^{(0)}_{\gamma_3\gamma_4}(z_3-z_4)\,\mathfrak{G}^{(0)}_{\gamma_4\beta}(z_4-y)\,\mathfrak{V}(z_1-z_4)\,\mathfrak{V}(z_2-z_3), \qquad \text{(V)}$$

$$\mp\int \mathrm{d}^4z_1\mathrm{d}^4z_2\mathrm{d}^4z_3\mathrm{d}^4z_4\, \mathfrak{G}^{(0)}_{\alpha\gamma_1}(x-z_1)\,\mathfrak{G}^{(0)}_{\gamma_1\gamma_2}(z_1-z_2)\,\mathfrak{G}^{(0)}_{\gamma_2\gamma_3}(z_2-z_3)$$
$$\times\,\mathfrak{G}^{(0)}_{\gamma_3\beta}(z_3-y)\,\mathfrak{G}^{(0)}_{\gamma_4\gamma_4}(0)\,\mathfrak{V}(z_1-z_3)\,\mathfrak{V}(z_2-z_4), \qquad \text{(VI)}$$

$$\mp\int \mathrm{d}^4z_1\mathrm{d}^4z_2\mathrm{d}^4z_3\mathrm{d}^4z_4\, \mathfrak{G}^{(0)}_{\alpha\gamma_1}(x-z_1)\,\mathfrak{G}^{(0)}_{\gamma_1\beta}(z_1-y)\,\mathfrak{G}^{(0)}_{\gamma_2\gamma_3}(z_2-z_3)$$
$$\times\,\mathfrak{G}^{(0)}_{\gamma_3\gamma_4}(z_3-z_4)\,\mathfrak{G}^{(0)}_{\gamma_4\gamma_2}(z_4-z_2)\,\mathfrak{V}(z_1-z_2)\,\mathfrak{V}(z_3-z_4), \qquad \text{(VII)}$$

$$\int \mathrm{d}^4z_1\mathrm{d}^4z_2\mathrm{d}^4z_3\mathrm{d}^4z_4\, \mathfrak{G}^{(0)}_{\alpha\gamma_1}(x-z_1)\,\mathfrak{G}^{(0)}_{\gamma_1\gamma_2}(z_1-z_2)\,\mathfrak{G}^{(0)}_{\gamma_2\gamma_3}(z_2-z_3)$$
$$\times\,\mathfrak{G}^{(0)}_{\gamma_3\gamma_4}(z_3-z_4)\,\mathfrak{G}^{(0)}_{\gamma_4\beta}(z_4-y)\,\mathfrak{V}(z_1-z_3)\,\mathfrak{V}(z_2-z_4), \qquad \text{(VIII)}$$

$$\mp\int \mathrm{d}^4z_1\mathrm{d}^4z_2\mathrm{d}^4z_3\mathrm{d}^4z_4\, \mathfrak{G}^{(0)}_{\alpha\gamma_1}(x-z_1)\,\mathfrak{G}^{(0)}_{\gamma_1\gamma_2}(z_1-z_2)\,\mathfrak{G}^{(0)}_{\gamma_2\beta}(z_2-y)$$
$$\times\,\mathfrak{G}^{(0)}_{\gamma_3\gamma_4}(z_3-z_4)\,\mathfrak{G}^{(0)}_{\gamma_4\gamma_3}(z_4-z_3)\,\mathfrak{V}(z_1-z_3)\,\mathfrak{V}(z_2-z_4), \qquad \text{(IX)}$$

$$\mp\int \mathrm{d}^4z_1\mathrm{d}^4z_2\mathrm{d}^4z_3\mathrm{d}^4z_4\, \mathfrak{G}^{(0)}_{\alpha\gamma_1}(x-z_1)\,\mathfrak{G}^{(0)}_{\gamma_1\gamma_2}(z_1-z_2)$$
$$\times\,\mathfrak{G}^{(0)}_{\gamma_2\gamma_3}(z_2-z_3)\,\mathfrak{G}_{\gamma_3\beta}(z_3-y)\,\mathfrak{G}_{\gamma_4\gamma_4}(0)\,\mathfrak{V}(z_1-z_2)\,\mathfrak{V}(z_3-z_4). \qquad \text{(X)}$$

Perturbation theory for the case of two-particle interactions can be given another more symmetrical form, which proves particularly convenient, when the forces acting between the particles depend on the spins as well as the distance. The Hamiltonian of such an interaction is

$$\hat{H}_{int}(\tau) = \frac{1}{2}\int\int \mathrm{d}^3\boldsymbol{r}_1\mathrm{d}^3\boldsymbol{r}_2\bar{\psi}_\alpha(\boldsymbol{r}_1,\tau)\bar{\psi}_\beta(\boldsymbol{r}_2,\tau)\,U_{\alpha\beta;\delta\gamma}(\boldsymbol{r}_1-\boldsymbol{r}_2)\,\psi_\gamma(\boldsymbol{r}_2,\tau)\psi_\delta(\boldsymbol{r}_1,\tau). \tag{13.6}$$

We write the integral

$$\int_0^{1/T} H_{int}(\tau)\,\mathrm{d}\tau,$$

appearing in the expression for $\mathfrak{S}$, in a form symmetric with respect to all the variables:

$$\frac{1}{4}\int_0^{1/T}\cdots\int_0^{1/T} d\tau_1\cdots d\tau_4 \int\cdots\int d^3\boldsymbol{r}_1\cdots d^3\boldsymbol{r}_2\,\bar{\psi}_{\gamma_1}(\boldsymbol{r}_1,\tau_1)\,\bar{\psi}_{\gamma_2}(\boldsymbol{r}_2,\tau_2)$$

$$\times\mathscr{T}^{(0)}_{\gamma_1\gamma_2;\gamma_3\gamma_4}(\boldsymbol{r}_1,\tau_1,\boldsymbol{r}_2,\tau_2;\boldsymbol{r}_3,\tau_3,\boldsymbol{r}_4,\tau_4)\,\psi_{\gamma_4}(\boldsymbol{r}_4,\tau_4)\,\psi_{\gamma_3}(\boldsymbol{r}_3,\tau_3),$$

or, introducing a "four-dimensional" notation,

$$\frac{1}{4}\int d^4z_1 d^4z_2 d^4z_3 d^4z_4\,\bar{\psi}_{\gamma_1}(z_1)\bar{\psi}_{\gamma_2}(z_2)\,\mathscr{T}^{(0)}_{\gamma_1\gamma_2;\gamma_3\gamma_4}(z_1,z_2;z_3,z_4)\,\psi_{\gamma_4}(z_4)\,\psi_{\gamma_3}(z_3). \tag{13.7}$$

In view of the fact that the operators $\bar{\psi}_{\gamma_1}(z_1)$, $\bar{\psi}_{\gamma_2}(z_2)$ (or $\psi_{\gamma_3}(z_3)$, $\psi_{\gamma_4}(z_4)$) anti-commute or commute, depending on the statistics, we can regard $\mathscr{T}^{(0)}$ in the case of fermions as anti-symmetric with respect to the permutation $z_1, \gamma_1 \rightleftarrows z_2, \gamma_2$ or $z_3, \gamma_3 \rightleftarrows z_4, \gamma_4$, and as symmetric in these variables in the case of Bose statistics.

The function $\mathscr{T}^{(0)}$ thus defined is obtained from

$$U_{\gamma_1\gamma_2;\gamma_3\gamma_4}(\boldsymbol{r}_1-\boldsymbol{r}_2)\,\delta(\tau_1-\tau_2)\,\delta(\boldsymbol{r}_1-\boldsymbol{r}_3)\,\delta(\tau_1-\tau_3)\,\delta(\boldsymbol{r}_2-\boldsymbol{r}_4)\,\delta(\tau_2-\tau_4)$$

by anti-symmetrisation with respect to the variables $z_1\gamma_1$, $z_2\gamma_2$ (and $z_3\gamma_3$, $z_4\gamma_4$) in the case of Fermi statistics, and symmetrisation with respect to these variables in the case of Bose statistics.

Let us find the first order correction to the Green function. We have

$$\frac{1}{4}\int d^4z_1\cdots d^4z_4\,\mathscr{T}^{(0)}_{\gamma_1\gamma_2;\gamma_3\gamma_4}(z_1,z_2;z_3,z_4)$$

$$\times\langle T_\tau\{\psi_\alpha(x)\bar{\psi}_\beta(y)\bar{\psi}_{\gamma_1}(z_1)\bar{\psi}_{\gamma_2}(z_2)\psi_{\gamma_4}(z_4)\psi_{\gamma_3}(z_3)\}\rangle. \tag{13.8}$$

On applying Wick's theorem and using the symmetry properties of $\mathscr{T}^{(0)}$, (13.8) is easily shown to be expressible as the sum of two terms:

$$-\frac{1}{2}\,\mathfrak{G}^{(0)}_{\alpha\beta}(x-y)\int d^4z_1\cdots d^4z_4\,\mathfrak{G}^{(0)}_{\gamma_3\gamma_1}(z_3-z_1)$$

$$\times\,\mathfrak{G}^{(0)}_{\gamma_4\gamma_2}(z_4-z_2)\,\mathscr{T}^{(0)}_{\gamma_1\gamma_2;\gamma_3\gamma_4}(z_1,z_2;z_3,z_4), \tag{I}$$

$$-\int d^4z_1\cdots d^4z_4\,\mathfrak{G}^{(0)}_{\alpha\gamma_1}(x-z_1)\,\mathscr{T}^{(0)}_{\gamma_1\gamma_2;\gamma_3\gamma_4}(z_1,z_2;z_3,z_4)$$

$$\times\,\mathfrak{G}^{(0)}_{\gamma_4\beta}(z_4-y)\,\mathfrak{G}^{(0)}_{\gamma_3\gamma_2}(z_3-z_2). \tag{II}$$

We shall denote $\mathscr{T}^{(0)}$ by a square on the diagram. We can now associate expressions (I), (II) with the diagrams of Fig. 35. Diagram I is unconnected, and its contribution may be neglected. Hence, to the first order of perturbation theory, we have a single diagram, whose contribution is given by expression (II).

To the second order of perturbation theory, there are altogether three connected topologically distinct diagrams (Fig. 36). The expressions corresponding to the diagrams may readily be found:

$$\int dz_1 \cdots dz_8\, \mathfrak{G}^{(0)}_{\alpha\gamma_1}(x-z_1)\, \mathcal{T}^{(0)}_{\gamma_1\gamma_2;\gamma_3\gamma_4}(z_1, z_2; z_3, z_4)\, \mathfrak{G}^{(0)}_{\gamma_4\gamma_5}(z_4-z_5)$$
$$\times \mathcal{T}^{(0)}_{\gamma_5\gamma_6;\gamma_7\gamma_8}(z_5, z_6; z_7, z_8)\, \mathfrak{G}^{(0)}_{\gamma_8\beta}(z_8-y)\, \mathfrak{G}^{(0)}_{\gamma_7\gamma_6}(z_7-z_6)\, \mathfrak{G}^{(0)}_{\gamma_3\gamma_2}(z_3-z_2), \quad \text{(I)}$$

$$\int dz_1 \cdots dz_8\, \mathfrak{G}^{(0)}_{\alpha\gamma_1}(x-z_1)\, \mathcal{T}^{(0)}_{\gamma_1\gamma_2;\gamma_3\gamma_4}(z_1, z_2; z_3, z_4)\, \mathfrak{G}^{(0)}_{\gamma_4\beta}(z_4-y)$$
$$\times \mathfrak{G}^{(0)}_{\gamma_3\gamma_6}(z_3-z_6)\, \mathfrak{G}^{(0)}_{\gamma_7\gamma_2}(z_7-z_2)\, \mathcal{T}^{(0)}_{\gamma_5\gamma_6;\gamma_7\gamma_8}(z_5, z_6; z_7, z_8)\, \mathfrak{G}^{(0)}_{\gamma_8\gamma_5}(z_8-z_5), \quad \text{(II)}$$

$$\frac{1}{2}\int dz_1 \cdots dz_2\, \mathfrak{G}^{(0)}_{\alpha\gamma_1}(x-z_1)\, \mathcal{T}^{(0)}_{\gamma_1\gamma_2;\gamma_3\gamma_4}(z_1, z_2; z_3, z_4)\, \mathfrak{G}^{(0)}_{\gamma_3\gamma_5}(z_3-z_5)$$
$$\times \mathfrak{G}^{(0)}_{\gamma_4\gamma_6}(z_4-z_6)\, \mathfrak{G}^{(0)}_{\gamma_7\gamma_2}(z_7-z_2)\, \mathcal{T}^{(0)}_{\gamma_5\gamma_6;\gamma_7\gamma_8}(z_5, z_6; z_7, z_8)\, \mathfrak{G}^{(0)}_{\gamma_8\beta}(z_8-y). \quad \text{(III)}$$

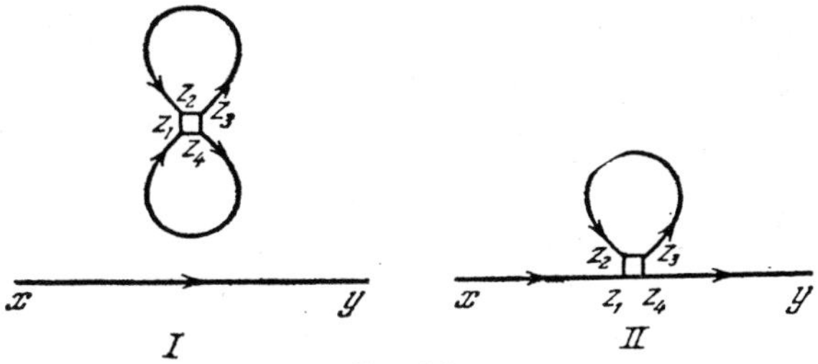

Fig. 35

The correction of any order to the Green functions is found in accordance with the following rules:

(1) We have to draw all the topologically non-equivalent diagrams, containing n squares (in our case the topologically equivalent diagrams include those that differ only in a commutation of the coordinates of the square vertices);

(2) each line of the diagram is associated with the particle Green function;

(3) each square is associated with a function $\mathcal{T}^{(0)}$;

(4) integration is performed over the coordinates of the square vertices;

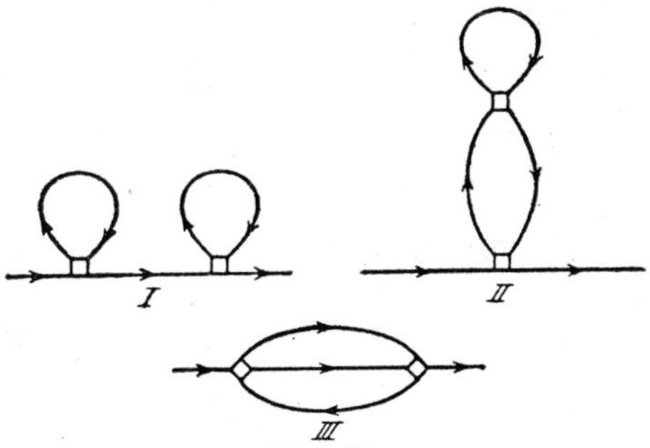

Fig. 36

(5) the resulting expression is multiplied by $m/2^n$, where m is the number of different diagrams in the non-symmetrised technique corresponding to the diagram under consideration. The sign is also obtained from a comparison with the non-symmetrised technique.

By using these rules, we can easily write down the expressions corresponding to the diagram of Fig. 37:

$$\mp \frac{1}{4} \int \mathrm{d}^4 z_1 \cdots \mathrm{d}^4 z_{12}\, \mathfrak{G}^{(0)}_{\alpha\gamma_1}(x - z_1)\, \mathcal{T}^{(0)}_{\gamma_1\gamma_2;\gamma_3\gamma_4}(z_1, z_2; z_3, z_4)\, \mathfrak{G}^{(0)}_{\gamma_3\gamma_5}(z_3 - z_5)$$
$$\times \mathfrak{G}^{(0)}_{\gamma_4\gamma_6}(z_4 - z_6)\, \mathcal{T}^{(0)}_{\gamma_5\gamma_6;\gamma_7\gamma_8}(z_5, z_6; z_7, z_8)\, \mathfrak{G}^{(0)}_{\gamma_7\gamma_9}(z_7 - z_9)\, \mathfrak{G}^{(0)}_{\gamma_8\gamma_{10}}(z_8 - z_{10})$$
$$\times \mathcal{T}^{(0)}_{\gamma_9\gamma_{10};\gamma_{11}\gamma_{12}}(z_9, z_{10}; z_{11}, z_{12})\, \mathfrak{G}^{(0)}_{\gamma_{12}\beta}(z_{12} - y)\, \mathfrak{G}^{(0)}_{\gamma_{11}\gamma_2}(z_{11} - z_2).$$

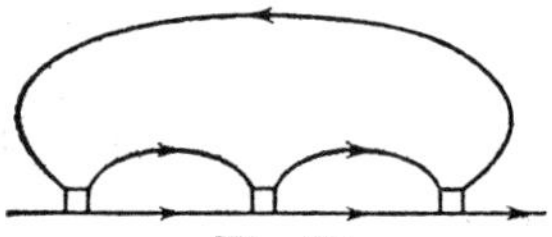

Fig. 37

The rather unwieldy expressions that are obtained by this technique (their complexity is to some extent redeemed by their symmetry) become very much simpler when it comes to dealing with point interactions described by the potential

$$U_{\alpha\beta;\gamma\delta}(\boldsymbol{r}_1 - \boldsymbol{r}_2) = \lambda\, \delta_{\alpha\delta}\, \delta_{\beta\gamma}\, \delta(\boldsymbol{r}_1 - \boldsymbol{r}_2).$$

In this case $\mathcal{T}^{(0)}$ has the simple form:

$$\mathcal{T}^{(0)}_{\gamma_1\gamma_2;\gamma_3\gamma_4} = \lambda(\delta_{\gamma_1\gamma_3}\delta_{\gamma_2\gamma_4} - \delta_{\gamma_1\gamma_4}\delta_{\gamma_2\gamma_3})\, \delta(z_1 - z_2)\, \delta(z_1 - z_3)\, \delta(z_1 - z_4)$$
$$= \lambda L_{\gamma_1\gamma_2;\gamma_3\gamma_4}\, \delta(z_1 - z_2)\, \delta(z_1 - z_3)\, \delta(z_1 - z_4).$$

Due to the presence of three δ-functions in $\mathcal{T}^{(0)}$, of the four integrations over the vertices of squares in the expressions for the corrections, only one remains. This enables us to replace the squares by points (vertices) on the diagrams. For instance, diagrams II of Fig. 35 and III of Fig. 36 are more conveniently drawn as shown in Fig. 38;

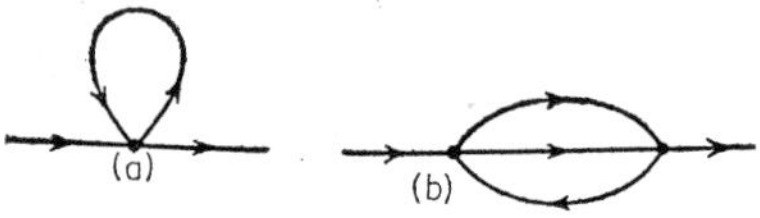

Fig. 38

the corresponding corrections are transformed to

$$-\lambda L_{\gamma_1\gamma_2;\gamma_3\gamma_4} \int \mathrm{d}^4 z\, \mathfrak{G}^{(0)}_{\alpha\gamma_1}(x - z)\, \mathfrak{G}^{(0)}_{\gamma_4\beta}(z - y)\, \mathfrak{G}^{(0)}_{\gamma_3\gamma_2}(0), \tag{a}$$

$$\frac{\lambda^2}{2} L_{\gamma_1\gamma_2;\gamma_3\gamma_4} L_{\gamma_5\gamma_6;\gamma_7\gamma_8} \int \mathrm{d}^4 z_1 \mathrm{d}^4 z_2\, \mathfrak{G}^{(0)}_{\alpha\gamma_1}(x - z_1)\, \mathfrak{G}^{(0)}_{\gamma_3\gamma_5}(z_1 - z_2)$$
$$\times \mathfrak{G}^{(0)}_{\gamma_4\gamma_6}(z_1 - z_2)\, \mathfrak{G}^{(0)}_{\gamma_7\gamma_2}(z_2 - z_1)\, \mathfrak{G}^{(0)}_{\gamma_8\beta}(z_2 - y). \tag{b}$$

The general rules for calculation using this diagram technique will be obvious from the foregoing.

B. Interaction of particles with phonons. The interaction of particles with phonons (for instance, the interaction of particles of a liquid with sound waves or the interaction of electrons in a metal with lattice vibrations) is described by the Hamiltonian

$$\hat{H}_{int}(\tau) = g\, \bar{\psi}_\alpha(\boldsymbol{r}, \tau)\psi_\alpha(\boldsymbol{r}, \tau)\varphi(\boldsymbol{r}, \tau)\,\mathrm{d}^3\boldsymbol{r},$$

where g is the coupling constant.

It is easily shown that the only non-zero approximations to the Green function for a particle $\mathfrak{G}$ and for a phonon $\mathfrak{D}$ are those of even orders of perturbation theory. (The expressions for the odd order approximations contain an odd number of phonon operators φ.) By evaluating the expressions for the approximations to the particle Green function $\mathfrak{G}$ it can be shown that they are precisely the same as the expressions for the approximations to $\mathfrak{G}$ in the first statement of perturbation theory for two-particle interactions, provided we replace the potential $\mathfrak{V}(z_1 - z_2)$ in the latter by $g^2\, \mathfrak{D}^{(0)}(z_1 - z_2)$. The approximations will naturally be described by precisely the same diagrams as in Figs. 33 and 34. However, to fix our ideas, we shall represent the phonon Green function $\mathfrak{D}^{(0)}$ by a dotted instead of a wavy line.

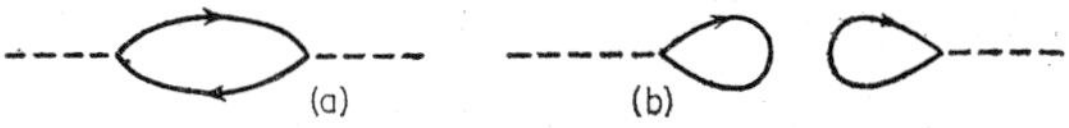

Fig. 39

The second order approximations to the phonon Green function are described by two connected diagrams (Fig. 39 *a, b*). The expression in the case of Fig. 39 *a* is found to be

$$\pm g^2 \int d^4z_1 d^4z_2 \mathfrak{D}^{(0)}(x - z_1)\, \mathfrak{G}^{(0)}_{\alpha\beta}(z_1 - z_2)\, \mathfrak{G}^{(0)}_{\beta\alpha}(z_2 - z_1)\, \mathfrak{D}^{(0)}(z_2 - y),$$

and in the case of Fig. 39 *b*:

$$g^2 \int d^4z_1 \mathfrak{D}^{(0)}(x - z_1)\, \mathfrak{G}_{\alpha\alpha}(0) \int d^4z_2\, \mathfrak{D}^{(0)}(z_2 - y)\, \mathfrak{G}_{\beta\beta}(0).$$

It can be shown as in § 9 that the expression corresponding to diagram 39 *b* vanishes. It is this fact that enables us to ignore, in general, the diagrams whose expressions contain the integral $\int \mathfrak{D}^{(0)}(z)\, d^4z$. These latter include all the diagrams for $\mathfrak{D}$ that split into two disconnected parts, each part being approached along a single external line, as also the diagrams for $\mathfrak{G}$ of the type of Fig. 40 (these always contain a part with no external lines, joined to the rest by a single phonon line).

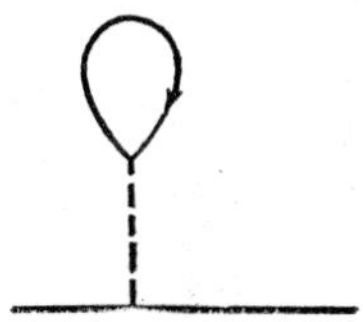
Fig. 40

If we consider the corrections to $\mathfrak{D}$ of subsequent orders and take notice of what has been said regarding $\mathfrak{G}$, the following general rules can be stated for using the diagram technique to compute the corrections of order $2n$:

(1) each full-drawn line of the diagram is associated with the free particle Green function $\mathfrak{G}^{(0)}_{\alpha\beta}(x - y)$, and each dotted line with a function $\mathfrak{D}^{(0)}(x - y)$;

(2) integration is performed over the coordinates of all the vertices (over $\boldsymbol{r}$ and over τ);

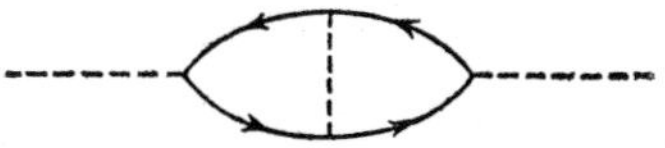
Fig. 41

(3) the expression obtained is multiplied by $g^{2n}(-1)^{n+F}$, where F is the number of closed fermion loops in the diagram.

For example, the fourth order correction to the phonon Green function, described by Fig. 41, is

$$\mp g^4 \int d^4z_1 \cdots d^4z_4 \mathfrak{D}^{(0)}(x - z_1)\, \mathfrak{G}^{(0)}_{\gamma_1\gamma_2}(z_1 - z_2)\, \mathfrak{G}^{(0)}_{\gamma_3\gamma_1}(z_3 - z_1)$$
$$\times\, \mathfrak{D}^{(0)}(z_2 - z_3)\, \mathfrak{G}^{(0)}_{\gamma_2\gamma_4}(z_2 - z_4)\, \mathfrak{G}^{(0)}_{\gamma_4\gamma_3}(z_4 - z_3)\, \mathfrak{D}^{(0)}(z_4 - y).$$

§ 14. THE DIAGRAM TECHNIQUE IN MOMENTUM SPACE

1. Transformation to the momentum representation

The diagram technique in coordinate space developed in the previous section proves to be far from convenient in practical calculations. The fact is that the success of the methods of field theory at absolute zero is due in the last instance to the highly automatic nature of the calculations, this being a consequence of the expansion of all the quantities featured in the theory into Fourier integrals over all four coordinates; this was in fact the method adopted at $T = 0$. In the Matsubara technique described above, this automatic quality is absent, due to the fact that the variable τ varies between the finite limits 0 and $1/T$, and the consequent impossibility of passing to a Fourier representation (in τ).

In the coordinate form, $\mathfrak{G}^{(0)}$ and $\mathfrak{D}^{(0)}$ are discontinuous functions of τ; every integration over τ splits up into integrals over a large number of domains, and this number increases very rapidly with the order n of the approximation. As a result, an application of the Matsubara technique is rendered extremely difficult.

A substantial improvement in technique results if we expand the quantities dependent on τ into Fourier series in this variable (Abrikosov, Gor'kov, Dzyaloshinskii [31], Fradkin [32]).

The temperature-dependent Green function $\mathfrak{G}$ (or $\mathfrak{D}$) is a function of $\tau_1 - \tau_2$, and, as such, is specified in the interval $(-1/T, 1/T)$. We expand $\mathfrak{G}(\tau)$ into a Fourier series:

$$\mathfrak{G}(\tau) = T \sum_n e^{-i\omega_n \tau} \mathfrak{G}(\omega_n),$$
$$\mathfrak{G}(\omega_n) = \frac{1}{2} \int_{-1/T}^{1/T} e^{i\omega_n \tau} \mathfrak{G}(\tau)\, d\tau, \quad \omega_n = \pi n T. \tag{14.1}$$

Our problem is to find a means of passing to the Fourier representation in the expressions of § 13 for the approximations to the Green functions. At the same time, we want above all to avoid introducing any further complexities into the formulae (e.g. we want no extra factors to depend on the "frequencies" ω_n).

It turns out that, in practice, we encounter the same sort of situation at finite temperatures as we had at $T = 0$. This refers to a general property of Green functions, remarked upon in § 11. We showed, in fact, that $\mathfrak{G}(\tau)$ at $\tau < 0$ is connected by simple relations with $\mathfrak{G}(\tau)$ at $\tau > 0$ (cf. (11.8) and (11.8a)). It follows from these relations that the Fourier components $\mathfrak{G}(\omega_n)$ of the Green function for a boson or a phonon only differ from zero at "even" frequencies $\omega_n = 2\pi n T$, whereas the $\mathfrak{G}$ for a fermion only has components with $\omega_n = (2n+1)\pi T$. Indeed,

$$\mathfrak{G}(\omega_n) = \frac{1}{2}\int_{-1/T}^{1/T} e^{i\omega_n\tau}\mathfrak{G}(\tau)\,\mathrm{d}\tau = \frac{1}{2}\int_{0}^{1/T} e^{i\omega_n\tau}\mathfrak{G}(\tau)\,\mathrm{d}\tau + \frac{1}{2}\int_{-1/T}^{0} e^{i\omega_n\tau}\mathfrak{G}(\tau)\,\mathrm{d}\tau.$$

We use (11.8) to replace the $\mathfrak{G}(\tau < 0)$ in the second integral and then carry out the change of variable $\tau' = \tau + 1/T$. We have:

$$\mathfrak{G}(\omega_n) = \frac{1}{2}\int_{0}^{1/T} e^{i\omega_n\tau}\mathfrak{G}(\tau)\,\mathrm{d}\tau \mp \frac{1}{2}\int_{-1/T}^{0} e^{i\omega_n\tau}\mathfrak{G}\left(\tau + \frac{1}{T}\right)\mathrm{d}\tau$$
$$= \frac{1}{2}\left(1 \mp e^{-i\omega_n/T}\right)\int_{0}^{1/T} e^{i\omega_n\tau}\mathfrak{G}(\tau)\,\mathrm{d}\tau,$$

whence our assertion follows at once. Notice that we always have

$$\mathfrak{G}(\omega_n) = \int_{0}^{1/T} e^{i\omega_n\tau}\mathfrak{G}(\tau)\,\mathrm{d}\tau, \qquad \omega_n = \begin{cases}(2n+1)\pi T \text{ for fermions,}\\ 2n\pi T \text{ for bosons.}\end{cases} \tag{14.2}$$

We carry out Fourier transformations in all the terms of the perturbation theory series, by substituting the Fourier expansion (14.1) in the relevant expressions. At the same time, we carry out Fourier transformation with respect to the position variables:

$$\mathfrak{G}(\boldsymbol{r}) = \frac{1}{(2\pi)^3}\int e^{i(\boldsymbol{p}\cdot\boldsymbol{r})}\mathfrak{G}(\boldsymbol{p})\,\mathrm{d}^3\boldsymbol{p}, \qquad \mathfrak{G}(\boldsymbol{p}) = \int e^{-i(\boldsymbol{p}\cdot\boldsymbol{r})}\mathfrak{G}(\boldsymbol{r})\,\mathrm{d}^3\boldsymbol{r}. \tag{14.3}$$

The transformations in the position variables are performed here exactly as at $T = 0$.

Note that every point, over whose coordinates integration is perfomed is a point of convergence of an even number of fermion lines; as a result of this, in the integral over the time coordinates of a vertex,

$$\int_{0}^{1/T} \mathrm{d}\tau \exp\left(i\tau \textstyle\sum \omega_n\right), \tag{14.4}$$

the sum of the "frequencies" $\Sigma\,\omega_n$ in the power of the exponential is always "even": $\Sigma\,\omega_n = 2N\pi T$ (N is an integer). In the case of even frequencies, integral (14.4) is equal to

$$\frac{1}{T}\delta_{\Sigma\omega_n}, \qquad \delta_{\omega_n} = \begin{cases}1, & \omega_n = 0,\\ 0, & \omega_n \neq 0.\end{cases} \tag{14.5}$$

In essence, therefore, we have the same situation here as at $T = 0$. In the latter case, the integrals over the coordinates and the time t of the vertices were found to contain δ-functions of the frequencies and momenta, expressing the laws of conservation of energy and momentum in the virtual processes. When $T \neq 0$, the δ-function of the frequencies is replaced by the Kronecker delta δ_{ω_n}, expressing the law of conservation of the discrete "frequency" ω_n.

All this enables us to retain the ordinary Feynman diagrams, with which we were concerned at $T = 0$, for describing the perturbation theory series in momentum space. The only important difference (apart from differences in the coefficients) is the appearance, in the expressions for the matrix elements, of sums over discrete frequences ω_n instead of integrals over frequencies ω.

Before turning to actual examples, we shall quote expressions for the Fourier components of the free Green functions. These functions were calculated in coordinate space in § 11. By (11.13a), the Green function for a free fermion becomes, at $\tau > 0$, after carrying out the coordinate Fourier transformation (14.3):

$$\mathfrak{G}^{(0)}_{\alpha\beta}(\boldsymbol{p}, \tau) = -\delta_{\alpha\beta}\big(1 - n(\boldsymbol{p})\big)\exp\left[-\tau\big(\varepsilon_0(\boldsymbol{p}) - \mu\big)\right],$$
$$n(\boldsymbol{p}) = \left[e^{(\varepsilon_0(\boldsymbol{p})-\mu)/T} + 1\right]^{-1}.$$

On substituting this into (14.2), we get ($\omega_n = (2n+1)\pi T$):

$$\mathfrak{G}^{(0)}_{\alpha\beta}(\boldsymbol{p}, \omega_n) = -\delta_{\alpha\beta}\big(1 - n(\boldsymbol{p})\big)\int\limits_0^{1/T} \exp\left[i\omega_n\tau - \tau\big(\varepsilon_0(\boldsymbol{p}) - \mu\big)\right]\mathrm{d}\tau$$
$$= -\frac{\delta_{\alpha\beta}}{i\omega_n - \varepsilon_0(\boldsymbol{p}) + \mu}\big(1 - n(\boldsymbol{p})\big)\left\{\exp\left[(2n+1)\pi i - \big(\varepsilon_0(\boldsymbol{p}) - \mu\big)/T\right] - 1\right\},$$

i.e. finally,

$$\mathfrak{G}^{(0)}_{\alpha\beta}(\boldsymbol{p}, \omega_n) = \delta_{\alpha\beta}\frac{1}{i\omega_n - \varepsilon_0(\boldsymbol{p}) + \mu}, \quad \omega_n = (2n+1)\pi T. \tag{14.6}$$

Similar calculations give us, for bosons:

$$\mathfrak{G}^{(0)}(\boldsymbol{p}, \omega_n) = \frac{1}{i\omega_n - \varepsilon_0(\boldsymbol{p}) + \mu}, \quad \omega_n = 2\pi n T, \tag{14.7}$$

and for phonons:

$$\mathfrak{D}^{(0)}(\boldsymbol{k}, \omega_n) = -\frac{\omega_0^2(\boldsymbol{k})}{\omega_n^2 + \omega_0^2(\boldsymbol{k})}, \quad \omega_n = 2\pi n T. \tag{14.8}$$

The free Green functions for fermions and bosons therefore differ only in the "parity" of the frequency ω_n. The functions (14.6)—(14.8) are obtainable from the Green functions (7.7) and (7.16) at $T = 0$ by the substitution $\omega \to i\omega_n$. We shall see later that a similar connection (admittedly with certain restrictions) exists for the exact Green functions.

2. *Examples*

We have shown that the temperature-dependent Green functions can be found from the Feynman diagram technique in momentum space. Each line of the diagram is associated with the zero-order Green function $\mathfrak{G}^{(0)}(\boldsymbol{p}, \omega_n)$ for a particle or $\mathfrak{D}^{(0)}(\boldsymbol{k}, \omega_n)$ for a phonon, and each vertex with $\delta(\Sigma \boldsymbol{p})\, \delta_{\Sigma \omega_n}$, expressing the laws of conservation of momentum and discrete "frequency" ω_n. Along every interior line, we integrate over the momenta and sum over the "frequencies" ω_n.

The actual nature of the diagrams and of the associated expressions depends on the type of interaction. We start with two-particle interaction.

A. Two-particle interactions. Let us consider the approximation to the Green function given by diagram IV of Fig. 33. We obtained the expression for it in § 13:

$$-\int\int d^4 z_1 d^4 z_2\, \mathfrak{G}^{(0)}_{\alpha\gamma_1}(x - z_1)\, \mathfrak{G}^{(0)}_{\gamma_1\gamma_2}(z_1 - z_2)\, \mathfrak{G}^{(0)}_{\gamma_2\beta}(z_2 - y)\, \mathfrak{V}(z_1 - z_2).$$

We carry out a Fourier transformation in it with respect to the coordinates and the "time" τ:

$$\delta\mathfrak{G}^{(1)}(\boldsymbol{p}, \omega_n) = \frac{1}{2}\int d^3(\boldsymbol{x} - \boldsymbol{y}) \times \int_{-1/T}^{1/T} d(\tau_x - \tau_y)\, \delta\mathfrak{G}^{(1)}(\boldsymbol{x} - \boldsymbol{y}, \tau_x - \tau_y)\, e^{-i(\boldsymbol{p}\cdot\boldsymbol{x}-\boldsymbol{y}) + i\omega_n(\tau_x - \tau_y)}.$$

We introduce the Fourier component of the potential $\mathfrak{V}(z_1 - z_2)$:

$$\mathfrak{V}(\boldsymbol{r}, \tau) = \frac{T}{(2\pi)^3}\sum_{\omega_n}\int d^3 \boldsymbol{q}\, e^{i(\boldsymbol{q}\cdot\boldsymbol{r}) - i\omega_n\tau}\, \mathfrak{V}(\boldsymbol{q}, \omega_n).$$

In view of the fact that

$$T\sum_{n=-\infty}^{\infty} e^{2\pi i n T\tau} = \delta(\tau),$$

$$\mathfrak{V}(\boldsymbol{q}, \omega_n) = U(\boldsymbol{q}).$$

We have:

$$\delta\mathfrak{G}^{(1)}_{\alpha\beta}(\boldsymbol{p}, \omega_n) = -\frac{1}{2}\sum_{\substack{\omega_{n1}, \omega_{n2} \\ \omega_{n3}, \omega_{n4}}}\int d^3\boldsymbol{p}_1 d^3\boldsymbol{p}_2 d^3\boldsymbol{p}_3 d^3\boldsymbol{p}_4 \left[\frac{T}{(2\pi)^3}\right]^4 \mathfrak{G}^{(0)}_{\alpha\gamma_1}(\boldsymbol{p}_1, \omega_{n1})$$

$$\times\, \mathfrak{G}^{(0)}_{\gamma_1\gamma_2}(\boldsymbol{p}_2, \omega_{n2})\, \mathfrak{G}^{(0)}_{\gamma_2\beta}(\boldsymbol{p}_3, \omega_{n3})\, \mathfrak{V}(\boldsymbol{q}, \omega_{n4}) \int d^3(\boldsymbol{x} - \boldsymbol{y})\, d^3\boldsymbol{z}_1 d^3\boldsymbol{z}_2 \int d(\tau_x - \tau_y)$$

$$\times\, d\tau_1 d\tau_2 \exp[-i(\boldsymbol{p}\cdot\boldsymbol{x} - \boldsymbol{y})] \exp[i\omega_n(\tau_x - \tau_y)] \exp[i(\boldsymbol{p}_1\cdot\boldsymbol{x} - \boldsymbol{z}_1)$$

$$+\, i(\boldsymbol{p}_2\cdot\boldsymbol{z}_1 - \boldsymbol{z}_2) + i(\boldsymbol{p}_3\cdot\boldsymbol{z}_3 - \boldsymbol{y})] \exp[-i\omega_{n1}(\tau_x - \tau_1) - i\omega_{n2}(\tau_1 - \tau_2)]$$

$$\times\, \exp[-i\omega_{n3}(\tau_2 - \tau_y)] \exp[i(\boldsymbol{q}\cdot\boldsymbol{z}_1 - \boldsymbol{z}_2) - i\omega_{n4}(\tau_1 - \tau_2)].$$

We make the transformation $\boldsymbol{x}-\boldsymbol{y}\to\boldsymbol{x}$, $\tau_x-\tau_y\to\tau$ in the integrals over space and time. Now,

$$\frac{1}{2}\int d^3\boldsymbol{x}\,d^3\boldsymbol{z}_1\,d^3\boldsymbol{z}_2\int_{-1/T}^{1/T}d\tau\int_0^{1/T}\int_0^{1/T}d\tau_1\,d\tau_2\exp[i(\boldsymbol{x}\cdot-\boldsymbol{p}+\boldsymbol{p}_1)$$
$$+i(\boldsymbol{z}_1\cdot-\boldsymbol{p}_1+\boldsymbol{p}_2+\boldsymbol{q})+i(\boldsymbol{z}_2\cdot-\boldsymbol{p}_2+\boldsymbol{p}_3-\boldsymbol{q})]$$
$$\times\exp[i\tau(\omega_n-\omega_{n1})+i\tau_1(\omega_{n1}-\omega_{n2}-\omega_{n4})+i\tau_2(\omega_{n2}-\omega_{n3}+\omega_{n4})]$$
$$\times\exp[i(\boldsymbol{y}\cdot\boldsymbol{p}_1-\boldsymbol{p}_3)+i\tau_y(-\omega_{n1}+\omega_{n3})]=\left(\frac{(2\pi)^3}{T}\right)^3\delta(\boldsymbol{p}-\boldsymbol{p}_1)$$
$$\times\,\delta(\boldsymbol{p}_1-\boldsymbol{p}_2-\boldsymbol{q})\,\delta(\boldsymbol{p}_2-\boldsymbol{p}_3+\boldsymbol{q})\delta_{\omega_n-\omega_{n1}}\delta_{\omega_{n1}-\omega_{n2}-\omega_{n4}}\delta_{n_2\omega-\omega_{n3}+\omega_{n4}},$$

whence

$$\delta\mathfrak{G}^{(1)}_{\alpha\beta}(\boldsymbol{p},\omega_n)=-\frac{T}{(2\pi)^3}\sum_{\omega_{n1}}\int d^3\boldsymbol{p}_1\,\mathfrak{G}^{(0)}_{\alpha\gamma_1}(\boldsymbol{p},\omega_n)$$
$$\times\,\mathfrak{G}^{(0)}_{\gamma_1\gamma_2}(\boldsymbol{p}_1,\omega_{n1})\,\mathfrak{G}^{(0)}_{\gamma_2\beta}(\boldsymbol{p},\omega_n)\,\mathfrak{V}(\boldsymbol{p}-\boldsymbol{p}_1,\omega_n-\omega_{n1}).$$

On substituting here expressions (14.6) and (14.7) for the zero-order Green functions, we finally get

$$\delta\mathfrak{G}^{(1)}_{\alpha\beta}=-\frac{\delta_{\alpha\beta}}{[i\omega_n-\varepsilon_0(\boldsymbol{p})+\mu]^2}\frac{T}{(2\pi)^3}\sum_{\omega_{n1}}\int d^3\boldsymbol{p}_1\frac{\mathfrak{V}(\boldsymbol{p}-\boldsymbol{p}_1,\omega_n-\omega_{n1})}{i\omega_{n1}-\varepsilon_0(\boldsymbol{p}_1)+\mu}.\qquad(14.9)$$

Similar calculations give us, for the contribution of diagram III of Fig. 33:

$$\pm\frac{\delta_{\alpha\beta}}{(i\omega_n-\varepsilon_0(\boldsymbol{p})+\mu)^2}\mathfrak{V}(0,0)\,(2s+1)\frac{T}{(2\pi)^3}\sum_{\omega_{n1}}\int d^3\boldsymbol{p}_1\frac{e^{i\omega_{n1}\tau}}{i\omega_{n1}-\varepsilon_0(\boldsymbol{p}_1)+\mu},\qquad(14.10)$$

where $\tau\to0$ and s is the particle spin, equal to $^1/_2$ for fermions and zero for bosons. We have here introduced $\exp(i\omega_n\tau)$ $(\tau\to+0)$ under the sign of summation, in accordance with the proviso mentioned in § 13 that the $\mathfrak{G}$-function be defined in coordinate space for coincident time arguments as

$$\mathfrak{G}^{(0)}(0,0)=\lim_{\tau\to+0}\mathfrak{G}^{(0)}(0,-\tau).$$

Expressions (14.9) and (14.10) can be associated with the Feynman diagrams of Fig. 42 *a*, *b* respectively. The external lines of these diagrams carry the external momentum $\boldsymbol{p}$ and frequency ω_n; the momenta and frequencies at each vertex satisfy the laws of conservation: the sum of the momenta "entering" the vertex is equal to the sum of those "leaving".

Let us consider the diagram for $\mathfrak{G}(\boldsymbol{p},\omega_n)$ of any order k of perturbation theory. It will have $2k$ vertices, $2k+1$ full-drawn lines, and k wavy lines. When carrying out the Fourier transformation we have $2k$ integrations over the space and "time" coordinates of the vertices and one integration over the difference between the coordinates of the end-points, leading to $2k+1$ quantities of the type $\delta(\Sigma \boldsymbol{p})\,\delta_{\Sigma\omega_n}$, expressing $2k+1$ laws of conservation. It may easily be seen that two laws of conservation

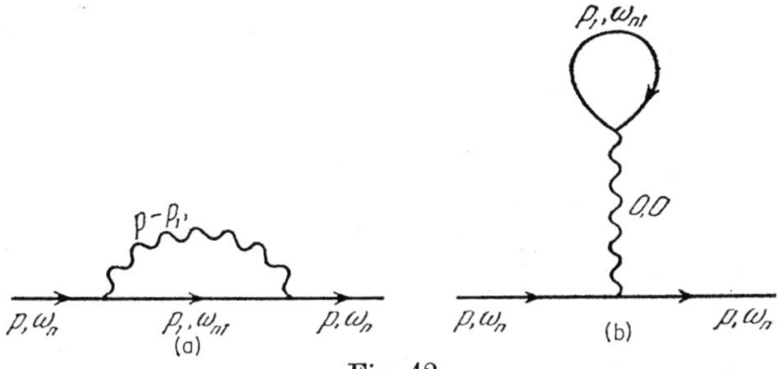

Fig. 42

express the fact that the external outer lines have momentum $\boldsymbol{p}$ and frequence ω_n. The remaining $2k-1$ laws of conservation imply that, of the $3k-1$ integrations over momenta and summations over frequencies of interior lines (both full-drawn and wavy), there actually remain only k integrations and summations.

Let us now state the rules for writing down the expression corresponding to a given diagram for a Green function.

(1) We first of all have to associate the lines of the diagram with momenta and frequencies; the outer lines must carry the external momentum and frequency, whilst the momenta and frequencies of the inner lines must satisfy at each vertex the laws of conservation $\Sigma \boldsymbol{p}' = 0$, $\Sigma \omega_n' = 0$. The frequencies of Bose lines are always even ($\omega_n = 2n\pi T$), and of Fermi lines always odd ($\omega_n = (2n+1)\pi T$).

(2) Summations and integrations are carried out over all the independent momenta and frequencies of the diagram.

(3) Each continuous interior line (momentum and frequency $\boldsymbol{p}'$, ω_n') is associated with

$$\frac{1}{i\omega_n' - \varepsilon_0(\boldsymbol{p}') + \mu},$$

and each wavy line ($\boldsymbol{q}$, ω_n') with

$$\mathfrak{V}(\boldsymbol{q}, \omega_n') \equiv U(\boldsymbol{q}).$$

(4) Both outer lines (momentum and frequency $\boldsymbol{p}$, ω_n) are associated with

$$\frac{\delta_{\alpha\beta}}{(i\omega_n - \varepsilon_0(\boldsymbol{p}) + \mu)^2}.$$

(5) The factor

$$(-1)^k \frac{T^k}{(2\pi)^{3k}} (2s+1)^F (\mp 1)^F$$

has to be put in front of the expression obtained, where F is the number of closed loops formed by particle lines on the diagram.

Using these rules, it is fairly easy to write down the correction corresponding to a diagram, no matter how complicated. For instance, the contribution of the diagram of Fig. 43 is

$$\pm \frac{\delta_{\alpha\beta}}{(i\omega - \varepsilon_0(\mathbf{p}) + \mu)^2} \frac{T^3}{(2\pi)^9} (2s+1) \sum_{\omega_1, \omega_2, \omega_3} \int d^3\mathbf{p}_1 d^3\mathbf{p}_2 d^3\mathbf{p}_3$$

$$\times \frac{1}{i(\omega - \omega_1) - \varepsilon_0(\mathbf{p} - \mathbf{p}_1) + \mu} \times \frac{1}{i(\omega_3 - \omega_1) - \varepsilon_0(\mathbf{p}_3 - \mathbf{p}_1) + \mu}$$

$$\times \frac{1}{i\omega_3 - \varepsilon_0(\mathbf{p}_3) + \mu} \times \frac{1}{i\omega_2 - \varepsilon_0(\mathbf{p}_2) + \mu} \times \frac{1}{i(\omega_1 + \omega_2) - \varepsilon_0(\mathbf{p}_1 + \mathbf{p}_2) + \mu}$$

$$\times \left(U(\mathbf{p}_1)\right)^2 U(\mathbf{p} - \mathbf{p}_3), \qquad \omega_1 = 2\pi n T; \qquad \omega_2, \omega_3 = (2n+1)\pi T.$$

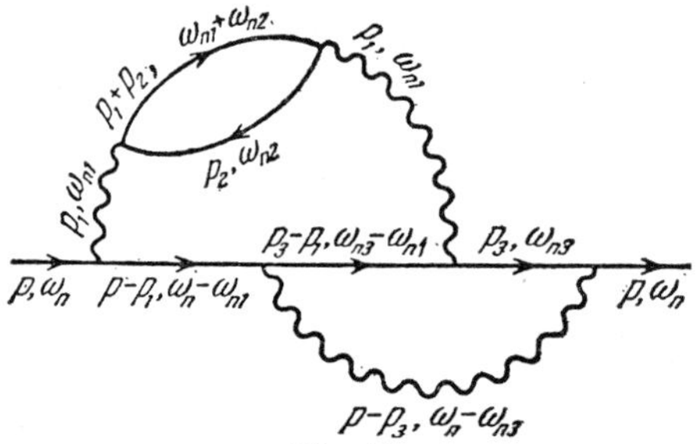

Fig. 43

We now turn to the second variant of the diagram technique for the case of two-particle interactions. The following formal method may conveniently be used for changing to Fourier transforms in the relevant expressions. We introduced above the quantity $\mathcal{T}^{(0)}_{\gamma_1\gamma_2;\gamma_3\gamma_4}(z_1, z_2; z_3, z_4)$. The $\mathcal{T}^{(0)}$ depends on four "times" τ_i, where each τ_i varies from 0 to $1/T$. We continue $\mathcal{T}^{(0)}$ into the interval $-1/T$, $1/T$ using relations such as (11.8) for the $\mathfrak{G}$-functions:

$$\mathcal{T}^{(0)}(\tau_1 < 0, \tau_2; \tau_3, \tau_4) = \mp \mathcal{T}^{(0)}(\tau_1 + 1/T, \tau_2; \tau_3, \tau_4)$$

together with similar relationships for τ_2, τ_3, τ_4. We define the Fourier transform as

$$\frac{1}{16} \int_{-1/T}^{1/T} d\tau_1 \cdots d\tau_4 e^{i(\omega_1\tau_1 + \omega_2\tau_2 - \omega_3\tau_3 - \omega_4\tau_4)} \mathcal{T}^{(0)}(\tau_1, \tau_2; \tau_3, \tau_4).$$

Obviously all four frequencies will be "odd" in the case of Fermi statistics, and "even" in the case of Bose statistics.

Notice further that, since $\mathcal{T}^{(0)}(z_1, z_2; z_3, z_4)$ is by definition a function of the differences of coordinates and "times" τ only, the Fourier components of $\mathcal{T}^{(0)}$ with respect to the space and time variables will contain a δ-function of a sum of momenta $\delta(\mathbf{p}_1 + \mathbf{p}_2 - \mathbf{p}_3 - \mathbf{p}_4)$ and the Kronecker delta of a sum of frequencies $\delta_{\omega_1+\omega_2-\omega_3-\omega_4}$. We therefore determine the

Fourier component of $\mathcal{T}^{(0)}$ at once as

$$\frac{(2\pi)^3}{T}\delta(\mathbf{p}_1+\mathbf{p}_2-\mathbf{p}_3-\mathbf{p}_4)\,\delta_{\omega_1+\omega_2-\omega_3-\omega_4}\mathcal{T}^{(0)}(\mathbf{p}_1\omega_1,\mathbf{p}_2\omega_2;\mathbf{p}_3\omega_3,\mathbf{p}_4\omega_4)$$

$$=\frac{1}{16}\int_{-1/T}^{1/T}\mathrm{d}\tau_1\cdots\mathrm{d}\tau_4\int\mathrm{d}^3\mathbf{r}_1\cdots\mathrm{d}^3\mathbf{r}_4\exp-i[(\mathbf{p}_1\cdot\mathbf{r}_1)+(\mathbf{p}_2\cdot\mathbf{r}_2)-(\mathbf{p}_3\cdot\mathbf{r}_3)$$

$$-(\mathbf{p}_4\cdot\mathbf{r}_4)]\exp i(\omega_1\tau_1+\omega_2\tau_2-\omega_3\tau_3-\omega_4\tau_4)\,\mathcal{T}^{(0)}(z_1,z_2;z_3,z_4). \quad (14.11)$$

As an example, let us perform a Fourier transformation in the expression for the first order correction corresponding to diagram II of Fig. 35. We find easily

$$-\frac{1}{[i\omega-\varepsilon_0(\mathbf{p})+\mu]^2}\frac{T}{(2\pi)^3}\sum_{\omega_1}\int\mathrm{d}^3\mathbf{p}_1\mathcal{T}^{(0)}_{\alpha\gamma;\gamma\beta}$$

$$\times(\mathbf{p},\omega;\mathbf{p}_1,\omega_1;\mathbf{p}_1,\omega_1;\mathbf{p},\omega)\frac{1}{i\omega_1-\varepsilon_0(p_1)+\mu},$$

which corresponds to the diagram of Fig. 44.

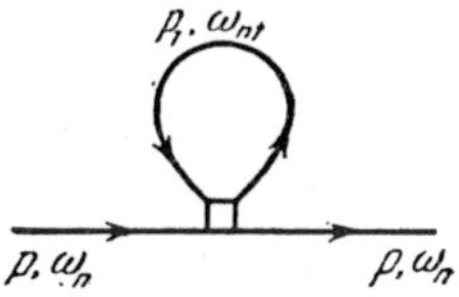

Fig. 44

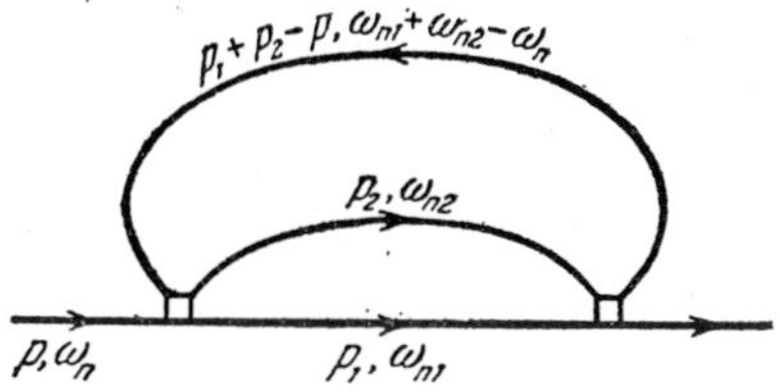

Fig. 45

Similar calculations, applied to the correction gives by the diagram of Fig. 45, lead to the formula (†)

$$\frac{1}{2}\frac{1}{[i\omega-\varepsilon_0(\mathbf{p})+\mu]^2}\frac{T^2}{(2\pi)^6}\sum_{\omega_1\omega_2}\int\mathrm{d}^3\mathbf{p}_1\mathrm{d}^3\mathbf{p}_2\mathcal{T}^{(0)}_{\alpha\gamma_1;\gamma_2\gamma_3}(p,p_1+p_2-p;p_1,p_2)$$

$$\times\frac{1}{i\omega_1-\varepsilon_0(\mathbf{p}_1)+\mu}\times\frac{1}{i\omega_2-\varepsilon_0(\mathbf{p}_2)+\mu}$$

$$\times\frac{1}{i(\omega_1+\omega_2-\omega)-\varepsilon_0(\mathbf{p}_1+\mathbf{p}_2-\mathbf{p})+\mu}\mathcal{T}^{(0)}_{\gamma_2\gamma_3;\gamma_1\beta}(p_1,p_2;p_1+p_2-p,p).$$

The nth order diagram for the $\mathfrak{G}$-function contains n squares (vertices) and $2n+1$ lines; of these latter, $2n-1$ interior lines are connected at each vertex by the conservation laws $\Sigma\,\mathbf{p}'=0$, $\Sigma\,\omega_n'=0$. It may easily be verified that there are altogether n independent integrations and summations over the momenta and frequencies of the interior

(†) Four-dimensional notation has been used here: $p=(\mathbf{p},\omega_n)$.

lines. The following operations are required for calculating the contribution of a diagram:

(1) each interior line is associated with

$$\frac{1}{i\omega' - \varepsilon_0(\boldsymbol{p}') + \mu};$$

(2) each external line is associated with

$$\frac{1}{i\omega - \varepsilon_0(\boldsymbol{p}) + \mu};$$

(3) each vertex is associated with the function

$$\mathcal{T}^{(0)}_{\alpha\beta;\gamma\delta}(p_1, p_2; p_3, p_1 + p_2 - p_3);$$

(4) integrations and summations are carried out over all the independent interior momenta and frequencies;

(5) summations are carried out over the indices $\alpha, \beta, \ldots$ of the $\mathcal{T}^{(0)}$, joined by the $\mathfrak{G}^{(0)}$-lines;

(6) the result is multiplied by $\pm T^n m/2^n(2\pi)^{3n}$ (m and the sign are determined from a comparison with the non-symmetrised technique).

For instance, the diagram of Fig. 46 corresponds to

$$-\frac{1}{4}\frac{1}{[i\omega - \varepsilon_0(\boldsymbol{p}) + \mu]^2}\frac{T^3}{(2\pi)^9}\sum_{\omega_1\omega_2\omega_3}\int d^3\boldsymbol{p}_1 d^3\boldsymbol{p}_2 d^3\boldsymbol{p}_3$$

$$\times \mathcal{T}^{(0)}_{\alpha\gamma_1;\gamma_2\gamma_3}(p, p_1 + p_2 - p; p_1, p_2)\frac{1}{i\omega_1 - \varepsilon_0(\boldsymbol{p}_1) + \mu}$$

$$\times \frac{1}{i\omega_2 - \varepsilon_0(\boldsymbol{p}_2) + \mu}\mathcal{T}^{(0)}_{\gamma_2\gamma_3;\gamma_4\gamma_5}(p_1, p_2; p_3, p_1 + p_2 - p_3)$$

$$\times \frac{1}{i\omega_3 - \varepsilon_0(\boldsymbol{p}_3) + \mu}\frac{1}{i(\omega_1 + \omega_2 - \omega_3) - \varepsilon_0(\boldsymbol{p}_1 + \boldsymbol{p}_2 - \boldsymbol{p}_3) + \mu}$$

$$\times \mathcal{T}^{(0)}_{\gamma_4\gamma_5;\gamma_1\beta}(p_3, p_1 + p_2 - p_3; p_1 + p_2 - p, p)$$

$$\times \frac{1}{i(\omega_1 + \omega_2 - \omega) - \varepsilon_0(\boldsymbol{p}_1 + \boldsymbol{p}_2 - \boldsymbol{p}) + \mu}.$$

In the case of a point interaction, $\mathcal{T}^{(0)}$ is independent of the momenta and frequencies.

B. Interactions with phonons. In this case the only non-zero diagrams are of even order. Any diagram of order $2n$ contains $3n + 1$ interior (electron and phonon) lines and $2n$ vertices, which corresponds to

$$3n - 1 - (2n - 1) = n$$

independent integrations. The following operations are performed when evaluating its contribution:

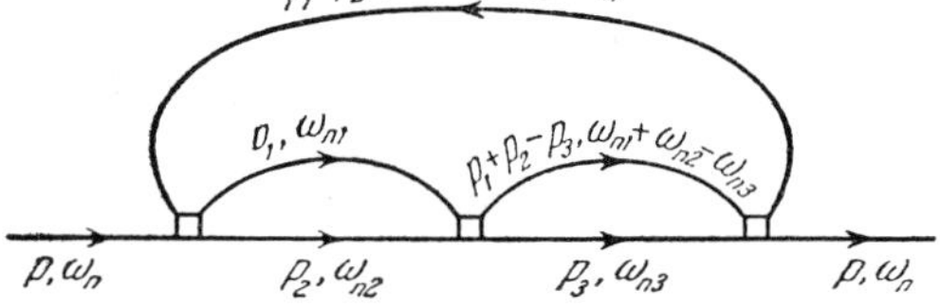

Fig. 46

(1) each interior full-drawn line is associated with

$$\frac{1}{i\omega' - \varepsilon_0(\boldsymbol{p}') + \mu'},$$

whilst both external full-drawn lines (in diagrams for the corrections to the particle 𝔊-functions) are associated with

$$\frac{\delta_{\alpha\beta}}{(i\omega - \varepsilon_0(\boldsymbol{p}) + \mu)^2};$$

(2) each phonon (dotted) line is associated with

$$-\frac{\omega_0^2(\boldsymbol{k})}{\omega^2 + \omega_0^2(\boldsymbol{k})};$$

(3) the result is multiplied by

$$g^{2n}\frac{T^n}{(2\pi)^{3n}}(-1)^n(2s+1)^F(\mp 1)^F,$$

where g is the coupling constant and F the number of closed loops on the diagram of spin s particles.

Suppose we write down, say, the expression for the second order correction to the 𝔇-function for a phonon, corresponding to the diagram of Fig. 47:

$$\pm\left[\frac{\omega_0^2(\boldsymbol{k})}{\omega^2 + \omega_0^2(\boldsymbol{k})}\right]^2 g^2(2s+1)\frac{T}{(2\pi)^3}$$
$$\times\sum_{\omega'}\int \mathrm{d}^3\boldsymbol{p}'\frac{1}{i\omega' - \varepsilon_0(\boldsymbol{p}') + \mu}\,\frac{1}{i(\omega'-\omega) - \varepsilon_0(\boldsymbol{p}'-\boldsymbol{k}) + \mu}.$$

The rules given in this section are related in a very simple way to the corresponding rules for finding the corrections to Green functions at $T = 0$. As is easily verified, the approximation to the temperature-dependent Green function 𝔊 can be obtained from the expression for the correction to the Green function G at $T = 0$ by replacing all the frequencies ω in the latter by $i\omega_n(\omega_n = 2n\pi T$ for bosons, and $(2n+1)\pi T$ for fermions) and all the integrals over frequencies by the sums:

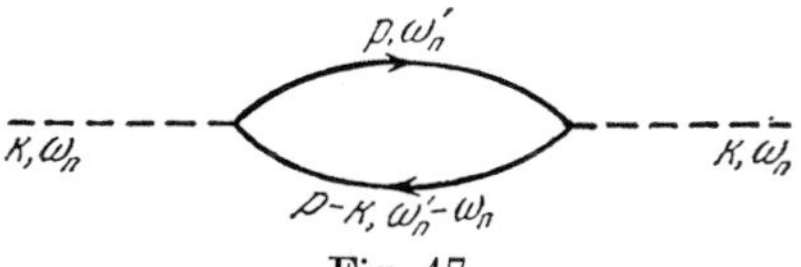

Fig. 47

$$\int \mathrm{d}\omega \cdots \to 2\pi T i \sum_{\omega_n}\cdots.$$

Let us see in conclusion how we can pass to the limit to the case $T = 0$ in the new technique. When T tends to zero, the chief role is played in the sums over frequencies by large n, so that we can replace these sums by integrals. Noting that $\Delta\omega = \omega_{n+1} - \omega_n = 2\pi T$, we get

$$T\sum_{\omega_n} \to \frac{1}{2\pi}\int \mathrm{d}\omega.$$

It must be emphasised that, at $T = 0$, $𝔊(\omega)$ is not at all the same as $G(\omega)$. The connection between these quantities will be established later.

§ 15. THE PERTURBATION THEORY SERIES FOR THE THERMODYNAMIC POTENTIAL

There are cases when it proves more convenient to find the thermodynamic potential Ω directly, rather than indirectly, via an evaluation of the Green functions.

The correction to the thermodynamic potential is given in terms of the average of the $\mathfrak{S}$-matrix by (see (12.11))

$$\Delta\Omega = -T \ln \langle \mathfrak{S} \rangle, \quad \mathfrak{S} = \exp\left\{-\int_0^{1/T} \hat{H}_{int}(\tau)\, d\tau\right\}. \tag{15.1}$$

The logarithm can in fact be taken in the ordinary form in (15.1); or, more precisely, a diagram technique can be developed for dealing directly with Ω.

It is obvious from the foregoing that the diagrams which describe the perturbation theory series for Ω are made up of closed loops. Typical diagrams are shown for both variants of a two-particle interaction in Fig. 48 *a*, *b*, and for interaction with phonons in Fig. 49 (the diagram in Fig. 49, I, is in fact equal to zero).

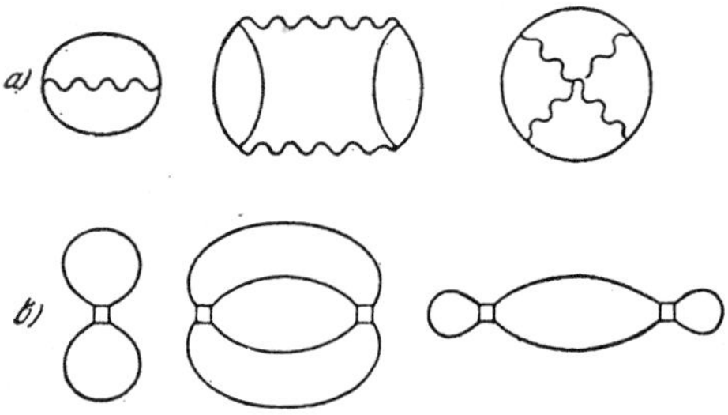

Fig. 48

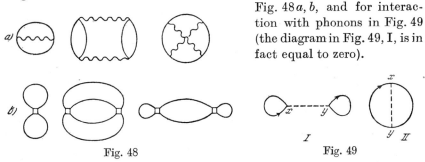

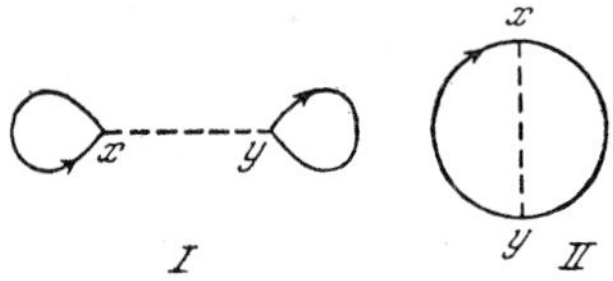

Fig. 49

The diagrams of a given order of the perturbation theory series will include both types — connected and unconnected. The latter consist of two or more closed loops, with no lines joining them. Connected diagrams are obtained if, when describing a term of the series for $\langle \mathfrak{S} \rangle$ in accordance with Wick's theorem,

$$\frac{(-1)^n}{n!} \int_0^{1/T} \cdots \int_0^{1/T} d\tau_1 \cdots d\tau_n \langle T_\tau \{\hat{H}_{int}(\tau_1) \cdots \hat{H}_{int}(\tau_n)\}\rangle, \tag{15.2}$$

we can start the pairing with an operator appearing in $\hat{H}_{int}(\tau_1)$ then return to $\hat{H}_{int}(\tau_1)$ without passing over any of the $\hat{H}_{int}$. In any other case, the diagram must be unconnected.

Suppose an nth order unconnected diagram consists of k closed loops, and, for a start, that every one of these loops contains a different number of vertices. The corresponding expression will be

$$\frac{(-1)^n}{n!}\int d\tau_1^{(1)}\cdots d\tau_{m_1}^{(1)}\langle T_\tau\{\hat{H}_{int}(\tau_1^{(1)})\cdots\hat{H}_{int}(\tau_{m_1}^{(1)})\}\rangle_c$$
$$\times\int d\tau_1^{(2)}\cdots d\tau_{m_2}^{(2)}\langle T_\tau\{\hat{H}_{int}(\tau_1^{(2)})\cdots\hat{H}_{int}(\tau_{m_2}^{(2)})\}\rangle_c$$
$$\cdots\int d\tau_1^{(k)}\cdots d\tau_{m_k}^{(k)}\langle T_\tau\{\hat{H}_{int}(\tau_1^{(k)})\cdots\hat{H}_{int}(\tau_{m_k}^{(k)})\}\rangle_c, \quad (15.3)$$

where

$$m_1+m_2+\cdots+m_k=n\ \ (m_1\neq m_2\neq\cdots\neq m_k),$$

and the symbol $\langle\cdots\rangle_c$ denotes the averaging corresponding to a given connected diagram. We now sum all the topologically equivalent diagrams containing k loops of the type selected. Obviously this can be done simply by multiplying (15.3) by the number of such diagrams F_k. This number is the same as the number of methods by which we can accomodate n operators $\hat{H}_{int}$ in k different "cells" $\langle\cdots\rangle_c$, containing respectively $m_1, m_2, \ldots, m_k$ places, i.e.

$$F_k=\frac{n!}{m_1!\,m_2!\cdots m_k!}.$$

We get as a result:

$$\frac{(-1)^m}{m_1!}\int d\tau_1^{(1)}\cdots d\tau_{m_1}^{(1)}\langle T_\tau\{\hat{H}_{int}(\tau_1^{(1)})\cdots\hat{H}_{int}(\tau_{m_1}^{(1)})\}\rangle_c$$
$$\times\frac{(-1)^{m_2}}{m_2!}\int d\tau_1^{(2)}\cdots d\tau_{m_2}^{(2)}\langle T_\tau\{\hat{H}_{int}(\tau_1^{(2)})\cdots\hat{H}_{int}(\tau_{m_2}^{(2)})\}\rangle_c$$
$$\cdots\frac{(-1)^{m_k}}{m_k!}\int d\tau_1^{(k)}\cdots d\tau_{m_k}^{(k)}\langle T_\tau\{\hat{H}_{int}(\tau_1^{(k)})\cdots\hat{H}_{int}(\tau_{m_k}^{(k)})\}\rangle_c. \quad (15.4)$$

Notice that we did not really need the assumption that each averaging $\langle\cdots\rangle_c$ corresponds to a connected diagram of a definite type; instead, we could have simply assumed that $\langle\cdots\rangle_c$ is the sum of all the connected diagrams with a given number of vertices. It may be concluded from this that the sum of all the unconnected diagrams containing k closed loops with $m_1, m_2, \ldots, m_k$ vertices respectively is of the form

$$\Xi_{m_1}\Xi_{m_2}\cdots\Xi_{m_k},$$

where

$$\Xi_m=\frac{(-1)^m}{m!}\int d\tau_1\cdots d\tau_m\langle T_\tau\{\hat{H}_{int}(\tau_1)\cdots\hat{H}_{int}(\tau_m)\}\rangle_c \quad (15.5)$$

is nothing but the sum of all the connected diagrams of order m for $\langle\mathfrak{S}\rangle$. Obviously,

$$1+\Xi_1+\Xi_2+\cdots=\langle\mathfrak{S}\rangle_c. \quad (15.6)$$

If some of the numbers $m_1, m_2, \ldots$ are the same, so that the diagram splits up into $p_1+p_2+\cdots+p_k$ closed loops, containing respectively

$m_1, m_2, \ldots, m_k$ ($m_1 \neq \cdots m_k$) vertices, it can be shown that expression (15.5) must be replaced by(†)

$$\frac{1}{p_1!}\Xi_{m_1}^{p_1}\frac{1}{p_2!}\Xi_{m_2}^{p_2}\cdots\frac{1}{p_k!}\Xi_{m_k}^{p_k} \tag{15.7}$$

or, what amounts to the same thing, by

$$\frac{1}{p_1!}\Xi_1^{p_1}\frac{1}{p_2!}\Xi_2^{p_2}\cdots\frac{1}{p_l!}\Xi_l^{p_l}, \tag{15.8}$$

where the p_l ($p_l = 0, 1, 2, \ldots$) indicate how many closed loops of order l are contained in all the unconnected diagrams. On summing (15.8) over all p_l (the summations over different p_l are obviously independent), we get

$$\begin{aligned}\langle\mathfrak{S}\rangle &= \sum_{p_1,p_2,\ldots}\frac{1}{p_1!}\Xi_1^{p_1}\frac{1}{p_2!}\Xi_2^{p_2}\cdots = \sum_{p_1}\frac{1}{p_1!}\Xi_1^{p_1}\sum_{p_2}\frac{1}{p_2!}\Xi_2^{p_2}\cdots\\ &= e^{\Xi_1}e^{\Xi_2}\cdots = \exp\{\Xi_1+\Xi_2+\cdots\}.\end{aligned} \tag{15.9}$$

Finally, on substituting (15.9) in (15.1), we get

$$\Delta\Omega = -T(\Xi_1+\Xi_2+\cdots) = -T\{\langle\mathfrak{S}\rangle_c - 1\}. \tag{15.10}$$

A very important result has been obtained: to find the corrections to the thermodynamic potential, we only need to find the contribution of the connected diagrams for $\langle\mathfrak{S}\rangle$.

As already remarked, the diagrams for $\langle\mathfrak{S}\rangle$ are in the form of closed loops, and these can be evaluated in accordance with essentially the same rules as for $\mathfrak{G}$-functions. The only difference lies in the factor in front of the diagram.

We mentioned in § 12 that the factor $1/n!$ in the perturbation theory series (12.13) for the $\mathfrak{G}$-functions cancels out if we take into account the contribution of all the topologically equivalent diagrams, the number of which is in fact equal to $n!$ The situation is different when we evaluate $\langle\mathfrak{S}\rangle_c$. The number of equivalent diagrams that give the nth term of series (12.12) will be equal(‡) to $(n-1)!$, so that a factor $1/n$ appears

(†) This can be proved as follows. When some of the $m_1, \ldots, m_k$ are the same, the F_k mentioned above is the same as the number of ways in which $p_1m_1 + p_2m_2 + \cdots + p_km_k = n$ operators $\hat{H}_{int}$ can be allocated to $p_1 + p_2 + \cdots + p_k$ cells $\langle\cdots\rangle_c$, containing $m_1, m_2, \ldots, m_k$ places where, respectively, $p_1, p_2, \ldots, p_k$ of the cells are the same. In this case F_k is equal to

$$F_k = \frac{n!}{p_1!(m_1!)^{p_1}p_2!(m_2!)^{p_2}\cdots p_k!(m_k!)^{p_k}}.$$

(‡) All the equivalent diagrams follow by taking all possible permutations of the $n-1$ operators H_{int} in (12.12). One of the H_{int} must be regarded as fixed. When evaluating $\mathfrak{G}$, the beginning and end of the external lines were fixed, i.e. the operators $\psi_\alpha(\mathbf{r}_1, \tau_1)$, $\overline{\psi}_\beta(\mathbf{r}_2, \tau_2)$ in (12.13).

in front of each diagram (assuming that only topologically non-equivalent diagrams are distinct). The presence of a factor dependent on the order n makes the perturbation theory series for Ω very awkward, especially in cases when we cannot confine ourselves to a finite number of terms but have to sum infinite sequences of diagrams.

We quote some examples of finding the corrections $\Delta\Omega$, and confine ourselves for brevity to interaction with phonons. Only the connected diagram of Fig. 49, II, is non-zero in the second order of perturbation theory. We find its contribution from Wick's theorem, putting $\Omega_2 = -T\Xi_2$ (we use a four-dimensional notation):

$$\Omega_2 = \pm\frac{1}{2} T g^2 \int \mathrm{d}^4x \mathrm{d}^4y \, \mathfrak{G}^{(0)}_{\alpha\beta}(x-y)\, \mathfrak{G}^{(0)}_{\beta\alpha}(y-x)\, \mathfrak{D}^{(0)}(x-y).$$

It turns out that Ω_2 is proportional to the volume V of the system, as may easily be verified by introducing the new variable $x' = x - y$ in the integral. The situation is the same in any approximation; this is to be expected, since the potential Ω is well known to have the form

$$\Omega = -V P(\mu, T),$$

where P is the pressure expressed as a function of the chemical potential and the temperature. In future, therefore, we shall always give the formulae for ΔP ($P = P_0(\mu, T) + \Delta P$, where P_0 is the pressure in the system of free particles).

We have for ΔP_2:

$$\Delta P_2 = \mp\frac{1}{2} g^2 \int \mathrm{d}^4x \, \mathfrak{G}^{(0)}_{\alpha\beta}(x)\, \mathfrak{G}^{(0)}_{\beta\alpha}(-x)\, \mathfrak{D}^{(0)}(x). \tag{15.11}$$

If we change to the momentum representation, we have

$$\Delta P_2 = \pm\frac{1}{2} g^2 \frac{T^2}{(2\pi)^6}(2s+1) \sum_{\omega_1,\omega_2} \int \mathrm{d}^3\boldsymbol{p}\, \mathrm{d}^3\boldsymbol{k}$$
$$\times \frac{1}{i\omega_1 - \varepsilon_0(\boldsymbol{p}) + \mu} \; \frac{1}{i(\omega_1+\omega_2) - \varepsilon_0(\boldsymbol{p}+\boldsymbol{k}) + \mu} \; \frac{\omega_0^2(\boldsymbol{k})}{\omega_2^2 + \omega_0^2(\boldsymbol{k})}.$$

The corresponding diagram is shown in Fig. 50.

Let us take any diagram of order $2n$. It contains $3n$ lines and $2n$ vertices. However, one of the $2n$ conservation laws turns out to be an identity, provided the remaining $2n-1$ laws are fulfilled. Thus there are altogether $n+1$ independent integrations in a $2n$th order diagram. The extra law of conservation leads to the appearance in the diagram for $\langle\mathfrak{S}\rangle$ of an extra factor $\delta(\boldsymbol{p} = 0)$, proportional to the volume V of the system (†).

(†) By definition,

$$\delta(\boldsymbol{p} = 0) = \frac{1}{(2\pi)^3}\int \mathrm{d}^3\boldsymbol{r} = \frac{V}{(2\pi)^3}.$$

The rules by which the individual elements of the diagrams are associated with the Green functions (and the vertex parts, in the case of

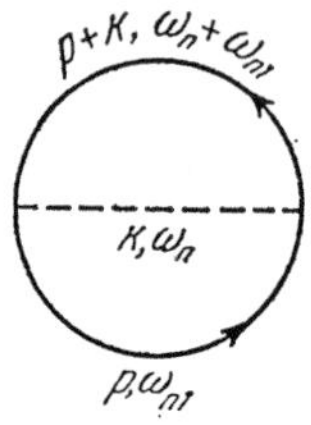

Fig. 50

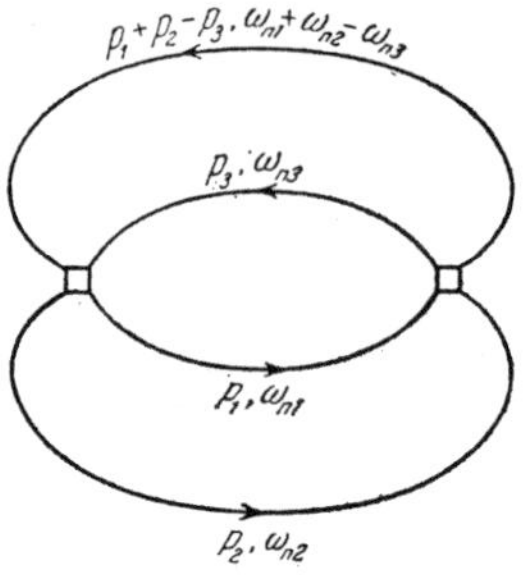

Fig. 51

other interactions) remain the same as for the diagrams for $\mathfrak{G}$. The factor in front of the diagram of order $2n$ for the correction ΔP is equal to

$$M_n = \frac{(-1)^{n+1}}{2n} g^{2n} \left(\frac{T}{(2\pi)^3}\right)^{2n} (\mp 1)^F (2s+1)^F,$$

where F is the number of closed loops, formed by single $\mathfrak{G}$-lines of the particles.

We quote in addition the expression for ΔP for the case of binary interactions. When the interaction has the form (13.7), the second order correction to the pressure, corresponding to the diagram of Fig. 51, is

$$-\frac{1}{4}\frac{T^3}{(2\pi)^9}\sum_{\omega_1\omega_2\omega_3}\int d^3p_1 d^3p_2 d^3p_3 \frac{1}{i\omega_1 - \varepsilon_0(p_1) + \mu}\,\frac{1}{i\omega_2 - \varepsilon_0(p_2) + \mu}$$
$$\times \frac{1}{i\omega_3 - \varepsilon(p_3) + \mu}\,\frac{1}{i(\omega_1 + \omega_2 - \omega_3) - \varepsilon_0(p_1 + p_2 - p_3) + \mu}$$
$$\times \mathcal{T}^{(0)}_{\alpha\beta;\gamma\delta}(p_1, p_2; p_1 + p_2 - p_3, p_3)\,\mathcal{T}^{(0)}_{\gamma\delta;\beta\alpha}(p_1 + p_2 - p_3, p_3; p_2, p_1).$$

§ 16. DYSON EQUATION. MANY-PARTICLE GREEN FUNCTIONS

1. *Dyson equation*

As in the absolute zero case, statistical problems at $T \neq 0$ virtually always involve finding several of the first terms of the perturbation theory series as corrections to the Green functions. Within the framework of almost any physical problem which is correctly stated, the formal parameter of the expansion of the diagram technique, namely the interaction Hamiltonian $\hat{H}_{int}$, is not small; as a result, several infinite sequences of terms of the perturbation theory series will give contributions of the same order of magnitude.

We saw in the previous chapter that a summation of infinite series is carried out by the diagram method in the field theory technique.

The sum can be presented in this method as a diagram whose elements — lines and vertices — are in turn the sums of an infinite number of diagrams. The association of definite expressions with the elements of such a diagram is performed in accordance with the same rules as for perturbation theory diagrams. This fact enables us to form different equations for the Green functions. We have already encountered one such equation in Chap. II; this was the Dyson equation, which expresses the Green function in terms of the mass operator of the system.

Two properties of the diagram technique are fundamental to the formation of such equations: the topological structure of the diagrams, and the rules by which a diagram is associated with a definite expression. The diagrams in the technique at absolute zero and in Matsubara's technique are in general the same, whilst the rules for association only differ in that the integrations over frequencies for the first case are replaced by summations over discrete imaginary "frequencies" $i\omega_n$ for the second; to be more precise, the expression for the correction to the

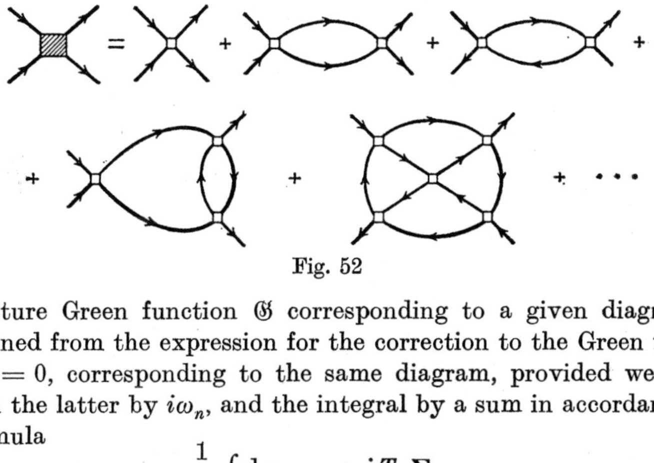

Fig. 52

temperature Green function $\mathfrak{G}$ corresponding to a given diagram can be obtained from the expression for the correction to the Green function G at $T = 0$, corresponding to the same diagram, provided we replace the ω in the latter by $i\omega_n$, and the integral by a sum in accordance with the formula

$$\frac{1}{2\pi}\int d\omega \cdots \to iT\sum_n \cdots$$

(see end of § 14).

The above-mentioned fact enables us to carry over at once all the results of § 10 to the $T \neq 0$ case simply by changing the notation. In particular, the Dyson equation is retained in Matsubara's technique. It becomes, for a system of particles with binary interactions:

$$\begin{aligned}\mathfrak{G}_{\alpha\beta}(p) = \mathfrak{G}^{(0)}_{\alpha\beta}(p) - \frac{T}{(2\pi)^3}\sum_{\omega_1}\int d^3\boldsymbol{p}_1 \mathcal{T}^{(0)}_{\gamma_1\gamma_2;\gamma_3\gamma_4}(p, p_1; p_1, p)\\ \times \mathfrak{G}^{(0)}_{\alpha\gamma_1}(p)\,\mathfrak{G}_{\gamma_3\gamma_2}(p_1)\,\mathfrak{G}_{\gamma_4\beta}(p) + \frac{1}{2}\,\mathfrak{G}^{(0)}_{\alpha\gamma_1}(p)\,\frac{T^2}{(2\pi)^6}\sum_{\omega_1\omega_2}\int d^3\boldsymbol{p}_1 d^3\boldsymbol{p}_2\\ \times \mathcal{T}^{(0)}_{\gamma_1\gamma_2;\gamma_3\gamma_4}(p, p_1+p_2-p; p_1, p_2)\,\mathfrak{G}_{\gamma_3\gamma_5}(p_1)\,\mathfrak{G}_{\gamma_4\gamma_6}(p_2)\\ \times \mathfrak{G}_{\gamma_7\gamma_2}(p_1+p_2-p)\,\mathcal{T}_{\gamma_5\gamma_6;\gamma_7\gamma_8}(p_1, p_2; p_1+p_2-p, p)\,\mathfrak{G}_{\gamma_8\beta}(p).\end{aligned} \tag{16.1}$$

Here, $\mathscr{T}$ is the exact vertex part, with the same meaning as in the $T = 0$ method. It is equal to the sum of all the compact diagrams with four external lines, each line being an exact Green function $\mathfrak{G}$ (examples of such diagrams are shown in Fig. 52).

The Dyson equation (16.1) is represented graphically in the same way as the equation in Chap. II (Fig. 53). A heavy line denotes $\mathfrak{G}$, a light line $\mathfrak{G}^{(0)}$, and a shaded square the vertex part $\mathscr{T}$.

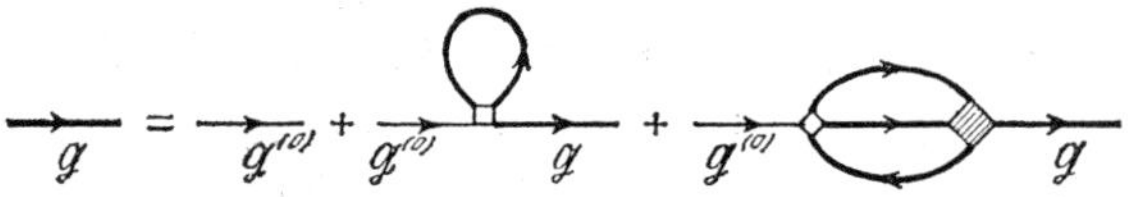

Fig. 53

If we introduce the inverse $\mathfrak{G}_{\alpha\beta}^{-1}$ of the matrix $\mathfrak{G}_{\alpha\beta}$, equation (16.1) can be written as

$$\mathfrak{G}_{\alpha\beta}^{-1}(p) = [i\omega - \varepsilon(\mathbf{p}) + \mu]\delta_{\alpha\beta} + \frac{T}{(2\pi)^3}\sum_{\omega_1}\int d^3\mathbf{p}_1$$

$$\times \mathscr{T}^{(0)}_{\alpha\gamma_1;\gamma_2\beta}(p, p_1; p_1, p)\,\mathfrak{G}_{\gamma_2\gamma_1}(p_1) - \frac{T^2}{2(2\pi)^6}\sum_{\omega_1\omega_2}\int d^3\mathbf{p}_1 d^3\mathbf{p}_2$$

$$\times \mathscr{T}^{(0)}_{\alpha\gamma_1;\gamma_2\gamma_3}(p, p_1 + p_2 - p; p_1, p_2)\,\mathfrak{G}_{\gamma_2\gamma_4}(p_1)\,\mathfrak{G}_{\gamma_3\gamma_5}(p_2)$$

$$\times \mathfrak{G}_{\gamma_6\gamma_1}(p_1 + p_2 - p)\,\mathscr{T}_{\gamma_4\gamma_5;\gamma_6\beta}(p_1, p_2; p_1 + p_2 - p; p). \qquad (16.2)$$

We can write in a similar way a system of equations for $\mathfrak{G}$ and $\mathfrak{D}$ in the case of interactions with phonons (Fig. 54):

$$\mathfrak{G}_{\alpha\beta}^{-1}(\mathbf{p}, \omega_n) = [i\omega_n - \varepsilon(\mathbf{p}) + \mu]\delta_{\alpha\beta}$$

$$+ \frac{gT}{(2\pi)^3}\sum_{\omega_n'}\int d^3\mathbf{p}'\,\mathfrak{G}_{\alpha\beta}(\mathbf{p}', \omega_n')\,\mathfrak{D}(\mathbf{p}' - \mathbf{p}, \omega_n' - \omega_n)\,\mathscr{T}(\mathbf{p}, \mathbf{p}'; \omega_n, \omega_n'),$$

$$\mathfrak{D}^{-1}(\mathbf{k}, \omega_n) = -\omega_0^{-2}(\mathbf{k})\left(\omega_n^2 + \omega_0^2(\mathbf{k})\right) \mp \frac{gT}{(2\pi)^3}\sum_{\omega_n'}\int d^3\mathbf{p}'\,\mathfrak{G}_{\alpha\beta}(\mathbf{p}', \omega_n)$$

$$\times\,\mathfrak{G}_{\beta\alpha}(\mathbf{p}' - \mathbf{k}, \omega_n' - \omega_n)\,\mathscr{T}(\mathbf{p}', \mathbf{p}' - \mathbf{k}; \omega_n', \omega_n' - \omega_n). \qquad (16.3)$$

The function $\mathscr{T}$ in (16.3) is the total vertex part. It is described by the sum of all the compact diagrams with two external particle lines and one external phonon line (Fig. 55).

As in the absolute zero case, the total vertex parts at $T \neq 0$ are connected by definite relations with the many-particle temperature-dependent Green functions. In Matsubara's method these latter are given by (11.1)—(11.4), which have precisely the same form as the corresponding expressions at $T = 0$. If we also observe that the statement of Wick's theorem is the same in both cases, we soon arrive at the conclusion that the many-particle Green functions can be found by using many-tailed diagrams and retaining (in the $\boldsymbol{r}$, τ space) all the rules of association described in § 14.

A formal method is useful for passing to the momentum representation.

Let us take, say, the two-particle Green function

$$\mathfrak{G}^{\mathrm{II}}_{\alpha\beta;\gamma\delta}(1,\,2;\,3,\,4) = \langle T_\tau\{\tilde{\psi}_\alpha(1)\tilde{\psi}_\beta(2)\,\tilde{\bar{\psi}}_\gamma(3)\tilde{\bar{\psi}}_\delta(4)\}\rangle \tag{16.4}$$

(the $\tilde{\psi}$ are the "Heisenberg" operators (11.3)); it depends on four "times" $\tau_1, \tau_2, \tau_3, \tau_4$, each of which varies from 0 to $1/T$. We continue (16.4) in the τ_1 domain from $-1/T$ to 0, on the assumption that $\mathfrak{G}^{\mathrm{II}}(1,\,2;\,3,\,4)$ with $\tau_1 < 0$ is connected with its values for $\tau_1 > 0$ by (11.8); similar continuations are made as regards τ_2, τ_3, τ_4. On now carrying out the Fourier transformation (14.2) with respect to each of the τ, we discover at once that the frequencies corresponding to each "Fermi" variable (Fermi operator in (16.4)) can only be odd: $(2n+1)\pi T$, whilst the frequencies corresponding to a "Bose" variable are only even: $2n\pi T$.

Of course any many-particle Green function can be subjected to a similar continuation procedure.

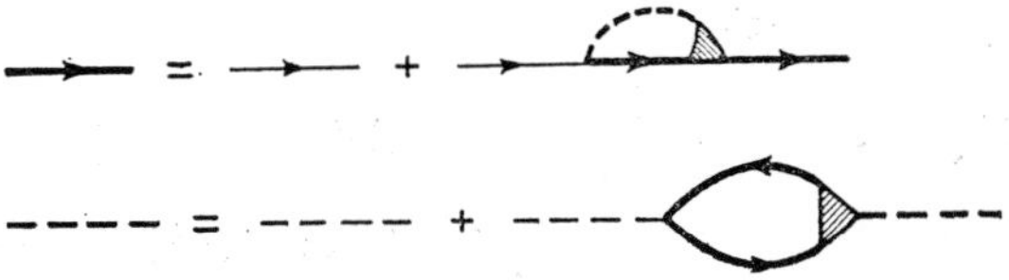

Fig. 54

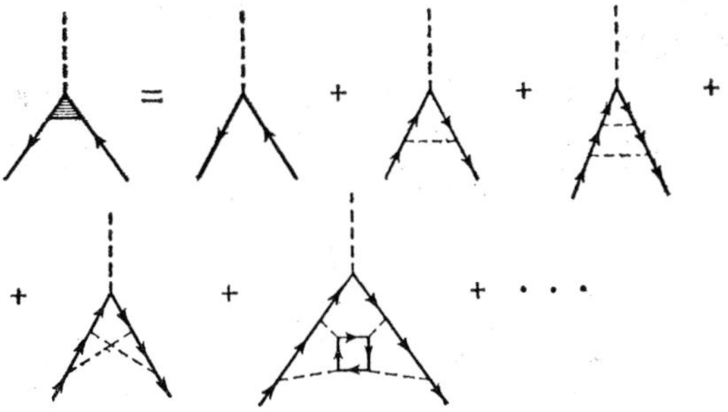

Fig. 55

As in § 14, it can easily be shown that we can perform a Fourier transformation with respect to τ in every term of the perturbation theory series. The resulting rules of association are the same as the rules of § 14 for the single-particle Green function. We can also retain the connection, mentioned at the start of this section, between the diagram techniques at $T = 0$ and $T \neq 0$. The expression for the correction to $\mathfrak{G}^{\mathrm{II}}(1,\,2;\,3,\,4)$ is obtained from the expression for $G^{\mathrm{II}}(1,\,2;\,3,\,4)$, corresponding to the same diagram, with the aid of the above-mentioned substitutions $\omega \to i\omega_n$, $(2\pi)^{-1}\int d\omega \to iT\sum_n$. The existence of this connection enables us to repeat word for word everything said in § 10 about the diagrams for many-particle functions. The perturbation theory series for $\mathfrak{G}^{\mathrm{II}}(1,\,2;\,3,\,4)$ can be reduced to a sum of all the compact diagrams consisting of heavy lines only, corresponding to the exact Green function $\mathfrak{G}(1,\,2)$. These diagrams are the same as those for the total vertex part $\mathscr{T}$, whence it follows that a relation must exist between $\mathfrak{G}^{\text{II}}(1, 2; 3, 4)$ and $\mathcal{T}$. This relation is easily shown to be(†)

$$\mathfrak{G}^{\text{II}}_{\alpha\beta;\gamma\delta}(p_1, p_2; p_3, p_4) = \frac{(2\pi)^3}{T}\left\{\frac{(2\pi)^3}{T}[\mathfrak{G}_{\alpha\delta}(p_1)\mathfrak{G}_{\beta\gamma}(p_2)\delta_{\omega_1\omega_4}\delta(\boldsymbol{p}_1-\boldsymbol{p}_4)\right.$$

$$\mp\mathfrak{G}_{\alpha\gamma}(p_1)\mathfrak{G}_{\beta\delta}(p_2)\delta_{\omega_1\omega_3}\delta(\boldsymbol{p}_1-\boldsymbol{p}_3)]\pm\mathfrak{G}_{\alpha\lambda}(p_1)\mathfrak{G}_{\beta\mu}(p_2)\mathcal{T}_{\lambda\mu;\nu\tau}(p_1, p_2; p_3, p_4)$$

$$\left.\times\mathfrak{G}_{\nu\gamma}(p_3)\mathfrak{G}_{\tau\delta}(p_4)\right\}\delta_{\omega_1+\omega_2-\omega_3-\omega_4}\delta(\boldsymbol{p}_1+\boldsymbol{p}_2-\boldsymbol{p}_3-\boldsymbol{p}_4). \tag{16.5}$$

In the case of interactions with phonons, the vertex part $\mathcal{T}(p_1, p_2)$ is connected with the Fourier components of the Green function

$$\mathfrak{G}_{\alpha\beta}(1, 2; 3) = \langle T_\tau\{\tilde{\psi}_\alpha(1)\,\tilde{\bar{\psi}}_\beta(2)\tilde{\varphi}(3)\}\rangle$$

by the relation

$$\mathfrak{G}_{\alpha\beta}(p_1, p_2; k) = \frac{(2\pi)^3}{T}\mathfrak{G}_{\alpha\gamma}(p_1)\mathfrak{G}_{\gamma\beta}(p_2)\mathfrak{D}(k)$$

$$\times\mathcal{T}(p_1, p_1+k)\delta(\boldsymbol{p}_1-\boldsymbol{p}_2+\boldsymbol{k})\delta_{\omega_1-\omega_2+\omega}. \tag{16.6}$$

As might be expected, equations (16.5) and (16.6) only differ in their numerical coefficients from the corresponding equations (10.17) and (10.21).

Let us emphasise that the method of graphical summation can only be applied to diagrams for $\mathfrak{G}$-functions. It cannot be used for the perturbation theory series for the potential Ω, discussed in § 15, on account

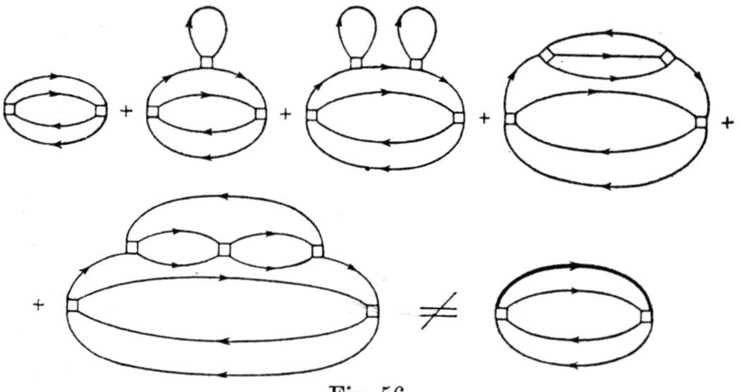

Fig. 56

of the factor $1/n$ appearing in front of the nth order diagram. Diagrams of this type, unlike those for $\mathfrak{G}$, will not break down into separate blocks, and the result of the summation of an infinite sequence of diagrams is not obtained simply by replacing light by heavy lines. In particular, the graphical process illustrated in Fig. 56 is quite impermissible.

(†) The coefficients in (16.5), (16.6) are most easily verified by evaluating both sides of the equation to the first order of perturbation theory.

2. Connection between the Green functions and the thermodynamic potential Ω

We shall conclude by deducing some relations between the thermodynamic potential Ω and the temperature-dependent Green functions. We start with the case of binary interactions between particles. Instead of the potential $\mathscr{T}^{(0)}$, we introduce the potential $\lambda\mathscr{T}^{(0)}$, $0 < \lambda < 1$, and differentiate the expression for Ω:

$$\Omega = \Omega_0 - T \ln \langle \mathfrak{S} \rangle$$

with respect to λ. We get

$$\frac{\partial \Omega}{\partial \lambda} = \frac{T}{4} \int \mathrm{d}^4x_1 \cdots \mathrm{d}^4x_4 \mathscr{T}^{(0)}_{\alpha\beta;\gamma\delta}(x_1, x_2; x_3, x_4)$$
$$\times \langle T_\tau \{\bar{\psi}_\alpha(x_1)\bar{\psi}_\beta(x_2)\psi_\delta(x_4)\psi_\gamma(x_3)\,\mathfrak{S}(\lambda)\}\rangle / \langle \mathfrak{S}(\lambda)\rangle$$

or, from (16.4),

$$\frac{\partial \Omega}{\partial \lambda} = \frac{T}{4} \int \mathrm{d}^4x_1 \cdots \mathrm{d}^4x_4 \mathscr{T}^{(0)}_{\alpha\beta;\gamma\delta}(x_1, x_2; x_3, x_4)\, \mathfrak{G}^{\mathrm{II}}_{\delta\gamma;\alpha\beta}(x_4, x_3; x_1, x_2).$$

On performing a Fourier transformation and using (16.5), we get ($\mathfrak{G}(p, \lambda)$ is the Green function for $\lambda \neq 1$):

$$\frac{\partial \Omega}{\partial \lambda} = \frac{VT}{2\lambda(2\pi)^3} \sum_\omega \int \mathrm{d}^3\boldsymbol{p} \left\{ \frac{T}{(2\pi)^3} \sum_{\omega_1} \int \mathrm{d}^3\boldsymbol{p}\, \lambda \mathscr{T}^{(0)}_{\gamma_1\gamma_2;\gamma_3\gamma_4}(p, p_1; p, p_1)\, \mathfrak{G}_{\gamma_3\gamma_1}(p, \lambda) \right.$$
$$\times \mathfrak{G}_{\gamma_4\gamma_2}(p_1, \lambda) \pm \frac{T^2}{2(2\pi)^6} \sum_{\omega_1,\omega_2} \int \mathrm{d}^3\boldsymbol{p}_1 \mathrm{d}^3\boldsymbol{p}_2 \lambda \mathscr{T}^{(0)}_{\gamma_1\gamma_2;\gamma_3\gamma_4}(p, p_1 + p_2 - p; p_1, p_2)$$
$$\times \mathfrak{G}_{\gamma_3\gamma_6}(p_1, \lambda)\, \mathfrak{G}_{\gamma_4\gamma_5}(p_2, \lambda)\, \mathfrak{G}_{\gamma_8\gamma_2}(p_1 + p_2 - p, \lambda)$$
$$\left. \times \mathscr{T}_{\gamma_5\gamma_6;\gamma_7\gamma_8}(p_2, p_1; p, p_1 + p_2 - p)\, \mathfrak{G}_{\gamma_7\gamma_1}(p, \lambda) \right\}. \quad (16.7)$$

Notice that the expression in curly brackets in (16.7) is the same as the right-hand side of the Dyson equation (16.1) with the interaction potential $\lambda\mathscr{T}^{(0)}$, i.e.

$$\frac{\partial \Omega}{\partial \lambda} = \pm \frac{VT}{2\lambda(2\pi)^3} \sum_\omega \int \mathrm{d}^3\boldsymbol{p}\, \mathfrak{G}^{(0)-1}_{\alpha\beta}(p) \{\mathfrak{G}_{\beta\alpha}(p, \lambda) - \mathfrak{G}^{(0)}_{\beta\alpha}(p)\}. \quad (16.8)$$

On integrating (16.8) over λ and using the condition $\Omega(\lambda = 0) = \Omega_0$, we get the required relation

$$\Omega = \Omega_0 \pm \frac{1}{2} V \int_0^1 \frac{\mathrm{d}\lambda}{\lambda} \frac{T}{(2\pi)^3} \sum_\omega \int \mathrm{d}^3\boldsymbol{p}\, \mathfrak{G}^{(0)-1}_{\alpha\beta}(p) \{\mathfrak{G}_{\beta\alpha}(p, \lambda) - \mathfrak{G}^{(0)}_{\beta\alpha}(p)\}. \quad (16.9)$$

Similar relations exist in the case of interactions with phonons; they express Ω in terms of an integral of the $\mathfrak{G}$-function of the system over the charge g. Proceeding in a way similar to that used for deriving (16.9) we obtain:

$$\Omega = \Omega_0 \pm V \int\limits_0^g \frac{\mathrm{d}g}{g} \frac{T}{(2\pi)^3} \sum_{\omega_n} \int \mathrm{d}^3\boldsymbol{p}[i\omega_n - \varepsilon(\boldsymbol{p}) + \mu] \big(\mathfrak{G}_{\alpha\alpha}(\boldsymbol{p}, \omega_n)$$

$$- \mathfrak{G}^{(0)}_{\alpha\alpha}(\boldsymbol{p}, \omega_n)\big) = \Omega_0 + V \int\limits_0^g \frac{\mathrm{d}g}{g} \frac{T}{(2\pi)^3} \sum_{\omega_n} \int \mathrm{d}^3\boldsymbol{k} \frac{\omega_n^2 + \omega_0^2(\boldsymbol{k})}{\omega_0^2(\boldsymbol{k})}$$

$$\times \big(\mathfrak{D}(\boldsymbol{k}, \omega_n) - \mathfrak{D}^{(0)}(\boldsymbol{k}, \omega_n)\big). \tag{16.10}$$

Another useful formula follows from the familiar relation between the derivative of Ω with respect to the mass m of the particles and the derivative with respect to m of the total Hamiltonian $\hat{H}$ of the system (see Landau and Lifshitz [1]):

$$\left(\frac{\partial \Omega}{\partial m}\right)_{T,V,\mu} = \left\langle \frac{\partial \hat{H}}{\partial m} \right\rangle.$$

Since

$$\frac{\partial \hat{H}}{\partial m} = \frac{1}{2m^2} \int \psi_\alpha^+(\boldsymbol{r}) \nabla^2 \psi_\alpha(\boldsymbol{r}) \mathrm{d}^3\boldsymbol{r},$$

we obtain at once, from the definition (11.1) of the $\mathfrak{G}$-function:

$$\frac{\partial \Omega}{\partial m} = \pm \frac{1}{2m^2} \int [\nabla_r^2 \mathfrak{G}_{\alpha\alpha}(\boldsymbol{r}, \boldsymbol{r}'; -0)]_{\boldsymbol{r}=\boldsymbol{r}'} \mathrm{d}^3\boldsymbol{r},$$

or in Fourier components(†)

$$\frac{\partial \Omega}{\partial m} = \mp \frac{VT}{m(2\pi)^3} \sum_{\omega_n} \int \mathrm{d}^3\boldsymbol{p} \frac{\boldsymbol{p}^2}{2m} \mathfrak{G}_{\alpha\alpha}(\boldsymbol{p}, \omega_n) e^{i\omega_n \tau}, \quad \tau \to +0. \tag{16.11}$$

On using (16.9)—(16.11), together with (11.6), which becomes in Fourier components

$$\frac{\partial \Omega}{\partial \mu} = -N = \mp \frac{VT}{(2\pi)^3} \sum_{\omega_n} \int \mathrm{d}^3\boldsymbol{p}\, \mathfrak{G}_{\alpha\alpha}(\boldsymbol{p}, \omega_n) e^{i\omega_n \tau} \; (\tau \to +0), \tag{16.12}$$

we can work out the derivatives of Ω with respect to the different parameters. These relations express the thermodynamic potential Ω in terms of the Green functions.

We conclude by deriving a further formula expressing the potential Ω in terms of the exact Green function $\mathfrak{G}$. As already remarked, the presence of the factor $1/n$ in the perturbation theory diagrams for Ω makes it impossible to apply the methods of diagrams summation to the series of § 15. In spite of this, there is an interesting possible approach, namely, to carry out a partial summation and to write Ω as the sum of an infinite number of diagrams consisting only of heavy lines, representing the exact $\mathfrak{G}$-functions (Luttinger and Ward [37]).

(†) We shall show in § 17 how the sum over the ω_n is evaluated.

To simplify the treatment, we confine ourselves to the case of binary interactions between the particles and assume a non-ferromagnetic system ($\mathfrak{G}_{\alpha\beta} = \delta_{\alpha\beta}\,\mathfrak{G}$). We select from the perturbation theory diagrams for the potential Ω all the compact diagrams, i.e. those diagrams of which no interior part represents a correction to a $\mathfrak{G}$-function (cf. § 10, sec. 2), and replace all the light lines by heavy lines. For instance, of the three diagrams of Fig. 48 *b* we should only take the first and second. The sum of all the diagrams thus obtained (with the coefficient $1/n$ for each diagram!) will be denoted by Ω'.

By definition, Ω' is a functional of the exact Green function $\mathfrak{G}$. By varying in turn each diagram for Ω' with respect to $\delta\mathfrak{G}(p)$ it is easily seen that the resulting series represents (up to a factor) the series for the correction to the exact Green function $\mathfrak{G}(p)$, if we replace in the latter the product of the Green functions $\mathfrak{G}^{(0)}(p)$ and $\mathfrak{G}(p)$, corresponding respectively to the extreme left-hand and extreme right-hand lines, by $\delta\mathfrak{G}(p)$. To be more precise,

$$\delta\Omega' = \pm\, 2TV \sum_n \int \frac{d^3\boldsymbol{p}}{(2\pi)^3} \Sigma\,(p)\,\delta\mathfrak{G}\,(p)\,, \tag{16.13}$$

where $\Sigma(p)$ is the self-energy part of the exact Green function $\mathfrak{G}(p)$. It is determined in precisely the same way as the self-energy part in the diagram technique at $T = 0$ (see § 10, sec. 1) and is connected with $\mathfrak{G}$ by the relationship

$$\mathfrak{G}^{-1} = \mathfrak{G}^{(0)-1} - \Sigma. \tag{16.14}$$

If Σ is understood as the sum of the corresponding diagrams consisting of heavy lines, (16.14) is precisely the Dyson equation (16.2).

We now form the expression

$$\overline{\Omega} = \Omega_0 \mp 2TV \sum_n \int \frac{d^3\boldsymbol{p}}{(2\pi)^3} \{\ln\,(1 - \mathfrak{G}^{(0)}(p)\,\Sigma(p)) + \Sigma(p)\,\mathfrak{G}\,(p)\} + \Omega'\,. \tag{16.15}$$

We show that $\overline{\Omega} = \Omega$. To do this, we observe that $\overline{\Omega}$, considered as a functional of Σ (or what amounts to the same thing, of $\mathfrak{G}$), possesses the stationary property, namely $\delta\overline{\Omega}/\delta\Sigma = 0$, if $\mathfrak{G}$ satisfies the Dyson equation (16.14). For, on varying (16.15) and taking (16.13) into account, we at once obtain:

$$\begin{aligned}\delta\overline{\Omega} &= \mp\, 2TV \sum_n \int \frac{d^3\boldsymbol{p}}{(2\pi)^3} \left\{-\frac{1}{\mathfrak{G}^{(0)-1} - \Sigma}\,\delta\Sigma + \mathfrak{G}\,\delta\Sigma + \Sigma\,\delta\mathfrak{G}\right\}\\ &\quad \pm 2TV \sum_n \int \frac{d^3\boldsymbol{p}}{(2\pi)^3}\,\Sigma\,\delta\mathfrak{G}\\ &= \pm\, 2TV \sum_n \int \frac{d^3\boldsymbol{p}}{(2\pi)^3} \left\{\mathfrak{G} - \frac{1}{\mathfrak{G}^{(0)-1} - \Sigma}\right\} \delta\Sigma,\end{aligned}$$

whence follows our assertion.

As at the start of this section, we consider now, instead of the interaction potential $\mathcal{T}^{(0)}$, the potential $\lambda\mathcal{T}^{(0)}$, and evaluate $\delta\bar{\Omega}(\lambda)/\partial\lambda$. Here, in view of the stationary property just proved of the functional $\bar{\Omega}$, we can disregard the dependence of $\mathfrak{G}$ and Σ on λ. Hence

$$\frac{\partial\bar{\Omega}(\lambda)}{\partial\lambda}=\frac{\partial\Omega'(\lambda)}{\partial\lambda},$$

where $\mathfrak{G}(\lambda)$ is not to be differentiated when evaluating $\partial\bar{\Omega}(\lambda)/\partial\lambda$.

On the other hand, it follows from the structure of the diagrams that the functional $\Omega'(\lambda)$ is the same as Ω' at $\lambda = 0$, if we substitute $\sqrt{\lambda}\,\mathfrak{G}(\lambda)$ in the latter instead of $\mathfrak{G}(\lambda = 0)$. We obtain from this, on taking (16.13) into account:

$$\frac{\partial\Omega'(\lambda)}{\partial\lambda}=\pm 2TV\sum_n\int\frac{d^3p}{(2\pi)^3}\frac{1}{\sqrt{\lambda}}\Sigma(\lambda)\frac{\partial}{\partial\lambda}\left(\sqrt{\lambda}\,\mathfrak{G}(\lambda)\right),$$

and on putting $\partial\mathfrak{G}/\partial\lambda = 0$ here, we find that

$$\frac{\partial\bar{\Omega}}{\partial\lambda}=\pm\frac{TV}{\lambda}\sum_n\int\frac{d^3p}{(2\pi)^3}\Sigma(\lambda)\,\mathfrak{G}(\lambda).$$

On comparing this expression with (16.8) (and taking the definition of Σ into account), we can conclude that

$$\frac{\partial\bar{\Omega}(\lambda)}{\partial\lambda}=\frac{\partial\Omega(\lambda)}{\partial\lambda}. \tag{16.16}$$

On the other hand, by definition (16.15), $\bar{\Omega}(0) = \Omega_0$, whence, on integrating (16.16) and substituting $\lambda = 1$, it may be seen that $\bar{\Omega} = \Omega$.

Formula (16.15) for Ω in fact holds for arbitrary (not merely binary) particle interactions. It is only necessary to understand by Ω' the functional possessing the property (16.13). The existence of such a functional may be proved by a word-for-word repetition of the arguments given in § 19, sec. 4, for the case $T = 0$ as applied to the case of the temperature-dependent technique. However, the actual construction of such a functional is rather laborious, and we shall not dwell on this.

Let us mention another expression. It can be shown that (see, e.g. the paper by Luttinger and Ward [37] cited above)

$$\Omega_0=\mp 2TV\sum_n\int\frac{d^3p}{(2\pi)^3}\ln\mathfrak{G}^{(0)}e^{i\omega_n\delta},\qquad \delta\to+0.$$

Hence (16.15) can also be rewritten as

$$\Omega=\pm 2TV\sum_n\int\frac{d^3p}{(2\pi)^3}e^{i\omega_n\delta}\{\ln\mathfrak{G}-\Sigma\,\mathfrak{G}\}+\Omega'. \tag{16.17}$$

§ 17. TIME-DEPENDENT GREEN FUNCTIONS G AT FINITE TEMPERATURES. ANALYTIC PROPERTIES OF THE GREEN FUNCTIONS

The time-dependent Green functions G considered in the previous chapter still have an importance at finite temperatures, along with the temperature-dependent Green functions $\mathfrak{G}$. Later examples will show that the functions G determine the transport properties of a system, and in particular, the electrical resistance and complex dielectric constant ε as a function of the field frequency; in addition, they describe non-elastic particle scattering processes in solids.

The single-particle Green function $G(\boldsymbol{r}_1 - \boldsymbol{r}_2, t_1 - t_2)$ has to be defined at non-zero temperatures as

$$G_{\alpha\beta}(\boldsymbol{r}_1 - \boldsymbol{r}_2, t_1 - t_2; E_n, N) = -i\langle E_n, N \mid T_t\{\tilde{\psi}_\alpha(\boldsymbol{r}_1, t_1)\tilde{\psi}^+_\beta(\boldsymbol{r}_2, t_2)\} \mid E_n, N\rangle, \tag{17.1}$$

where $\tilde{\psi}$, $\tilde{\psi}^+$ are the Heisenberg operators of the system. The averaging in (17.1) is over the states of the system with energies E_n and number of particles N. Definition (17.1) includes as a particular case the definition of G at $T = 0$, when the averaging is over the ground state. The Green function (17.1) depends on the total energy E of the system and the number of particles in it. In quantum statistics, it is more convenient to regard all the quantities as functions of the temperature and the chemical potential μ, this being equivalent to passing from a microcanonical to a grand canonical ensemble (see [1]). On averaging (17.1) over the ensemble, we get

$$\begin{aligned} G_{\alpha\beta}(\boldsymbol{r}_1 - \boldsymbol{r}_2, t_1 - t_2; T, \mu) &= \sum_{N,n} e^{(\Omega+\mu N - E_n)/T}\, G_{\alpha\beta}(\boldsymbol{r}_1 - \boldsymbol{r}_2, t_1 - t_2, E_n, N) \\ &= -i\mathrm{Tr}\{e^{(\Omega+\mu\hat{N}-\hat{H})/T}\, T_t[\tilde{\psi}_\alpha(\boldsymbol{r}_1, t_1)\tilde{\psi}^+_\beta(\boldsymbol{r}_2, t_2)]\}. \end{aligned} \tag{17.2}$$

The many-particle Green functions are defined by similar formulae. The phonon Green function is

$$D(1, 2) = -i\,\mathrm{Tr}\{e^{(\Omega-\hat{H})/T}\, T_t[\tilde{\varphi}(1)\tilde{\varphi}(2)]\} \tag{17.3}$$

and the two-particle Green function is

$$G^{\mathrm{II}}(1, 2; 3, 4) = \mathrm{Tr}\{e^{(\Omega+\mu\hat{N}-\hat{H})/T}\, T_t[\tilde{\psi}(1)\tilde{\psi}(2)\tilde{\psi}^+(3)\tilde{\psi}^+(4)]\}. \tag{17.4}$$

The Fourier components of the Green function $G(\omega, \boldsymbol{p})$ satisfy an extremely general relationship (Landau [33)]. To derive this, we note that the time dependence of the matrix elements of the Heisenberg operators $\tilde{\psi}$, $\tilde{\psi}^+$ is given by

$$\begin{aligned} \tilde{\psi}_{nm}(\boldsymbol{r}, t) &= \psi_{nm}(\boldsymbol{r})\exp(i\,\omega_{nm}t), \\ \tilde{\psi}^+_{nm}(\boldsymbol{r}, t) &= \psi^+_{nm}(\boldsymbol{r})\exp(i\,\omega_{nm}t), \\ \omega_{nm} &= E_n - E_m - \mu(N_n - N_m) \end{aligned} \tag{17.5}$$

($N_n = N_m \pm 1$ throughout).

In turn, provided the system is homogeneous and infinite, the coordinate dependence of the matrix elements of the operators $\psi(\boldsymbol{r})$ is(†)

$$\begin{aligned}\psi_{nm}(\boldsymbol{r}) &= \psi_{nm}(0)\exp[-i(\boldsymbol{p}_{nm}\cdot\boldsymbol{r})],\\ \psi^{+}_{nm}(\boldsymbol{r}) &= \psi^{+}_{nm}(0)\exp[-i(\boldsymbol{p}_{nm}\cdot\boldsymbol{r})],\\ \boldsymbol{p}_{nm} &= \boldsymbol{P}_n - \boldsymbol{P}_m,\end{aligned} \tag{17.6}$$

where $\boldsymbol{P}_n$, $\boldsymbol{P}_m$ are the momenta of the system in the states n, m. The $\psi_{nm}(0)$ and $\psi^{+}_{nm}(0)$ in (17.6) are not dependent on the coordinates. On substituting (17.5), (17.6) into (17.1), we get

$$G_{\alpha\beta}(\boldsymbol{r}, t > 0) = -i\sum_{n,m}\exp[(\Omega+\mu N_n - E_n)/T]$$
$$\times\exp[i\omega_{nm}t - i(\boldsymbol{p}_{nm}\cdot\boldsymbol{r})]\big(\psi_\alpha(0)\big)_{nm}\big(\psi^{+}_\beta(0)\big)_{mn},$$
$$G_{\alpha\beta}(\boldsymbol{r}, t < 0) = \pm i\sum_{n,m}\exp[(\Omega+\mu N_m - E_m)/T]$$
$$\times\exp[-i\omega_{mn}t + i(\boldsymbol{p}_{nm}\cdot\boldsymbol{r})]\big(\psi_\alpha(0)\big)_{nm}\big(\psi^{+}_\beta(0)\big)_{mn},$$

We change from the spatial representation of the Green function to its Fourier components:

$$G(\boldsymbol{p}, \omega) = \int\int G(\boldsymbol{r}, t)e^{-i(\boldsymbol{p}\cdot\boldsymbol{r})+i\omega t}\mathrm{d}^3\boldsymbol{r}\,\mathrm{d}t.$$

Integration over space gives δ-functions of $\boldsymbol{p}+\boldsymbol{p}_{nm}$. Integration over t is carried out in separate stages, from $-\infty$ to 0 and from 0 to ∞. We have to use here the familiar formula

$$\int_0^\infty e^{i\alpha x}\mathrm{d}x = \pi\delta(\alpha) + \frac{i}{\alpha}.$$

Integration over t from 0 to ∞ gives $(N_n = N_m - 1)$:

$$(2\pi)^3\sum_{n,m}\exp[(\Omega+\mu N_n - E_n)/T]\,\big(\psi_\alpha(0)\big)_{nm}\big(\psi^{+}_\beta(0)\big)_{mn}$$
$$\times\delta(\boldsymbol{p}+\boldsymbol{p}_{nm})\left[\frac{1}{\omega+\omega_{nm}} - i\pi\delta(\omega+\omega_{nm})\right].$$

The integral over the $t < 0$ region gives

$$\pm(2\pi)^3\sum_{n,m}\exp[(\Omega+\mu N_m - E_m)/T]\,\big(\psi_\alpha(0)\big)_{nm}\big(\psi^{+}_\beta(0)\big)_{mn}$$
$$\times\delta(\boldsymbol{p}-\boldsymbol{p}_{mn})\left[\frac{1}{\omega-\omega_{mn}} + i\pi\delta(\omega-\omega_{mn})\right],$$

whence

$$G_{\alpha\beta}(\boldsymbol{p}, \omega) = -(2\pi)^3\sum_{n,m}\exp[(\Omega+\mu N_n - E_n)/T]\,\big(\psi_\alpha(0)\big)_{nm}\big(\psi^{+}_\beta(0)\big)_{mn}$$
$$\times\delta(\boldsymbol{p}-\boldsymbol{p}_{mn})\left\{\frac{1}{\omega_{mn}-\omega}\,[1\pm e^{-\omega_{mn}/T}] + i\pi\delta(\omega-\omega_{mn})\,[1\mp e^{-\omega_{mn}/T}]\right\}. \tag{17.7}$$

(†) See the remark on p. 54.

The course of the future arguments is linked with the dependence of $G_{\alpha\beta}$ on the spin variables. If the system is not ferromagnetic (and this is the only case we consider), it follows from symmetry considerations that $G_{\alpha\beta}$ must be proportional to the unit tensor $\delta_{\alpha\beta}$:

$$G_{\alpha\beta} = \delta_{\alpha\beta} G, \tag{17.8}$$

$$G(\boldsymbol{p}, \omega) = \frac{1}{2s+1} G_{\alpha\alpha}(\boldsymbol{p}, \omega)$$

$$= -(2\pi)^3 \sum_{n,m} \exp[(\Omega + \mu N_n - E_n)/T] \frac{1}{2s+1} \sum_{\alpha} |(\psi_\alpha(0))_{nm}|^2 \delta(\boldsymbol{p} - \boldsymbol{p}_{mn})$$

$$\times \left\{ \frac{1}{\omega_{mn} - \omega} [1 \pm e^{-\omega_{mn}/T}] + i\pi\delta(\omega - \omega_{mn}) [1 \mp e^{-\omega_{mn}/T}] \right\} \tag{17.9}$$

(s is the spin of the particles).

On comparing the two terms in the curly brackets in (17.9), it will be seen that a definite relationship exists between the real and imaginary parts G', G'' of the Green function (Landau [33]). We have in fact, for Fermi statistics:

$$G'(\boldsymbol{p}, \omega) = \frac{1}{\pi} \int_{-\infty}^{\infty} \coth \frac{x}{2T} \frac{G''(\boldsymbol{p}, x)}{x - \omega}, \tag{17.10}$$

where the principal value of the integral is taken; in the case of Bose statistics,

$$G'(\boldsymbol{p}, \omega) = \frac{1}{\pi} \int_{-\infty}^{\infty} \tanh \frac{x}{2T} \frac{G''(\boldsymbol{p}, x)}{x - \omega}. \tag{17.11}$$

In addition, it follows from (17.9) that G'' is always negative for bosons. On the contrary, the imaginary part of the G-function for a Fermi system changes sign at $\omega = 0$; it is positive for $\omega < 0$ and negative for $\omega > 0$.

It follows from (17.10) and (17.11) that G, as a function of the complex variable ω, is not analytic. However, there are simple relationships between G and two functions, G^R, G^A, which are analytic in the upper and lower half-planes of ω, respectively. The function G^R is written in terms of the real and imaginary parts G' and G'' of G as

$$G^R(\boldsymbol{p}, \omega) = G'(\boldsymbol{p}, \omega) + i \coth \frac{\omega}{2T} G''(\boldsymbol{p}, \omega) \tag{17.12}$$

for fermions, and

$$G^R(\boldsymbol{p}, \omega) = G'(\boldsymbol{p}, \omega) + i \tanh \frac{\omega}{2T} G''(\boldsymbol{p}, \omega) \tag{17.13}$$

for bosons. Similarly,

$$G^A(\boldsymbol{p}, \omega) = G'(\boldsymbol{p}, \omega) - i \coth \frac{\omega}{2T} G''(\boldsymbol{p}, \omega),$$

$$G^A(\boldsymbol{p}, \omega) = G'(\boldsymbol{p}, \omega) - i \tanh \frac{\omega}{2T} G''(\boldsymbol{p}, \omega).$$

The functions G^R, G^A satisfy the dispersion relations:

$$\begin{aligned} \operatorname{Re} G^R(\omega) &= \frac{1}{\pi} \int_{-\infty}^{\infty} \frac{\operatorname{Im} G^R(x)}{x-\omega} \mathrm{d}x, \\ \operatorname{Re} G^A(\omega) &= -\frac{1}{\pi} \int_{-\infty}^{\infty} \frac{\operatorname{Im} G^A(x)}{x-\omega} \mathrm{d}x, \end{aligned} \tag{17.14}$$

whence their analyticity follows by a familiar theorem of the theory of functions of a complex variable.

It is easily shown that G^R is the same as the so-called retarded Green function:

$$G^R(1,2) = \begin{cases} -i \operatorname{Tr}\{e^{(\Omega+\mu\hat{N}-\hat{H})/T} (\tilde{\psi}(1)\tilde{\psi}^+(2) \pm \tilde{\psi}^+(2)\tilde{\psi}(1))\}, & t_1 > t_2, \\ 0, & t_1 < t_2, \end{cases} \tag{17.15}$$

and G^A the same as the advanced function:

$$G^A(1,2) = \begin{cases} 0, & t_1 > t_2, \\ i \operatorname{Tr}\{e^{(\Omega+\mu\hat{N}-\hat{H})/T} (\tilde{\psi}(1)\tilde{\psi}^+(2) \pm \tilde{\psi}^+(2)\tilde{\psi}(1))\}, & t_1 < t_2. \end{cases} \tag{17.16}$$

Indeed, we have, on carrying out precisely the same procedure for (17.15) as we applied to G:

$$\begin{aligned} G^R(\boldsymbol{p}, \omega) = &-(2\pi)^3 \sum_{n,m} e^{(\Omega+\mu N_n - E_n)/T} |\psi_{nm}(0)|^2 (1 \pm e^{-\omega_{mn}/T}) \\ &\times \delta(\boldsymbol{p} - \boldsymbol{p}_{mn}) \left\{ i\pi\delta(\omega-\omega_{mn}) - \frac{1}{\omega-\omega_{mn}} \right\}. \end{aligned} \tag{17.17}$$

The function (17.17) clearly satisfies the relations (17.14).

We can write formula (17.17) for G^R (and the analogous formula for G^A) in a rather different form:

$$\begin{aligned} G^R(\boldsymbol{p}, \omega) &= \int_{-\infty}^{\infty} \frac{\varrho(\boldsymbol{p}, x)}{x-\omega-i\delta} \mathrm{d}x, \\ G^A(\boldsymbol{p}, \omega) &= \int_{-\infty}^{\infty} \frac{\varrho(\boldsymbol{p}, x)}{x-\omega+i\delta} \mathrm{d}x; \quad \delta \to +0, \end{aligned} \tag{17.18}$$

where ϱ is the real function

$$\begin{aligned} \varrho(\boldsymbol{p}, \omega) = &-(2\pi)^3 \sum_{n,m} e^{(\Omega+\mu N_n - E_n)/T} |\psi_{nm}(0)|^2 \\ &\times [1 \pm e^{-\omega_{mn}/T}] \delta(\boldsymbol{p} - \boldsymbol{p}_{mn}) \delta(\omega - \omega_{mn}). \end{aligned} \tag{17.19}$$

Expressions of the type (17.18) were first obtained by Lehmann [28] for the Green functions of quantum electrodynamics. They enable us, in particular, to draw conclusions about the behaviour of G^R, G^A for large ω. We find, in fact, on noting that the integral $\int_{-\infty}^{\infty} \varrho \, \mathrm{d}x$ is finite:

$$G^R \approx G^A \approx -\frac{1}{\omega} \int_{-\infty}^{\infty} \varrho \, \mathrm{d}x.$$

On the other hand, we can find the integral of ϱ by using the definition (17.15) of G^R. Indeed, we have, by the rules for the commutation of Heisenberg operator at $t_1 = t_2$:

$$G^R(\boldsymbol{r}_1, \boldsymbol{r}_2; t_1 = t_2 + 0) = -i\delta(\boldsymbol{r}_1 - \boldsymbol{r}_2), \quad G^R(\boldsymbol{p}; t_1 = t_2 + 0) = -i.$$

We express $G^R(\boldsymbol{p}; t_1 = t_2 + 0)$ in terms of $G^R(\boldsymbol{p}, \omega)$:

$$G^R(\boldsymbol{p}, t_1 = t_2 + 0) = \frac{1}{2\pi}\int_{-\infty}^{\infty} \mathrm{d}\omega G^R(\boldsymbol{p}, \omega) e^{-i\omega\alpha}, \quad \alpha \to +0,$$

and substitute into this the expression (17.18) for G^R:

$$-i = \frac{1}{2\pi}\int_{-\infty}^{\infty} \mathrm{d}x\, \varrho(x, \boldsymbol{p}) \int_{-\infty}^{\infty} \mathrm{d}\omega \frac{e^{-i\omega\alpha}}{x - \omega - i\delta} = i\int_{-\infty}^{\infty} \varrho(x, \boldsymbol{p})\,\mathrm{d}x.$$

Hence,

$$\int_{-\infty}^{\infty} \varrho(x, \boldsymbol{p})\,\mathrm{d}x = -1. \tag{17.20}$$

We thus have, in the case of large ω,

$$G^R \approx G^A \approx \frac{1}{\omega}, \tag{17.21}$$

i.e. G^R, G^A behave like the corresponding functions for non-interacting particles.

The retarded and advanced functions satisfy an infinite system of coupled equations (Bogolyubov and Tyablikov [34]). There is no diagram technique for evaluating them, however, as there is for the temperature-dependent Green functions $\mathfrak{G}$. It is therefore of interest to establish a connection between G^R and $\mathfrak{G}$. This is done by finding an integral form for $\mathfrak{G}$, analogous to (17.18).

We can use definition (11.1) to write ($\tau > 0$):

$$\mathfrak{G}(\boldsymbol{r}, \tau) = -\sum_{n,m} e^{(\Omega + \mu N_n - E_n)/T} e^{\omega_{nm}\tau - i(\boldsymbol{p}_{nm}\cdot\boldsymbol{r})} |\psi_{nm}(0)|^2. \tag{17.22}$$

Changing to Fourier components in (17.22) in accordance with the formula

$$\mathfrak{G}(\boldsymbol{p}, \omega_k) = \int_0^{1/T} \mathrm{d}\tau \int \mathrm{d}^3\boldsymbol{r} \exp[i\omega_k\tau - i(\boldsymbol{p}\cdot\boldsymbol{r})]\,\mathfrak{G}(\boldsymbol{r}, \tau),$$

where $\omega_k = (2k+1)\pi T$ for fermions and $\omega_k = 2k\pi T$ for bosons, we get

$$\mathfrak{G}(\boldsymbol{p}, \omega_k) = -(2\pi)^3 \sum_{n,m} e^{(\Omega + \mu N_n - E_n)/T} |\psi_{mn}(0)|^2 \delta(\boldsymbol{p} - \boldsymbol{p}_{mn}) \frac{1 \pm e^{-\omega_{mn}/T}}{\omega_{mn} - i\omega_k}. \tag{17.23}$$

The function (17.23) can be written as(†)

$$\mathfrak{G}(\boldsymbol{p}, \omega_k) = \int_{-\infty}^{\infty} \frac{\varrho(\boldsymbol{p}, x)}{x - i\omega_n} \tag{17.24}$$

(†) It can be shown that, in the Bose case, with $\omega_n = 0$, the integral (17.24) has a singularity at $x = 0$. It follows from (17.19), however, that $\varrho \sim x$ for small x in this case.

with the same ϱ as in (17.19), whence a relationship is obtained between $\mathfrak{G}$ for $\omega_n > 0$ and $G^R(\omega)$:

$$\mathfrak{G}(\omega_n) = G^R(i\omega_n), \quad \omega_n > 0. \tag{17.25}$$

On the other hand, it follows from (17.24) that

$$\mathfrak{G}(\omega_n) = \mathfrak{G}^*(-\omega_n). \tag{17.26}$$

A knowledge of the function $G^R(\omega)$, analytic in the upper half-plane, thus enables us to form, with the aid of (17.25) and (17.26), the temperature-dependent Green function $\mathfrak{G}$ for all "frequencies" ω_n.

Much more interest attaches, however, to the inverse problem of finding the function G^R, given $\mathfrak{G}$. Suppose that $\mathfrak{G}$ is known for all frequencies ω_n and that we have managed to find a function $F(\omega)$, analytic in the upper half-plane, with the properties:

$$F(i\omega_n) = \mathfrak{G}(\omega_n), \quad \omega_n > 0.$$

We could then fairly soon show, by using a familiar theorem from the theory of functions of a complex variable(†), that $F(\omega)$ is the same as $G^R(\omega)$ everywhere on the upper half-plane.

The problem of finding $G^R(\omega)$ thus amounts to the analytic continuation of $\mathfrak{G}(\omega_n)$ with a discrete set of points throughout the upper half-plane (Abrikosov, Gor'kov, Dzyaloshinskii [31], Fradkin [32]). Although no general solution is possible, the analytic continuation can be performed in certain particular cases.

A special case is $T = 0$. In that case all sums over ω must be replaced by integrals, according to the rule

$$T \sum_{\omega} \to \frac{1}{2\pi} \int_{-\infty}^{\infty} d\omega.$$

In contradistinction to the time-dependent technique for $T = 0$ which was given in the preceding chapter, all integrals are here taken along the imaginary axes in the complex planes of the true frequencies. The external frequencies are also imaginary.

The analytical continuation onto the real axis can be performed as follows. Let us take an arbitrary diagram for one of the many-particle Green functions $\mathfrak{G}^{(n)}$ or the thermodynamic potential Ω and let us split up the expression corresponding to that diagram into parts in which the integrals are taken in such a way that all arguments of $\mathfrak{G}^{(0)}$-lines have a well-defined sign.

(†) The theorem in question says that two analytic functions coincide if they take the same values over an infinite sequence of points having a limit-point in the domain of analyticity. The sequence in our case is formed by the integral points $i\omega_n$, the limit-point being the point at infinity.

We replace each $\mathfrak{G}^{(0)}$ occurring in the diagram by $G_R^{(0)}$ if $\omega > 0$ or by $G_A^{(0)}$ if $\omega < 0$. Let one of the frequencies ω (external or integration variable) be such that when it is replaced by zero none of the frequency arguments of the functions changes sign. Using the notation $G_R^{(0)}$ and $G_A^{(0)}$ enables us in this case to make the transformation $\omega \to -i\omega$, since none of the arguments of $G_R^{(0)}$ and $G_A^{(0)}$ passes through the real axis. If we assume that after this transformation we can find another frequency which possesses similar properties, and so on, we can by successive transformations of this kind transform all frequencies onto the real axis. It is then clear that the arguments of all $G_A^{(0)}$ functions will be negative, while those of all $G_R^{(0)}$ functions will be positive, and those functions can thus be replaced by $G^{(0)}$. We arrive in this way at the time-dependent technique for $T = 0$. This argument is naturally also completely relevant for diagrams which consist of complete $\mathfrak{G}$-lines.

We must also note that if the diagram considered refers to a temperature-dependent Green function $\mathfrak{G}^{(n)}$, the function obtained by the procedure which we just described will just be the time-dependent Green function $G^{(n)}$ and not some other function. One sees this most easily from the example of the single-particle function $\mathfrak{G}(\omega)$. If we have $\omega > 0$, it changes to $G_R(\omega)$, while if $\omega < 0$, it changes to $G_A(\omega)$. Hence, $\mathfrak{G}(\omega)$ changes into $G(\omega)$.

The problem thus reduces to the question whether we can find such a sequence of frequencies that they enable us to realise the successive transformations $\omega \to -i\omega$. In practice it is sufficient for this to show that there is necessarily one such frequency, since after the transformation of that frequency the problem reduces to a completely analogous one, but with a smaller number of variables (the diagram is obtained from the original one by deleting one of the $\mathfrak{G}^{(0)}$-lines).

Let us consider an arbitrary diagram where the frequency ω occurs in the arguments of some of the $\mathfrak{G}^{(0)}$-functions. Let us denote these arguments by $\omega, \omega + \omega_1, \ldots, \omega + \omega_s$. Let $\omega > 0$, and let there be negative frequencies among the ω_i. We shall now assume that $\omega + \omega_0$ has the smallest absolute magnitude of the negative arguments of the $\mathfrak{G}^{(0)}$-lines. It is clear that if all other negative frequencies have a larger absolute magnitude than ω_0, the substitution $\omega \to -i\omega$ would be possible. If this is not the case, we perform a change of variables: $\omega' = \omega + \omega_0$. The arguments then become of the form $\omega' - \omega_0, \omega' + \omega_1 - \omega_0, \ldots, \omega', \ldots, \omega' + \omega_s - \omega_0$. Since $\omega' < 0$, those arguments which were positive remain the same when $\omega' = 0$. Those which were negative remain negative since the corresponding $\omega_i - \omega_0 > 0$. In that case we can thus perform the transformation $\omega' \to -i\omega'$.

The discussion given here shows that the time-dependent technique for $T = 0$ which was given in Chapter II can also be obtained without applying the very artificial method of the adiabatic switching on of the interaction.

If we know the retarded functions $G^R(\omega)$, $\mathfrak{D}^R(\omega)$ or the electromagnetic field Green function $D^R_{\alpha\beta}(\omega)$ (see § 29), a number of transport characteristics of the system can be determined. For instance, having found the pole of $G^R(\omega)$ for an electron in a metal, the time of free flight of the electron can be determined as a function of its energy, whilst the pole of $\mathfrak{D}^R(\omega)$ for a phonon yields the acoustic absorption coefficient. The function $\mathfrak{D}^R_{\alpha\beta}(\omega)$ determines the dielectric permeability $\varepsilon(\omega)$ of the system, and from this the low frequency conductivity of the metal (from the relationship $\varepsilon(\omega) \to 4\pi i\sigma/\omega$ as $\omega \to 0$). The method of analytic continuation in the technique of temperature-dependent Green functions thus enables us to go beyond the framework of the purely statistical problem of evaluating the thermodynamic potential; in essence, concurrently with the evaluation of Ω, we can also find the transport coefficients.

But a whole group of transport coefficients (such as viscosity, heat conductivity, etc.) is connected with the many-particle Green functions. In principle, relationships similar to (17.24) can also be obtained for the latter. However, in view of the fact that these functions depend on a large number of frequencies (for instance, on three for the two-particle functions), the general relationships prove unwieldy. Moreover, in the majority of cases of practical importance Green functions are required in which several of the arguments coincide (we shall encounter examples of this in the chapter on superconductivity). In this case the analytic properties are the same as those for the single-particle functions, and relationships connecting the temperature-dependent and time-dependent quantities may easily be established for them.

Let us take, say, the scattering of slow neutrons in a liquid, which we shall assume for simplicity to consist of spin-zero bosons. It is well known that the interaction of a slow neutron with an atom of a liquid can be described by a point interaction (see e.g. [16]):

$$V(\boldsymbol{r}-\boldsymbol{R}) = 2\pi\frac{m+m_n}{m m_n}a\,\delta(\boldsymbol{r}-\boldsymbol{R}), \tag{17.27}$$

where $\boldsymbol{r}$, $\boldsymbol{R}$ are the radius vectors of the incident neutron and the atom in the liquid, m_n and m are their respective masses, and a is the scattering amplitude. On summing (17.27) over all the atoms of the liquid, we obtain for the energy of interaction of the slow neutron with the liquid:

$$V(\boldsymbol{r}) = 2\pi\frac{m+m_n}{m m_n}a\sum_k\delta(\boldsymbol{r}-\boldsymbol{R}_k). \tag{17.28}$$

In the second quantisation representation for the liquid particles, $V(\boldsymbol{r})$ is

$$V(\boldsymbol{r}) = 2\pi\frac{m+m_n}{m m_n}a\psi^+(\boldsymbol{r})\psi(\boldsymbol{r}),$$

where ψ, ψ^+ are the field operators for the particles of the liquid in the Schrödinger representation.

The transition matrix element for neutron scattering with momentum transfer $\boldsymbol{q}$ is proportional to

$$a \int e^{-i(\boldsymbol{q}\cdot\boldsymbol{r})} \langle i|\psi^+(\boldsymbol{r})\psi(\boldsymbol{r})|f\rangle \mathrm{d}^3\boldsymbol{r}$$

(i is the initial, f the final state of the liquid). We thus have, for the differential scattering section (Δ is the energy transfer):

$$\mathrm{d}\sigma \sim a^2 \int \mathrm{d}^3\boldsymbol{r}_1 \mathrm{d}^3\boldsymbol{r}_2 e^{-i(\boldsymbol{q}\cdot\boldsymbol{r}_1-\boldsymbol{r}_2)} \langle i|\psi^+(\boldsymbol{r}_1)\psi(\boldsymbol{r}_1)|f\rangle \times \langle f|\psi^+(\boldsymbol{r}_2)\psi(\boldsymbol{r}_2)|i\rangle \delta(E_i - E_f + \Delta).$$

We sum this expression over the final state f and then take the ensemble average over the initial state i:

$$\mathrm{d}\sigma \sim a^2 \sum_{i,f} \int \mathrm{d}^3\boldsymbol{r}_1 \mathrm{d}^3\boldsymbol{r}_2 e^{-i(\boldsymbol{q}\cdot\boldsymbol{r}_1-\boldsymbol{r}_2)} e^{(\Omega+\mu N_i-E_i)/T} \langle i|\psi^+(\boldsymbol{r}_1)\psi(\boldsymbol{r}_1)|f\rangle \times \langle f|\psi^+(\boldsymbol{r}_2)\psi(\boldsymbol{r}_2)|i\rangle\, \delta(E_f - E_i - \Delta).$$

On substituting expression (17.6) for the operators $\psi(\boldsymbol{r})$, we finally get

$$\mathrm{d}\sigma \sim a^2 (2\pi)^3 V \sum_{i,f} e^{(\Omega+\mu N_i-E_i)/T} |\langle i|\psi^+(0)\psi(0)|f\rangle|^2 \delta(\boldsymbol{p}_{fi} - \boldsymbol{q})\delta(\omega_{fi} - \Delta), \tag{17.29}$$

where V is the volume of the system, $\omega_{fi} = E_f - E_i$, $\boldsymbol{p}_{fi} = \boldsymbol{P}_f - \boldsymbol{P}_i$.

Apart from a numerical factor, (17.29) is easily shown to be the same as the imaginary part of the Fourier component of the function

$$K(\boldsymbol{r}_1 - \boldsymbol{r}_2, t_1 - t_2) = -i\mathrm{Tr}\{e^{(\Omega+\mu\hat{N}-\hat{H})/T} T_t\{\tilde{\psi}^+(\boldsymbol{r}_1, t_1)\tilde{\psi}(\boldsymbol{r}_1, t_1)\tilde{\psi}^+(\boldsymbol{r}_2, t_2)\tilde{\psi}(\boldsymbol{r}_2, t_2)\}\},$$

and in fact:

$$\mathrm{d}\sigma \sim -Va^2 \frac{\mathrm{Im}\, K(\boldsymbol{q}, \Delta)}{1 + e^{-\Delta/T}}. \tag{17.30}$$

K is the two-particle Green function with pairwise coincident arguments, and its analytic properties are in no way different from those of the single-boson Green function G. If, in analogy with G^R and $\mathfrak{G}$, we introduce functions K^R and $\mathfrak{K}$, we can repeat for them everything that has been said about G, G^R and $\mathfrak{G}$, after replacing the operators $\psi(1)$, $\psi^+(2)$ in all the formulae from (17.1) to (17.21) by $\psi^+(1)\psi(1)$ and $\psi^+(2)\psi(2)$ respectively.

To find $\mathrm{d}\sigma$, therefore, we only need to find the temperature-dependent Green function $\mathfrak{K}$ and construct its analytic continuation in the upper half-plane K^R. We can then find the cross-section from the relations

$$\mathrm{d}\sigma \sim -Va^2 \frac{\mathrm{Im}\, K^R(\boldsymbol{q}, \Delta)}{1 - e^{-\Delta/T}}. \tag{17.30'}$$

Formula (17.30′) is sometimes written in another way:

$$d\sigma = A \frac{S(\boldsymbol{q}, \Delta)}{1 - \exp(-\Delta/T)}$$

where A is a constant, and $S(\boldsymbol{q}, \Delta)$ is the so-called structure function

$$S(\boldsymbol{q}, \Delta) = 2\pi \operatorname{Im} K^R(\boldsymbol{q}, \Delta).$$

It may easily be verified that $S(\boldsymbol{q}, \Delta)$ is the Fourier component of the average of the commutator of the density operators:

$$\langle [\varrho(0), \varrho(x)]_- \rangle = (2\pi)^{-4} \int S(\boldsymbol{q}, \omega) \exp[i(\boldsymbol{q} \cdot \boldsymbol{z}) - i\omega t]\, d^3\boldsymbol{q}\, d\omega .$$

A useful formula has been published for S (Cohen and Feynman [72], Nozières and Pines [73]), which holds both for bosons and for fermions:

$$\frac{1}{2\pi} \int_0^\infty \omega\, d\omega\, S(\boldsymbol{q}, \omega) = \frac{\bar{n} q^2 \hbar}{2m} \tag{17.31}$$

($\bar{n}$ is the density of the number of particles in the system: $\bar{n} = \langle \varrho(x) \rangle$, m is the mass of a free atom).

It may conveniently be derived by using the equation of continuity, which is satisfied by the density and the number density current operators:

$$\frac{\partial \varrho(x)}{\partial t} + \operatorname{div} \boldsymbol{j}(x) = 0, \tag{17.32}$$

where $\varrho(x) = \psi^+(x)\psi(x)$, and where the flux density operator, if the particle interactions in the liquid are independent of the particle velocities, has the form

$$\boldsymbol{j}(x) = -\frac{i\hbar}{2m} [\psi^+ \nabla\psi - (\nabla\psi^+)\psi].$$

Equation (17.32) expresses the law of conservation of matter. On applying it to the commutator $[\varrho(1), \varrho(2)]_-$ at $t_1 = t_2$:

$$\frac{\partial}{\partial t_1} [\varphi(1), \varphi(2)]^-_{t_1 = t_2} = -\operatorname{div} [\boldsymbol{j}(1), \varphi(2)]^-_{t_1 = t_2},$$

and using the commutation rule (6.2) for operators with identical time arguments, we find that

$$\frac{\partial}{\partial t_1} [\varphi(1), \varphi(2)]^-_{t_1 = t_2} = \frac{i\hbar}{m} \operatorname{div}_1 [\varrho(1) \nabla_1 \delta(\boldsymbol{r}_1 - \boldsymbol{r}_2) - \frac{1}{2} \delta(\boldsymbol{r}_1 - \boldsymbol{r}_2) \nabla_1 \varrho(1)].$$

On taking the average of this operator equation and using the spatial uniformity of the system, by virtue of which $\nabla \langle \varrho(1) \rangle = \nabla \bar{n} \equiv 0$, we obtain in Fourier components:

$$\frac{1}{2\pi} \int_{-\infty}^{+\infty} \omega S(\boldsymbol{q}, \omega)\, d\omega = \frac{\bar{n} q^2 \hbar}{m}.$$

But $S(\boldsymbol{q}, -\omega) = -S(\boldsymbol{q}, \omega)$, whence (17.31) follows.

It also follows from (17.29) that

$$\frac{1}{2\pi}\int_{-\infty}^{+\infty}\frac{S(\boldsymbol{q},\omega)\,\mathrm{d}\omega}{1-\exp(-\omega/T)}=\int e^{-i(\boldsymbol{q}\cdot\boldsymbol{r})}\langle\varrho(0)\varrho(\boldsymbol{r})\rangle\,\mathrm{d}\boldsymbol{r}=S(\boldsymbol{q})\bar{n},$$

where $S(\boldsymbol{q})$ is the correlation function which we introduced in § 1 (for $\boldsymbol{q}\neq 0$). This last relationship may be conveniently transformed to

$$\frac{1}{2\pi}\int_{0}^{\infty}S(\boldsymbol{q},\omega)\coth\frac{\omega}{2T}\,\mathrm{d}\omega=S(\boldsymbol{q})\bar{n}.$$

We conclude by showing how the sums over ω_n such as (16.12) are evaluated. We note that, for large ω_n,

$$\mathfrak{G}\approx\frac{1}{i\omega_n} \tag{17.33}$$

(this follows at once from (17.24) and (17.20)), so that

$$T\sum_{\omega_n}\mathfrak{G}(\omega_n)\exp(-i\omega_n\tau) \tag{17.34}$$

is divergent for $\tau=0$. This implies in fact, as is clear from the definition of $\mathfrak{G}(\tau)$, that the latter has a discontinuity at $\tau=0$.

We shall take τ to be arbitrarily small, but finite. The series (17.34) is now convergent. Using (17.26), we rewrite it as

$$2T\sum_{\omega_n\geqslant 0}{}'\cos\omega_n\tau\,\mathrm{Re}\,\mathfrak{G}(\omega_n)+2T\sum_{\omega_n\geqslant 0}{}'\sin\omega_n\tau\,\mathrm{Im}\,\mathfrak{G}(\omega_n)$$

(the prime on the summation sign means that the $\omega_n=0$ term is taken with half its weight). Since, by (17.33), Re $\mathfrak{G}$ tends to zero more rapidly than $1/\omega_n$ as $\omega_n\to\infty$, we can simply put $\tau=0$ in the first sum. To find the second sum, we note that the ω_n for which $\omega_n\tau\sim 1$ play the main role in the sum over ω_n, i.e. large ω_n as $\tau\to 0$. In view of this, we can replace the sum over ω_n by the integral ($T\Sigma\to(1/2\pi)\int$), whilst at the same time we must, of course, use the asymptotic value (17.33) for $\mathfrak{G}(\omega_n)$. We have:

$$\lim_{\tau\to 0}2T\sum_{\omega_n\geqslant 0}{}'\sin\omega_n\tau\,\mathrm{Im}\,\mathfrak{G}(\omega_n)=-\frac{1}{\pi}\int_{0}^{\infty}\frac{\sin\tau x}{x}\,\mathrm{d}x=-\frac{1}{2}\,\mathrm{sign}\,\tau.$$

We thus get the following rule for evaluating the sum (17.34):

$$\lim_{\tau\to 0}T\sum_{\omega_n}\mathfrak{G}(\omega_n)\exp(-i\omega_n\tau)=2T\sum_{\omega_n\geqq 0}{}'\,\mathrm{Re}\,\mathfrak{G}(\omega_n)-\frac{1}{2}\,\mathrm{sign}\,\tau. \tag{17.35}$$

CHAPTER IV

FERMI LIQUID THEORY(†)

§ 18. PROPERTIES OF THE VERTEX PART FOR SMALL MOMENTUM TRANSFER. ZERO SOUND(‡)

WE SHALL start the present chapter by indicating how the methods of quantum field theory can provide us with a basis for a general theory of Fermi liquids. To do this, we consider a system of fermions with arbitrary short-range interaction forces at $T = 0$. We considered the properties of the Green function for this case in § 7. In particular, we established there that an excitation of the "particle" type corresponds to a pole of the function G_R in the lower half-plane close to the real positive semi-axis of the complex variable ε(††), whilst holes correspond to poles of G_A in the upper half-plane close to the semi-axis $\varepsilon < 0$. Since both these functions represent analytic continuations of the G-function from different real semi-axes of the variable ε, we can assert that, in the neighbourhood of the point $\varepsilon = 0$, $|\boldsymbol{p}| = p_0$, G is of the form

$$G(\boldsymbol{p}, \varepsilon) = \frac{a}{\varepsilon - v(|\boldsymbol{p}| - p_0) + i\delta \operatorname{sign}(|\boldsymbol{p}| - p_0)}, \tag{18.1}$$

where a is the coefficient whose meaning was explained in § 7 (see (7.40)); $\delta \to +0$; $v(|\boldsymbol{p}| - p_0)$ is $\varepsilon(\boldsymbol{p}) - \mu$, expanded in the neighbourhood of $|\boldsymbol{p}| = p_0$ (remember that p_0 is given by the equation $\varepsilon(p_0) = \mu$). The coefficient v in the expansion is the velocity of the excitations on the Fermi boundary, equal to p_0/m^*, where m^* is the effective mass of the excitations.

We consider the properties of the vertex part Γ. This function, along with G, plays an essential role in the theory of Fermi liquids. We shall consider the behaviour of the vertex part for p_1 close to p_3, and p_2 close to p_4. We introduce the notation

$$\Gamma(p_1, p_2; p_1 + k, p_2 - k) \equiv \Gamma(p_1, p_2, k), \tag{18.2}$$

(†) It will be convenient to denote the frequency variable in the Green function by ε in this chapter.

(‡) This section is largely based on the results of L. D. LANDAU [35].

(††) This corresponds to $\varepsilon > \mu$ for the Green functions with fixed N, as defined in § 7.

where the energy-momentum transfer $k = (\boldsymbol{k}, \omega)$ is a small 4-vector (i.e. $|\boldsymbol{k}| \ll p_0$, $|\omega| \ll \mu$). The simplest diagrams for such a vertex part are illustrated in Fig. 57. The expressions for these diagrams contain integrals of two Green functions. Whereas there is nothing distinctive about the case $k = 0$ for diagrams (*a*) and (*b*), in case (*c*) the poles of the two Green functions approach one another as $k \to 0$. As we shall see below, this leads to the appearance of singularities in Γ. It should be remarked that, although the diagrams of Fig. 57 formally refer to the

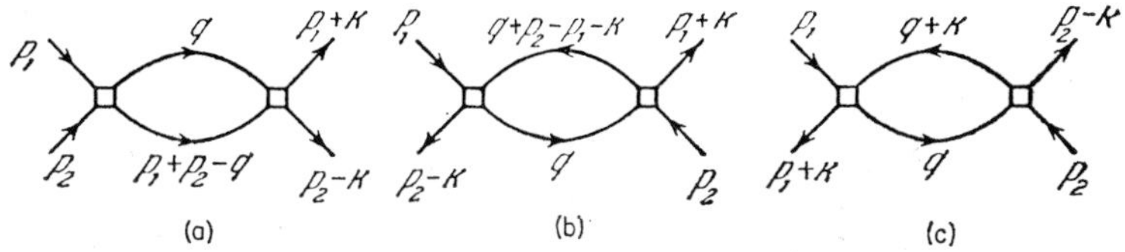

Fig. 57

case of pair forces, the diagram of type (*c*) in fact remains distinctive whatever the interaction forces.

Let us denote by $\Gamma^{(1)}$ the set of all possible diagrams for Γ, not containing "singular elements" (lines $G(p)G(p+k)$). It may readily be seen that the complete Γ is obtained by summation of the "ladders" illustrated in Fig. 58, where the vertices are the functions $\Gamma^{(1)}$ and all the lines are singular. We can express such a summation with the aid of the integral equation

$$\Gamma_{\alpha\beta,\gamma\delta}(p_1, p_2, k) = \Gamma^{(1)}_{\alpha\beta,\gamma\delta}(p_1, p_2) - i \int \Gamma^{(1)}_{\alpha\xi,\gamma\eta}(p_1, q) G(q) \times G(q+k) \Gamma_{\eta\beta,\xi\delta}(q, p_2, k) \frac{d^4 q}{(2\pi)^4}. \quad (18.3)$$

In view of the fact that $\Gamma^{(1)}$ has no singularities when $k = 0$ (short-range forces), we can put $k = 0$ in it.

We now consider the integral in (18.3). It consists of a term arising from the domain remote from the point $\varepsilon = 0$, $|\boldsymbol{p}| = p_0$, and an integral over the neighbourhood of this point, which determines the singularity of the entire expression. If k is small, we can take this neighbourhood to

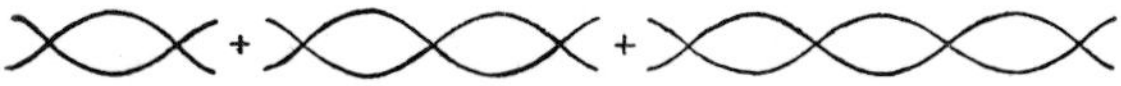

Fig. 58

be extremely small, and only the circuit round the poles of the G-functions is important in the corresponding integral. Since the arguments of the two G are close together, we can assume that all the remaining quantities under the integral vary slowly with q. In this case there will be a contribution from the poles only when these lie on opposite sides of the real axis. For this, we must have $|\boldsymbol{q}| < p_0$, $|\boldsymbol{q} + \boldsymbol{k}| > p_0$, or vice versa. Recalling that k is small, it is easily seen that $|\boldsymbol{q}| \approx p_0$ and $\varepsilon \approx 0$. Thus in the part of the integral over q which is connected with the circuit round the poles, the product $G(q)G(q+k)$ can be replaced by $A\,\delta(\varepsilon)\,\delta(|\boldsymbol{q}|-p_0)$.

The coefficient A can be determined by integration of $G(q)G(q+k)$ over ε and $|\boldsymbol{q}|$; it is equal to

$$\frac{2\pi i a^2}{v} \times \frac{(\boldsymbol{k}\cdot\boldsymbol{v})}{\omega-(\boldsymbol{k}\cdot\boldsymbol{v})},$$

where $\boldsymbol{v}$ is a vector directed along $\boldsymbol{q}$ and equal to v in absolute value. The product $G(q)G(q+k)$ can therefore be written as

$$G(q)G(q+k) = \frac{2\pi i a^2}{v} \times \frac{(\boldsymbol{k}\cdot\boldsymbol{v})}{\omega-(\boldsymbol{k}\cdot\boldsymbol{v})}\,\delta(\varepsilon)\,\delta(|\boldsymbol{q}|-p_0)+\varphi(q), \quad (18.4)$$

where $\varphi(q)$ represents the regular part of $G(q)G(q+k)$, which is only important in the integral over the remote domain (so that we put $k = 0$ in it).

The limit of (18.4) as $\boldsymbol{k}$, $\omega \to 0$ depends essentially on the relation between ω and $\boldsymbol{k}$. The same applies to Γ in the limit as $\omega \to 0$, $\boldsymbol{k} \to 0$.

We first consider the limit of Γ, Γ^{ω}, as $\omega \to 0$, $|\boldsymbol{k}|/\omega \to 0$.

We obtain in this limit using (18.3) and (18.4):

$$\Gamma^{\omega}_{\alpha\beta,\gamma\delta}(p_1, p_2) = \Gamma^{(1)}_{\alpha\beta,\gamma\delta}(p_1, p_2) - i\int \Gamma^{(1)}_{\alpha\xi,\gamma\eta}(p_1, q)\varphi(q)\,\Gamma^{\omega}_{\eta\beta,\xi\delta}(q, p_2)\frac{d^4q}{(2\pi)^4}. \quad (18.5)$$

We can eliminate $\Gamma^{(1)}$ from the two equations (18.3) and (18.5). To do this, we write the equations in the operator form (the product is understood to be an integral):

$$\begin{aligned}\Gamma^{\omega} &= \Gamma^{(1)} - i\Gamma^{(1)}\varphi\Gamma^{\omega},\\ \Gamma &= \Gamma^{(1)} - i\Gamma^{(1)}(i\Phi+\varphi)\Gamma.\end{aligned} \quad (18.6)$$

Here, $i\Phi$ denotes the first term in (18.4). We get from the first equation:

$$\Gamma^{\omega} = (1+i\Gamma^{(1)}\varphi)^{-1}\Gamma^{(1)}.$$

We take the term containing φ over to the left-hand side in the second of equations (18.6) and operate on it with $(1+i\Gamma^{(1)}\varphi)^{-1}$; this gives us

$$\Gamma = \Gamma^{\omega} + \Gamma^{\omega}\Phi\Gamma.$$

We find on writing this explicitly:

$$\Gamma_{\alpha\beta,\gamma\delta}(p_1 p_2, k) = \Gamma^{\omega}_{\alpha\beta,\gamma\delta}(p_1, p_2) + \frac{a^2 p_0^2}{(2\pi)^3 v}\int \Gamma^{\omega}_{\alpha\xi,\gamma\eta}(p_1, q)\,\Gamma_{\eta\beta,\xi\delta}(q, p_2)\frac{(\boldsymbol{v}\cdot\boldsymbol{k})}{\omega-(\boldsymbol{v}\cdot\boldsymbol{k})}\,d\Omega. \quad (18.7)$$

We now take the other limit, namely $|\boldsymbol{k}| \to 0$, $\omega/|\boldsymbol{k}| \to 0$. Denoting this by Γ^k, we find that, by virtue of (18.7), it is connected with Γ^ω as follows

$$\Gamma^k_{\alpha\beta,\gamma\delta}(p_1, p_2) = \Gamma^\omega_{\alpha\beta,\gamma\delta}(p_1, p_2) - \frac{p_0^2 a^2}{v(2\pi)^3} \int \Gamma^\omega_{\alpha\xi,\gamma\eta}(p_1, q)\, \Gamma^k_{\eta\beta,\xi\delta}(q, p_2)\, \mathrm{d}\Omega. \tag{18.8}$$

Let us investigate the poles of the function $\Gamma(p_1, p_2, k)$ for small $\boldsymbol{k}$ and ω. In view of the fact that $\Gamma(p_1, p_2; k) \gg \Gamma^\omega(p_1, p_2)$ in the neighbourhood of the pole, we can neglect the term Γ^ω on the right-hand side of (18.7). It may be observed in addition that the variable p_2, as also the indices β and δ, play the role of parameters in the equation. We can therefore write Γ near the pole as the product of two functions $\chi_{\alpha\gamma}(p_1; k)$ $\chi'_{\beta\delta}(p_2; k)$, the latter of which cancels on the two sides of (18.7). We introduce the notation

$$\nu_{\alpha\gamma}(\boldsymbol{n}) = \frac{(\boldsymbol{n} \cdot \boldsymbol{k})}{\omega - v(\boldsymbol{n} \cdot \boldsymbol{k})} \chi_{\alpha\gamma}(p_1; k),$$

where $\boldsymbol{n}$ is the unit vector in the direction of $\boldsymbol{p}_1$. We obtain for $\nu_{\alpha\gamma}$ the equation:

$$\big(\omega - v(\boldsymbol{n} \cdot \boldsymbol{k})\big)\, \nu_{\alpha\gamma}(\boldsymbol{n}) = (\boldsymbol{n} \cdot \boldsymbol{k}) \frac{p_0^2 a^2}{(2\pi)^3} \int \Gamma^\omega_{\alpha\xi,\gamma\eta}(\boldsymbol{n}, \boldsymbol{l})\, \nu_{\eta\xi}(\boldsymbol{l})\, \mathrm{d}\Omega. \tag{18.9}$$

Equation (18.9) has the same form as the equation for zero sound and spin waves (see § 2, equation (2.24)). We shall show in the next section that this result is quite justified, since the poles of Γ determine the spectrum of the Bose excitations of a Fermi liquid.

The quantity $a^2\Gamma^\omega$ in (18.9) plays the role of the function f introduced in the theory of Fermi liquids (§ 2). This quantity, has in itself no direct physical meaning; but it is connected with $a^2\Gamma^k$ by means of (18.8), and we shall shortly prove that $a^2\Gamma^k$ can be interpreted, up to a constant factor, as the scattering amplitude of two quasi-particles with $|\boldsymbol{p}_1| = |\boldsymbol{p}_2| = p_0$ at zero angle.

Let us consider the auxiliary problem of the scattering of two particles in vacuo. Let the wave function of this system at $t = -\infty$ be $a^+_{p_3\gamma} a^+_{p_4\delta} \Phi_0$, where Φ_0 is the vacuum wave function. At $t = \infty$ the system has gone over into the states $S(\infty) a^+_{p_3\gamma} a^+_{p_4\delta} \Phi_0$. The scattering amplitude for a transition of the particles to states $\boldsymbol{p}_1\alpha$, $\boldsymbol{p}_2\beta$ is proportional to

$$-i\langle a_{p_1\alpha} a_{p_2\beta} S(\infty)\, a^+_{p_3\gamma} a^+_{p_4\delta} \rangle_0, \tag{18.10}$$

where $\langle \cdots \rangle_0$ denotes the vacuum average. The operator $S(\infty)$ is defined by (8.8), i.e. when expanded in powers of H_{int}, it represents the sum of the integrals of the T-products of the operators ψ.

According to (8.10), each of these T-products can be written as a set of N-products. It is obvious that the only important terms in our matrix element are those which contain N-products of four ψ-operators with all possible contractions. Since the contractions are simply numbers, they do not take part in the averaging. There are two particles in the present case, so that

$$N\left(\psi^+(x_3)\psi^+(x_4)\psi(x_1)\psi(x_2)\right) = \psi^+(x_3)\psi^+(x_4)\psi(x_1)\psi(x_2).$$

On averaging the matrix element (18.10) over the vacuum, this N-product gives a factor

$$\exp\{i(\mathbf{r}_1\cdot\mathbf{p}_1) + (\mathbf{r}_2\cdot\mathbf{p}_2) - (\mathbf{r}_3\cdot\mathbf{p}_3) - (\mathbf{r}_4\cdot\mathbf{p}_4) \\ - i\left(\varepsilon_0(\mathbf{p}_1)t_1 + \varepsilon_0(\mathbf{p}_2)t_2 - \varepsilon_0(\mathbf{p}_3)t_3 - \varepsilon_0(\mathbf{p}_4)t_4\right)\}.$$

In (18.10) this factor is integrated over the coordinates along with an expression that only contains the contractions. We obtain as a result the set of all diagrams with four vertices, the energy and momentum of every end-point being connected by the relationship $\varepsilon = \varepsilon_0(\mathbf{p})$.

This quantity corresponds to the last term of the two-particle Green function (10.17) introduced earlier, without the final $G^{(0)}$-functions. In our present problem of the scattering of two particles in a vacuum, these $G^{(0)}$-functions are the same as the complete G-functions. For, by (7.3),

$$G^{(0)}(\mathbf{r}, t) = 0 \text{ for } t < 0.$$

At the same time, a diagram for a correction to G always contains at least one pair of lines in opposite directions (†) (i.e. one $G^{(0)}$ for $t > 0$ and one $G^{(0)}$ for $t < 0$), as a result of which any correction to $G^{(0)}$ is equal to zero.

For the same reason, of the diagrams for the vertex part only those remain in which all the $G^{(0)}$ lines are in the same direction (to the left or the right), i.e. only diagrams of the type of Fig. 57 *a*.

It follows from all this that the scattering amplitude (18.10) is equal to (‡) $\Gamma'_{\alpha\beta,\gamma\delta}(p_1p_2, p_3p_4)\big|_{\varepsilon_i=\varepsilon_0(p_i)}$, where Γ' is the vertex part for our problem.

Formula (9.17), expressing the connection between the two-particle Green function and the vertex part, takes the form here:

$$G^{II'}_{\alpha\beta\gamma\delta}(p_1p_2; p_3p_4) = G^{(0)}(p_1)G^{(0)}(p_2)\,(2\pi)^4\,[\delta(p_1-p_3)\,\delta_{\alpha\gamma}\,\delta_{\beta\delta} \\ - \delta(p_1-p_4)\,\delta_{\beta\gamma}\,\delta_{\alpha\delta}] + iG^{(0)}(p_1)\,G^{(0)}(p_2)\,G^{(0)}(p_3)\,G^{(0)}(p_4)\,\Gamma'_{\alpha\beta,\gamma\delta}(p_1p_2; p_3p_4). \tag{18.11}$$

The momenta in this formula are connected by the conservation laws. We can regard $G^{II'}$ as the Green function of two particles (whence its

(†) Diagrams of the type Fig. 4 *a* are exceptions, but, as already remarked in § 8, $G^{(0)}(0)$ must be regarded as $\lim_{t\to+0} G^{(0)}(-t)$ in such diagrams, and this limit is here equal to zero.

(‡) A constant factor is omitted in all the formulae; comparison with perturbation theory (see Chap. I) shows that it is equal to $4\pi/m$.

name derives). The first term in (18.11) corresponds to independent motion of the particles, and the second to their scattering by each other.

Let us return to Fermi liquids, and compare formulae (10.17) and (18.11). By (18.1), in the domain of small ε and $|\mathbf{p}|$ close to p_0, the Green functions are very similar in form to the functions for free particles. In order to be able to regard G^{II} as the Green function of two interacting quasi-particles, it has to be divided by a^2. The free particle term will have in this case precisely the same normalisation as for real particles with energies $\varepsilon(\mathbf{p})$.

The second term in (10.17) corresponds to the scattering of quasi-particles. On comparing it with the expression for real particles, we can conclude that

$$a^2 \Gamma_{\alpha\beta,\gamma\delta}(p_1 p_2; p_3 p_4)\,|_{\varepsilon_i = \varepsilon(\mathbf{p}_i) - \mu} \tag{18.12}$$

plays the role of the scattering amplitude for the quasi-particles.

In particular, if $|\mathbf{p}_1| = |\mathbf{p}_2| = p_0$, $\varepsilon_i = 0$ for all i. The scattering amplitude for small momentum transfer is now equal to $a^2\Gamma(p_1 p_2, k)$ with $\omega = 0$, whilst the scattering amplitude at zero angle is $a^2\Gamma^k$.

Equation (18.8), connecting $a^2\Gamma^k$ with the function $f = a^2\Gamma^\omega$, can be solved if we assume that the spin-dependent particle interaction has a purely exchange origin. We can write $a^2\Gamma^k$ in this case as

$$\frac{p_0^2}{\pi^2 v} a^2 \Gamma^k = A(\mathbf{n}_1, \mathbf{n}_2) + B(\mathbf{n}_1, \mathbf{n}_2)\,(\boldsymbol{\sigma}_1 \cdot \boldsymbol{\sigma}_2), \tag{18.13}$$

whilst $(p_0^2/\pi^2 v)f = E$ in accordance with (2.28). The equations for Φ and Z (see (2.28)) separate:

$$\begin{aligned} A(\mathbf{n}_1, \mathbf{n}_2) &= \Phi(\mathbf{n}_1, \mathbf{n}_2) - \int \Phi(\mathbf{n}_1, \mathbf{n}')\,A(\mathbf{n}', \mathbf{n}_2)\,\frac{\mathrm{d}\Omega}{4\pi}, \\ B(\mathbf{n}_1, \mathbf{n}_2) &= Z(\mathbf{n}_1, \mathbf{n}_2) - \int Z(\mathbf{n}_1, \mathbf{n}')\,B(\mathbf{n}', \mathbf{n}_2)\,\frac{\mathrm{d}\Omega}{4\pi}. \end{aligned} \tag{18.14}$$

In an isotropic liquid, all the quantities on the Fermi surface depend only on $\cos(\widehat{\mathbf{n}_1 \mathbf{n}_2}) = \cos\chi$. We expand them in Legendre polynomials, say $A(\chi) = \Sigma A_l P_l(\cos\chi)$, then at once obtain the following relationships between the coefficients of the expansions:

$$A_l = \frac{\Phi_l}{1 + \Phi_l/(2l+1)}, \qquad B_l = \frac{Z_l}{1 + Z_l/(2l+1)}. \tag{18.15}$$

§ 19. EFFECTIVE MASS. CONNECTION BETWEEN FERMI MOMENTUM AND NUMBER OF PARTICLES(†). BOSE BRANCHES OF THE SPECTRUM. SPECIFIC HEAT

1. *Auxiliary relations*

We shall first of all obtain some useful relationships for the G-functions. Suppose that our system is in an infinitely small field $\delta U(t)$, uniform in space and varying weakly in time. The corresponding interaction Hamiltonian has the form $H_{int} = \int \psi_\alpha^+(\mathbf{r})\, \delta U(t) \psi_\alpha(\mathbf{r})\, \mathrm{d}^3\mathbf{r}$. If we pass to the interaction representation in respect of H_{int}, expansion of the G-function in powers of δU gives us, up to first order terms:

$$\delta G_{\alpha\beta}(x, x') = -\int \mathrm{d}^4 y\, \delta U(t_y) \{\langle T(\tilde{\psi}_\alpha(x) \tilde{\psi}_\gamma^+(y) \tilde{\psi}_\gamma(y) \tilde{\psi}_\beta^+(x'))\rangle - \langle T(\tilde{\psi}_\alpha(x) \tilde{\psi}_\beta^+(x'))\rangle \langle \tilde{\psi}_\gamma^+(y) \tilde{\psi}_\gamma(y)\rangle\}.$$

Here $\tilde{\psi}$ are the Heisenberg operators of the interacting particles in the absence of the field δU. On using (10.17), we get

$$\delta G_{\alpha\beta}(x, x') = \delta_{\alpha\beta} \int \mathrm{d}^4 y\, \delta U(t_y) G(x-y) G(y-x') - i \int \mathrm{d}^4 y\, \mathrm{d}^4 x_1 \cdots \mathrm{d}^4 x_4\, \delta U(t_y) G(x-x_1) G(y-x_2) \times G(x_3 - x') G(x_4 - y) \Gamma_{\alpha\gamma;\beta\gamma}(x_1, x_2; x_3, x_4).$$

Fourier transforming we have:

$$\delta G_{\alpha\beta} = \delta_{\alpha\beta} G(p) \delta U(\omega) G(p+k_1) - i G(p) G(p+k_1) \times \Gamma_{\alpha\gamma,\beta\gamma}(p, q; k_1) G(q) \delta U(\omega) G(q+k_1) \frac{\mathrm{d}^4 q}{(2\pi)^4},$$

where $k_1 = (0, \omega)$.

In view of the fact that the field δU does not affect the spins, $\delta G_{\alpha\beta}$ must be proportional to $\delta_{\alpha\beta}$. On taking one-half of the trace we get

$$\delta G = G(p) \delta U G(p+k_1) - i G(p) G(p+k_1) \times \frac{1}{2} \int \Gamma_{\alpha\beta,\alpha\beta}(p, q; k_1) G(q) \delta U(\omega) G(q+k_1) \frac{\mathrm{d}^4 q}{(2\pi)^4}.$$

On the other hand, if we add to the Hamiltonian the term

$$\delta U(t) \int \psi_\alpha^+(\mathbf{r}) \psi_\alpha(\mathbf{r})\, \mathrm{d}^3\mathbf{r} = \delta U(t) \hat{N},$$

in the limit $\delta U \to \text{const}$ the function G is simply multiplied by $\exp[-i\delta U(t-t')]$, which corresponds to the addition of $-\delta U$ to ε. Hence $\delta G/\delta U \to -\delta G/\delta\varepsilon$ in the limit as $\omega \to 0$. We therefore find that

$$\frac{\partial G}{\partial \varepsilon} = -\{G^2(p)\}_\omega \left[1 - \frac{i}{2} \int \Gamma^\omega_{\alpha\beta,\alpha\beta}(p, q)\, \{G^2(q)\}_\omega \frac{\mathrm{d}^4 q}{(2\pi)^4}\right],$$

(†) These results were obtained by L. D. Landau and L. P. Pitaevskii and were published in part in reference [36].

where $\{G^2(p)\}_\omega = \varphi$ (see (18.4)) denotes the limit of $G(p)G(p+k_1)$ as $\omega \to 0$. Let us consider this relationship close to the pole of $G(p)$. In this case we can write $G(p)$ in form (18.1). On dividing by $-\{G^2(p)\}_\omega$, we get the first relation

$$\frac{\partial G^{-1}(p)}{\partial \varepsilon} = \frac{1}{a} = 1 - \frac{i}{2} \int \Gamma^{\omega}_{\alpha\beta,\alpha\beta}(p, q) \{G^2(q)\}_\omega \frac{d^4 q}{(2\pi)^4}. \tag{19.1}$$

The second relation is obtained as follows. Suppose the particles have infinitely small charge δe and the system is located in a magnetic field which is weakly non-uniform in space and constant in time. The term $-(\delta e/c)\boldsymbol{A}$ is added in this case to the momentum operator in the Hamiltonian. If the charge δe is extremely small, the variaton of the Hamiltonian is given by the term $(-\delta e/mc) \int \psi^+_\alpha(\boldsymbol{r})[\hat{\boldsymbol{p}} \cdot \boldsymbol{A}(\boldsymbol{r})]\psi_\alpha(\boldsymbol{r})\, d^3\boldsymbol{r}$, where $\hat{\boldsymbol{p}}$ is the momentum operator. As before we now get for the variation of the Green function:

$$\delta G = -G(p)\frac{\delta e}{mc}(\boldsymbol{p} \cdot \boldsymbol{A})G(p+k_2) + \frac{i}{2} G(p)G(p+k_2)$$
$$\times \int \Gamma_{\alpha\beta,\alpha\beta}(p, q; k_2)G(q)\frac{\delta e}{mc}(\boldsymbol{q} \cdot \boldsymbol{A})G(q+k_2)\frac{dq^4}{(2\pi)^4},$$

where $k_2 = (\boldsymbol{k}, 0)$ (we assume $\boldsymbol{k}$ small here). On the other hand, it follows from gauge invariance that, in the limit as $\boldsymbol{k} \to 0$, all the momentum-dependent functions must transform to functions of $\boldsymbol{p} - \delta e\boldsymbol{A}/c$. Hence $\delta G/(\delta e\, \boldsymbol{A}/c) = -\partial G/\partial p$ as $\boldsymbol{k} \to 0$.

Consequently, in the limit as $\delta e \to 0$, $\boldsymbol{k} \to 0$, we get our second relation for $G(p)$ close to the pole:

$$\frac{\partial G^{-1}}{\partial \boldsymbol{p}} = -\frac{\boldsymbol{v}}{a} = -\frac{\boldsymbol{p}}{m^* a} = -\frac{\boldsymbol{p}}{m} + \frac{i}{2}\int \Gamma^k_{\alpha\beta,\alpha\beta}(p, q)\frac{\boldsymbol{q}}{m}\{G^2(q)\}_k \frac{d^4 q}{(2\pi)^4}. \tag{19.2}$$

We get a third relation by considering the variation of the G-function when the system moves as a whole with a small, slowly varying velocity $\delta\boldsymbol{u}(t)$.

The variation of the Hamiltonian of the system amounts to the addition of the term

$$-(\delta\boldsymbol{u} \cdot \hat{\boldsymbol{P}}) = -\left(\delta\boldsymbol{u} \cdot \int \psi^+_\alpha(\boldsymbol{r})\hat{\boldsymbol{p}}\psi_\alpha(\boldsymbol{r})\, d^3\boldsymbol{r}\right),$$

where $\hat{\boldsymbol{P}}$ is the operator of the total momentum of the system. The variation of the G-function is given by

$$\delta G = -G(p)\big(\boldsymbol{p} \cdot \delta\boldsymbol{u}(\omega)\big)G(p+k_1)$$
$$+\frac{i}{2}G(p)G(p+k_1)\int \Gamma_{\alpha\beta,\alpha\beta}(p, q; k_1)\big(\boldsymbol{q} \cdot \delta\boldsymbol{u}(\omega)\big)G(q)G(q+k_1)\frac{d^4 q}{(2\pi)^4},$$

where $k_1 = (0, \omega)$. On the other hand, when $\omega = 0$, this transformation implies the passage to a new coordinate system moving with constant velocity $\delta \boldsymbol{u}$. The energy of the system must in this case change by an amount $-(\delta \boldsymbol{u} \cdot \boldsymbol{P})$. The frequency ε is here replaced by $\varepsilon + (\delta \boldsymbol{u} \cdot \boldsymbol{p})$, and consequently the Green function varies by an amount $(\partial G/\partial \varepsilon)\,(\boldsymbol{p} \cdot \delta \boldsymbol{u})$.

We therefore obtain close to the pole, as $\omega \to 0$, $\delta \boldsymbol{u} \to 0$,

$$\boldsymbol{p}\frac{\partial G^{-1}}{\partial \varepsilon} = \frac{\boldsymbol{p}}{a} = \boldsymbol{p} - \frac{i}{2}\int \Gamma^{\omega}_{\alpha\beta,\alpha\beta}(p, q)\,\boldsymbol{q}\,\{G^2(q)\}_{\omega}\frac{\mathrm{d}^4 q}{(2\pi)^4}. \tag{19.3}$$

Finally, we get our last relation by considering the variation of the G-function under the influence of a small field $\delta U(\boldsymbol{r})$, constant in time and slightly non-uniform in space. The variation of G is in this case

$$\delta G = G(p)\,\delta U(\boldsymbol{k})\,G(p + k_2)$$
$$- \frac{i}{2} G(p) G(p + k_2) \int \Gamma_{\alpha\beta,\alpha\beta}(p, q;\, k_2) G(q)\,\delta U(\boldsymbol{k}) G(q + k_2)\frac{\mathrm{d}^4 q}{(2\pi)^4},$$

where $k_2 = (\boldsymbol{k}, 0)$.

On the other hand, the equilibrium condition

$$\mu + \delta U = \mathrm{const}$$

must be fulfilled in a constant external field. In the limit, as $\boldsymbol{k} \to 0$, the chemical potential changes by a small constant $-\delta U$.

We therefore get, as $\boldsymbol{k} \to 0$, $\delta U \to 0$,

$$\frac{\partial G^{-1}}{\partial \mu} = 1 - \frac{i}{2}\int \Gamma^{k}_{\alpha\beta,\alpha\beta}(p, q)\,\{G^2(q)\}_k \frac{\mathrm{d}^4 q}{(2\pi)^4}. \tag{19.4}$$

This formula holds for all momenta.

2. *Proof of the fundamental relations of the Fermi liquid theory*

The fundamental relation of the Fermi liquid theory can be derived using (19.1)–(19.4), together with (18.8), connecting Γ^k with Γ^{ω}. It is worth noticing that the last formula is suitable for Γ for arbitrary momenta p_1, p_2, regardless of whether or not they lie close to the Fermi surface.

We start by substituting (18.8) in (19.2); this gives us

$$-\frac{\boldsymbol{p}}{m^* a} + \frac{\boldsymbol{p}}{m} = \frac{i}{2}\int \Gamma^{\omega}_{\alpha\beta,\alpha\beta}(p, q)\,\frac{\boldsymbol{q}}{m}\{G^2(q)\}_k \frac{\mathrm{d}^4 q}{(2\pi)^4}$$
$$-\frac{1}{2}\,\frac{p_0^2 a^2}{(2\pi)^3 v}\int \Gamma^{\omega}_{\alpha\beta,\alpha\beta}(p, q)\left(-\frac{\boldsymbol{q}}{m^* a} + \frac{\boldsymbol{q}}{m}\right)\mathrm{d}\Omega.$$

It follows from (18.4) that

$$\{G^2(p)\}_k = \{G^2(p)\}_{\omega} - \frac{2\pi i a^2}{v}\,\delta(\varepsilon)\,\delta(|\boldsymbol{p}| - p_0). \tag{19.5}$$

We substitute this in the first integral of the previous equation and use (19.3). We obtain after some cancellations when $|\boldsymbol{p}| = p_0$, $\varepsilon = 0$,

$$\frac{1}{m} = \frac{1}{m^*} + \frac{p_0}{2(2\pi)^3}\int a^2 \Gamma^{\omega}_{\alpha\beta,\alpha\beta}(\chi)\cos\chi\,\mathrm{d}\Omega. \tag{19.6}$$

It is easily seen that this is the same as (2.12), where $a^2\Gamma^{\omega}_{\alpha\beta,\alpha\beta} = \mathrm{Tr}_{\sigma\sigma'} f(\chi, \boldsymbol{\sigma}; \boldsymbol{\sigma}')$.

Let us prove (2.1). We consider expression (18.1) for G close to the pole, i. e. when $|\boldsymbol{p}| \to p_0$, $\varepsilon \to 0$. The coefficients a, v and momentum p_0 in this expression depend on the chemical potential μ. By differentiating G with respect to μ it may easily be seen that the terms originating from differentiation of a and v with respect to μ are small close to the pole ($\sim(|\boldsymbol{p}| - p_0)/\mu$ or ε/μ) compared to the term produced by differentiation of p_0. We thus get

$$\frac{\partial G}{\partial \mu} \approx -G^2 \frac{v}{a}\frac{\mathrm{d}p_0}{\mathrm{d}\mu}.$$

Hence,

$$v\frac{\mathrm{d}p_0}{\mathrm{d}\mu} = a\left(\frac{\partial G^{-1}}{\partial \mu}\right)_{\substack{\varepsilon=0\\ |\boldsymbol{p}|=p_0}}.$$

We substitute (19.4) into this and write Γ^k in accordance with (18.8). This gives

$$\frac{v}{a}\frac{\mathrm{d}p_0}{\mathrm{d}\mu} = 1 - \frac{i}{2}\int \Gamma^{\omega}_{\alpha\beta,\alpha\beta}(p, q)\,\{G^2(q)\}_k \frac{\mathrm{d}^4 q}{(2\pi)^4}$$
$$-\frac{1}{2}\frac{p_0^2 a^2}{(2\pi)^3 v}\int \Gamma^{\omega}_{\alpha\beta,\alpha\beta}(p, q)\left(\frac{v}{a}\frac{\mathrm{d}p_0}{\mathrm{d}\mu} - 1\right)\mathrm{d}\Omega.$$

On substituting $\{G^2\}_k$ in this in accordance with (19.5) and using (19.1), we find after some cancellations that

$$v\frac{\mathrm{d}p_0}{\mathrm{d}\mu} = \left(1 + \frac{p_0^2}{2(2\pi)^3 v}\int a^2\Gamma^{\omega}_{\alpha\beta,\alpha\beta}(p, q)\,\mathrm{d}\Omega\right)^{-1}. \tag{19.7}$$

The total number of particles in the system is defined by (7.37). We differentiate this relation with respect to μ, and get(†)

$$\frac{\mathrm{d}(N/V)}{\mathrm{d}\mu} = -2i\int \frac{\partial G(p)}{\partial \mu}\frac{\mathrm{d}^4 p}{(2\pi)^4} = 2i\int \frac{\partial G^{-1}(p)}{\partial \mu}\{G^2(p)\}_k \frac{\mathrm{d}^4 p}{(2\pi)^4}.$$

We substitute (19.4) in this and write Γ^k in terms of Γ^{ω} using (18.8). This gives us

$$\frac{\mathrm{d}(N/V)}{\mathrm{d}\mu} = 2i\int \{G^2(p)\}_k \frac{\mathrm{d}^4 p}{(2\pi)^4}$$
$$+\int \{G^2(p)\}_k \Gamma^{\omega}_{\alpha\beta,\alpha\beta}(p, q)\,\{G^2(q)\}_k \frac{\mathrm{d}^4 q}{(2\pi)^4}\frac{\mathrm{d}^4 p}{(2\pi)^4}$$
$$-\frac{ip_0^2 a^2}{(2\pi)^3 v}\int \{G^2(p)\}_k \Gamma^{\omega}_{\alpha\beta,\alpha\beta}(p, q)\left(\frac{v}{a}\frac{\mathrm{d}p_0}{\mathrm{d}\mu} - 1\right)\frac{\mathrm{d}^4 p}{(2\pi)^4}\mathrm{d}\Omega.$$

(†) For brevity, the factor $e^{i\varepsilon t}$ $(t \to +0)$ is omitted here under the integral.

We substitute (19.5) in this and use (19.1), obtaining as a result:

$$\frac{\mathrm{d}(N/V)}{\mathrm{d}\mu} = 2i \int \{G^2(p)\}_\omega \frac{\mathrm{d}^4 p}{(2\pi)^4} + \int \{G^2(p)\}_\omega \Gamma^\omega_{\alpha\beta,\alpha\beta}(p, q)$$

$$\times \{G^2(q)\}_\omega \frac{\mathrm{d}^4 q}{(2\pi)^4} \frac{\mathrm{d}^4 p}{(2\pi)^4} + \frac{8\pi a p_0^2}{(2\pi)^3 v} - \frac{8\pi p_0^2 (a-1)}{(2\pi)^3 v} v \frac{\mathrm{d}p_0}{\mathrm{d}\mu}$$

$$-8\pi \left[\frac{p_0^2 a^2}{(2\pi)^3 v}\right]^2 \frac{1}{2} \int \Gamma^\omega_{\alpha\beta,\alpha\beta}(p, q)\, \mathrm{d}\Omega \frac{v}{a} \frac{\mathrm{d}p_0}{\mathrm{d}\mu}.$$

By (19.1), the first two terms on the right-hand side of this relation are none other than

$$-2i \int \frac{\partial G}{\partial \varepsilon} \frac{\mathrm{d}^4 p}{(2\pi)^4} = -2i \int [G(\varepsilon = \infty) - G(\varepsilon = -\infty)] \frac{\mathrm{d}^3 \boldsymbol{p}}{(2\pi)^4},$$

and this last expression vanishes, since, by (7.11), G vanishes when $\varepsilon \to \pm\infty$. This is also obvious from the fact that the expression is equal to the variation of the number of particles in the system when we shift the zero of the energy scale.

In the last of the remaining terms, we express $\int a^2 \Gamma^\omega \mathrm{d}\Omega$ in accordance with (19.7); this gives us

$$\frac{\mathrm{d}(N/V)}{\mathrm{d}\mu} = \frac{8\pi p_0^2}{(2\pi)^3} \frac{\mathrm{d}p_0}{\mathrm{d}\mu}. \tag{19.8}$$

Integrating over μ we finally get

$$\frac{N}{V} = \frac{8\pi}{3} \frac{p_0^3}{(2\pi)^3}.$$

Formula (19.7) enables us to verify expression (2.19) for the sound velocity. All we need to do is observe that, by virtue of (19.8), (19.7) is the same as (2.18).

3. *Bose branches of spectrum*

We now consider the subject of acoustic excitations, and for this purpose carry out an analysis similiar to formulae (7.32)—(7.33). We consider the time variation of the state of a system which is described at the instant $t = t'$ by the function

$$\Psi_0(t') = \sum_{\boldsymbol{p},\alpha} \psi_{\boldsymbol{p}\alpha}(t') \psi^+_{\boldsymbol{p}+\boldsymbol{k}\alpha}(t') \Phi_i(t'). \tag{19.9}$$

Here $|\boldsymbol{k}| \ll p_0$; $\psi_{\boldsymbol{p}\alpha}$ are the operators of particles with momentum $\boldsymbol{p}$ in the interaction representation. On performing the operations in the same sequence as in (7.32)—(7.33), we get the probability amplitude

$$\langle \Psi_0^*(t) \Psi(t) \rangle = \frac{1}{V^2} \sum_{\boldsymbol{p}_1 \boldsymbol{p}_2 \alpha\beta} \langle \tilde{\psi}_{\boldsymbol{p}_2\beta}(t) \tilde{\psi}^+_{\boldsymbol{p}_2 - \boldsymbol{k}\beta}(t) \psi_{\boldsymbol{p}_1\alpha}(t') \tilde{\psi}^+_{\boldsymbol{p}_1 + \boldsymbol{k}\alpha}(t') \rangle$$

$$= -\int G^{\mathrm{II}}_{\alpha\beta,\alpha\beta}(\boldsymbol{p}_1, t', \boldsymbol{p}_2, t; \boldsymbol{p}_1 + \boldsymbol{k}, t', \boldsymbol{p}_2 - \boldsymbol{k}, t) \frac{\mathrm{d}^3 \boldsymbol{p}_1}{(2\pi)^3} \frac{\mathrm{d}^3 \boldsymbol{p}_2}{(2\pi)^3}$$

$$= -\int G^{\mathrm{II}}_{\alpha\beta,\alpha\beta}(p_1, p_2; p_1 + k, p_2 - k) \frac{\mathrm{d}^4 p_1}{(2\pi)^4} \frac{\mathrm{d}^4 p_2}{(2\pi)^4} e^{-i\omega(t-t')} \frac{\mathrm{d}\omega}{2\pi},$$

$$k = (\boldsymbol{k}, \omega).$$

We substitute (10.17) for G^{II} into this formula, and obtain

$$\langle \Psi_0^*(t)\Psi(t)\rangle = \int \Big[2\int G(p)G(p+k)\frac{\mathrm{d}^4p}{(2\pi)^4} - i\int G(p_1)G(p_1+k)\Gamma_{\alpha\beta,\alpha\beta}(p_1,p_2;k) \times G(p_2)G(p_2-k)\frac{\mathrm{d}^4p_1}{(2\pi)^4}\frac{\mathrm{d}^4p_2}{(2\pi)^4}\Big] e^{-i\omega(t-t')}\frac{\mathrm{d}\omega}{2\pi}. \quad (19.10)$$

The expression in square brackets (which we shall denote symbolically by $ia\Pi$) can be transformed with the aid of (18.4), (18.7) and (19.1). The aim of these transformations is to eliminate the terms containing integrations far from $\varepsilon = 0$, $|\boldsymbol{p}| = p_0$. The method of doing this follows the same pattern as earlier in this section.

The vertex part Γ is expressed with the aid of (18.7) in terms of Γ^ω, whilst an integral such as $\int \Gamma^\omega \varphi$ is replaced in accordance with (19.1).

The net result is that the function Π is given by

$$\Pi = \frac{2p_0^2}{(2\pi)^3 v}\int \frac{(\boldsymbol{k}\cdot\boldsymbol{v})}{\omega-(\boldsymbol{k}\cdot\boldsymbol{v})}\mathrm{d}\Omega + \left(\frac{p_0^2}{(2\pi)^3 v}\right)^2 \int \frac{(\boldsymbol{k}\cdot\boldsymbol{v}_1)}{\omega-(\boldsymbol{k}\cdot\boldsymbol{v}_1)} \times a^2\Gamma_{\alpha\beta,\alpha\beta}(p_1,p_2;k)\frac{(\boldsymbol{k}\cdot\boldsymbol{v}_2)}{\omega-(\boldsymbol{k}\cdot\boldsymbol{v}_2)}\mathrm{d}\Omega_1\mathrm{d}\Omega_2.$$

It may easily be verified directly that the function Π can be written in the form $(p_2^0/v(2\pi)^3)\int \Pi_{1\alpha\alpha}\mathrm{d}\Omega$, where $\Pi_{1\alpha\gamma}$ satisfies the equation

$$\big(\omega-(\boldsymbol{k}\cdot\boldsymbol{v})\big)\Pi_{1\alpha\gamma}(\omega,\boldsymbol{k},\boldsymbol{n}) - \frac{p_0^2(\boldsymbol{k}\cdot\boldsymbol{v})}{(2\pi)^3 v}\frac{1}{2}\int a^2\Gamma^\omega_{\gamma\eta,\alpha\xi}(\boldsymbol{n},\boldsymbol{n}')\Pi_{1\eta\xi}(\omega,\boldsymbol{k},\boldsymbol{n}')\mathrm{d}\Omega' = (\boldsymbol{k}\cdot\boldsymbol{v})\,\delta_{\alpha\gamma}. \quad (19.11)$$

It follows from (19.10) that, as in § 7, the value of $\langle\Psi_0^*(t)\Psi(t)\rangle$ as $t\to\infty$ is determined by the poles of the function Π of the complex variable ω in the lower half-plane. The equation for the poles is obtained from (19.11) by neglecting the right-hand side. On comparing the equation obtained with (18.9), we find that the function Π_1 corresponds to $\boldsymbol{\nu}$. Since our interest is in $\mathrm{Tr}_\sigma\int\Pi_1\mathrm{d}\Omega$ rather than in Π_1, the only solution of equation (18.9) that we consider is the one which is isotropic in the plane perpendicular to the vector $\boldsymbol{k}$. By choosing other functions $\Psi_0(t)$, we can similarly obtain equations for all the components (with respect to both angles and spins) of the $\nu_{\alpha\gamma}$.

It has thus been shown that there can exist in Fermi liquids excitations whose spectrum is determined by the poles of the Fourier transform of the function Γ, i.e. by equation (18.9). These excitations obey Bose statistics, since the corresponding operators are bilinear with respect to the Fermi operators (see (19.9)).

We showed in § 2 that the different branches of the zero sound spectrum represent excitations of this type. In other words, we have determined the physical meaning of the poles of the function Γ in the domain of small energy and momentum transfer and have proved the identity of equations (18.9) and (2.24).

It follows from (19.9) that the acoustic excitations can be regarded as bound pairs of quasi-particles and holes with neighbouring values of momentum.

4. *Another derivation of the connection between the Fermi momentum p_0 and the number of particles* (†)

We shall now describe another derivation of (2.1).

We write formula (7.37) for the density as follows:

$$\frac{N}{V} = 2i \int \left[\frac{\partial}{\partial \varepsilon} \ln G(p) - G(p) \frac{\partial}{\partial \varepsilon} \Sigma(p) \right] e^{i\varepsilon t} \frac{d^4 p}{(2\pi)^4}, \tag{19.12}$$

where $t \to +0$, and Σ is the total irreducible self-energy part ($G = (\varepsilon - \xi - \Sigma)^{-1}$). The integral of the second term in the square brackets may be shown to vanish. To do this, we first of all prove that $\Sigma(p)$ can be regarded as the variational derivative with respect to $G(p)$ of a functional in G, i.e.

$$\delta X = \int \Sigma(p)\, \delta G(p) \frac{d^4 p}{(2\pi)^4}. \tag{19.13}$$

Let us find the variational derivative $\delta\Sigma(p)/\delta G(q)$; to do this we consider the diagrams composing Σ, and successively vary all the G-lines in each diagram. Let us take, for instance, the self-energy part in Fig. 10*c*. If we separate in turn each of the three G-lines in it and denote the corresponding frequency and momentum each time by ε_q, q, it may easily be seen that the result will be equal to the product of this G-function with the sum of two diagrams, which represent simply the second approximation to the function $\Gamma^{(1)}(p, q)$ introduced in § 18, i.e. that part of the function $\Gamma(p, q)$, which does not contain "singular elements" with two identical G-functions. On applying this procedure to all the diagrams forming Σ, we get

$$\delta\Sigma(p) = -\frac{i}{2} \int \Gamma^{(1)}_{\alpha\beta,\alpha\beta}(p, q)\, \delta G(q) \frac{d^4 q}{(2\pi)^4}.$$

Hence it follows that

$$\frac{\delta\Sigma(p)}{\delta G(q)} = -\frac{i}{2} \Gamma^{(1)}_{\alpha\beta,\alpha\beta}(p, q). \tag{19.14}$$

This quantity is symmetrical with respect to the exchange $p \rightleftarrows q$, which, as is well known, is a sufficient condition for the existence of the functional X.

(†) The derivation in this section is mainly due to LUTTINGER and WARD [37].

The functional X can be represented as a set of diagrams that include only complete G-lines. It is clear from (19.13) that these diagrams are obtained from the diagrams for Σ (the skeleton diagrams with complete G-lines) if we "close" each diagram by a complete G-line. In order for (19.13) to be obtained with the correct normalization, we have to introduce a numerical coefficient, dependent on the type of diagram (for instance, if there is only one type of interaction, the coefficient is equal to $1/n$, where n is the number of vertices).

The diagrams composing X do not change if all the frequencies in the G-lines are shifted by a small amount $\delta\varepsilon^{(0)}$, since the limits of integration over the frequency are $(-\infty, +\infty)$, whilst the δ-functions at the vertices contain the same number of ε_i with the "plus" sign as with the "minus" sign. We obtain from this:

$$\frac{\delta X}{\delta\varepsilon^{(0)}} = \int \Sigma(p)\frac{\partial G(p)}{\partial\varepsilon}\frac{d^4p}{(2\pi)^4} = 0.$$

Let us return to (19.12). The integral of the second term in the square brackets can be written as

$$-2i\int G(p)\frac{\partial}{\partial\varepsilon}\Sigma(p)e^{i\varepsilon t}\frac{d^4p}{(2\pi)^4}$$
$$= -2i\int G(p)\Sigma(p)\Big|_{-\infty}^{\infty}\frac{d^3\boldsymbol{p}}{(2\pi)^4} + 2i\int\Sigma(p)\frac{\partial G(p)}{\partial\varepsilon}\frac{d^4p}{(2\pi)^4}.$$

The second term here is equal to zero. As regards the first term, by (7.21), $G(\boldsymbol{p}, \varepsilon) \approx G^{(0)}(\boldsymbol{p}, \varepsilon) \approx 1/\varepsilon$ as $\varepsilon \to \infty$, i.e. $\Sigma(p)$ cannot increase proportionally to ε and $G(p)\Sigma(p) \to 0$ as $\varepsilon \to \infty$. We have thus proved that the integral of the second term in (19.12) vanishes. Hence

$$\frac{N}{V} = 2i\int\frac{\partial}{\partial\varepsilon}\ln G(p)e^{i\varepsilon t}\frac{d^4p}{(2\pi)^4}. \qquad (19.15)$$

As already mentioned in § 7, although the G-function is not analytic, the function $G_R(\varepsilon)$, equal to $G(\varepsilon)$ for $\varepsilon > 0$ and $G^*(\varepsilon)$ for $\varepsilon < 0$, is analytic in the upper half-plane. It can also be shown that G_R has no zeros in the upper half-plane(†). Hence it follows that

$$\frac{N}{V} = 2i\int_{-\infty}^{\infty}\frac{\partial}{\partial\varepsilon}\ln G_R(p)e^{i\varepsilon t}\frac{d^4p}{(2\pi)^4} + 2i\int_{-\infty}^{0}\frac{d\varepsilon}{2\pi}\int\frac{d^3\boldsymbol{p}}{(2\pi)^3}\frac{\partial}{\partial\varepsilon}\ln\frac{G(p)}{G^*(p)}$$
$$= 2i\int_{-\infty}^{0}\frac{d\varepsilon}{2\pi}\int\frac{d^3\boldsymbol{p}}{(2\pi)^3}\frac{\partial}{\partial\varepsilon}\ln\frac{G(p)}{G^*(p)} = \frac{2i}{2\pi}\int\frac{d^3\boldsymbol{p}}{(2\pi)^3}\ln\frac{G(p)}{G^*(p)}\Big|_{-\infty}^{0}$$

(in the integral containing G_R, the contour can be shifted into the region $\operatorname{Im}\varepsilon = \infty$, after which the integral vanishes).

(†) The proof is the same as for the absence of zeros of the complex dielectric constant $\varepsilon(\omega)$ in the upper half-plane of the variable ω (see [47], § 62).

If we denote the phase of the G-function by φ, we have

$$\frac{N}{V}=-\frac{2}{\pi}\int\frac{\mathrm{d}^3\boldsymbol{p}}{(2\pi)^3}\,[\varphi(0)-\varphi(-\infty)].$$

We consider the variation of the phase φ on passing from $\varepsilon=0$ to $\varepsilon=-\infty$. We know from § 7 that $\operatorname{Im} G>0$ for $\varepsilon<0$, whilst $\operatorname{Im} G=0$ at $\varepsilon=0$. As $\varepsilon\to-\infty$, $\operatorname{Im} G$ decreases more rapidly than $\operatorname{Re} G$, and $\operatorname{Re} G\approx 1/\varepsilon<0$. Since $\operatorname{Im} G$ has a definite sign, the point in the complex G plane, corresponding to $\operatorname{Im} G$ and $\operatorname{Re} G$ for a given ε, only moves in the upper half-plane, i.e. the phase can only vary from 0 to π. Since $\operatorname{Im} G/\operatorname{Re} G\to-0$ when $\varepsilon\to-\infty$, we have $\varphi(-\infty)=\pi$. The value of the phase at $\varepsilon=0$ is determined by the sign of $\operatorname{Re} G(0,\boldsymbol{p})\equiv G(0,\boldsymbol{p})$. If $G(0,\boldsymbol{p})>0$, then $\varphi(0)=0$. If $G(0,\boldsymbol{p})<0$, then $\varphi(0)=\pi$.

We therefore obtain from (19.15):

$$\frac{N}{V}=2\int\limits_{G(0,\boldsymbol{p})>0}\frac{\mathrm{d}^3\boldsymbol{p}}{(2\pi)^3}\,. \tag{19.16}$$

The region $G(0,\boldsymbol{p})>0$ is bounded by a surface on which G either vanishes or becomes infinite. Vanishing of $G(0,\boldsymbol{p})$ ($\Sigma\to\infty$) would appear to correspond to superconductivity (see § 34). As regards $G(0,\boldsymbol{p})$ becoming infinite, this is the case in an ordinary Fermi liquid and occurs on the Fermi surface.

In the neighbourhood of the Fermi surface (in the present case $|\boldsymbol{p}|=p_0$), $G(0,\boldsymbol{p})=-a/\xi$, where $a>0$, i.e. the $G(0,\boldsymbol{p})>0$ region corresponds to $\xi<0$ (the interior of the Fermi sphere). On integrating in (19.16), we get (2.1).

5. *Specific heat*

We have so far considered Fermi liquids at $T=0$. The properties of Fermi liquids at non-zero temperatures are also of interest. It seems reasonable to expect that, in the case of low temperatures, all the functions will be determined by the values of the fundamental characteristics of the Fermi liquid at $T=0$. This will be demonstrated by working out the specific heat. The method to be employed for obtaining the temperature correction may prove valuable for other calculations.

We shall start from expression (16.12) for the total number of particles in the system as a function of μ and T:

$$\frac{N(\mu,T)}{V}=2T\sum_{\varepsilon}\int\limits_{\tau\to+0}\frac{\mathrm{d}^3\boldsymbol{p}}{(2\pi)^3}\,\mathfrak{G}(\varepsilon,\boldsymbol{p})\,e^{i\varepsilon\tau}. \tag{19.17}$$

A knowledge of this function enables us to determine the entropy from the thermodynamic relation

$$\left(\frac{\partial N}{\partial T}\right)_{\mu}=\left(\frac{\partial S}{\partial\mu}\right)_{T}. \tag{19.18}$$

Let us consider the $\mathfrak{G}$-function appearing in the expression for N. If we recall that this is the complete $\mathfrak{G}$-function, obtained as a result of adding all the diagrams containing sums over the frequencies, it becomes clear that $\mathfrak{G}(\varepsilon, \boldsymbol{p})$ depends on the temperature not only through the discrete variable $\varepsilon = \pi T(2n+1)$. We shall therefore denote it here by $\mathfrak{G}(T; \varepsilon, \boldsymbol{p})$. As we know from Chap. III, $\mathfrak{G}$ is connected with the self-energy part Σ by the relation

$$\mathfrak{G}^{-1}(T; \varepsilon, \boldsymbol{p}) = \mathfrak{G}^{(0)-1}(\varepsilon, \boldsymbol{p}) - \Sigma(T; \varepsilon, \boldsymbol{p}). \qquad (19.19)$$

Notice that $\mathfrak{G}^{(0)}(\varepsilon, \boldsymbol{p})$ depends only on the temperature through ε. Now let $T \to 0$, but at the same time $\varepsilon = \text{const}$. Equation (19.19) now becomes

$$\mathfrak{G}^{-1}(0; \varepsilon, \boldsymbol{p}) = \mathfrak{G}^{(0)-1}(\varepsilon, \boldsymbol{p}) - \Sigma(0; \varepsilon, \boldsymbol{p}). \qquad (19.20)$$

We find from (19.19) and (19.20) that

$$\mathfrak{G}^{-1}(T; \varepsilon, \boldsymbol{p}) = \mathfrak{G}^{-1}(0; \varepsilon, \boldsymbol{p}) - [\Sigma(T; \varepsilon, \boldsymbol{p}) - \Sigma(0; \varepsilon, \boldsymbol{p})]. \qquad (19.21)$$

The quantity $\Sigma(0; \varepsilon, \boldsymbol{p})$ differs from $\Sigma(T; \varepsilon, \boldsymbol{p})$ in that all the sums over frequencies in it are replaced by integrals, in accordance with the formula

$$T \sum_{\varepsilon} \to \frac{1}{2\pi} \int_{-\infty}^{\infty} d\varepsilon .$$

At low temperatures, the difference $\Sigma(t) - \Sigma(0)$ can be found as follows. We consider two similar diagrams for $\Sigma(T)$ and $\Sigma(0)$, say $\Sigma_1(T)$ and $\Sigma_1(0)$. We shall regard these diagrams as composed of complete $\mathfrak{G}$-lines. If we were to replace all the sums in $\Sigma_1(T)$ by integrals, this diagram would still differ from $\Sigma_1(0)$ by virtue of the difference of one of its components $\mathfrak{G}(T)$ from $\mathfrak{G}(0)$. To a first approximation it is sufficient to take into account the difference of one of the $\mathfrak{G}(T)$ from $\mathfrak{G}(0)$, and to regard the rest as equal to $\mathfrak{G}(0)$. Another contribution to the difference $\Sigma_1(T) - \Sigma_1(0)$ is due to the presence of sums instead of integrals. If the corresponding frequency is small, this difference is of the order T. To a first approximation these frequency regions will be significant, in which one of the $\mathfrak{G}$-lines has a frequency of the order T, whilst all the rest have much higher frequencies. In view of this we can split off the $\mathfrak{G}$-line with low frequency, and in all the remaining places replace the sums by integrals and all the $\mathfrak{G}(T)$ by $\mathfrak{G}(0)$.

We therefore need to vary the diagram for Σ_1 over all the $\mathfrak{G}$-lines appearing in it. The working here is precisely the same as in the previous section. On applying this procedure to all the diagrams composing Σ, we obtain

$$\Sigma(T, \varepsilon, \boldsymbol{p}) - \Sigma(0, \varepsilon, \boldsymbol{p})$$
$$= \frac{1}{2} \int \frac{d^3 q}{(2\pi)^3} \frac{d\varepsilon_1}{2\pi} \mathcal{T}^{(1)}_{\alpha\beta,\alpha\beta}(0; \varepsilon, \boldsymbol{p}; \varepsilon_1, \boldsymbol{q}) \, [\mathfrak{G}(T, \varepsilon_1, \boldsymbol{q}) - \mathfrak{G}(0, \varepsilon_1, \boldsymbol{q})]$$
$$+ \frac{1}{2} \left[T \sum_{\varepsilon_1} - \frac{1}{2\pi} \int d\varepsilon_1 \right] \int \frac{d^3 q}{(2\pi)^3} \mathcal{T}^{(1)}_{\alpha\beta,\alpha\beta}(0; \varepsilon, \boldsymbol{p}; \varepsilon_1, \boldsymbol{q}) \, \mathfrak{G}(0, \varepsilon_1, \boldsymbol{q}).$$

On applying formula (19.21) and solving this equation for $\Sigma(T) - \Sigma(0)$ (to do this, it is sufficient to write the equation for $\mathcal{T}$, analogous to (18.3)), we find that

$$\Sigma(T; \varepsilon, \boldsymbol{p}) - \Sigma(0; \varepsilon, \boldsymbol{p}) \approx \frac{1}{2}\left[T \sum_{\varepsilon_1} - \frac{1}{2\pi}\int \mathrm{d}\varepsilon_1\right] \int \frac{\mathrm{d}^3\boldsymbol{q}}{(2\pi)^3} \mathcal{T}_{\alpha\beta,\alpha\beta}(0; \varepsilon, \boldsymbol{p}; \varepsilon_1, \boldsymbol{q})\, \mathfrak{G}(0; \varepsilon_1, \boldsymbol{q}). \tag{19.22}$$

This gives us, using (19.21), for the first approximation to $\mathfrak{G}(T)$:

$$\mathfrak{G}(T; \varepsilon, \boldsymbol{p}) = \mathfrak{G}(0; \varepsilon, \boldsymbol{p}) + \frac{1}{2}\mathfrak{G}^2(0; \varepsilon, \boldsymbol{p})\left[T \sum_{\varepsilon_1} - \frac{1}{2\pi}\int \mathrm{d}\varepsilon_1\right] \times \int \frac{\mathrm{d}^3\boldsymbol{q}}{(2\pi)^3} \mathcal{T}_{\alpha\beta,\alpha\beta}(0; \varepsilon, \boldsymbol{p}; \varepsilon_1, \boldsymbol{q})\, \mathfrak{G}(0; \varepsilon_1, \boldsymbol{q}).$$

We substitute this expression into (19.17) and subtract $(1/V)N(\mu, 0)$ from the result. We can obviously write, up to first order terms:

$$\begin{aligned}
\frac{1}{V}[N(\mu, T) - N(\mu, 0)] &= 2\left[T \sum_{\varepsilon} - \frac{1}{2\pi}\int \mathrm{d}\varepsilon\right] \mathfrak{G}(0; \varepsilon, \boldsymbol{p}) \\
&\quad + \frac{1}{2\pi}\int \mathrm{d}\varepsilon \frac{\mathrm{d}^3\boldsymbol{p}}{(2\pi)^3} \mathfrak{G}^2(0; \varepsilon, \boldsymbol{p})\left[T \sum_{\varepsilon_1} - \frac{1}{2\pi}\int \mathrm{d}\varepsilon_1\right] \\
&\quad \times \int \frac{\mathrm{d}^3\boldsymbol{q}}{(2\pi)^3} \mathcal{T}_{\alpha\beta,\alpha\beta}(0; \varepsilon, \boldsymbol{p}; \varepsilon_1, \boldsymbol{q})\, \mathfrak{G}(0; \varepsilon_1, \boldsymbol{q}) \\
&= 2\left[T \sum_{\varepsilon} - \frac{1}{2\pi}\int \mathrm{d}\varepsilon\right] \int \frac{\mathrm{d}^3\boldsymbol{p}}{(2\pi)^3} \mathfrak{G}(0; \varepsilon, \boldsymbol{p}) \\
&\quad \times \left[1 + \frac{1}{2}\int \frac{\mathrm{d}\varepsilon_1}{2\pi}\int \frac{\mathrm{d}^3\boldsymbol{q}}{(2\pi)^3} \mathcal{T}_{\alpha\beta,\alpha\beta}(0; \varepsilon, \boldsymbol{p}; \varepsilon_1, \boldsymbol{q})\, \mathfrak{G}^2(0; \varepsilon_1, \boldsymbol{q})\right].
\end{aligned} \tag{19.23}$$

We have used in this last equation the symmetry property

$$\mathcal{T}_{\alpha\beta,\alpha\beta}(\varepsilon, \boldsymbol{p}; \varepsilon_1, \boldsymbol{q}) = \mathcal{T}_{\alpha\beta,\alpha\beta}(\varepsilon_1, \boldsymbol{q}; \varepsilon, \boldsymbol{p}).$$

This expression can be modified to some extent. By proceeding as in deriving (19.4), we can obtain a formula for the finite temperature case which is analogous to (19.4):

$$\frac{\partial \mathfrak{G}^{-1}(T; \varepsilon, \boldsymbol{p})}{\partial \mu} = 1 + \frac{1}{2} T \sum_{\varepsilon_1} \int \frac{\mathrm{d}^3\boldsymbol{q}}{(2\pi)^3} \mathcal{T}_{\alpha\beta,\alpha\beta}(T; \varepsilon, \boldsymbol{p}; \varepsilon_1, \boldsymbol{q})\, \mathfrak{G}^2(T; \varepsilon_1, \boldsymbol{q}).$$

We obtain in the limit as $T \to 0$ when $\varepsilon = \text{const}$:

$$\frac{\partial}{\partial \mu} \mathfrak{G}^{-1}(0; \varepsilon, \boldsymbol{p}) = 1 + \frac{1}{2}\int \frac{\mathrm{d}\varepsilon_1}{2\pi}\int \frac{\mathrm{d}^3\boldsymbol{q}}{(2\pi)^3} \mathcal{T}_{\alpha\beta,\alpha\beta}(0; \varepsilon, \boldsymbol{p}; \varepsilon_1, \boldsymbol{q})\, \mathfrak{G}^2(0; \varepsilon_1, \boldsymbol{q}).$$

On substituting in (19.23), we get

$$\begin{aligned}
\frac{1}{V}[N(\mu, T) - N(\mu, 0)] &= 2\left[T \sum_{\varepsilon} - \frac{1}{2\pi}\int \mathrm{d}\varepsilon\right] \int \frac{\mathrm{d}^3\boldsymbol{p}}{(2\pi)^3} \mathfrak{G}(0; \varepsilon, \boldsymbol{p}) \\
&\quad \times \frac{\partial}{\partial \mu}\mathfrak{G}^{-1}(0; \varepsilon, \boldsymbol{p}) = -2\left[T \sum_{\varepsilon} - \frac{1}{2\pi}\int \mathrm{d}\varepsilon\right] \int \frac{\mathrm{d}^3\boldsymbol{p}}{(2\pi)^3} \frac{\partial}{\partial \mu} \ln \mathfrak{G}(0; \varepsilon, \boldsymbol{p}).
\end{aligned}$$

Differentiating this with respect to the temperature with $\mu = \text{const}$, and comparing it with (19.18), we have

$$\frac{S}{V} = -2\frac{\partial}{\partial T}\left[T\sum_{\varepsilon}\int\frac{d^3\boldsymbol{p}}{(2\pi)^3}\ln\mathfrak{G}(0;\varepsilon,\boldsymbol{p})\right]. \tag{19.24}$$

The rest of the calculations proceeds as follows. We use the connection between the temperature-dependent function $\mathfrak{G}$ and the retarded and advanced Green functions, and write (19.24) for the entropy as the sum of two contour integrals:

$$\frac{S}{V} = -2\frac{\partial}{\partial T}\left\{\int\frac{d^3\boldsymbol{p}}{(2\pi)^3}\left[\frac{1}{4\pi i}\int\limits_{C_1}\tanh\frac{\varepsilon}{2T}\ln G_R(\varepsilon,\boldsymbol{p})\,d\varepsilon + \frac{1}{4\pi i}\int\limits_{C_2}\tanh\frac{\varepsilon}{2T}\ln G_A(\varepsilon,\boldsymbol{p})\,d\varepsilon\right]\right\}$$

$$= 2\int\frac{d^3\boldsymbol{p}}{(2\pi)^3}\frac{1}{2\pi i T}\left[\int\limits_{C_1}\varepsilon\left(-\frac{\partial n_F(\varepsilon)}{\partial\varepsilon}\right)\ln G_R(\varepsilon,\boldsymbol{p})\,d\varepsilon + \int\limits_{C_2}\varepsilon\left(-\frac{\partial n_F}{\partial\varepsilon}\right)\ln G_A(\varepsilon,\boldsymbol{p})\,d\varepsilon\right],$$

where n_F is the Fermi function, and the contours C_1, C_2 are given in Fig. 59 *a*. The function G_R has no zeros in the upper half-plane, and G_A none in the lower half-plane (†). Using this property, and also the analyticity of G_R, G_A in the appropriate half-planes and the rapid decreases of $\partial n_F/\partial\varepsilon$ as $\varepsilon \to \pm\infty$, we can arrange C_1, C_2 along the real axis (see Fig. 59 *b*); this gives us

$$\frac{S}{V} = 2\int\frac{d^3\boldsymbol{p}}{(2\pi)^3}\frac{1}{2\pi i T}\int\limits_{-\infty}^{\infty}\varepsilon\left(-\frac{\partial n_F(\varepsilon)}{\partial\varepsilon}\right)[\ln G_R(\varepsilon,\boldsymbol{p}) - \ln G_A(\varepsilon,\boldsymbol{p})]\,d\varepsilon.$$

On applying the usual rule for finding integrals containing Fermi functions (‡), we find that

$$\frac{S}{V} = \frac{2\pi^2 T}{3}\frac{1}{2\pi i}\int\frac{d^3\boldsymbol{p}}{(2\pi)^3}\left[G_R^{-1}\frac{\partial G_R}{\partial\varepsilon} - G_A^{-1}\frac{\partial G_A}{\partial\varepsilon}\right]_{\varepsilon=0}$$

$$= \frac{2\pi^2 T}{3}\frac{1}{2\pi}\int\frac{d^3\boldsymbol{p}}{(2\pi)^3}2\,\mathrm{Im}\left[G_R^{-1}\frac{\partial G_R}{\partial\varepsilon}\right]_{\varepsilon=0}.$$

The last equality is connected with the fact that $G_R^* = G_A$ on the real axis. It is easily seen that the integration over $\boldsymbol{p}$ is performed close to the Fermi surface. On substituting $G_R \approx a/(\varepsilon - \xi + i\delta)$, we get

$$\frac{S}{V} = \frac{p_0 m^*}{3}T. \tag{19.25}$$

The specific heat is obviously equal to the entropy.

(†) See the footnote on p. 165.

(‡) $\int\limits_{-\infty}^{\infty} f(\varepsilon)\frac{\partial n_F(\varepsilon)}{\partial\varepsilon}\,d\varepsilon = -f(0) - \frac{\pi^2}{6}T^2\left[\frac{\partial^2 f}{\partial\varepsilon^2}\right]_{\varepsilon=0} + \cdots.$

Notice that our derivation of the temperature correction only made use of the neighbourhood of the point $\varepsilon = 0$, $\xi = 0$, i.e. the real poles of G_R(or G) at $T = 0$. This situation would seem to hold for any first order temperature corrections. In other words, to a first approximation the temperature corrections must always be determined by the poles of the function G (or Γ) at $T = 0$, i.e. they are determined by the spectrum of the elementary excitations.

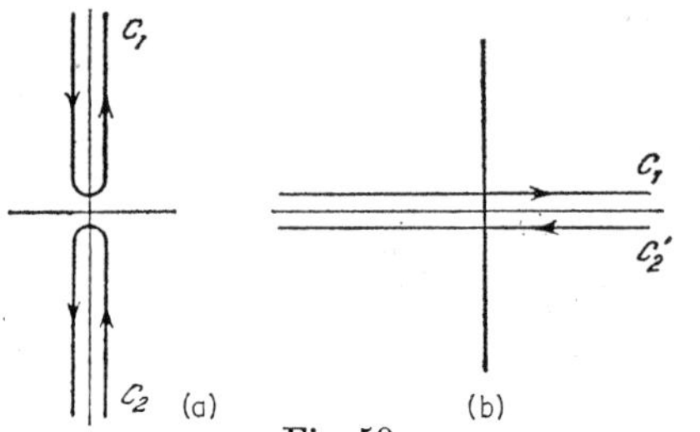

Fig. 59

6. Damping of quasi-particles in a Fermi liquid (†)

The basic parameters characterising the energy spectrum of a Fermi liquid, i.e. the constants a and v, vary weakly as functions of the temperature. It is clear from general considerations that the corrections must be proportional to $(T/\varepsilon_F)^2$. It is not possible to say this about the damping. The example of a rarefied gas discussed in § 2 (the footnote on p. 17) showed that the damping must be proportional to ε^2 for $\varepsilon \gg T$ and proportional to T^2 for $\varepsilon \ll T$. In view of this, it is of interest to consider the case of finite temperatures when evaluating the damping.

The damping of the quasi-particles is due to various collision processes in which two or more quasi-particles can participate. Degeneration limits the domains of phase space accessible to the quasi-particles on scattering, and it can therefore be assumed that the most probable processes are those in which two quasi-particles participate.

The quasi-particle damping is determined by the imaginary part of the retarded Green function $G_R(\varepsilon, \boldsymbol{p})$ at small frequencies ε and momenta $\boldsymbol{p}$, close to the limiting Fermi momentum. In the previous section the first term of the temperature expansion of G_R was evaluated. It was obtained by the replacement, in each diagram for the temperature-dependent self-energy part $\Sigma(\varepsilon, \boldsymbol{p})$, of all the summations over the discrete frequencies ε_n, except one, by integration. In other words, we extracted one line from each diagram for Σ and assumed that this line correspond to a low frequency. The remaining frequencies were assumed large and the corresponding sums were replaced by integrals.

However, the temperature correction to G_R thus obtained turns out to be purely real. This means that we must evaluate the next term in the expansion in ε and T. To do this, we have to suppose that there are two low frequencies instead of one in each diagram, and in view of this there must be two summations of discrete frequencies instead of one. This corresponds to splitting off three $\mathfrak{G}$-lines in the diagrams for the

(†) This section is based on G. M. Eliashberg's paper [74].

temperature-dependent self-energy part. A diagram now has the form illustrated in Fig. 60. As earlier, there are several possible methods of carrying out this splitting off. We shall assume that the remaining fre-

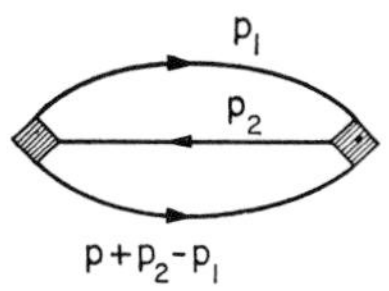

Fig. 60

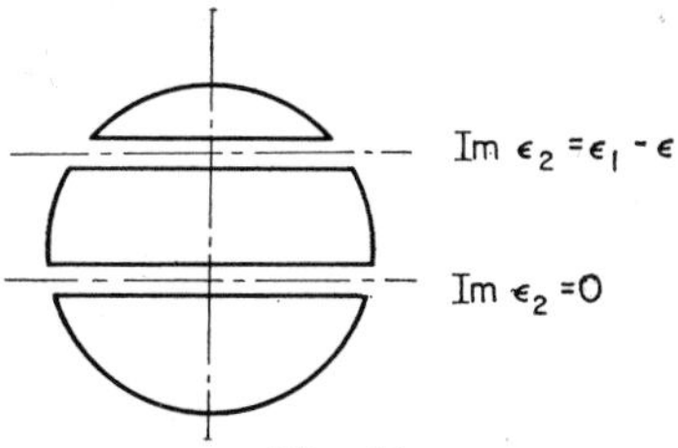

Fig. 61

quencies are high, so that we can put $T = 0$ and $\varepsilon_1 = \varepsilon_2 = \varepsilon_3 = 0$ in the vertex parts represented by shaded squares in Fig. 60. We thus obtain

$$-\frac{T^2}{(2\pi)^6}\sum_{\varepsilon_1,\varepsilon_2}\int d^3p_1 d^3p_2 \Gamma_1(p, p_2; p_1, p+p_2-p_1)$$
$$\times \Gamma_2(p+p_2-p_1, p_1; p_2, p)\,\mathfrak{G}(p_1)\mathfrak{G}(p_2)\mathfrak{G}(p+p_2-p_1), \quad (19.26)$$

where Γ_1 and Γ_2 correspond to a definite diagram and a definite method of splitting off the three lines (summation over the spins is also understood here).

We now pass from summation over the imaginary parts to integration over the real parts. We shall first do this for ε_2. In accordance with the analytic properties of $\mathfrak{G}$ and the G-functions, we can replace the sum over ε_2 by a contour integral:

$$-\frac{T}{2(2\pi)^7}\, i\int d^3p_1 d^3p_2 \Gamma_1\Gamma_2\Big\{\sum_{\varepsilon_1>\varepsilon}\mathfrak{G}(\varepsilon_1)\int_C d\varepsilon_2 \tanh(\varepsilon_2/2T)$$
$$\times G_{RA}(\varepsilon_2)G_{RA}(\varepsilon_2+i\varepsilon-i\varepsilon_1) + \sum_{\varepsilon_1<\varepsilon}\mathfrak{G}(\varepsilon_1)\int_{C'} d\varepsilon_2 \tanh(\varepsilon_2/2T)$$
$$\times G_{RA}(\varepsilon_2)G_{RA}(\varepsilon_2+i\varepsilon-i\varepsilon_1) + \mathfrak{G}(\varepsilon)\int_{C''} d\varepsilon_2 \tanh(\varepsilon_2/2T)\, G_{RA}(\varepsilon_2)\Big\}.$$

The contour C is illustrated in Fig. 61. The contour C' differs from C in that the line $\text{Im}\,\varepsilon_2 = \varepsilon_1 - \varepsilon$ lies below the line $\text{Im}\,\varepsilon_2 = 0$. The contour C'' contains one horizontal piece. $G_{RA}(\varepsilon)$ denotes $G_R(\varepsilon)$ if $\text{Im}(\varepsilon) > 0$ and $G_A(\varepsilon)$ if $\text{Im}(\varepsilon) < 0$. The integral over the large circle vanishes, so that there only remain the integrals over the horizontal lines. This gives

$$-\frac{T}{(2\pi)^7}\int d^3p_1 d^3p_2 \Gamma_1\Gamma_2\Bigg\{\sum_{\varepsilon_1>\varepsilon}\mathfrak{G}(\varepsilon_1)\int_{-\infty}^{\infty} d\varepsilon_2 \tanh(\varepsilon_2/2T)\,[\text{Im}\, G_R(\varepsilon)$$
$$\times G_A(\varepsilon_2+i\varepsilon-i\varepsilon_1) + \text{Im}\, G_R(\varepsilon_2)G_R(\varepsilon_2+i\varepsilon_1-i\varepsilon)] + \sum_{\varepsilon_1<\varepsilon}\mathfrak{G}(\varepsilon_1)\int_{-\infty}^{\infty} d\varepsilon_2$$
$$\times \tanh(\varepsilon_2/2T)\,[\text{Im}\, G_R(\varepsilon_2)G_R(\varepsilon_2+i\varepsilon-i\varepsilon_1) + \text{Im}\, G_R(\varepsilon_2)G_A(\varepsilon_2+i\varepsilon_1-i\varepsilon)]$$
$$+\,\mathfrak{G}(\varepsilon)\int_{-\infty}^{\infty} d\varepsilon_2 \tanh(\varepsilon_2/2T)\,\text{Re}\, G_R(\varepsilon_2)\,\text{Im}\, G_R(\varepsilon_2)\Bigg\}. \quad (19.27)$$

We have made use here of the relationship $G_A(\varepsilon) = G_R^*(\varepsilon)$, so that $G_R(\varepsilon) - G_A(\varepsilon) = 2i \operatorname{Im} G_R(\varepsilon)$, and also of the periodicity $\tanh(x + i\pi n) = \tanh x$.

We now perform the same replacement of the sum by a contour integral for ε_1. In this case we obtain ($\varepsilon > 0$):

$$-\frac{1}{(2\pi)^8}\int d^3\mathbf{p}_1 d^3\mathbf{p}_2 \Gamma_1\Gamma_2 \left\{ \oint_{-\infty}^{\infty} d\varepsilon_1 \coth(\varepsilon_1/2T)\, G_R(\varepsilon_1 + i\varepsilon) \int_{-\infty}^{\infty} d\varepsilon_2 \tanh(\varepsilon_2/2T) \right.$$

$$\times [\operatorname{Im} G_R(\varepsilon_2) \operatorname{Im} G_R(\varepsilon_2 + \varepsilon_1) - \operatorname{Im} G_R(\varepsilon_2) \operatorname{Im} G_R(\varepsilon_2 - \varepsilon_1)]$$

$$+ \int_{-\infty}^{\infty} d\varepsilon_1 \tanh(\varepsilon_1/2T) \operatorname{Im} G_R(\varepsilon_1) \int_{-\infty}^{\infty} d\varepsilon_2 \tanh(\varepsilon_2/2T)$$

$$\left. \times [\operatorname{Im} G_R(\varepsilon_2)\, G_R(\varepsilon_2 + i\varepsilon - \varepsilon_1) + \operatorname{Im} G_R(\varepsilon_2)\, G_A(\varepsilon_2 + \varepsilon_1 - i\varepsilon)] \right\}. \qquad (19.28)$$

We remark here that the circuit round the pole of $\coth(\varepsilon_1/2T)$ at $\varepsilon_1 = 0$ cancels with the last term in (19.27).

In order to find the contribution to $\operatorname{Im} \Sigma_R(\varepsilon_1 \mathbf{p})$ it is sufficient to replace $i\varepsilon$ by $\varepsilon + i\delta$ and take the imaginary part. The result can be written in the form

$$\frac{1}{(2\pi)^8}\int d^3\mathbf{p}_1 d^3\mathbf{p}_2 \Gamma_1\Gamma_2 \int d\varepsilon_1 d\varepsilon_2 \operatorname{Im} G_R(\varepsilon_1, \mathbf{p}_1) \operatorname{Im} G_R(\varepsilon_2, \mathbf{p}_2)$$

$$\times \operatorname{Im} G_R(\varepsilon_2 + \varepsilon - \varepsilon_1, \mathbf{p}_2 + \mathbf{p} - \mathbf{p}_1) \big(\tanh(\varepsilon_1/2T) - \coth[(\varepsilon_1 - \varepsilon)/2T]\big)$$

$$\times \big(\tanh(\varepsilon_2/2T) - \tanh[(\varepsilon_2 + \varepsilon - \varepsilon_1)/2T]\big). \qquad (19.29)$$

Since we are interested in small ε, small ε_1 and ε_2 and momenta close to p_0 are important in the integrals. Now, $\operatorname{Im} G_R(\varepsilon, \mathbf{p}) \approx -a\pi\delta[\varepsilon - \xi(\mathbf{p})]$. This method thus actually enables us to take account of the binary quasi-particle interactions. On summing over all different ways of splitting off three lines, then over all the diagrams, we obtain an expression that differs from (19.29) by the factor

$$\frac{1}{2}\Gamma(\mathbf{p}_1, \mathbf{p}_2; \mathbf{p}_1, \mathbf{p} + \mathbf{p}_2 - \mathbf{p}_1)\, \Gamma(\mathbf{p} + \mathbf{p}_2 - \mathbf{p}_1, \mathbf{p}_1; \mathbf{p}_2, \mathbf{p})$$

$$= \frac{1}{2}\left|\Gamma(\mathbf{p}, \mathbf{p}_2; \mathbf{p}_1, \mathbf{p} + \mathbf{p}_2 - \mathbf{p}_1)\right|^2$$

instead of $\Gamma_1\Gamma_2$. The factor $\frac{1}{2}$ owes its origin to the fact that two of the three lines split off are equivalent. As we shall see in a moment, expression (19.29) is of the order $\varepsilon^2/p_0 v$ or $T^2/p_0 v$. We mention without offering a proof that the splitting off of more than three real lines in the diagrams for Σ leads to higher order terms.

On substituting a δ-function for $\operatorname{Im} G_R(\varepsilon, \mathbf{p})$ and noting that the integration over the momenta proceeds close to the Fermi surface. i.e.

$\mathrm{d}\boldsymbol{p} \to (p_0^2/v)\,\mathrm{d}\xi\,\mathrm{d}\Omega$, we obtain from (19.29):

$$a\,\mathrm{Im}\,\Sigma_R(\varepsilon) = -\frac{\pi}{32}\cosh\,(\varepsilon/2T)\int\frac{\mathrm{d}\Omega_1\mathrm{d}\Omega_2}{(4\pi)^2}\left|\frac{a^2p_0^2}{\pi^2 v}\Gamma(\boldsymbol{p}_1,\boldsymbol{p}_2;\boldsymbol{p}_1,\boldsymbol{p}+\boldsymbol{p}_2-\boldsymbol{p}_1)\right|^2$$

$$\times\int \mathrm{d}\xi_1\,\mathrm{d}\xi_2\frac{\delta(\varepsilon+\xi_2-\xi_1-\xi_3)}{\cosh\dfrac{\xi_1}{2T}\cosh\dfrac{\xi_2}{2T}\cosh\dfrac{\xi_3}{2T}}, \qquad (19.30)$$

where $\xi_3 = \xi(\boldsymbol{p}+\boldsymbol{p}_2-\boldsymbol{p}_1)$.

The δ-function appearing in the integral merely has the practical effect of imposing the condition on the momenta: $|\boldsymbol{p}+\boldsymbol{p}_2-\boldsymbol{p}_1| = p_0$. In view of this, (19.30) can be integrated over ξ_1 and ξ_2. As a result we finally obtain:

$$a\,\mathrm{Im}\,\Sigma_R(\varepsilon) = -\frac{\pi^3}{16}A\frac{T^2}{p_0 v}\left[1+\left(\frac{\varepsilon}{\pi T}\right)^2\right],$$

$$A = \int\frac{\mathrm{d}\Omega_1\mathrm{d}\Omega_2}{(4\pi)^2}\left|\frac{a^2p_0^2}{\pi^2 v}\Gamma(\boldsymbol{p},\boldsymbol{p}_2;\boldsymbol{p}_1,\boldsymbol{p}+\boldsymbol{p}_2-\boldsymbol{p}_1)\right|^2\delta(|\boldsymbol{n}+\boldsymbol{n}_2-\boldsymbol{n}_1|-1), \qquad (19.31)$$

where the $\boldsymbol{n}_i$ are unit vectors in the directions of the $\boldsymbol{p}_i$.

It is clear from this expression that $(a^2p_0^2/\pi^2 v)\,\Gamma(\boldsymbol{p},\boldsymbol{p}_2,\boldsymbol{p}_1,\boldsymbol{p}+\boldsymbol{p}_2-\boldsymbol{p}_1)$ plays the role of the quasi-particle mutual scattering amplitude. It depends only on the angles and can be evaluated at $T = 0$. As already mentioned above, summation over the spins is understood in the expression (19.31) for A.

The function Σ_R possesses the same analytic properties as G_R, and hence the imaginary additional term to the time function has the form

$$\mathrm{Im}\,\Sigma(\varepsilon) = \mathrm{Im}\,\Sigma_R(\varepsilon)\tanh\,(\varepsilon/2T). \qquad (19.32)$$

§ 20. SPECIAL PROPERTIES OF THE VERTEX PART IN THE CASE WHERE THE TOTAL MOMENTUM OF THE COLLIDING PARTICLES IS SMALL(†)

The vertex part has another special property, apart from those occurring for small energy and momentum transfers; this will be seen later to be of interest in the theory of superconductivity (Chap. VII). The case to be considered is the one where the sum $p_1 + p_2$ is small, whilst the energies ε_1, ε_2 and the sum of the momenta $\boldsymbol{p}_1 + \boldsymbol{p}_2 = \boldsymbol{s}$ are small.

We consider the diagrams of Fig. 57. It may easily be seen that diagram 57*a* is the singular one in the present case. In this diagram the poles of the two G-functions under the integral approach one another.

We proceed with this vertex part as in § 18. We use the notation

$$\Gamma_{\alpha\beta,\gamma\delta}(p_1, p_3, s) = \Gamma_{\alpha\beta,\gamma\delta}(p_1, -p_1+s; p_3, -p_3+s).$$

(†) This section is based on unpublished results of A. A. Abrikosov, L. P. Gor'kov, L. D. Landau, and I. M. Khalatnikov.

Further, we denote by $\Gamma^{(2)}$ the sum of all the "non-singular" diagrams. We can put $s = 0$ in $\Gamma^{(2)}$. To obtain the total Γ, we have to sum "ladder" diagrams such as Fig. 58. Before doing this, we differentiate Γ with respect to the fourth component of s, which we denote by λ. Each ladder diagram now yields a sum of terms, in each of which one of the "rungs" is differentiated. If we fix the differentiated rung, it is easily seen that all the diagrams split up into two independent ladders from the left and the right, the sum of the ladders to either side being the total vertex part. We therefore get the equation

$$\frac{\partial}{\partial \lambda}\Gamma_{\alpha\beta,\gamma\delta}(p_1, p_3; s)$$
$$= \frac{i}{2}\int \Gamma_{\alpha\beta,\xi\eta}(p_1, q; s)\, G(q)\frac{\partial}{\partial\lambda}G(-q+s)\,\Gamma_{\xi\eta,\gamma\delta}(q, p_3; s)\frac{d^4q}{(2\pi^4)}. \quad (20.1)$$

The expression $G(q)\,\partial/\partial\lambda\, G(-q+s)$ under the integral has close to $\varepsilon = 0$, $|\boldsymbol{q}| = p_0$ the form:

$$-\frac{a^2}{[\varepsilon - v(|\boldsymbol{q}| - p_0) + i\delta\,\mathrm{sign}\,(|\boldsymbol{q}| - p_0)]}$$
$$\times \frac{1}{[\varepsilon - \lambda + v\,(|\boldsymbol{q}-\boldsymbol{s}| - p_0) - i\delta\,\mathrm{sign}\,(|\boldsymbol{q}-\boldsymbol{s}| - p_0)]^2}. \quad (20.2)$$

It is clear from this that the integration with respect to $d^4q/(2\pi)^4 \to p_0 d|\boldsymbol{q}|\, d\Omega\, d\varepsilon/(2\pi)^4$ is mainly over the neighbourhood of $|\boldsymbol{q}| = p_0$, $\varepsilon = 0$. If we assume that the Γ in the integral do not vary rapidly in this region, they can be regarded as constant when integrating with respect to $d|\boldsymbol{q}|$ and $d\varepsilon$, and only expression (20.2) has to be integrated. We obtain as a result:

$$\frac{\partial}{\partial\lambda}\Gamma_{\alpha\beta,\gamma\delta}(p_1, p_3; s)$$
$$= \frac{a^2 p_0^2}{2(2\pi)^3 v}\int \Gamma_{\alpha\beta,\xi\eta}(p_1 q; s)\Gamma_{\xi\eta,\gamma\delta}(q p_3; s)\frac{\lambda}{\lambda^2 - (\boldsymbol{v}\cdot\boldsymbol{s}^2 + i\delta)}\,d\Omega. \quad (20.3)$$

We now find the gradient of Γ with respect to $\boldsymbol{s}$. It must not be forgotten, when differentiating the second of the G-functions, that $(\boldsymbol{v}\cdot\boldsymbol{s})$, as well as appearing directly in the denominator, determines the sign of the imaginary part. In view of this, close to $\varepsilon = 0$, $|\boldsymbol{q}| = p_0$, the expression $G(q)\,\partial G(-q+s)/\partial\boldsymbol{s}$ has the form

$$-\frac{a^2\boldsymbol{v}}{[\varepsilon - v(|\boldsymbol{q}| - p_0) + i\delta\,\mathrm{sign}\,(|\boldsymbol{q}| - p_0)]}$$
$$\times\frac{1}{\left[\varepsilon - \lambda + v(|\boldsymbol{q}| - p_0) - (\boldsymbol{v}\cdot\boldsymbol{s}) - i\delta\,\mathrm{sign}\left(|\boldsymbol{q}| - p_0 - \frac{(\boldsymbol{v}\cdot\boldsymbol{s})}{v}\right)\right]^2}$$
$$+\frac{2\pi i a^2\delta(\varepsilon-\lambda)\,\delta((\boldsymbol{v}\cdot\boldsymbol{s}) - v(|\boldsymbol{q}| - p_0))\,\boldsymbol{v}}{\lambda - (\boldsymbol{v}\cdot\boldsymbol{s}) + i\delta\,\mathrm{sign}\,(\boldsymbol{v}\cdot\boldsymbol{s})}.$$

The integral of this expression with respect to $d\varepsilon\, d|\boldsymbol{q}|$ leads to the following relation for Γ:

$$\frac{\partial}{\partial s}\Gamma_{\alpha\beta,\gamma\delta}(p_1, p_3; s)$$
$$= -\frac{a^2 p_0^2}{2(2\pi)^3 v}\int \Gamma_{\alpha\beta,\xi\eta}(p_1, q; s)\,\Gamma_{\xi\eta,\gamma\delta}(q, p_3; s)\,\frac{\boldsymbol{v}(\boldsymbol{v}\cdot\boldsymbol{s})}{\lambda^2-(\boldsymbol{v}\cdot\boldsymbol{s})^2+i\delta}\,d\Omega. \quad (20.4)$$

Combining equations (20.3) and (20.4), we get

$$\lambda\frac{\partial}{\partial\lambda}\Gamma_{\alpha\beta,\gamma\delta}(p_1, p_3; s) + \left(\boldsymbol{s}\cdot\frac{\partial}{\partial \boldsymbol{s}}\Gamma_{\alpha\beta,\gamma\delta}(p_1, p_3; s)\right)$$
$$= \frac{a^2 p_0^2}{2(2\pi)^3 v}\int \Gamma_{\alpha\beta,\xi\eta}(p_1, q; s)\,\Gamma_{\xi\eta,\gamma\delta}(q, p_3; s)\,d\Omega. \quad (20.5)$$

If magnetic interactions are not considered, the forces between the particles depend only on the mutual orientation of the spins. If we take into account the fact that the total spin is unchanged when the particles interact, (20.5) splits into two independent equations. One of them corresponds to the interaction of two particles with opposite spins (i.e. for instance, $\alpha = 1/2$, $\beta = -1/2$), and the other to the interaction of particles with parallel spins. These equations look completely identical. The difference lies in the initial conditions (i.e. in $\Gamma^{(2)}$). In view of this, we shall simply write Γ in future, understanding by this either one of the two different components.

We shall make the assumption, to be justified later, that $\Gamma(p_1, p_3; s)$ does not depend on the angles $\widehat{(\boldsymbol{p}_1, \boldsymbol{s})}$, $\widehat{(\boldsymbol{p}_3, \boldsymbol{s})}$. It can then be expanded in Legendre polynomials, dependent on $\cos\theta$, where θ is the angle between $\boldsymbol{p}_1$ and $\boldsymbol{p}_3$:

$$\Gamma(p_1, p_3; s) = \sum_l \Gamma_l P_l(\cos\theta). \quad (20.6)$$

Notice that $\Gamma_{\alpha\beta,\xi\eta}(p_1, p_3; s)$ is antisymmetric in the spins when these are antiparallel and symmetric when they are parallel. Interchanging the momenta of the initial particles corresponds to replacing $\cos\theta$ by $-\cos\theta$. In view of the fact that $\Gamma_{\alpha\beta,\xi\eta}(p_1, p_2; p_3, p_4)$ must be antisymmetric with respect to the interchange $p_1\alpha \rightleftarrows p_2\beta$, the expansion (20.6) contains only odd harmonics in the case of parallel spins, and only even ones in the antiparallel case. We get separate equations for the individual Γ_l from (20.5) of the form

$$\lambda\frac{\partial\Gamma_l}{\partial\lambda} + \left(\boldsymbol{s}\cdot\frac{\partial\Gamma_l}{\partial \boldsymbol{s}}\right) = \frac{4\pi a^2 p_0^2}{(2\pi)^3 v}\,\frac{\Gamma_l^2}{2l+1} \quad (20.7)$$

(the factor 2 results from summing over the spins in (20.5)).

The solution of (20.7) is

$$\Gamma_l(\lambda, |\boldsymbol{s}|) = -\frac{(2\pi)^3 v(2l+1)}{2\pi a^2 p_0^2}\cdot\frac{1}{\ln\lambda|\boldsymbol{s}| + f_l\left(\frac{\lambda}{|\boldsymbol{s}|}\right)}, \quad (20.8)$$

where f_l is an arbitrary function.

Equations (20.3) and (20.4) enable us to obtain limiting expressions for $f_l(x)$ as $x \to 0$ and $x \to \infty$. Let us consider equation (20.3) in the limit as $|s| \to 0$. On expanding Γ in spherical harmonics, we get

$$\lambda \frac{\mathrm{d}\Gamma_l}{\mathrm{d}\lambda} = \frac{4\pi a^2 p_0^2}{(2\pi)^3 v (2l+1)} \Gamma_l^2. \tag{20.9}$$

The solution of this equation is

$$\Gamma_l(\lambda, 0) = -\frac{(2\pi)^3 v (2l+1)}{4\pi a^2 p_0^2} \cdot \frac{1}{\ln \dfrac{\lambda}{c_1^l}}, \tag{20.10}$$

where c_1 is constant. Hence, as $x \to \infty$, $f_l(x) \to \ln [x/(c_1^l)^2]$. We similarly obtain from (20.4), as $\lambda \to 0$:

$$\Gamma_l(0, |s|) = -\frac{(2\pi)^3 v (2l+1)}{4\pi a^2 p_0^2} \cdot \frac{1}{\ln \dfrac{v\,|s|}{c_2^l}}. \tag{20.11}$$

Hence $f_l(x) \to \ln [v^2/x(c_2^l)^2]$ as $x \to 0$.

The constants c_1^l and c_2^l have the dimensions of energies and may be complex. If, for example, the constant c_1^l has the order of the Fermi energy, it follows from (20.10) that, as $\lambda \to 0$, $\Gamma_l(\lambda, 0)$ vanishes as $1/\ln (c_1^l/\lambda)$. But a singular case is possible, when c_1^l is small: $\Gamma_l(\lambda, 0)$ now has a pole for some (in general, complex) value of λ. The meaning of this fact, which is closely connected with the phenomenon of superconductivity, will be explained in Chap. VII.

* * *

So far we have considered an isotropic Fermi liquid. Only one isotropic Fermi liquid is known at the present time, namely liquid He^3. The class of anisotropic Fermi liquids, namely electrons in metals, is much wider. However, apart from anisotropy, an electron liquid in metals has specific features such as the long-range Coulomb forces, interactions with lattice vibrations, and so on. Some of these features will be discussed in the next two sections, taking as examples the isotropic model of electrons interacting with phonons and a degenerate plasma.

A general consideration of an electron Fermi liquid in a metal undoubtedly represents one of those subjects that should be capable of being treated by the methods of quantum field theory. In particular, it should be possible to make a deeper study of the phenomenon of superconductivity, which has so far only been treated by using an elementary model (see Chap. VII).

§ 21. ELECTRON-PHONON INTERACTIONS (†)

Electron-phonon interactions in a metal will be considered in this section by using an isotropic model. We shall assume that the temperature of the metal is in the range

$$T_c \ll T \ll \omega_D, \tag{21.1}$$

where T_c is the superconductivity transition temperature, and ω_D is the Debye frequency of the phonons. The first of these inequalities is connected with the fact that, in the model in question, superconductivity occurs at fairly low temperatures (see Chap. VII).

Naturally our model differs substantially from a normal metal. This is because a real metal is anisotropic, and in addition, Coulomb interactions occur between the electrons. We shall mention at the end of this section the qualitative changes that result from the Coulomb interaction of the electrons.

1. The vertex part

Dyson's equations were obtained for electrons and phonons in the temperature technique in § 16. We shall start by considering the vertex part $\mathcal{T}$ that appears in these equations. We shall show that $\mathcal{T}$ differs from its zero value, equal to g, by small quantities of the order $\sqrt{m/M}$, where m is the mass of an electron, and M the mass of the nucleus. Let us evaluate the first approximation correction to $\mathcal{T}$, illustrated in Fig. 55. We shall assume here that the electron momenta are of the order p_0, and the phonon momentum and energy restricted by the conditions $|\boldsymbol{k}| \lesssim k_D$, $\omega \lesssim \omega(\boldsymbol{k})/|\boldsymbol{k}| = 2p_0 \sim \omega_D$ (since $k_D \sim p_0$). It is precisely these values of the energies and momenta that will be of most importance below.

In accordance with § 14, the first order diagram is equal to

$$\mathcal{T}^{(1)}(p+q, p; q) = -g^3\, T \int \mathfrak{D}^{(0)}(p-p_1)\, \mathfrak{G}^{(0)}(p_1+q)\, \mathfrak{G}^{(0)}(p_1).\ \mathrm{d}^3\boldsymbol{p}_1/(2\pi)^3,$$

$$q = (\boldsymbol{k}, \omega)\ (‡), \tag{21.2}$$

where, in accordance with § 14, $\mathfrak{D}^{(0)}$ and $\mathfrak{G}^{(0)}$ are given by

$$\mathfrak{G}^{(0)}(p) = \frac{1}{i\varepsilon - \varepsilon_0(\boldsymbol{p}) + \mu}, \tag{21.3}$$

$$\varepsilon = \pi T(2n+1), \qquad \varepsilon_0(\boldsymbol{p}) = \boldsymbol{p}^2/2m, \qquad \mu = p_0^2/2m,$$

$$\mathfrak{D}^{(0)}(q) = -\frac{\omega_0^2(\boldsymbol{k})}{\omega^2 + \omega_0^2(\boldsymbol{k})}. \tag{21.4}$$

(†) This section is based on papers by A. B. MIGDAL [38] and G. M. ELIASHBERG [75].

(‡) In this section we shall denote the 4-momentum of the phonons in the temperature technique by the letter $q = (\boldsymbol{k}, \omega)$. We denote $|\boldsymbol{k}|$ by k.

We shall assume that the phonon spectrum is bounded, i.e. $k < k_D < p_0$. For large momenta $\mathfrak{D}^{(0)}$ vanishes. This restriction implies physically the absence of vibrations with wavelength less than the interatomic spacings.

The function $\mathfrak{D}^{(0)}$ appearing in (21.2) can be replaced by a constant $\mathfrak{D}^{(0)} \approx -1$ for $|\boldsymbol{p}-\boldsymbol{p}_1| \sim p_0$ and $|\varepsilon-\varepsilon_1| \ll \omega_D$. In the region $|\varepsilon-\varepsilon_1| \ll \omega_D$, $\mathfrak{D}^{(0)}$ decreases according to the law $(\varepsilon-\varepsilon_1)^{-2}$. This enables us to estimate the integral for $\mathcal{T}^{(1)}$:

$$\mathcal{T}^{(1)} \approx -g^3 T \sum_{|\varepsilon-\varepsilon_1|<\omega_D} \int_{|\boldsymbol{p}-\boldsymbol{p}_1|<k_D} \mathfrak{G}^{(0)}(p_1+q)\,\mathfrak{G}^{(0)}(p_1)\frac{d^3\boldsymbol{p}_1}{(2\pi)^3}. \tag{21.5}$$

The summation over ε_1 yields a factor $\omega_D/2\pi \sim u p_0 \sim \mu\sqrt{m/M}$, where u is the velocity of sound. Thus, $\mathcal{T}^{(1)}$ is of the order (see § 9):

$$\mathcal{T}^{(1)} \sim g^3 \frac{p_0^2}{v}\sqrt{\frac{m}{M}} \sim g\,\zeta\sqrt{\frac{m}{M}}$$

($\zeta \sim 1$), provided the integration over $\boldsymbol{p}_1$ does not introduce factors of the order $1/\omega_D$.

The only case which is dubious in this sense is that of small transfers $k \ll p_0$ and $\omega \ll \omega_D$. In this case the poles of the two $\mathfrak{G}$-functions in the integral for $\mathcal{T}^{(1)}$ approach one another. It may easily be verified that the integration over the momenta will be carried out over a region close to $|\boldsymbol{p}_1| = p_0$, and the summation over energies for which $\varepsilon_1 \ll \mu$. In this case $\mathfrak{G}^{(0)}$ can be replaced by

$$\mathfrak{G}^{(0)} = \frac{1}{i\varepsilon - v(|\boldsymbol{p}|-p_0)}, \tag{21.6}$$

where $v = p_0/m$. The situation recalls that considered in § 18; yet there is an essential difference. As in § 18, the product of the two $\mathfrak{G}^{(0)}$-functions has a sharp maximum close to $|\boldsymbol{p}_1| = p_0$ for small ε_1. The sum and the integral of the two $\mathfrak{G}$-functions, taken between infinite limits, are formally divergent and, by virtue of this, depend essentially on the order of the integration. In § 18 we first performed the integration over the frequencies, and then over $\xi = v(|\boldsymbol{p}|-p_0)$. This was connected with the fact that the integral over ε is actually taken between infinite limits, whilst the integration over ξ is essentially restricted by the limits $|\xi| \ll \mu$. In the present case the presence of the $\mathfrak{D}$-function in the integral makes it convergent, in accordance with (21.5). Hence the order in which the summation over the frequencies and the integration over ξ are carried out is of no consequence, and we shall in fact first integrate over ξ_1, then sum over ε_1. As a result we obtain (the condition $T \ll \omega_D$ is essential):

$$\frac{-p_0^2}{(2\pi)^3 v}\int \frac{i\omega}{(\boldsymbol{v}\cdot\boldsymbol{k}) - i\omega}\,d\Omega.$$

This quantity is always small, except when $\omega \gtrsim vk$, and as we shall see, this particular case is of no interest for what follows.

The above estimate is unaffected if we take into account higher order diagrams. Hence

$$\mathcal{T} = g\left[1 + O\left(\sqrt{\frac{m}{M}}\right)\right]. \tag{21.7}$$

2. *The phonon Green function*

Let us now find the phonon Green function. The Dyson equation is given by (16.3). We consider the irreducible self-energy part appearing in this equation:

$$\mathcal{H}(q) = 2g^2 T \sum_{\varepsilon} \int \mathfrak{G}(p)\, \mathfrak{G}(p-q) \frac{\mathrm{d}^3 \boldsymbol{p}}{(2\pi)^3}. \tag{21.8}$$

We carry out an analytic continuation of $\mathcal{H}$ in the region of real frequencies. To do this, we proceed as when evaluating $\operatorname{Im} \Sigma_R$ at the and of § 19. Let the phonon frequency $\omega = 2\pi n T$ be positive. We can now write the sum over ε in (21.8) as the contour integral

$$\mathcal{T}(q) = \frac{g^2}{2\pi i} \int_C \mathrm{d}\varepsilon \tanh \frac{\varepsilon}{2T} \int G_{RA}(\varepsilon, \boldsymbol{p})\, G_{RA}(\varepsilon - i\omega, \boldsymbol{p} - \boldsymbol{k}) \frac{\mathrm{d}^3 \boldsymbol{p}}{(2\pi)^3},$$

where the contour of integration C is the same as in Fig. 61. The function $G_{RA}(\varepsilon)$ appearing in the integral must be understood as $G_R(\varepsilon)$ if $\operatorname{Im} \varepsilon > 0$ and as $G_A(\varepsilon)$ if $\operatorname{Im} \varepsilon < 0$. Since the integral over the outer circumference vanishes, we are left only with the integrals over the lines $\operatorname{Im} \varepsilon = 0$ and $\operatorname{Im} \varepsilon = \omega$. In the latter integral we carry out the change of variables $\varepsilon - i\omega \to \varepsilon$, whilst making use of the fact that $\tanh [(\varepsilon/2T) + i\pi n] = \tanh [\varepsilon/2T]$. In addition, we observe that $G_R(\varepsilon) = G_A^*(\varepsilon)$. We obtain then:

$$\mathcal{T} = \frac{g^2}{\pi (2\pi)^3} \int_{-\infty}^{\infty} \mathrm{d}\varepsilon \int \mathrm{d}^3 \boldsymbol{p} \tanh \frac{\varepsilon}{2T} [\operatorname{Im} G_R(\boldsymbol{p}, \varepsilon)\, G_A(\boldsymbol{p} - \boldsymbol{k}, \varepsilon - i\omega)$$
$$+ \operatorname{Im} G_R(\boldsymbol{p} - \boldsymbol{k}, \varepsilon)\, G_R(\boldsymbol{p}, \varepsilon + i\omega)]. \tag{21.9}$$

We can now quite easily perform the analytic continuation to real ω. A retarded function Π_R is then obtained. On separating real and imaginary parts in this function, we find that

$$\operatorname{Re} \Pi_R = \frac{2g^2}{(2\pi)^4} \int \mathrm{d}^4 p \tanh \frac{\varepsilon}{2T} [\operatorname{Im} G_R(\boldsymbol{p}, \varepsilon) \operatorname{Re} G_R(\boldsymbol{p} - \boldsymbol{k}, \varepsilon - \omega)$$
$$+ \operatorname{Im} G_R(\boldsymbol{p} - \boldsymbol{k}, \varepsilon) \operatorname{Re} G_R(\boldsymbol{p}, \varepsilon + \omega)] \tag{21.10}$$

$$\operatorname{Im} \Pi_R = \frac{-2g^2}{(2\pi)^4} \int \mathrm{d}^4 p \left(\tanh \frac{\varepsilon}{2T} - \tanh \frac{\varepsilon - \omega}{2T}\right) \operatorname{Im} G_R(\boldsymbol{p}, \varepsilon)$$
$$\times \operatorname{Im} G_R(\boldsymbol{p} - \boldsymbol{k}, \varepsilon - \omega).$$

It will be shown below that the function G_R has the form

$$G_R = \frac{1}{\varepsilon - \xi(\boldsymbol{p}) - \Sigma_R(\varepsilon)}, \tag{21.11}$$

where Σ depends only on ε. When $\varepsilon < \omega_D$, $\Sigma(\varepsilon) \approx -b\varepsilon$, where b is a constant of the order of unity, whilst when $\varepsilon > \omega_D$, $\Sigma(\varepsilon) \approx \text{const} \sim \omega_D$. We first take the case of long-wave phonons $k \ll p_0$. In this case we can write approximately $\xi(\boldsymbol{p}-\boldsymbol{k}) \approx \xi(\boldsymbol{p}) - (\boldsymbol{v}\cdot\boldsymbol{k})$ and inasmuch as the region $\xi \sim vk$ is important in the integral, we can assume that $\xi(\boldsymbol{p}) = \varepsilon_0(\boldsymbol{p}) - \mu \approx v(|\boldsymbol{p}| - p_0)$. We shall make no use of this approximation for phonons with $k \sim p_0$, which we shall discuss below.

After substituting for G in (21.10), these equations can be integrated over ξ. The integral for $\operatorname{Re} \Pi_R$ is formally divergent, so that we take account of the fact that the integration over ξ is carried out between finite limits, which we denote by $-L_1$, L_2. These limits are of the order $\mu \sim p_0 v$, i.e. are substantially greater than Σ. In view of this we have

$$\operatorname{Re} \Pi_R = -\frac{p_0^2}{(2\pi)^3 v} \int d\Omega \int_{-\infty}^{\infty} d\varepsilon \tanh\frac{\varepsilon}{2T} \left\{ \frac{\theta(L_2-\varepsilon)\,\theta(\varepsilon+L_1)}{-\omega + (\boldsymbol{v}\cdot\boldsymbol{k}) - \Sigma(\varepsilon-\omega) + \Sigma(\varepsilon)} + \frac{\theta(L_2 - (\boldsymbol{v}\cdot\boldsymbol{k}) - \varepsilon)\,\theta(\varepsilon + L_1 + (\boldsymbol{v}\cdot\boldsymbol{k}))}{\omega - (\boldsymbol{v}\cdot\boldsymbol{k}) - \Sigma(\varepsilon+\omega) + \Sigma(\varepsilon)} \right\},$$

where $\theta(x) = \begin{cases} 1, & x > 0, \\ 0, & x < 0. \end{cases}$

Since $\Sigma(\varepsilon) \lesssim \omega_D$, the principal regions in the integral over ε are the neighbourhoods $\varepsilon \approx -L_1, L_2$, and $\Sigma(\varepsilon \pm \omega) \approx \Sigma(\varepsilon)$ in these regions. On taking account of this, and assuming $\omega \sim uk \ll vk$, we obtain

$$\operatorname{Re} \Pi_R = -\frac{2g^2 p_0^2}{(2\pi)^3 v} \int d\Omega \frac{(\boldsymbol{v}\cdot\boldsymbol{k})}{(\boldsymbol{v}\cdot\boldsymbol{k}) - \omega} \approx -\frac{g^2 m p_0}{\pi^2}. \tag{21.12}$$

Integral (21.10) for $\operatorname{Im} \Pi_R$ is convergent. In view of this, the limits of integration over ξ can be assumed infinite. We therefore obtain

$$\operatorname{Im} \Pi_R = \frac{\pi g^2 p_0^2}{(2\pi)^3 v} \int d\Omega \int_{-\infty}^{\infty} d\varepsilon \left(\tanh\frac{\varepsilon}{2T} - \tanh\frac{\varepsilon-\omega}{2T}\right) \times \delta[(\boldsymbol{v}\cdot\boldsymbol{k}) + \omega + \Sigma(\varepsilon) - \Sigma(\varepsilon-\omega)].$$

In the case $\omega \ll T$, the region $\varepsilon \sim T$ is important in the integral, whilst in the case $\omega > T$, the region $\varepsilon \sim \omega$ is important. In the present case $\omega \sim uk \ll \omega_D$ ($k \ll p_0 \sim k_D$), so that we always have $\varepsilon \ll \omega_D$. In view of this, $\Sigma(\varepsilon) = -b\varepsilon$ in the argument of the δ-function, where b is a constant (see below); consequently this argument is equal to $\delta[(\boldsymbol{v}\cdot\boldsymbol{k}) - \omega(1+b)]$. The integration over ε yields 2ω. We thus have

$$\operatorname{Im} \Pi_R = \frac{g^2 p_0 m}{2\pi} \frac{\omega}{vk}. \tag{21.13}$$

Notice that both $\operatorname{Re} \Pi_R$ and $\operatorname{Im} \Pi_R$ are the same as in the first approximation of perturbation theory, i.e. when $\mathfrak{G}^{(0)}$ is substituted for $\mathfrak{G}$ in (21.8). In the case of $\operatorname{Re} \Pi_R$, this circumstance is explained by the fact that the region of integration corresponds to high frequencies, where the self-energy is independent of frequency, whilst in the case of $\operatorname{Im} \Pi_R$ it is due to the linearity of $\Sigma(\varepsilon)$ for $\varepsilon \ll \omega_D$.

We now take the case $k \sim k_D \sim p_0$. We now always have $vk \gg \omega_D$. Since $\Sigma \lesssim \omega_D$, the G-function in (21.10) can be replaced by the free functions $G^{(0)}$. In view of the fact that $\omega \sim \omega_D \gg T$, we can substitute sign ε for $\tanh \varepsilon/2T$. We can put $\omega = 0$ in the integral for $\operatorname{Re} \Pi_R$. Here, however, it is not possible to make the substitution $\xi = \boldsymbol{p}^2/2m - p_0^2/2m \to v(|\boldsymbol{p}| - p_0)$. We find as a result:

$$\operatorname{Re} \Pi_R = -\frac{2g^2\pi}{(2\pi)^4} \mathrm{d}^3\boldsymbol{p}\, [\operatorname{Re} G^{(0)}(\boldsymbol{p}-\boldsymbol{k}, \xi(\boldsymbol{p}))\operatorname{sign}\xi(\boldsymbol{p}) + \operatorname{Re} G^{(0)}(\boldsymbol{p}, \xi(\boldsymbol{p}-\boldsymbol{k}))\operatorname{sign}\xi(\boldsymbol{p}-\boldsymbol{k})]$$

$$= -\frac{4g^2}{(2\pi)^3} \int\limits_{\substack{\xi\left(\boldsymbol{p}+\frac{\boldsymbol{k}}{2}\right)>0 \\ \xi\left(\boldsymbol{p}-\frac{\boldsymbol{k}}{2}\right)<0}} \mathrm{d}^3p \frac{1}{\xi\left(\boldsymbol{p}+\dfrac{\boldsymbol{k}}{2}\right) - \xi\left(\boldsymbol{p}-\dfrac{\boldsymbol{k}}{2}\right)} = -\frac{g^2 m p_0}{2\pi^2} h\left(\frac{u}{2p_0}\right) \tag{21.14}$$

where

$$h(x) = \left(1 + \frac{1-x^2}{2x} \ln\left|\frac{1+x}{1-x}\right|\right).$$

We now consider $\operatorname{Im} \Pi_R$. If we only retain the ω originating from $\tanh[(\varepsilon-\omega)/2T]$, we obtain

$$\operatorname{Im} \Pi_R = \frac{g^2\omega}{(2\pi)^2} \int \mathrm{d}^3\boldsymbol{p}\, \delta(\xi(\boldsymbol{p}) - \xi(\boldsymbol{p}-\boldsymbol{k})) = \frac{g^2 m p_0}{2\pi} \frac{\omega}{vk} \theta(2p_0 - k). \tag{21.15}$$

These expressions include as particular cases (21.12) and (21.13), valid for $k \ll p_0$, and they can therefore be regarded as universal. On using the analytic continuation of the Dyson temperature equation, we find that

$$D_R(\boldsymbol{k}, \omega) = D_{0R}^{-1} - \Pi_R$$

$$= \frac{1}{\omega_0^2(\boldsymbol{k})} \left\{\omega^2 - \omega_0^2(\boldsymbol{k}) \left[1 - \zeta h\left(\frac{k}{2p_0}\right) - \zeta \frac{i\pi m\omega}{p_0 k} \theta(2p_0 - k)\right]\right\}, \tag{21.16}$$

where the constant $\zeta = g^2 p_0 m/2\pi^2 \sim 1$ is introduced instead of g^2.

The pole of the function D_R determines the true energy of the phonons and their damping:

$$\omega(\boldsymbol{k}) = \omega_0(\boldsymbol{k}) \sqrt{1 - \zeta h\left(\frac{k}{2p_0}\right)}, \tag{21.17}$$

$$\gamma_1(\boldsymbol{k}) = \frac{\pi}{2} \frac{\omega_0^2(\boldsymbol{k}) m}{p_0 k} \theta(2p_0 - k). \tag{21.18}$$

According to (21.17), when $k \ll p_0$, $\omega(\boldsymbol{k}) = \omega_0(\boldsymbol{k}) \sqrt{1 - 2\zeta}$.

Close to $k = 2p_0$ the derivative $d\omega(\boldsymbol{k})/dk$ has a logarithmic singularity:

$$\frac{d\omega(\boldsymbol{k})}{dk} = \frac{\omega(2p_0)}{4p_0} \frac{\zeta}{1-\zeta} \ln \frac{4p_0}{|2p_0 - k|}. \qquad (21.19)$$

The damping vanishes discontinuously at this point. It is worth remarking that this behaviour of $\omega(\boldsymbol{k})$ and γ_1 is the result of the approximation made when deriving the formulae for Π_R. If, for example, we do not neglect ω in the $G^{(0)}$-functions appearing in (21.10), these singularities fall out. In particular,

$$\gamma_1 \sim \arctan\left[\omega m / p_0 (k - 2p_0)\right] \qquad (k > 2p_0),$$

i.e. it vanishes gradually, and not with a jump (†). Obviously, the corrections to the formulae obtained become important in the region $|k - 2p_0| \sim \omega/v \sim \omega_D/v \ll p_0$.

An accurate evaluation of the behaviour of $\omega(\boldsymbol{k})$ and γ_1 in this region requires not only that we take into account ω in (21.10), but also that we cannot replace G by $G^{(0)}$. Notice also that the difference between $\mathscr{T}$ and g becomes significant here. In view of all this, the working becomes extremely complicated. Since the region of which we are speaking is very small, the results that can be obtained along these lines do not justify the effort.

It is clear from (21.18) and (21.17) that, in the case when ζ is not too close to 1/2, the damping $\gamma_1(k)$ is relatively weak. In fact,

$$\frac{\gamma_1(k)}{\omega(k)} = \frac{\pi}{2} \frac{\zeta \omega_0^2(k)\, m}{p_0\, k\, \omega(k)} \sim \frac{\pi}{2} \zeta \frac{k}{v} \frac{1}{1-2\zeta} \sim \sqrt{\frac{m}{M}} (1-2\zeta)^{-1}. \qquad (21.20)$$

The phonon Green function $D_R(\boldsymbol{k}, \omega)$ can be written in the following form:

$$D(\boldsymbol{k}, \omega) = \frac{\omega_0^2(\boldsymbol{k})}{2\omega(\boldsymbol{k})} \left(\frac{1}{\omega - \omega(\boldsymbol{k}) + i\gamma_1(\boldsymbol{k})} - \frac{1}{\omega + \omega(\boldsymbol{k}) + i\gamma_1(\boldsymbol{k})} \right). \qquad (21.21)$$

It is clear from this that the function $D_R(\boldsymbol{k}, \omega)$ differs from $D_R^{(0)}(\boldsymbol{k}, \omega)$ by a change in eigenfreqency and a constant factor.

3. *The electron Green function*

We now turn to the electron Green function. The Dyson equation (16.3) for $\mathfrak{G}$ contains the self-energy Σ, which satisfies the equation

$$\Sigma(\boldsymbol{p}, \varepsilon) = -\frac{g^2 T}{(2\pi)^3} \sum_{\varepsilon_1} \int d^3 p_1\, \mathfrak{G}(\varepsilon_1, \boldsymbol{p}_1)\, \mathfrak{D}(\varepsilon - \varepsilon_1, \boldsymbol{p} - \boldsymbol{p}_1). \qquad (21.22)$$

As in the previous section, we carry out an analytic continuation into the region of real frequencies. To do this, we write (21.22) as a contour

(†) The maximum attained by $d\omega/dk$ is of the order v.

integral:

$$\Sigma(p, \varepsilon) = \frac{i g^2 T}{2(2\pi)^4} \int d^3 p_1 \int_C d\varepsilon_1 \tanh \frac{\varepsilon_1}{2T} G_{RA}(\varepsilon_1, p_1) D_{RA}(i\varepsilon - \varepsilon_1, p - p_1)$$

$$- \frac{g^2 T}{(2\pi)^3} \int d^3 p_1 \, \mathfrak{G}(\varepsilon, p) \, \mathfrak{D}(0, p - p_1). \quad (21.23)$$

The contour of integration is the same as in Fig. 61 ($\varepsilon > 0$). G_{RA} and D_{RA} have the same meaning as before. The second term is due to the fact that the contour integral does not include a term corresponding to $\varepsilon_1 = \varepsilon$ in (21.22). Remember that the frequencies in the $\mathfrak{D}$-function are even ($2\pi n T$).

The integral over the outer circumference vanishes as before and we are only left with the integrals over the horizontal lines $\varepsilon_1 = 0$ and $\varepsilon_1 = i\varepsilon$. On observing that $G_A = G_R^*$ and $\tanh[(\varepsilon + i\pi T(2n+1))/2T] = \coth(\varepsilon/2T)$, we can transform expression (21.23) in the same way as above for $\mathcal{H}$. The circuit round the point $\varepsilon_1 = i\varepsilon$ here yields a term which cancels the second term in (21.23). After this, it is easy to carry out a continuation into the upper half-plane of the variable ε. We obtain thus:

$$\Sigma_R(p, \varepsilon) = - \frac{g^2}{(2\pi)^4} \int d^3 p_1 \int_{-\infty}^{\infty} d\varepsilon_1 \tanh(\varepsilon_1/2T) \, \mathrm{Im}\, G_R(\varepsilon_1, p_1)$$

$$\times D_R(\varepsilon - \varepsilon_1, p - p_1) - \frac{g^2}{(2\pi)^4} \int d^3 p_1 \int_{-\infty}^{\infty} d\varepsilon_1 \coth(\varepsilon_1/2T)$$

$$\times G_R(\varepsilon - \varepsilon_1, p_1) \, \mathrm{Im}\, D_R(\varepsilon_1, p - p_1). \quad (21.24)$$

In accordance with formulae (17.14), (17.18):

$$G_R^*(\varepsilon, p) = \frac{1}{\pi} \int_{-\infty}^{\infty} \frac{\mathrm{Im}\, G_R(\varepsilon_1, p)}{\varepsilon_1 - \varepsilon - i\delta} d\varepsilon_1, \quad (21.25)$$

and the same holds for D_R. We substitute (21.25) in (21.24) and obtain after carrying out a change of variables:

$$\Sigma_R(p, \varepsilon) = - \frac{g^2}{(2\pi)^4 \pi} \int d^3 p_1 \int_{-\infty}^{\infty} d\omega \int_{-\infty}^{\infty} d\varepsilon_1 \frac{\mathrm{Im}\, G_R(\varepsilon_1, p) \, \mathrm{Im}\, D_R(\omega, p - p_1)}{\omega + \varepsilon_1 - \varepsilon - i\delta}$$

$$\times \left(\tanh \frac{\varepsilon_1}{2T} + \coth \frac{\omega}{2T} \right). \quad (21.26)$$

We find first of all the integral over p_1. We introduce for this purpose the new variables $\xi_1 = (p_1^2/2m) - (p_0^2/2m)$, $k = |p - p_1|$:

$$\int_0^{\infty} |p_1|^2 d^3 p_1 \int_{-1}^{1} dx = \frac{m}{|p|} \int_0^{k_D} k\, dk \int_{\xi(|p|-k)}^{\xi(|p|+k)} d\xi. \quad (21.27)$$

We shall seek G_R in the form (21.11). The imaginary part of this expression has the form:

$$\operatorname{Im} G_R(\boldsymbol{p}_1, \varepsilon_1) = \frac{\operatorname{Im} \Sigma_R(\varepsilon_1)}{[\varepsilon_1 - \xi_1 - \operatorname{Re} \Sigma_R(\varepsilon_1)]^2 + [\operatorname{Im} \Sigma_R(\varepsilon_1)]^2}. \quad (21.28)$$

If we regard the external momentum as close to p_0, the region of variation of ξ_1 depends on k, by (21.27). In the case when the region $k \sim k_D$ is important in the integral, we can assume in practice that ξ_1 varies between the limits $-\infty, \infty$. We shall also encounter later the case when momenta $k \ll p_0$ are important in the integral; here, however, the scale of all the energies appearing in the integral will be of the order $\omega(\boldsymbol{k}) \ll \omega_D$. In both cases $\operatorname{Im} \Sigma_R(\varepsilon) \ll \varepsilon$, so that

$$\operatorname{Im} G_R(\boldsymbol{p}_1, \varepsilon_1) \approx \pi \delta[\varepsilon_1 - \operatorname{Re} \Sigma_R(\varepsilon_1) - \xi_1] \operatorname{sign} \operatorname{Im} \Sigma_R(\varepsilon_1)$$
$$= -\pi \delta(\varepsilon_1 - \operatorname{Re} \Sigma_R(\varepsilon_1) - \xi_1).$$

On substituting this in (21.4), we obtain

$$\Sigma_R(\boldsymbol{p}, \varepsilon) = \frac{g^2}{(2\pi)^3} \frac{m}{|\boldsymbol{p}|} \int_0^{k_D} k \, dk \int_{\varepsilon'}^{\varepsilon''} d\varepsilon_1 \int_{-\infty}^{\infty} d\omega \frac{\operatorname{Im} D_R(\omega, \boldsymbol{k})}{\omega + \varepsilon_1 - \varepsilon - i\delta}$$
$$\times [\tanh(\varepsilon_1/2T) + \coth(\omega/2T)], \quad (21.29)$$

where $\varepsilon'' - \operatorname{Re} \Sigma_R(\varepsilon'') = \xi(|\boldsymbol{p}| + k)$, $\varepsilon' - \operatorname{Re} \Sigma_R(\varepsilon') = \xi(|\boldsymbol{p}| - k)$. We shall see below that the main role is played by the region $\varepsilon_1 \lesssim \omega_D$. In order for these small values to fall within the region of variation of ε_1, it is necessary for k to be less than $2p_0$. The actual upper limit of the integral over k is therefore $\min(k_D, 2p_0)$.

Let us consider $\operatorname{Re} \Sigma_R$, on the assumption, which will be justified later, that the region $k \sim k_D \sim p_0$ is important in the integral over k. In this case the integration over ε_1 can be performed between infinite limits. We transform from $\int_{-\infty}^{\infty} d\omega$ to $\int_0^{\infty} d\omega$:

$$\operatorname{Re} \Sigma_R = \frac{g^2}{(2\pi)^3} \frac{m}{p_0} \int_0^{k_1} k \, dk \int_{-\infty}^{\infty} d\varepsilon_1 \int_0^{\infty} d\omega \times$$
$$\left[\frac{\operatorname{Im} D_R(\omega, k)}{\omega + \varepsilon_1 - \varepsilon}\left(\tanh\frac{\varepsilon_1}{2T} + \coth\frac{\omega}{2T}\right) - \frac{\operatorname{Im} D_R(\omega, k)}{-\omega + \varepsilon_1 - \varepsilon}\left(\tanh\frac{\varepsilon_1}{2T} - \coth\frac{\omega}{2T}\right)\right]$$
$$= \frac{g^2}{(2\pi)^3} \frac{m}{p_0} \int_0^{k_1} k \, dk \int_{-\infty}^{\infty} d\varepsilon_1 \fint_0^{\infty} d\omega \frac{\operatorname{Im} D_R(\omega, k)}{\omega + \varepsilon_1} \tanh\left(\frac{\varepsilon + \varepsilon_1}{2T} + \tanh\frac{\varepsilon - \varepsilon_1}{2T}\right).$$

($k_1 = \min(2p_0, k_D)$. The antisymmetry of $\operatorname{Im} D_R(\omega) = -\operatorname{Im} D_R(-\omega)$ has been used here. In view of the connection between $\operatorname{Im} D_R$ and $\operatorname{Re} D_R$ (see (21.25)) and the antisymmetry of $\operatorname{Im} D_R$, the formula for $\operatorname{Re} \Sigma_R$ can be rewritten in the form

$$\operatorname{Re} \Sigma_R = \frac{g^2}{16\pi^2} \frac{m}{p_0} \int_0^{k_1} k \, dk \fint_{-\infty}^{\infty} d\omega \operatorname{Re} D_R(\omega, k) \left(\tanh\frac{\varepsilon + \omega}{2T} + \tanh\frac{\varepsilon - \omega}{2T}\right). \quad (21.30)$$

On substituting (21.21), we obtain

$$\operatorname{Re}\Sigma_R = \frac{g^2}{16\pi^2}\frac{m}{p_0}\int_0^{k_1} k\,\mathrm{d}k \fint_{-\infty}^{\infty} \mathrm{d}\omega \frac{\omega_0^2(k)}{\omega - \omega_0^2(k)}\left(\tanh\frac{\varepsilon+\omega}{2T} + \tanh\frac{\varepsilon-\omega}{2T}\right).$$

It follows at once from this that the region $k \sim k_1$ is important, which supports our above assumption. As regards ω, in the case $\varepsilon \gg T$, the region $\omega \sim \varepsilon$ or $\omega \sim \omega_D$ are important; whereas if $\varepsilon \ll T$, the region $\omega \sim T \ll \omega_D$ is important. In all these cases we can write

$$\operatorname{Re}\Sigma_R = \frac{g^2}{8\pi^2}\frac{m}{p_0}\int_0^{k_1} k\,\mathrm{d}k \fint_{-\varepsilon}^{\varepsilon} \mathrm{d}\omega \frac{\omega_0^2(k)}{\omega^2 - \omega^2(k)}. \tag{21.31}$$

On integrating over ω, we obtain

$$\operatorname{Re}\Sigma_R = \frac{-g^2}{8\pi^2}\frac{m}{p_0}\int_0^{k_1} k\,\mathrm{d}k \frac{\omega_0^2(k)}{\omega(k)} \ln\left|\frac{\varepsilon+\omega(k)}{\varepsilon-\omega(k)}\right|. \tag{21.32}$$

In this integral the region $k \sim k_1$ is important, i.e. $\omega(k) \sim \omega_D$. If $\varepsilon \ll \omega_D$, then

$$\operatorname{Re}\Sigma_R \approx -\frac{g^2}{4\pi^2}\frac{m}{p_0}\int_0^{k_1} k\,\mathrm{d}k \frac{\omega_0^2(k)}{\omega^2(k)}\varepsilon = -b\varepsilon, \tag{21.32'}$$

where b is a positive constant of order 1. When $\varepsilon \gg \omega_D$,

$$\operatorname{Re}\Sigma_R \approx -\frac{g^2}{4\pi^2}\frac{m}{p_0}\int_0^{k_1} k\,\mathrm{d}k\,\omega_0^2(k)\frac{1}{\varepsilon} \sim -\frac{\omega_D^2}{\varepsilon} \to 0.$$

Let us now consider the imaginary part of Σ_R. It is obviously obtained from a circuit round the pole in the denominator of (21.29):

$$\operatorname{Im}\Sigma_R(\mathbf{p},\varepsilon) = \frac{g^2\pi}{(2\pi)^3}\frac{m}{p_0}\int_0^{k_1} k\,\mathrm{d}k \int_{\varepsilon'}^{\varepsilon''} \mathrm{d}\varepsilon_1 \operatorname{Im} D_R(\varepsilon-\varepsilon_1, k)$$
$$\times\left(\tanh\frac{\varepsilon_1}{2T} + \coth\frac{\varepsilon-\varepsilon_1}{2T}\right). \tag{21.33}$$

If we substitute here for $\operatorname{Im} D_R$, which, by (21.21), is equal to

$$\operatorname{Im} D_R(\omega, k) = -\frac{2\omega_0^2(k)\,\omega\gamma_1(k)}{[\omega^1 - \omega^2(k)]^2 + 4\,\omega^2\gamma_1^2(k)}, \tag{21.34}$$

there are in essence two possible cases. In the case when ω varies in an interval which is large compared with $\gamma_1(k)$, we have

$$\operatorname{Im} D_R(\omega, k) \approx -\pi\omega_0^2(k)\,\delta\left(\omega^2 - \omega^2(k)\right)\operatorname{sign}\omega. \tag{21.35}$$

On the other hand, if the interval of variation of ω is much less than $\gamma_1(k)$, we can neglect ω in the denominator of (21.33) by comparison with the remaining terms, and we now have

$$\operatorname{Im} D_R(\omega, k) \approx -\frac{2\omega_0^2(k)\,\omega\gamma_1(k)}{\omega^4(k)}. \tag{21.36}$$

We substitute (21.35) in (21.33), after carrying out the change of variables $\varepsilon_1 - \varepsilon = \omega$. In view of the factor in front of tanh and coth, the important values here are $\omega \sim \max(\varepsilon, T)$. On the other hand, the δ-function requires $|\omega| \sim \omega(k) \sim uk$. Consequently the important values are $k \sim \varepsilon/u$, i.e. $vk \gg \varepsilon$. This means that the limits of the integral can be taken as infinite in the present case. We therefore obtain

$$\operatorname{Im} \Sigma_R(\boldsymbol{p}, \varepsilon) = \frac{g^2 m}{16\pi p_0} \int_0^{k_1} \frac{k\,\mathrm{d}k\, \omega_0^2(k)}{\omega(k)} \left[\tanh \frac{\omega(k) + \varepsilon}{2T} - \tanh \frac{\varepsilon - \omega(k)}{2T} - 2 \coth \frac{\omega(k)}{2T}\right]. \quad (21.37)$$

In the case when $\max(\varepsilon, T) \ll \omega_D$, we obtain

$$\operatorname{Im} \Sigma_R(\varepsilon) = -\frac{\pi \zeta T^3}{4(1 - 2\zeta) p_0^2 k^2} f_1\left(\frac{\varepsilon}{T}\right), \quad (21.38)$$

where $u = \mathrm{d}\omega(k)/\mathrm{d}k$ is the velocity of sound,

$$f_1(x) = \int_0^\infty z^2 \mathrm{d}z \left[\coth \frac{z}{2} - \frac{1}{2} \tanh \frac{z + x}{2} - \frac{1}{2} \tanh \frac{z - x}{2}\right]. \quad (21.39)$$

We find from this, when $\varepsilon \ll T \ll \omega_D$:

$$\operatorname{Im} \Sigma_R(\varepsilon) = -\frac{7\pi\zeta(3)}{8} \frac{\zeta}{1 - 2\zeta} \frac{T^3}{p_0^2 u^2} \quad (21.40)$$

(here $\zeta(3)$ is the Riemann zeta function).

When $T \ll \varepsilon \ll \omega_D$ we have

$$\operatorname{Im} \Sigma_R(\varepsilon) = -\frac{\pi\zeta |\varepsilon|^3}{12(1 - 2\zeta) p_0^2 u^2}. \quad (21.41)$$

In the case when $\varepsilon \gg \omega_D$ ($T \gg \omega_D$ by hypothesis), we obtain

$$\operatorname{Im} \Sigma_R(\varepsilon) = \text{const} \sim \omega_D. \quad (21.42)$$

The important values in this derivation have been $\omega \sim \max(\varepsilon, T)$. Therefore, according to the above, the replacement of (21.34) by (21.35) is only valid for $\max(\varepsilon, T) \gg \max \gamma_1(k) \sim \omega_D \sqrt{m/M}$. In the opposite limiting case we have to use formula (21.36). Here, the important values in the integral are $k \sim k_1$, so that

$$\operatorname{Im} \Sigma_R(\varepsilon) = -\frac{g^2 m}{2\pi^2 p_0} \int_0^{k_1} k\,\mathrm{d}k \frac{\omega_0^2(k) \gamma_1(k)}{\omega^4(k)} \int_0^\infty \omega\,\mathrm{d}\omega \left[\coth \frac{\omega}{2T} - \frac{1}{2} \tanh \frac{\omega + \varepsilon}{2T} - \frac{1}{2} \tanh \frac{\omega - \varepsilon}{2T}\right]. \quad (21.43)$$

The integral over the frequencies can be found. We obtain as a result, when $\max(\varepsilon, T) \ll \omega_D \sqrt{m/M}$:

$$\operatorname{Im} \Sigma_R(\varepsilon) = -\frac{d}{p_0 v} (\pi^2 T^2 + \varepsilon^2), \quad (21.44)$$

where

$$d = \frac{g^2 p_0^2}{4\pi^2} \int_0^{k_1} k \, dk \, \frac{\omega_0^2(k)\, \gamma_1(k)}{\omega^4(k)} = \text{const} \sim 1 .$$

Formulae (21.38) and (21.44) correspond to different damping mechanisms. The first determines the damping due to radiation and absorption of phonons by electrons. However, when the energy of the quasi-particles is very close to the Fermi surface: $\varepsilon \ll \omega_D/\sqrt{m/M}$, the electron interactions due to exchange with phonons become important, and damping (21.44) becomes operative.

It follows from what was said earlier (see § 19) that this is precisely the form that the damping due to the interactions of the Fermi particles must have.

The energy of the electron excitations can be determined from the real part of the pole of the function G_R. We have by (21.11):

$$\varepsilon - \Sigma_R(\varepsilon) = \xi .$$

In the case $\varepsilon \ll \omega_D$, we obtain with the aid of (21.32):

$$\varepsilon = \frac{v}{1+b}(|\boldsymbol{p}| - p_0) . \qquad (21.45)$$

The velocity of the quasi-particles thus decreases on the Fermi surface ($b > 0$). In addition, the G-function acquires the form (18.1) close to its pole, where $a = (1+b)^{-1} < 1$. It follows from (21.41) that the quasi-particle damping is equal to their energy when $|\varepsilon| \sim \omega_D$. However, it may easily be seen that (see 21.42) on further increase of the excitation energy the damping ceases to increase and again becomes less than the quasi-particle energy. There are consequently two regions in which the concept of quasi-particles has a meaning: $|\varepsilon| \ll \omega_D$ and $|\varepsilon| \gg \omega_D$. In both regions the electron energy has the form $v(|\boldsymbol{p}| - p_0)$, but the velocities v are different.

We now observe qualitatively what the result would be if we took into account the direct Coulomb interactions of the electrons. As a result of the screening of the Coulomb forces at distances of the order of the lattice period (i.e. of the order $1/p_0$), these can still be regarded as short-range in the present case. If we take account of these forces we obtain a different velocity at the Fermi surface and a different coefficient a in the G-function close to the pole.

A qualitative difference arises in the magnitude of the damping. The Coulomb interaction yields a damping expressed by (19.30). In the region $(\varepsilon, T) \ll \omega_D\sqrt{m/M}$ it will be added to damping (21.44), which has the same structure and the same order of magnitude, whereas, in the region $(\varepsilon, T) \gg \omega_D\sqrt{m/M}$ the phonon damping becomes predominant. As already remarked above, when $\varepsilon \gg \omega_D$ the damping becomes constant, of the order ω_D. Coulomb damping is also present in this region and starts to become predominant when

$$|\varepsilon| \gg \sqrt{p_0 v_0 \omega_D} \sim \omega_D \sqrt[4]{M/m} > \omega_D.$$

It follows from this that, whilst the excitation spectrum in the region $|\varepsilon| \gg \omega_D$ is determined by the Coulomb interaction of the electrons, the phonon damping continues for some time to dominate over the electron damping.

4. *A correction to the linear term in the electronic specific heat*

We can draw an interesting conclusion from the results of the preceding sections, concerning the electronic specific heat. It appears at first sight that the correction to the linear term (19.25) (linear in the temperature) must be of the relative order $(T/\mu)^2$. In fact, the electron-phonon interaction leads to the appearance of a substantially greater correction of the order b.

Let us consider the general formula (21.32) for $\operatorname{Re} \Sigma$. When $\varepsilon \ll \omega_D$, we obtain to a first approximation expression (21.32′). It may easily be seen that an expansion of the logarithm in the integrand of (21.32) up to terms of order ε^3 leads to the appearance of a logarithmically divergent integral over k. If we confine ourselves to a term of the order of the logarithmic one, we have (†)

$$\delta \operatorname{Re} \Sigma_R = -\frac{1}{6} \frac{\zeta}{1-2\zeta} \frac{1}{p_0^2 u^2} \varepsilon^3 \ln \frac{\omega_D}{|\varepsilon|}.$$

We took this correction into account when obtaining the entropy (§ 19.5). We find that

$$\delta \frac{S}{V} = 2\int \frac{d^3 \boldsymbol{p}}{(2\pi)^3} \frac{1}{T} \int_{-\infty}^{\infty} \left(-\frac{\partial n_F(\varepsilon)}{\partial \varepsilon}\right) \varepsilon \operatorname{Im} G_R \delta \operatorname{Re} \Sigma_R(\varepsilon) \, d\varepsilon .$$

On substituting $\operatorname{Im} G_R = -\pi \delta(\xi - \varepsilon - \Sigma_R(\varepsilon))$ and taking the integral with the Fermi function, we obtain

$$\delta \frac{S}{V} = \frac{7\pi^2}{180} \frac{\zeta}{1-2\zeta} \frac{m}{p_0^2 u^2} T^3 \ln \frac{\omega_D}{T}.$$

On differentiating with respect to T, we have

$$\delta \frac{C}{v} = \frac{7\pi^2}{60} \frac{\zeta}{1-2\zeta} \frac{m}{p_0^2 u^2} T^3 \ln \frac{\omega_D}{T}. \tag{21.46}$$

(†) Notice that $\delta \operatorname{Re} \Sigma_R$ and the damping (21.41) are the real and imaginary parts of the same analytic function

$$\delta \Sigma_R = \frac{1}{12} \frac{\zeta}{1-2\zeta} \frac{\varepsilon^3}{p_0^2 u^2} \ln \frac{\omega_D^2}{(-\varepsilon + i\delta)^2}.$$

The correction to the linear term in the specific heat thus proves to be of the order $[\zeta/(1-2\zeta)]\,(T^2/\omega_D^2)\ln(\omega_D/T)$.

It is interesting to compare this correction with the cubic term coming from the lattice vibrations. The order of magnitude of the specific heat of the lattice is T^3/u^3. It follows from this that the correction obtained amounts to $[\zeta/(1-2\zeta)]\,(u/v)\ln(\omega_D/T)$ of the specific heat of the lattice, i.e. as a rule it must be regarded as small. Since, however, we are talking about one order of magnitude at most, this does not exclude the possibility that, in individual metals, the term may make a substantial contribution to the specific heat (†).

§ 22. SOME PROPERTIES OF A DEGENERATE PLASMA

1. Statement of the problem

We shall consider a plasma, i.e. a mixture of electron and ion gases, as an example of a system with Coulomb interactions. The interaction Hamiltonian is

$$\begin{aligned}H_{int} = \frac{e^2}{2}\int \psi_\alpha^+(\boldsymbol{r})\,\psi_\beta^+(\boldsymbol{r}')\,\frac{1}{|\boldsymbol{r}-\boldsymbol{r}'|}\,\psi_\beta(\boldsymbol{r}')\,\psi_\alpha(\boldsymbol{r})\,\mathrm{d}^3\boldsymbol{r}\,\mathrm{d}^3\boldsymbol{r}' \\ -Ze^2\int \psi_\alpha^+(\boldsymbol{r})\Phi^+(\boldsymbol{r}')\,\frac{1}{|\boldsymbol{r}-\boldsymbol{r}'|}\,\Phi(\boldsymbol{r}')\,\psi_\alpha(\boldsymbol{r})\,\mathrm{d}^3\boldsymbol{r}\,\mathrm{d}^3\boldsymbol{r}' \\ +\frac{Z^2e^2}{2}\int \Phi^+(\boldsymbol{r})\Phi^+(\boldsymbol{r}')\,\frac{1}{|\boldsymbol{r}-\boldsymbol{r}'|}\,\Phi(\boldsymbol{r}')\Phi(\boldsymbol{r})\,\mathrm{d}^3\boldsymbol{r}\,\mathrm{d}^3\boldsymbol{r}', \quad (22.1)\end{aligned}$$

where $\psi_\alpha(\boldsymbol{r})$ is the operator of the electron field, and Φ of the ionic field. We shall assume the electron gas to be degenerate, and the ion gas to be a Boltzmann gas. This requires that the temperature satisfy the inequalities

$$\frac{1}{m}\left(\frac{N}{V}\right)^{2/3} \gg T \gg \frac{1}{M}\left(\frac{N}{V}\right)^{2/3}. \quad (22.2)$$

We can arbitrarily regard the ions as a Fermi gas for our calculations, since the Boltzmann limit is the same for both statistics.

Suppose, further, that the effect of the Coulomb interaction is small. This requires that

$$\frac{e^2}{\bar{E}\,\bar{r}} \ll 1,$$

(†) The presence of this term can easily be discovered if the metal passes to the superconducting state (see Chap. VII). This transition does not affect the lattice specific heat, but it leads to an exponential drop in the electronic specific heat as $T \to 0$. If we compare the cubic term in the specific heat at a temperature above the superconducting transition temperature T_c and at $T \to 0$, it can be observed that the electronic term (21.46) is added to the cubic term above T_c.

where $\bar{E}$ is a mean energy, and $\bar{r}$ the mean distance between the particles. For the ions, $\bar{E} \sim T$, whilst for the electrons, $\bar{E} \sim p_0^2/2m$. Our condition is therefore equivalent to the following requirements:

$$\frac{T}{e^2} \gg \left(\frac{N}{V}\right)^{1/3} \gg e^2 m. \tag{22.3}$$

It may easily be seen that conditions (22.3) do not contradict the assumption of the degeneracy of the electron gas.

When conditions (22.3) are satisfied, the Coulomb interaction will almost always have little effect on the properties of the plasma. An exception is provided by the case when collisions of the particles with small momentum transfer are important. Due to the fact that the Fourier component of the Coulomb potential has the form(†)

$$U(k) = \frac{4\pi e^2}{k^2}, \tag{22.4}$$

the role of collisions with small transfer becomes extremely important.

We must first of all look at an apparent difficulty connected with Coulomb interactions. The diagrams of G-functions of § 8 include diagrams such as Fig. 5 *b*, containing $U(0)$. Similar diagrams are also obtained for the temperature-dependent $\mathfrak{G}$-functions. According to (22.4), such diagrams diverge. Let us change from a given chemical potential to a given number of particles and replace the Coulomb potential by the potential $U(r) = e^2(e^{-\alpha r}/r)$. We shall regard α as small, and in the end put it equal to zero. In order to avoid the difficulties that arise in specifying the number of particles, we shall carry out our discussion in $\boldsymbol{r}$, τ-space.

Let us take, for instance, the electron line $\mathfrak{G}_e$. We sum all the irreducible self-energy parts of the type illustrated in Figs. 4 *a* and 8 *e*, *f*, which are joined to the fundamental $\mathfrak{G}$-line by a single wavy line. It is easily seen that all these diagrams together yield

$$\begin{aligned} \Sigma'(\boldsymbol{r}-\boldsymbol{r}', \tau-\tau') &= 2e^2 \int d^3\boldsymbol{r}\,[\mathfrak{G}_e(0,-0) - Z\,\mathfrak{G}_i(0,-0)] \\ &\quad \times \frac{\exp(-\alpha\,|\boldsymbol{r}-\boldsymbol{r}_1|)}{|\boldsymbol{r}-\boldsymbol{r}_1|}\,\delta(\boldsymbol{r}-\boldsymbol{r}')\,\delta(\tau-\tau') \\ &= \frac{N_e - ZN_i}{V}\,\frac{4\pi e^2}{\alpha^2}\,\delta(\boldsymbol{r}-\boldsymbol{r}')\,\delta(\tau-\tau') \equiv 0 \end{aligned} \tag{22.5}$$

by virtue of the condition for electric neutrality: $N_e = ZN_i$.

The analogous corrections for the $\mathfrak{G}_i$-lines similarly vanish. This means that all the diagrams containing the integral of the Coulomb potential over the volume (the zero Fourier component) should simply be put equal to zero.

(†) We do not use 4-vectors in this section, so that ordinary type will denote the absolute values of the three-dimensional vectors.

After this, the following formal method can be used for returning to the representation with a given μ. We write all the $\mathfrak{G}^{(0)}$-functions in the diagrams in the form

$$\mathfrak{G}_N^{(0)} = \mathfrak{G}_\mu^{(0)} \exp[-\mu(\tau_1 - \tau_2)]. \tag{22.6}$$

It can easily be proved that the resultant $\mathfrak{G}$-function will be represented simply by the diagrams with $\mathfrak{G}_\mu^{(0)}$, multiplied by the same factor $\exp[-\mu(\tau_1 - \tau_2)]$. On dividing by this factor, we transform to $\mathfrak{G}_\mu$.

It is clear from this that the scheme for operating with the Coulomb potential amounts simply to throwing away all the diagrams containing $U(0)$. However, it must be borne in mind here that the results thus obtained will only be correct when the chemical potential is chosen in such a way that $N_e(\mu_e, \mu_i) = ZN_i(\mu_e, \mu_i)$, or what amounts to the same thing,

$$\frac{\partial \Omega}{\partial \mu_e} = Z \frac{\partial \Omega}{\partial \mu_i}. \tag{22.7}$$

2. *Vertex part for small momentum transfer*

Let us first of all consider the vertex part with small momentum transfer. In view of the fact that the ions are extremely important here, we apply the temperature method. The first order correction to expression (22.4) is illustrated by two diagrams such as Fig. 62*a*. In the first of these diagrams we have an electron loop and in the second an ion loop.

Fig. 62

Although this correction contains the extra factor e^2, a factor $(1/k^2)^2$ also appears in it. It can therefore become important at small transfer, and we have to sum a chain of diagrams as illustrated in Fig. 62*b*, with any number of electron and ion loops (†). As a result, it turns out that all the vertices — electron-electron, ion-ion and ion-electron — are multiplied by the same factor, i.e. by

$$\mathcal{T}_{12}(\mathbf{k}, \omega_m) = \frac{4\pi e^2 Z_1 Z_2}{k^2 \left[1 - \frac{4\pi e^2}{k^2} [\mathcal{K}_e(\mathbf{k}, \omega_m) + Z^2 \mathcal{K}_i(\mathbf{k}, \omega_m)]\right]} = \frac{4\pi e^2 Z_1 Z_2}{k^2 - 4\pi e^2 [\mathcal{K}_e(\mathbf{k}, \omega_m) + Z^2 \mathcal{K}_i(\mathbf{k}, \omega_m)]}; \tag{22.8}$$

here, $\omega_m = 2\pi T m$, where m is an integer, $\mathcal{K}_e$ corresponds to an electron loop, and $\mathcal{K}_i$ to an ion loop.

(†) Such a summation for a system with Coulomb interactions was first carried out by GELL-MANN and BRUECKNER [39].

As already mentioned, the ions can arbitrarily be regarded as a Fermi gas. The evaluation of $\mathcal{H}_e$ and $\mathcal{H}_i$ is therefore identical in the first stage:

$$\mathcal{H} = 2T \sum_n \int \frac{\mathrm{d}^3 \boldsymbol{p}}{(2\pi)^3} \frac{1}{[i\varepsilon_n - \varepsilon_0(\boldsymbol{p}) + \mu]\,[i\varepsilon_n + i\omega_m - \varepsilon_0(\boldsymbol{p}+\boldsymbol{k}) + \mu]}$$

$$= 2T \int \frac{\mathrm{d}^3 \boldsymbol{p}}{(2\pi)^3} \frac{1}{i\omega_m - \varepsilon_0(\boldsymbol{p}+\boldsymbol{k}) + \varepsilon_0(\boldsymbol{p})} \sum_n \left[\frac{1}{i\varepsilon_n - \varepsilon_0(\boldsymbol{p}) + \mu} - \frac{1}{i(\varepsilon_n + \omega_m) - \varepsilon_0(\boldsymbol{p}+\boldsymbol{k}) + \mu}\right]$$

$$= -2T \int \frac{\mathrm{d}^3 \boldsymbol{p}}{(2\pi)^3} \frac{1}{i\omega_m - \varepsilon_0(\boldsymbol{p}+\boldsymbol{k}) + \varepsilon_0(\boldsymbol{p})}$$

$$\times \sum_{n>0} \left[\frac{2(\varepsilon_0(\boldsymbol{p}) - \mu)}{\pi^2 T^2 (2n+1)^2 + (\varepsilon_0(\boldsymbol{p}) - \mu)^2} - \frac{2(\varepsilon_0(\boldsymbol{p}+\boldsymbol{k}) - \mu)}{\pi^2 T^2 (2n+1)^2 + (\varepsilon_0(\boldsymbol{p}+\boldsymbol{k}) - \mu)^2}\right]$$

$$= -\int \frac{\mathrm{d}^3 \boldsymbol{p}}{(2\pi)^3} \frac{1}{i\omega_m - \varepsilon_0(\boldsymbol{p}+\boldsymbol{k}) + \varepsilon_0(\boldsymbol{p})} \left[\tanh \frac{\varepsilon_0(\boldsymbol{p}) - \mu}{2T} - \tanh \frac{\varepsilon_0(\boldsymbol{p}+\boldsymbol{k}) - \mu}{2T}\right]$$

$$= -2 \int \frac{\mathrm{d}^3 \boldsymbol{p}}{(2\pi)^3} \frac{n(\boldsymbol{p}+\boldsymbol{k}) - n(\boldsymbol{p})}{i\omega_m - \varepsilon_0(\boldsymbol{p}+\boldsymbol{k}) + \varepsilon_0(\boldsymbol{p})}, \tag{22.9}$$

where $n(\boldsymbol{p}) = \{\exp[\varepsilon_0(p) - \mu]/T + 1\}^{-1}$ for the electrons, and $2n(\boldsymbol{p}) = \exp[\mu - \varepsilon_0(p)]/T$ for the ions. Use has been made here of the formula

$$\sum_{n>0} \frac{1}{(2n+1)^2 + x^2} = \frac{\pi}{4x} \tanh \frac{\pi x}{2}. \tag{22.10}$$

We shall assume that $|\boldsymbol{k}| \ll p_0$. The average momentum of an ion is of the order $\sqrt{MT}$, and by (22.2), is much greater than p_0. In view of this, we can expand (22.9) in powers of $\boldsymbol{k}$ and write it in the simplified form

$$\mathcal{H}(\boldsymbol{k}, \omega_m) = -2 \int \frac{\mathrm{d}^3 \boldsymbol{p}}{(2\pi)^3} \frac{\partial n}{\partial \varepsilon} \frac{(\boldsymbol{v} \cdot \boldsymbol{k})}{i\omega_m - (\boldsymbol{v} \cdot \boldsymbol{k})}. \tag{22.11}$$

The vertex part in question depends only on $\boldsymbol{k}$ and ω_m. It can therefore be associated with a $\mathfrak{D}$-function, transmitting the electromagnetic interaction between the particles. Indeed, let us consider, for instance, the following quantity:

$$\mathfrak{D}(\boldsymbol{r}_1 - \boldsymbol{r}_2, \tau_1 - \tau_2) = \langle T(\bar{\tilde{\psi}}_\alpha(\boldsymbol{r}_1\tau_1)\bar{\tilde{\psi}}_\beta(\boldsymbol{r}_2\tau_2)\tilde{\psi}_\alpha(\boldsymbol{r}_1\tau_1)\tilde{\psi}_\beta(\boldsymbol{r}_2\tau_2))\rangle,$$

where $\langle \cdots \rangle$ indicates here an ordinary statistical average. It may easily be verified that the Fouriertransform of this quantity with respect to the variables $\tau_1 - \tau_2$ and $\boldsymbol{r}_1 - \boldsymbol{r}_2$ has all the properties of the Bose temperature-dependent Green function. On the other hand, this Fourier transform is obviously equal to

$$\mathfrak{D}(\boldsymbol{k}, \omega_m) = \mathcal{H}_e(\boldsymbol{k}, \omega_m)\mathcal{T}_{ee}(\boldsymbol{k}, \omega_m)\mathcal{H}_e(\boldsymbol{k}, \omega_m).$$

The same applies to the other vertices with small transfer.

The method of changing from the temperature-dependent to the time-dependent functions becomes clear from this. We know that, in the case of the Green function, all that is required for this is to find the function which is analytic in the upper half-plane of the variable ω and is the same as the temperature-dependent Green function at the points $i\omega_m = i \, . \, 2\pi T m$. The retarded function D_R is determined in this way. The present Green function is equal to D_R when $\omega > 0$ and to D_R^* when $\omega < 0$.

This procedure can easily be carried out for the functions $\mathcal{H}_e$ and $\mathcal{H}_i$. (We denote the functions thus obtained by Π_e, Π_i.) It follows from integral (22.11) that this requires the change of variable $i\omega_m \to \omega + i\delta \operatorname{sign} \omega$. Since only the factors $\mathcal{H}^n$ are dependent on ω_m in the diagrams forming the function $\mathfrak{D}(\boldsymbol{k}, \omega_m)$, we can obtain the function $D(\boldsymbol{k}, \omega)$ by the same method. The same is obviously true for the functions $\Gamma(\boldsymbol{k}, \omega)$ (†). The time-dependent vertex parts $\Gamma(\boldsymbol{k}, \omega)$ are therefore expressible by the same formulae (22.8), (22.10), except for the obvious substitution $i\omega_m \to \omega + i\delta \operatorname{sign} \omega$. The appearance of a term additional to k^2 in the denominator of Γ is precisely the Debye screening of the Coulomb interaction. It makes the interaction retarded in the general case (i.e. Γ depends on ω).

Let us now consider the behaviour of Γ, depending on the relation between ω and k. When $\omega \ll v_i k$, (22.11) gives

$$\mathcal{H}_e = \Pi_e = -\frac{1}{V}\frac{\partial N_e}{\partial \mu_e}, \quad \mathcal{H}_i = \Pi_i = -\frac{1}{V}\frac{\partial N_i}{\partial \mu_i}. \tag{22.12}$$

We have $(1/V)(\partial N/\partial \mu) \sim N/V\mu$ in order of magnitude. For electrons $\mu_e \sim p_0^2/m$, and for ions $\mu_i \sim T \ln T$. Hence $\Pi_i \gg \Pi_e$.

We obtain from (22.8) and (22.11):

$$\Gamma_{12} = \frac{4\pi e^2 Z_1 Z_2}{k^2 + \varkappa_i^2}, \tag{22.13}$$

where $\varkappa_i = \sqrt{4\pi Z^2 e^2 N_i / VT}$ is the reciprocal of the Debye radius of the ions.

The next region is $v_e k \gg |\omega| \gg v_{ik}$. In this region

$$\Pi_i = -\frac{k^2}{\omega^2}\frac{N}{MV}, \quad \Pi_e = \frac{1}{V}\frac{\partial N_e}{\partial \mu} = \frac{p_0 m}{\pi^2}. \tag{22.14}$$

(†) A procedure that is correct for individual terms of a series may not be correct for its sum. However, it can be shown to lead to the correct result in the present case.

On substituting in (22.13), we have

$$\Gamma_{12} = \frac{4\pi e^2 Z_1 Z_2 \omega^2}{(k^2 + \varkappa_e^2)\,\omega^2 - k^2 \omega_{p1}^2}, \tag{22.15}$$

where $\varkappa_e = \sqrt{4 p_0 m e^2/\pi}$ is the inverse Debye radius of the electrons, and ω_{p1} is equal to

$$\omega_{p1} = \sqrt{\frac{4\pi N_i Z^2 e^2}{M V}}. \tag{22.16}$$

The function Γ_{12} has a pole when

$$\omega(k) = \frac{\omega_{p1} k}{\sqrt{k^2 + \varkappa_e^2}}. \tag{22.17}$$

When $k \ll \varkappa_e$ the function $\omega(k)$ is linear. This is the so-called "ionic sound". The velocity of the sound is

$$\frac{\omega(k)}{k} = \frac{\omega_{p1}}{\varkappa_e} = p_0 \sqrt{\frac{Z}{3 M m}}. \tag{22.18}$$

When $k \gg \varkappa_e$, ω approaches a constant value $\omega \approx \omega_{p1}$. The pole of Γ_{12} remains real as far as $k \sim \omega_{p1}/v$, so that $\omega(k)$ attains the value ω_{p1} in the case when $\omega_{p1} \gg v_i \varkappa_e$. This condition is observed by virtue of the first of inequalities (22.2).

The damping of these vibrations is determined by the imaginary parts of Π_i and Π_e. On substituting in (22.11) (with $i\omega_m \to \omega + i\delta \operatorname{sign} \omega$) the equilibrium distributions for the electrons and ions, we obtain

$$\operatorname{Im} \Pi_i = \frac{|\omega|}{k} \frac{N_i}{V} \sqrt{\frac{2\pi M}{T}} \exp\left(-M\omega^2/2k^2 T\right),$$
$$\operatorname{Im} \Pi_e = \frac{|\omega|}{k} \frac{m^2}{2\pi}. \tag{22.19}$$

The index of the exponential in $\operatorname{Im} \Pi_i$ is of the order p_0^2/mT or less, whilst the ratio of the factors in front of the exponentials in $\operatorname{Im} \Pi_i$ and $\operatorname{Im} \Pi_e$ is of the order $(p_0/mT)^{3/2}\sqrt{M/m}$. In view of this, any of these quantities may prove to be the main one. On determining the imaginary part of the pole of Γ_{12}, we obtain the damping of the vibrations:

$$\gamma = \frac{\omega^4}{k^3}\left[\sqrt{\frac{\pi}{8}}\left(\frac{M}{T}\right)^{3/2} \exp\left(-M\omega^2/2k^2 T\right) + \frac{m^2 e^2}{\omega_{p1}^2}\right], \tag{22.20}$$

where ω is determined by (22.19).

When $k v_e \gg \omega \gg \omega_{p1}$, we obtain by (22.15):

$$\Gamma_{12} = \frac{4\pi e^2 Z_1 Z_2}{k^2 + \varkappa_e^2}. \tag{22.21}$$

In the case $\omega \sim kv_e$ the electron loop predominates over the ionic. By (22.11), the total value of Π_e is equal to

$$\Pi_e = \frac{p_0^2}{\pi^2 v_e}\left[1 - \frac{\omega}{2kv_e}\ln\left|\frac{\omega + kv_e}{\omega - kv_e}\right| + \frac{i\pi|\omega|}{2kv_e}\theta(kv_e - |\omega|)\right]. \quad (22.22)$$

We obtain from this, in the region $\omega \gg kv_e$:

$$\Pi_e = -\frac{p_0^3}{\pi^2 m}\frac{k^2}{3\omega^2}\left[1 + \frac{3}{5}\frac{v_e^2 k^2}{\omega^2}\right].$$

On substituting this in (22.8), we have

$$\Gamma_{12}(\boldsymbol{k}, \omega) = \frac{4\pi e^2 Z_1 Z_2}{k^2\left[1 - \frac{4\pi e^2 N_e}{m\omega^2 V}\left(1 + \frac{3}{5}\frac{v_e^2 k^2}{\omega^2}\right)\right]}. \quad (22.23)$$

This expression has a pole at the point

$$\omega^2 = \omega_{p2}^2 + \frac{3}{5}v_e^2 k^2, \quad (22.24)$$

where

$$\omega_{p2}^2 = \frac{4\pi e^2 N_e}{mV}, \qquad v_e k \ll \omega_{p2}. \quad (22.25)$$

The pole corresponds to the so-called plasma vibrations of the electrons. The dispersion of the vibrations is expressed by a small additional term. The damping of the oscillations may be obtained as in the previous case, if we take into account the exponentially small contribution of the circuit round the pole in the transformed integral (22.11). It proves to be proportional to $\exp[-m\omega_{p2}^2/2k^2T]$. This expression proves to be incorrect at very low temperatures, since the damping contains larger terms that come from the subsequent approximations in e^2.

It follows from (22.23) that, in the limit as $k/\omega \to 0$, $\omega \to 0$, $\Gamma \to -\infty$. It is therefore clear from this example that Γ^ω contains an infinite constant in the case of Coulomb interaction.

3. *The electron spectrum*

We shall now find the electron Green function (†). The first order term added to the self-energy part is illustrated in Fig. 63. The analytic expression for this term is of the form:

$$\Sigma_1 = -4\pi e^2 T \sum_{\varepsilon_1} \int \frac{d^3 \boldsymbol{p}}{(2\pi)^3} e^{i\varepsilon_1 t}_{t\to+0}\, \mathfrak{G}_e^{(0)}(\varepsilon, \boldsymbol{p}_1)\frac{1}{(\boldsymbol{p} - \boldsymbol{p}_1)^2}. \quad (22.26)$$

(†) This section is based on ref. [77], and also on a calculation for $T \neq 0$ made by A. KOCHKIN.

On taking the sum over ε_1 in accordance with rules (17.33) and (22.10), we obtain

$$\Sigma_1 = -4\pi e^2 \int \frac{d^3 \boldsymbol{p}_1}{(2\pi)^3} n_F(\boldsymbol{p}_1) \frac{1}{(\boldsymbol{p}-\boldsymbol{p}_1)^2},$$

Fig. 63

where n_F is the Fermi function.

We add and subtract the same expression, but with n_F $(T = 0)$, and use the fact that the difference $n_F(\boldsymbol{p}, T) - n_F(\boldsymbol{p}, 0)$ tends to zero rapidly on moving away from the Fermi boundary. This enables us to replace $\xi_1 = (p_1^2/2m) - (p_0^2/2m)$ in the relevant integral by $v(p_1 - p_0)$. We thus obtain, after integration over the angles (†):

$$\Sigma_1 = -\frac{e^2}{\pi p} \int_0^{p_0} p_1 dp_1 \ln \left| \frac{p_1 + p}{p_1 - p} \right| - \frac{e^2}{\pi v} \int_0^{\infty} \frac{d\xi_1}{e^{\xi_1/T} + 1} \ln \frac{\xi_1 + \xi}{|\xi_1 - \xi|}.$$

On carrying out the first integration, we obtain after minor transformations:

$$\Sigma_1 = \frac{e^2 m}{\pi p} \left[\xi \left(\ln \frac{(p+p_0)^2}{2mT} - 1 \right) - T f\left(\frac{\xi}{T} \right) \right] - \frac{e^2 p_0}{\pi}, \tag{22.27}$$

where $\xi = (p^2/2m) - (p_0^2/2m)$,

$$f(x) = \frac{1}{2} \int_0^{\infty} dz \ln z \left[\tanh \frac{1}{2}(z + x) - \tanh \frac{1}{2}(z - x) \right].$$

In view of the fact that this expression does not depend on ε, it represents the correction to the energy of the quasi-particles. Since the Fermi limiting momentum is not affected by interactions and at the same time is connected with the chemical potential by the relationship $\varepsilon(p_0) = \mu$, the expression for Σ_1 at $p = p_0$ must be regarded as the variation of the chemical potential

$$\Delta\mu = -e^2 p_0/\pi. \tag{22.28}$$

The remaining part $\Sigma_1 - \Delta\mu$ is for $|p - p_0| \ll p_0$ equal to

$$\Sigma_1 - \Delta\mu = (e^2/\pi v) \{ \xi [\ln (2vp_0/T) - 1] - T f(\xi/T) \}. \tag{22.29}$$

We obtain from this:

$$\begin{aligned} \Sigma_1 - \Delta\mu &= (e^2/\pi v)\, \xi \ln (2p_0 v/\xi) \quad \text{for} \quad \xi \gg T, \\ \Sigma_1 - \Delta\mu &= (e^2/\pi v)\, \xi [\ln (4\gamma v p_0/\pi T) - 1] \quad \text{for} \quad \xi \ll T, \end{aligned} \tag{22.30}$$

where $\gamma = e^C = 1{\cdot}78$ (C is Euler's constant).

It follows from (22.30) that expression (22.27) is not entirely correct for the case of small ξ and T; the correction to the velocity of the quasi-particles $\partial\Sigma_1/\partial p$ close to the Fermi boundary as $T \to 0$, $\xi \to 0$ tends to infinity proportionally to $\ln(vp_0/T)$ or $\ln(vp_0/\xi)$. This result is connected with the fact that, as $p \to p_0$, small transfers of momentum at the vertex become important in integral (22.26). It is necessary here to take

(†) In this section we use v to denote the electron velocity on the Fermi boundary p_0/m.

account of all the loops threaded on the basic dotted loop, in other words to replace $4\pi e^2/(\boldsymbol{p}-\boldsymbol{p}_1)^2$ by the expression for $\Gamma_{ee}(\boldsymbol{p}_1-\boldsymbol{p}, \varepsilon_1-\varepsilon)$, corresponding to (22.8).

Instead of evaluating the whole of Σ, it will be more convenient for us to evaluate the difference between Σ and Σ_1, which latter has already been found. We denote this difference by Σ_2. It is represented by expression (22.26), in which $4\pi e^2/(\boldsymbol{p}-\boldsymbol{p}_1)^2$ is replaced by $\Gamma_{ee}-4\pi e^2/(\boldsymbol{p}-\boldsymbol{p}_1)^2$. It may be mentioned that no substantial corrections are introduced if we take into account higher order approximations to the $\mathfrak{G}$-function appearing in this integral, since in the corrected $\mathfrak{G}$-function p_0 corresponds as before to the Fermi boundary.

It was pointed out above that Γ_{ee} possesses, as a function of frequency, all the properties of the boson Green function. The same applies to the difference between Γ_{ee} and its zero-order approximation. Hence it follows that the problem has a great similarity to the evaluation of the self-energy of an electron interacting with phonons, which was discussed in the previous section. However, in the present case small momenta k are important. Let us denote $\Gamma_{ee}(\omega, \boldsymbol{k})-4\pi e^2/k^2$ by $D(\omega, \boldsymbol{k})$. Formula (21.29) of the previous section is now entirely applicable to our present case (with g^2 replaced by 1). On recalling that $\Sigma(\varepsilon) \ll \varepsilon$, and the fact that the important values are $k \ll p_0$, we obtain:

$$\Sigma_R = \frac{1}{(2\pi)^3 v}\int_{-\infty}^{\infty} d\eta \int_0^{\infty} k\,dk \int_{\xi-vk-\varepsilon}^{\xi+vk-\varepsilon} d\omega \frac{\operatorname{Im} D_R(\eta, k)}{\eta+\omega-i\delta} \times\left(\tanh\frac{\omega+\varepsilon}{2T}+\coth\frac{\eta}{2T}\right). \tag{22.31}$$

Since we are interested in the spectrum, which is expressed to a first approximation by the formula $\varepsilon=\xi$, we can assume that the integration over ω is between the limits $-vk$ and vk.

The real part of Σ_{2R} is equal to the principal value of integral (22.31). On carrying out a number of transformations, and taking account of the antisymmetry of $\operatorname{Im} D_R(\eta, k)$ and the connection between $\operatorname{Im} D_R(\eta,k)$ and $\operatorname{Re} D_R(\eta, k)$, we obtain

$$\operatorname{Re}\Sigma_{2R} = \frac{\pi}{(2\pi)^3 v}\int_0^{\infty} d\omega \int_{\omega/v}^{\infty} k\,dk \operatorname{Re} D_R(\omega, k)\left(\tanh\frac{\varepsilon+\omega}{2T}+\tanh\frac{\varepsilon-\omega}{2T}\right). \tag{22.32}$$

Similarly, a circuit round the pole in (22.31) yields

$$\operatorname{Im}\Sigma_{2R} = \frac{\pi}{(2\pi)^3 v}\int_0^{\infty} d\omega \int_{\omega/v}^{\infty} k\,dk \operatorname{Im} D_R(\omega, k) \times\left(2\coth\frac{\omega}{2T}-\tanh\frac{\omega+\varepsilon}{2T}-\tanh\frac{\omega-\varepsilon}{2T}\right). \tag{22.33}$$

The brackets containing coth and tanh show that, as earlier, only the greater of the two variables ε and T is important (except for the case when the pole of D_R is important (see below)). The value of D_R is determined by the region in which ω lies: in $\omega \ll kv_i$, $v_i k \ll \omega \ll v_e k$ or $v_e k \sim \omega$. It may easily be seen that there are correspondingly three regions of values of max (ε, T): (a) less than ω_{p1}, (b) between ω_{p1} and ω_{p2}, (c) greater than ω_{p2}. In accordance with inequalities (22.2) and (22.3), $T \gg \omega_{p1}$. In view of this it is sufficient for us to consider only regions (b) and (c).

Let us take the case $\omega_{p1} \ll \max(\varepsilon, T) \ll \omega_{p2}$. We split the integral over k for $\mathrm{Re}\,\Sigma_{2R}$ into the regions $\omega/v_e \lesssim k \lesssim \omega/v_i$, and $\omega/v_i < k < \infty$. On substituting

$$D_R = \frac{-(4\pi e^2)^2 (Z^2 \Pi_i + \Pi_e)}{k^2 [k^2 + 4\pi e^2 (Z^2 \Pi_i + \Pi_e)]}, \tag{22.34}$$

we find that the contribution from the region $\omega/v_i \lesssim k < \infty$ is of the order $(e^2/v)\,\omega_{p1} \tanh(\varepsilon/2T)$, i.e. it is small compared with Σ_1. On the contrary, the region $\omega/v_e \lesssim k \lesssim \omega/v_i$ yields an important contribution. If we notice that the significant values are $\omega \sim \max(\varepsilon, T) \gg \omega_{p1}$, we can assume $\Pi_i \ll \Pi_e$, $4\pi e^2 \Pi_e = \varkappa_e^2$. It follows from this that

$$\begin{aligned}
\mathrm{Re}\,\Sigma_{2R} &= -\frac{e^2}{2\pi v}\int_0^\infty \mathrm{d}\omega \int_{\omega/v}^\infty \frac{\varkappa_e^2\,\mathrm{d}k}{k(k^2+\varkappa_e^2)}\left(\tanh\frac{\varepsilon+\omega}{2T} + \tanh\frac{\varepsilon-\omega}{2T}\right) \\
&= -\frac{e^2}{2\pi v}\int_0^\infty \mathrm{d}\omega \ln\frac{v\varkappa_e}{\omega}\left(\tanh\frac{\varepsilon+\omega}{2T} + \tanh\frac{\varepsilon-\omega}{2T}\right) \\
&= -\frac{e^2}{\pi v}\left[\varepsilon \ln\frac{v\varkappa_e}{T} - Tf\left(\frac{\varepsilon}{T}\right)\right].
\end{aligned} \tag{22.35}$$

The regions of k can be split in the same way when finding $\mathrm{Im}\,\Sigma_{2R}$. In the region $\omega/v_e \lesssim k \lesssim \omega/v_i$ the function D_R has a pole corresponding to the ionic sound. A circuit round the pole gives the following contribution to $\mathrm{Im}\,\Sigma_{2R}(\omega, k)$:

$$\begin{aligned}
&-\frac{e^2\omega_{p1}}{4v}\int_0^\infty \mathrm{d}\omega \int_{\omega/v}^{\sim\omega/v_i} \frac{k^2\,\mathrm{d}k}{(k^2+\varkappa_e^2)^{3/2}}\,\delta\left(\omega - \omega_{p1}\frac{k}{\sqrt{k^2+\varkappa_e^2}}\right) \\
&\quad\times\left(2\coth\frac{\omega}{2T} - \tanh\frac{\omega+\varepsilon}{2T} - \tanh\frac{\omega-\varepsilon}{2T}\right) \approx -\frac{e^2 T}{v}\ln\frac{\varkappa_i}{\varkappa_e}
\end{aligned} \tag{22.36}$$

(this result holds to logarithmic accuracy).

In addition to this, the imaginary extra terms in Π_i and Π_e may prove to be important. Substitution yields

$$\begin{aligned}
&-\frac{e^2}{2\pi v}\int_0^\infty \mathrm{d}\omega \int_{\omega/v}^\infty \frac{k\,\mathrm{d}k\, 4\pi e^2\, \mathrm{Im}\,(\Pi_e + Z^2\Pi_i)}{[k^2 + 4\pi e^2(\Pi_e + Z^2\Pi_i)]^2} \\
&\qquad\times\left(2\coth\frac{\omega}{2T} - \tanh\frac{\omega+\varepsilon}{2T} - \tanh\frac{\omega-\varepsilon}{2T}\right).
\end{aligned}$$

In view of the exponential nature of $\mathrm{Im}\,\Pi_i$, the region $k \gtrsim \omega/v_i$ is important in the corresponding integral. An estimate shows that the contribution of this integral to $\mathrm{Im}\,\Sigma_{2R}$ is of the order $e^2 T/v$, i.e. to logarithmic accuracy it is small compared to (22.36).

The second imaginary correction to $\mathrm{Im}\,\Pi_e$ yields an integral over the region $\omega \sim \max(\varepsilon, T)$, $k \sim \varkappa_e$. The lower limit of the integral over k can therefore be put equal to zero:

$$-\frac{e^2}{4v^2}\int\limits_0^\infty \omega\,\mathrm{d}\omega\left[2\coth\frac{\omega}{2T} - \tanh\frac{\omega+\varepsilon}{2T} - \tanh\frac{\omega-\varepsilon}{2T}\right]\int\limits_0^\infty \frac{\mathrm{d}k\,\varkappa^2}{(k^2+\varkappa_e^2)^2}$$

$$= -\frac{\pi e^2}{16v^2\varkappa_e}[\varepsilon^2 + (\pi T)^2]. \qquad (22.37)$$

The second term of this expression is small compared with (22.36), but the term containing e^2 may in fact be greater than the ionic damping. We now turn to the region $\omega_{p2} \ll \max(\varepsilon, T)$. On considering the integral (22.35) for $\mathrm{Re}\,\Sigma_{2R}$, it may easily be seen that it yields an insignificant contribution of the order $(e^2/v)\,\omega_{p2}\tanh(\varepsilon/2T)$. Hence the real part of Σ_R is essentially equal to Σ_{R1}. As regards the imaginary part, the ionic acoustic pole of D_R yields in this case the same contribution (22.36). The same applies to the term from $\mathrm{Im}\,\Pi_i$, which is again insignificant.

The electronic, damping, which comes from $\mathrm{Im}\,\Pi_e$, yields an integral in which the most important values are $k \sim \varkappa_e$, $\omega \sim v\varkappa_e \sim \omega_{p2}$. If $T \gg \omega_{p2}$, it is of the order $e^2 T/v$, i.e. it can be neglected. Whereas if $T \ll \omega_{p2}$, we can take $T = 0$ in the integral. In view of the fact that $\omega \sim vk$ here, the entire expression (22.22) has to be substituted for Π_e. The simplest thing here is to use the fact that the temperature brackets are the same in (22.32) and (22.33) for $T = 0$, and are equal to 2. On combining these two formulae, we obtain

$$\Sigma_{R2}\Big|_{\substack{T=0\\ \varepsilon\gg\omega_{p2}}} = \frac{1}{(2\pi)^2 v}\int\limits_0^\infty \mathrm{d}\omega \int\limits_{\omega/v}^\infty k\,\mathrm{d}k\, D_R(\omega, k).$$

On substituting in this (22.34) and (22.22), we arrive at the expression

$$-e^2\varkappa_e(\beta_1 + i\beta_2), \qquad (22.38)$$

where β_1 and β_2 are constants, equal to the real and imaginary parts of the integral

$$\frac{1}{2}\int\limits_0^1 \mathrm{d}u\left[1 - u\left(\operatorname{artanh} u - \frac{1}{2}i\pi\right)\right]^{1/2} \qquad (22.39)$$

(the value of the square root with the positive imaginary part is taken here). The imaginary part of (22.38) is the required part of the damping.

We can find from the equation $\varepsilon - \xi - \Sigma + \Delta\mu = 0$ the energy spectrum and the damping of the electronic excitations in the various regions:

(a)
$$\omega_{p1} \ll \max(\xi, T) \ll \omega_{p2},$$
$$\varepsilon(p) = \xi(p)\left[1 + \frac{e^2}{\pi v}\left(\ln\frac{2p_0}{\varkappa_e} - 1\right)\right],$$
$$\gamma(p) = \frac{e^2}{v}\left[T\ln\frac{\varkappa_i}{\varkappa_e} + \frac{\pi\xi^2(p)}{16 v \varkappa_e}\right]; \qquad (22.40)$$

(b)
$$\omega_{p2} \ll \max(\xi, T),$$
$$\varepsilon(p) = \xi(p)\left[1 + \frac{e^2 m}{\pi p}\left(\ln\frac{(p+p_0)^2}{2mT} - 1\right)\right] - \frac{e^2 m T}{\pi p} f\left(\frac{\xi(p)}{T}\right),$$
$$\gamma(p) = \frac{e^2}{v} T \ln\frac{\varkappa_i}{\varkappa_e} + e^2 \varkappa_e \beta_2.$$

In particular,
$$\varepsilon(p) = \xi(p)\left[1 + \frac{e^2}{\pi v}\left(\ln\frac{4\gamma p_0 v}{\pi T} - 1\right)\right]$$
$$\text{for } p_0^2/2m \gg T \gg \max[\xi(p), \omega_{p2}], \qquad (22.41)$$
$$\varepsilon(p) = \xi(p)\left[1 + \frac{e^2 m}{\pi p}\ln\frac{p+p_0}{|p-p_0|}\right] \quad \text{for } \xi(p) \gg \max(T, \omega_{p2}).$$

4. *Thermodynamic functions*

Let us end by considering the thermodynamic functions for a degenerate plasma (†). We have, in accordance with (10.22):

$$\Delta\Omega = \Omega - \Omega_0 = \frac{1}{2}\int_0^{e^2} d(e^2) \int d^3r\, d^4x' \frac{\delta(\tau - \tau')}{|\mathbf{r} - \mathbf{r}'|} [\langle \tilde{\bar{\psi}}_\alpha(x) \tilde{\bar{\psi}}_\beta(x') \tilde{\psi}_\beta(x') \tilde{\psi}_\alpha(x)\rangle$$
$$- 2Z\langle \tilde{\bar{\psi}}_\alpha(x) \tilde{\bar{\Phi}}(x') \tilde{\Phi}(x') \tilde{\psi}_\alpha(x)\rangle + Z^2 \langle \tilde{\bar{\Phi}}(x) \tilde{\bar{\Phi}}(x') \tilde{\Phi}(x') \tilde{\Phi}(x)\rangle]. \qquad (22.42)$$

The expression in the brackets $\langle \cdots \rangle$ can be expressed in terms of the functions $\mathfrak{G}$ and $\mathcal{T}$; for example:

$$\langle \tilde{\bar{\psi}}_\alpha(x) \tilde{\bar{\psi}}_\beta(x') \tilde{\psi}_\beta(x') \tilde{\psi}_\alpha(x)\rangle = -2\mathfrak{G}_e^{(0)}(x - x')\mathfrak{G}_e^{(0)}(x' - x) + \left(\frac{N_e}{V}\right)^2$$
$$- \int d^4x_1 d^4x_2 d^4x_3 d^4x_4 \mathfrak{G}_e^{(0)}(x - x_1)\mathfrak{G}_e^{(0)}(x' - x_2)$$
$$\times \mathfrak{G}_e^{(0)}(x_3 - x)\mathfrak{G}_e^{(0)}(x_4 - x')\mathcal{T}^{ee}_{\alpha\beta\alpha\beta}(x_1x_2, x_3x_4).$$

The term $(N_e/V)^2$ can be thrown away, since it cancels in (22.42) with analogous terms from the electron-ion and ionic interactions as a consequence of the electrical neutrality of the plasma. When writing down the averages of the four field operators that come from the electron-ion

(†) A. A. Vedenov [40] has found the thermodynamic functions for a degenerate plasma.

and ion-ion interactions, it is not necessary to write the term with two $\mathfrak{G}$-functions as well. The fact is that this term has an exchange origin. Exchange of ions with electrons is impossible, while exchange between ions can produce a very small effect, since the ions form a Boltzmann gas.

The resulting expression therefore consists of two terms. One of them comes from the product of the two electron $\mathfrak{G}$-functions (it corresponds to exchange of electrons with electrons), whilst the other is the sum of terms with different G.

We start by considering the first term. It is equal in the momentum representation to

$$\frac{\Delta\Omega_1}{V} = -4\pi e^2 T^2 \sum_{\substack{n_1 n_2 \\ \tau_1,\tau_2 \to +0}} \int \frac{d^3\boldsymbol{p}_1 d^3\boldsymbol{p}_2}{(2\pi)^6} \frac{1}{(\boldsymbol{p}_1-\boldsymbol{p}_2)^2}$$

$$\times \frac{\exp(i\varepsilon_{n_1}\tau_1)}{i\varepsilon_{n_1}-\varepsilon_0(\boldsymbol{p}_1)+\mu} \frac{\exp(i\varepsilon_{n_2}\tau_2)}{i\varepsilon_{n_2}-\varepsilon_0(\boldsymbol{p}_2)+\mu}.$$

The factors $\exp i\varepsilon_n\tau$ take into account the order of the operators ψ in the Hamiltonian (18.1). The summations over n_1 and n_2 are independent. On recalling the definition of the Fourier component of the function $\mathfrak{G}$, we get

$$T \sum_{\substack{n \\ \tau\to+0}} \frac{\exp(i\varepsilon_n\tau)}{i\varepsilon_n-\varepsilon_0(\boldsymbol{p})+\mu} = \mathfrak{G}_e^{(0)}(-\tau,\boldsymbol{p})_{\tau\to+0} = n(\boldsymbol{p}).$$

Hence we find that

$$\frac{\Delta\Omega_1}{V} = -4\pi e^2 \int \frac{d^3\boldsymbol{p}_1 d^3\boldsymbol{p}_2 n(\boldsymbol{p}_1)n(\boldsymbol{p}_2)}{(2\pi)^6(\boldsymbol{p}_1-\boldsymbol{p}_2)^2}. \tag{22.43}$$

This term is small for a Boltzmann gas, since occupation numbers $n \ll 1$ appear in it. This justifies our neglecting the ion exchange.

We now consider the remaining terms. We obtain in the momentum representation:

$$\frac{\Delta\Omega_2}{V} = -\frac{4\pi}{2}\int_0^{e^2} d(e^2)\, T^3 \sum_{\varepsilon_1\varepsilon_2\omega} \int \frac{d^3\boldsymbol{k}\, d^3\boldsymbol{p}_1 d^3\boldsymbol{p}_2}{(2\pi)^9} \frac{1}{k^2}$$

$$\times\{4\mathfrak{G}^e(\boldsymbol{p}_1,\varepsilon_1)\mathfrak{G}^e(\boldsymbol{p}_1+\boldsymbol{k},\varepsilon_1+\omega)\mathfrak{G}^e(\boldsymbol{p}_2,\varepsilon_2)\mathfrak{G}^e(\boldsymbol{p}_2+\boldsymbol{k},\varepsilon_2+\omega)\mathcal{T}_{ee}(\boldsymbol{k},\omega_m)$$

$$-8Z\mathfrak{G}^i(\boldsymbol{p}_1,\varepsilon_1)\mathfrak{G}^i(\boldsymbol{p}_1+\boldsymbol{k},\varepsilon_1+\omega)\mathfrak{G}^e(\boldsymbol{p}_2,\varepsilon_2)\mathfrak{G}^e(\boldsymbol{p}_2+\boldsymbol{k},\varepsilon_2+\omega)\mathcal{T}_{ei}(\boldsymbol{k},\omega_m)$$

$$+4Z^2\mathfrak{G}^i(\boldsymbol{p}_1,\varepsilon_1)\mathfrak{G}^i(\boldsymbol{p}_1+\boldsymbol{k},\varepsilon_1+\omega)\mathfrak{G}^i(\boldsymbol{p}_2,\varepsilon_2)\mathfrak{G}^i(\boldsymbol{p}_2+\boldsymbol{k},\varepsilon_2+\omega)\mathcal{T}_{ii}(\boldsymbol{k},\omega_m)\}.$$

Since the vertex parts $\mathcal{T}$ are themselves of order e^2, this expression is formally of fourth order in e^2. What is important here, however, is that small k and ω matter in the integral over k and the sum over ω. Comparing with (22.9), we can write

$$\frac{\Delta\Omega^2}{V} = -2\pi \int_0^{e^2} d(e^2)\, T \sum_\omega \int \frac{d^3\boldsymbol{k}}{(2\pi)^3} \frac{1}{k^2} \{\mathcal{H}_e^2(\omega, \boldsymbol{k}) \mathcal{T}_{ee}(\omega, \boldsymbol{k})$$

$$+ 2Z \mathcal{H}_i(\omega, \boldsymbol{k}) \mathcal{H}_e(\omega, \boldsymbol{k}) \mathcal{T}_{ei}(\omega, \boldsymbol{k}) + Z^2 \mathcal{H}_i^2(\omega, \boldsymbol{k}) \mathcal{T}_{ii}(\omega, \boldsymbol{k})\}$$

$$= -8\pi^2 \int_0^{e^2} e^2 d(e^2)\, T \sum_\omega \int d^3\boldsymbol{k} \frac{1}{k^2} \frac{\mathcal{H}_e^2 + 2Z^2 \mathcal{H}_i \mathcal{H}_e + Z^4 \mathcal{H}_i^2}{k^2 - 4\pi e^2(\mathcal{H}_e + Z^2 \mathcal{H}_i)}$$

$$= -\frac{1}{\pi} \int_0^{e^2} e^2 d(e^2)\, T \sum_\omega \int dk \frac{1}{k^2} \frac{(\mathcal{H}_e + Z^2 \mathcal{H}_i)^2}{k^2 - 4\pi e^2(\mathcal{H}_e + Z^2 \mathcal{H}_i)}. \quad (22.44)$$

The important values of k^2 here are of order $4\pi e^2(\mathcal{H}_e + Z^2 \mathcal{H}_i)$. We want to find the relationship between the values $v_e k$, $v_i k$ and $\omega = 2m\pi T$. It is easily seen that the integral over k in (22.44) increases as $4\pi e^2(\mathcal{H}_e + Z^2 \mathcal{H}_i)$ increases. Let us consider (22.11) for $\mathcal{H}$. If we assume $vk \ll T$, the maximum value of $\mathcal{H}(\omega, k)$ is obtained when $\omega = 0$. The most important ion loop in this case is that for which $\mathcal{H}_i \sim N_i/TV$, so that $4\pi e^2(\mathcal{H}_e + Z^2 \mathcal{H}_i) \sim (e^2/V)(N_i Z^2/T)$. Hence it follows that $(v_i k)^2 \sim (T/M)(N_i/V)(e^2/T) \sim e^2 N_i/VM \ll T^2$. Our assumption is therefore justified. We need only take the term with $\omega = 0$ in (22.44) and put $\mathcal{H}_e + Z^2 \mathcal{H}_i \approx -(Z^2/V)(\partial N_i/\partial\mu)$. After this, integration over k^2 gives

$$\frac{\Delta\Omega_2}{V} = -\frac{2\sqrt{\pi}}{3}(Ze)^3 T \left(\frac{\partial N_i}{V\partial\mu}\right)^{3/2} = -\frac{2\sqrt{\pi}}{3}(Ze)^3 \frac{(N_i/V)^{3/2}}{T^{1/2}}. \quad (22.45)$$

Given our assumptions (22.2), (22.3), this term is small compared with (22.43). But this is the only term of order e^3, since the correction to the term $\Delta\Omega_1$ must be of order e^4.

If the condition of strong degeneracy is not imposed on the electron gas, $\Delta\Omega_2$ may become of order $\Delta\Omega_1$. Now, however, we must also take into account the electron loop. We thus get

$$\frac{\Delta\Omega}{V} = -4\pi e^2 \int \frac{d^3\boldsymbol{p}_1 d^3\boldsymbol{p}_2}{(2\pi)^6} \frac{n(\boldsymbol{p}_1) n(\boldsymbol{p}_2)}{(\boldsymbol{p}_1 - \boldsymbol{p}_2)^2} - \frac{2\sqrt{\pi}}{3} e^3 T \left(\frac{Z^2}{V}\frac{\partial N_i}{\partial\mu} + \frac{1}{V}\frac{\partial N_e}{\partial\mu}\right)^{3/2}.$$

This formula is suitable when the following conditions are fulfilled:

$$e^2 m \left(\frac{V}{N}\right) \ll 1, \qquad T \gg \max\left\{e^2\left(\frac{N}{V}\right)^{1/3}, \frac{1}{M}\left(\frac{N}{V}\right)^{2/3}\right\}. \quad (22.46)$$

CHAPTER V

SYSTEM OF INTERACTING BOSONS

§ 23. APPLICATION OF FIELD THEORY METHODS TO A SYSTEM OF BOSONS AT ABSOLUTE ZERO

Considerable difficulties are involved in extending the methods of quantum field theory to the case of a system of bosons at temperatures below the "Bose-condensation" temperature. Nevertheless a suitable formalism has been developed (Belyaev [41]), and the present chapter will be devoted to it. As usual, we start from a consideration of the absolute zero case.

Throughout the foregoing treatment our development of the diagram technique has been based on the fact that the average of the product of several non-interacting ψ-operators can be reduced to the product of Wick's averages of pairs of averages $\psi\psi^+$. This was a consequence of theorem, according to which the average of the time-ordered product of any number of field operators splits up into the sum of the products of normal products of pairs. The ground state for a system of fermions — the "vacuum" (we are at present only considering the case at absolute zero) — is such that, by varying the definition of the creation and annihilation operators, we can arrange for the average of the normal products to be equal to zero. The situation is quite different for a system of bosons. Because of the statistics in a Bose gas, any number of particles may be concentrated in the zero momentum state at absolute zero. In an ideal gas, at $T = 0$, the number of particles in the lowest level is simply equal to the total number of particles in the system. The state of Bose-condensation is therefore characterized by the fact that the density of the number of particles at the lowest level, with zero momentum, tends to a finite limit when the total number of particles N and the volume of the system tend to infinity. Hence the averages of the normal product of operators a_0^+ and a_0, of the form $(a_0^+)^n a_0^n$, are not only non-zero, but can be made arbitrarily large.

Let us suppose for a start that the system is at absolute zero. As we have just said, in an ideal Bose gas all the particles are at the zero momentum level. We separate out, from the operators $\psi(x)$ and $\psi^+(x)$ in the interaction representation, operators which correspond to the creation and annihilation of particles in the state with $p = 0$:

$$\psi(x) = \xi_0 + \psi'(x); \qquad \psi^+(x) = \xi_0^+ + \psi'^+(x) \tag{23.1}$$

(we shall use the notation $\xi_0 = a_0/\sqrt{V}$, $\xi_0^+ = a_0^+/\sqrt{V}$).

The total number of particles $N = V\xi_0^+\xi_0$ becomes arbitrarily large as $V \to \infty$. Hence, if we neglect the right-hand side in the commutation relation

$$\xi_0\xi_0^+ - \xi_0^+\xi_0 = \frac{1}{V}$$

the operators ξ_0 and ξ_0^+ can be regarded as c-numbers to a first approximation, as was done in Chap. II. It will be seen, however, that this is only meaningful when the interaction is sufficiently weak.

We shall write the total Hamiltonian of the system as

$$H = H_0 + H_{int},$$

where

$$H_0 = \frac{1}{2m}\int \nabla\psi^+(x)\nabla\psi(x)\,\mathrm{d}^3\boldsymbol{r},$$

and H_{int} is an interaction Hamiltonian, the form of which will be left unspecified for the present. All the familiar relations of field theory, connecting the operators in the Heisenberg representation with those in the interaction representation by means of the S-matrix, remain valid, as does the actual definition of the S-matrix:

$$S = T\exp\{-i\int H_{int}(x)\,\mathrm{d}^4x\}. \tag{23.2}$$

The single-particle Green function $G(x, x')$ is given in terms of the operators in the Heisenberg representation by

$$G(x - x') = -i\langle T(\tilde{\psi}(x)\tilde{\psi}^+(x'))\rangle \tag{23.3}$$

and in the interaction representation by

$$G(x - x') = -\frac{i\langle T(\psi(x)\psi^+(x')S)\rangle}{\langle S\rangle} \tag{23.3'}$$

(taking the average in the formulae with respect to the ground state of N interacting particles in (23.3), and with respect to the ground state of N non-interacting particles in (23.3')). It proves more convenient to consider, instead of (23.3), the following two parts of the Green function $G(x - x')$:

$$G'(x - x') = -i\langle T(\tilde{\psi}'(x), \tilde{\psi}'^+(x'))\rangle = \frac{-i\langle T(\psi'(x), \psi'^+(x')S)\rangle}{\langle S\rangle}, \tag{23.4}$$

and

$$G_0(t-t') = -i\langle T(\tilde{\xi}_0(t), \tilde{\xi}_0^+(t'))\rangle = \frac{-i\langle T(\xi_0(t)\xi_0^+(t')S)\rangle}{\langle S\rangle}; \tag{23.5}$$

$G'(x-x')$ is the Green function for particles "above the condensate", $G_0(t-t')$ is the Green function for particles in the condensate. Obviously, $G_0(t-t')$ does not depend on the difference of the positional coordinates and can therefore be defined as the zero momentum Fourier component of the complete Green function

$$G_0(t-t') = \int G(\mathbf{r}-\mathbf{r}', t-t')\,\mathrm{d}^3\mathbf{r}'.$$

The density of the number of particles in the condensate is

$$n_0 = iG_0(t-t'),$$
$$t' = t+0.$$

As regards the density of the total number of particles, it is always equal to

$$n = n' + n_0 = i[G'(0, t-t') + G_0(t-t')],\ t' = t+0. \qquad (23.6)$$

Notice again that, in accord with § 4, the number of particles in the condensate is different from the total number of particles when interactions are present.

Let us now turn our attention directly to the development of a diagram technique of perturbation theory for interacting particles. In accord with the special role played by the condensate particles, we shall assume that the substitution (23.1) has been carried out in the Hamiltonian H_{int} and that H_{int} has been reduced to the form in which the operators ξ_0 and ξ_0^+, ψ' and ψ'^+ figure separately. We shall similarly assume that H_{int} has this form in the definition of the S-matrix (23.2). Our subsequent treatment is applicable for a Hamiltonian $H_{int}(x)$ which is the product of any number of operators ψ and ψ^+ with any law of interaction between the particles.

After the division indicated, between the operation of time ordering T and the averaging over the ground state of non-interacting particles, each can be represented as a sequence of two operations, acting separately on the particles in the condensate and on the particles "above the condensate" ("uncondensed" particles)

$$T = T^0 \,.\, T', \qquad \langle\cdots\rangle = \langle\langle\cdots\rangle'\rangle^0, \qquad (23.7)$$

where T_0 and $\langle\cdots\rangle^0$ are applied to the operators ξ_0 and ξ_0^+. The expansion of the S-matrix in powers of the interaction contains a set of different products of operators ξ_0, ξ_0^+, ψ' and ψ'^+ in each term. As regards the independent operators ψ' and ψ'^+, Wick's general theorem can be applied in regard to them, since the means of the normal products of the uncondensed particles vanish. The time-ordered pair averages of $\psi'\psi'^+$, which we denote by $G^{(0)}(x-x')$, differ from zero and are equal to

$$G^{(0)}(x-x') = -i\langle T'\left(\psi'(x)\psi'^+(x')\right)\rangle' \equiv -i\langle T\left(\psi'(x)\psi'^+(x')\right)\rangle. \quad (23.8)$$

The corresponding Fourier components are

$$G^{(0)}(x-x') = (2\pi)^{-4} \int G^{(0)}(p) e^{ip(x-x')} d^4p,$$

$$G^{(0)}(p) = \frac{1}{\omega - \frac{p^2}{2m} + i\delta}. \tag{23.9}$$

If, therefore, we consider the operators ξ_0 and ξ_0^+ as (numerical) parameters, they will play the role of an external field in the different vertices of the diagrams.

Let us consider the question of finding the Green function for any number of uncondensed particles. The Green function is

$$G_n(x_1 \cdots x_n; x_1' \cdots x_n') = \frac{(-i)^n \langle T(\psi'(x_1) \cdots \psi'(x_n); \psi'^+(x_1') \cdots \psi'^+(x_n') S)\rangle}{\langle S \rangle}. \tag{23.10}$$

We split up the operations T and $\langle \cdots \rangle$ into operations T', T^0 and $\langle \cdots \rangle'$, $\langle \cdots \rangle^0$ in accordance with (23.7), and first investigate the perturbation theory series for

$$\overline{G}_n(x_1 \cdots x_n; x_1' \cdots x_n') = (-i)^n \langle T(\psi(x_1) \cdots \psi'(x_n); \psi'^+(x_1') \cdots \psi'^+(x_n') S)\rangle'. \tag{23.11}$$

Since the operations T' and $\langle \cdots \rangle'$ do not touch the operators ξ_0 and ξ_0^+, the latter are parameters with respect to these operations and have no effect on the time-ordering and averaging of the different products of operators of uncondensed particles. The corresponding matrix element can therefore be written in accordance with the usual rules of constructing Feynman diagrams and contains products of time-ordered averages (23.8) and powers of the operators ξ_0 and ξ_0^+. The number of the latter in a given order of the expansion of the S-matrix in powers of H_{int} depends on the form of the interaction Hamiltonian H_{int} and on the choice of the terms in H_{int} after the substitution (23.1). For example, the interaction (see § 25)

$$H_{int} = \frac{1}{2} \int \psi^+(x) \psi^+(x') U(\mathbf{r}-\mathbf{r}') \psi(x') \psi(x) d^3\mathbf{r} d^3\mathbf{r}' \tag{23.12}$$

splits up, after the substitution

$$\psi \to \xi_0 + \psi'; \; \psi^+ \to \xi_0^+ + \psi'^+$$

into eight terms, starting with a term of the fourth order in ξ_0 and ξ_0^+: $1/2 (\xi_0^+)^2 (\xi_0)^2 \int U(\mathbf{r}) d^3\mathbf{r}$, and ending with

$$\frac{1}{2} \int \psi'^+(x) \psi'^+(x') U(\mathbf{r}-\mathbf{r}') \psi'(x') \psi(x) d^3\mathbf{r} d^3\mathbf{r}'.$$

As an example, Fig. 64 shows one of the second order diagrams for the function $\overline{G}(x-x')$. The continuous line in the diagram corresponds to the function $G^{(0)}(x-x')$ (23.9), whilst the wavy line between the two points is the interaction potential $U(\boldsymbol{r}-\boldsymbol{r}')$; the free jagged lines represent here the operators ξ_0 and ξ_0^+, the line to the vertex being ξ_0^+, and from the vertex ξ_0. The matrix element of this diagram is

$$\overline{G}(x_1-x_2) = i\int G^{(0)}(x_1-x_3)G^{(0)}(x_3-x_5)\,\xi_0(t_5)\,U(\boldsymbol{r}_3-\boldsymbol{r}_4)$$
$$\times\xi_0^+(t_4)G^{(0)}(x_4-x_6)\,U(\boldsymbol{r}_6-\boldsymbol{r}_5)G^{(0)}(x_6-x_2)\,\mathrm{d}^4x_3\cdots\mathrm{d}^4x_6. \quad (23.13)$$

In the general case (23.10), the mth order matrix element in $\overline{G}_n(x_1,\ldots,x_n;x_1',\ldots,x_n')$ contains the product of an arbitrary number of operators ξ_0 and ξ_0^+. We shall only remark that the powers of ξ_0 and ξ_0^+ are necessarily the same. This is connected with the fact that the interaction H_{int} preserves the total number of particles. Hence, if the numbers of operators ξ_0, ξ_0^+ are not the same, the numbers of operators ψ and ψ'^+ in the average $\langle\cdots\rangle'$ are not the same, and the latter is consequently zero.

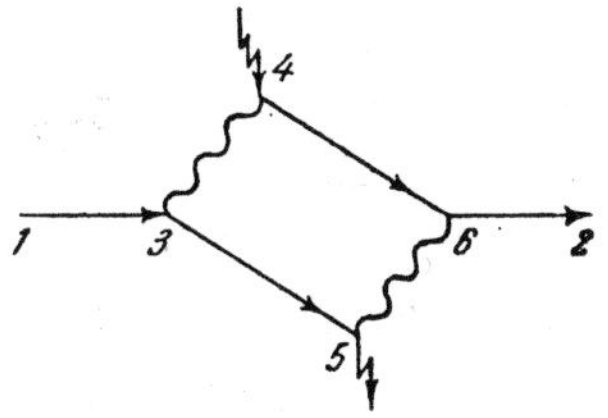

Fig. 64

Let $M_n(x_1\cdots x_n;x_1'\cdots x_n')$ be the connected diagram in (23.11), having $2m$ vertices corresponding to m operators ξ_0 and ξ_0^+. As usual, we understand by a connected diagram one which does not break down into several parts with no joining lines. Let us consider, along with M_n, all the diagrams that differ from it by the presence of "vacuum" loops, i.e. different disconnected diagrams. It is well known from field theory that the total set of such diagrams amounts to the multiplication of each matrix element by the mean value of the S-matrix. In our case, M_n is multiplied by $\langle S\rangle'$. It is sufficient, therefore, when forming the perturbation theory series for the functions (23.11), to take only connected diagrams into account, and to multiply the corresponding matrix element by $\langle S\rangle'$.

We now pass from evaluating $\overline{G}_n(x_1\cdots x_n;x_1'\cdots x_n')$ to finding

$$G_n(x_1\cdots x_n;x_1'\cdots x_n') \equiv \frac{\langle T^0\overline{G}_n(x_1\cdots x_n;x_1'\cdots x_n')\rangle^0}{\langle S\rangle}.$$

It becomes important at this stage to know the character of the operators ξ_0, ξ_0^+. This has so far been ignored, since the operations T' and $\langle\cdots\rangle'$ do not touch ξ_0 and ξ_0^+, which commute with ψ' and ψ'^+. Every matrix element M_n in $\overline{G}_n$, like (23.13), contains under the integral a definite number of operators ξ_0 and ξ_0^+, multiplied by averages such as (23.8). Suppose that M_n contains

$$\xi_0(t_1)\cdots\xi_0(t_m)\xi_0^+(t_1')\cdots\xi_0^+(t_m').$$

In order finally to obtain $G(x_1\cdots x_n;x_1'\cdots x_n')$, we have to find an average $\langle\cdots\rangle^0$ of the form

$$\frac{\langle T^0(\xi_0(t)\ \ldots\ \xi_0(t_m),\ \xi_0^+(t_1')\cdots\xi_0^+(t_m')\,\langle S\rangle')\rangle^0}{\langle S\rangle}.$$

Since in turn the operations T^0 and $\langle\cdots\rangle^0$ do not touch the uncondensed particles, the required averages are seen to be m-particle Green functions for condensate particles:

$$G_{0m}(t_1\cdots t_m;\, t_1'\cdots t_m') = \frac{\langle T(\xi_0(t_1)\cdots\xi_0(t_m)\xi_0^+(t_1')\cdots\xi_0^+(t_m')S)\rangle}{\langle S\rangle}. \quad (23.14)$$

Consequently, to find the Green functions for the uncondensed particles via the perturbation theory series, we have to know the exact m-particle Green function for condensate particles.

These functions can be found directly from formulae such as (23.14), in terms of the value of the density n of the number of particles in the condensate without interaction. This approach involves difficulties, however: the Wick theorem expansion into normal products of the product of operators ξ_0, ξ_0^+ has no meaning, because the average over the ground state of such normal products of the type $N(a_0^+, \ldots, a_0, \ldots)$ is not merely non-vanishing, but is in fact extremely large. At the same time, it is not possible to neglect the non-commutativeness of the operators ξ_0 and ξ_0^+ in (23.14). For we can write $\langle S\rangle'$ as(†)

$$\langle S\rangle' = e^{\sigma}, \quad (23.15)$$

where σ is the sum of all the singly-connected "vacuum" loops (i.e. those that do not break down into independent parts) and a functional of ξ_0, ξ_0^+. This sum is proportional to the volume (the density of the number of condensate particles $n_0 = \xi_0^+\xi_0$ is a finite quantity). Arbitrary powers of V are obtained in the formal expansion of $\langle S\rangle'$ into a series in powers of σ in (23.15), so that, although the right-hand side in the commutation relations

$$\xi_0\xi_0^+ - \xi_0^+\xi_0 = \frac{1}{V}$$

is of order $1/V$, we still cannot neglect it, because its smallness can be compensated by a suitable power of V in expansion (23.15).

Hence it is more satisfactory to look for another approach. It may be observed that expressions (23.14) can be written directly in terms of the Heisenberg operators:

$$G_{0m}(t_1\cdots t_m;\, t_1'\cdots t_m') = \langle T(\tilde{\xi}_0(t_1)\cdots\tilde{\xi}_0(t_m)\tilde{\xi}_0^+(t_1')\cdots\tilde{\xi}_0^+(t_m'))\rangle, \quad (23.16)$$

where the mean of the product is taken over the ground state of the interacting particles. We first consider the mean $V\langle\tilde{\xi}_0^+\tilde{\xi}_0\rangle$, which is the exact number of particles with zero momentum. In an ideal gas, this number

(†) In field theory, the possibility of writing $\langle S\rangle$ in the form (23.15) is proved on the assumption that ξ_0, ξ_0^+ are external parameters with no operator properties. We shall see, however, that the $\langle S\rangle'$ in (23.14) stands after the time-ordering sign with respect to operators ξ_0, ξ_0^+. The Bose operators under the sign of the T-product can be interchanged, in accordance with the actual meaning of this operation. Hence the assumption of field theory is fulfilled here.

is simply equal, at $T = 0$, to the total number of particles N — all the particles are at the level with $\boldsymbol{p} = 0$. Interaction between the particles (a repulsion at sufficiently small distances — an attraction everywhere would make the system unstable) leads to a decrease in the number of particles with momentum $\boldsymbol{p} = 0$. This does not mean, however (we emphasized this in Chap. I), that the condensate vanishes, i.e. the mean number of particles with zero momentum remains arbitrarily large for an arbitrarily large total number of particles in the system (the density n_0 of the number of particles in the condensate remains finite for any interaction of the particles, if $V \to \infty$). This fact is fairly obvious physically, though we cannot logically exclude the possibility that n_0 vanishes for a certain interaction. We shall not dwell here on a proof of our assertion, all the more, because the only substance occurring in nature of this kind is helium. The reader will find the proof in Belyaev's article [41].

The change in the total number of particles in the condensate as a result of interactions is in fact the reason why we could not simply regard the independent operators ξ_0, ξ_0^+ as c-numbers.

The perturbation theory series obtained for the uncondensed Green functions contains the averages of the exact Heisenberg operators $\tilde{\xi}_0$ and $\tilde{\xi}_0^+$. If the condensate does not disappear, the operators $\tilde{\xi}_0$, $\tilde{\xi}_0^+$ are in turn, to a first approximation, simply c-numbers, as regards their effect on the ground state of the interacting particles. We can use this fact to simplify the expressions for the Green functions for the condensate particles.

It must not be forgotten, of course, that the operator $\tilde{\xi}_0$ annihilates one particle, whilst $\tilde{\xi}_0^+$ creates one. Strictly speaking, then, importance will only attach throughout what follows to the matrix elements $\tilde{\xi}_0$, $\tilde{\xi}_0^+$ for a transitions from the ground state of the system with N particles ($N \to \infty$) to the ground state of the system with $N \pm 1$ particles: from the physical point of view, if we take a particle from an infinite number of them in the Bose condensate, or if we add a particle, we do not in practice change the ground state of the system, except for increasing its energy by the amount of the chemical potential μ. When speaking of the operators $\tilde{\xi}_0$ and $\tilde{\xi}_0^+$ as c-numbers, we shall always bear this last point in mind.

Let us discuss this in more detail, using as an example the single-particle Green function

$$G_0(t-t') = -i\langle T\big(\tilde{\xi}_0(t)\tilde{\xi}_0^+(t')\big)\rangle \approx -i\langle\tilde{\xi}_0(t)\tilde{\xi}_0^+(t')\rangle.$$

We write this last expression as a sum of products of matrix elements with respect to intermediate states:

$$\langle \Phi_N^* |\tilde{\xi}_0(t)\tilde{\xi}_0^+(t')| \Phi_N\rangle = \langle \Phi_N^* |\tilde{\xi}_0(t)| \Phi_{N+1}\rangle \langle \Phi_{N+1}^* |\tilde{\xi}_0(t')| \Phi_N\rangle$$
$$+ \sum_s \langle \Phi_N^* |\tilde{\xi}_0(t)| \Phi_{N+1}^s\rangle \langle \Phi_{N+1}^{s*} |\tilde{\xi}_0^+(t')|\Phi_N\rangle;$$

where Φ_N and Φ_{N+1} are the ground states of systems of N and $N+1$ interacting particles, and Φ^s_{N+1} are states other than ground states of the system of $N+1$ particles. The sum occurring in this expression is small, since, for example, $\xi_0^+ \Phi_N \simeq \Phi_{N+1}$, whilst Φ_{N+1} and Φ^s_{N+1} are orthogonal. As regards the matrix elements for passing from a ground state to a ground state, their dependence on time can be found from the usual formula of quantum mechanics:

$$-i\frac{\partial}{\partial t}\langle\Phi_N^*|\tilde{\xi}_0(t)|\Phi_{N+1}\rangle = \langle\Phi_N^*|[\hat{H}, \tilde{\xi}_0(t)]|\Phi_{N+1}\rangle$$

or

$$\tilde{\xi}_0(t) = \xi_0(0)e^{-it(E_{N+1}-E_N)}.$$

Using the definition of the chemical potential $\mu = \partial E/\partial N$, and replacing $\xi_0(0)$ by $n_0^{1/2}$, we find that

$$iG_0(t-t') = n_0 e^{-i\mu(t-t')}. \tag{23.17}$$

In other words, the function $G_0(t-t')$ has split up into a product of two independent factors; $\tilde{\xi}_0(t)$ corresponds to the factor $\sqrt{n_0}e^{-i\mu t}$, and $\tilde{\xi}_0^+(t')$ to $\sqrt{n_0}e^{i\mu t'}$. Obviously, the same situation holds for any Green function for a condensate particle: when replacing the operators $\tilde{\xi}_0(t)$, $\tilde{\xi}_0^+(t)$ by c-numbers, a factor of the type indicated has to be associated with each operator. The diagram method for finding the Green functions for uncondensed particles thus reduces to the usual diagram technique, in which the operators $\tilde{\xi}_0$ and $\tilde{\xi}_0^+$ play the role of an external field:

$$\tilde{\xi}_0(t) = \sqrt{n_0}\, e^{-i\mu t}, \qquad \tilde{\xi}_0^+(t) = \sqrt{n_0}\, e^{i\mu t}. \tag{23.18}$$

As usual, only the connected diagrams have to be taken into account when writing down the perturbation theory series. As we have shown, taking the unconnected diagrams into account amounts to replacing the density of the number of particles in the condensate of an ideal gas by the exact value of the density of the number of particles in the condensate of a gas with interacting particles, whilst it also leads to the appearance of frequency factors in (23.18). For the rest, all the diagrams are the same as though, after substituting in the interaction Hamiltonian H_{int} the operators in the form (23.1), we were to regard the operators ξ_0 and ξ_0^+ in the interaction representation as external parameters and, when evaluating expressions (23.10), carry out the averaging $\langle\cdots\rangle'$ and the time-ordering T' only with respect to the uncondensed particles (in the connected diagrams). In order to obtain the final expressions, we have to perform substitutions for ξ_0 and ξ_0^+ in accordance with (23.18). Let us remark once more that the density of the number of condensate particles n_0 in a gas of interacting particles differs from its value for an ideal gas.

Evaluation of the Green function for uncondensed particles by means of the Feynman diagram method developed above leads to expressions for the diagrams that feature two parameters, the number n_0 and the value of the chemical potential μ. Instead of directly evaluating the dependence of the chemical potential on the density of the total number of particles in the system using perturbation theory, we can apply general relations. First of all, the density n_0 of the total number of particles in the system is connected with μ by the obvious relationship

$$n = n_0 + iG'(x - x'), \ \boldsymbol{r} = \boldsymbol{r}'; t' = t + 0. \qquad (23.19)$$

The second relation follows from the condition that the ground state energy be a minimum with respect to n_0. Evaluation of the ground state energy $E = \langle \hat{H} \rangle$ using the approach described above leads to an expression for E as a function of the parameters n_0 and μ. On varying E with respect to n_0, with the total number of particles constant (n constant), we find the second condition as

$$\left(\frac{\partial E}{\partial n_0} \right)_n = 0. \qquad (23.20)$$

In principle the two conditions (23.19) and (23.20) solve the problem. It proves more convenient for practical calculations to use (24.17) instead of (23.20).

We shall end this section with a brief discussion of the choice of thermodynamic variables. To date, we have used the total number of particles in the system as the independent variable. This was connected with the fact that, when applying perturbation theory, we started out from the characteristic of an ideal Bose gas, in which there is no Bose condensation when the chemical potential is finite: as we know, the chemical potential of an ideal Bose gas is identically zero throughout the temperature interval from zero to the condensation temperature T_0. The chemical potential μ is not zero for a system of interacting particles and is therefore just as much a thermodynamic variable as the total number of particles. As usual, the value of μ can be found from the condition that the average number of particles in the system be equal to a given real number of particles. In essence, it is precisely this condition that is expressed by (23.19). Changing to the chemical potential μ as the independent variable enables us formally to avoid auxiliary time dependences in formulae (23.18), arising in the matrix elements from the vertices with $\tilde{\xi}_0(t)$ and $\tilde{\xi}_0^+(t)$.

For, as we have seen more than once, a change from the variable N to the variable μ is achieved by replacing the complete Hamiltonian of the system H by $H - \mu N$. Since the operators ψ and ψ^+ commute with the total number of particles N in accordance with

$$N\psi - \psi N = -\psi; \ N\psi^+ - \psi^+ N = \psi^+,$$

the change of Hamiltonian amounts to an extra time dependence of the operators ψ and ψ^+:

$$\psi \to e^{i\mu t}\psi; \; \psi^+ \to e^{-i\mu t}\psi^+. \tag{23.21}$$

The Green function is changed at the same time, e.g. we have for the complete single-particle Green function,

$$G(x - x') \to e^{i\mu(t-t')}G(x - x'). \tag{23.22}$$

In the Fourier components this transformation implies a change of all the frequencies ω in the old expressions to $\omega + \mu$. Hence, the Green functions for condensate particles are also independent of time in the new thermodynamic variables after the transformation (23.21). In view of this, the time factors (23.18) can be omitted at the corresponding vertices of the diagrams. The reader can obtain this result by redefining the Green function in accordance with (23.22) and investigating directly the perturbation theory series in connection with tranformations (23.21) and (23.22). It will be assumed everywhere in future that μ has been chosen as the independent thermodynamic variable.

§ 24. GREEN FUNCTIONS

1. *Structure of the equations*

Let us dwell in rather more detail on the structure of the perturbation theory series for a single-particle Green function for uncondensed particles. A diagram of any order can be split up into several irreducible parts, joined by a single line, corresponding to the function $G^{(0)}(x - x')$. Any diagram for a Green function is therefore a chain of self-energy diagrams, connected by zero order Green functions. Fig. 65 shows a few examples, where the circles denote schematically irreducible self-energy parts of any required structure. The presence of the condensate amounts to the appearance among the self-energy diagrams of new diagrams which have never been featured in the problems considered in earlier chapters.

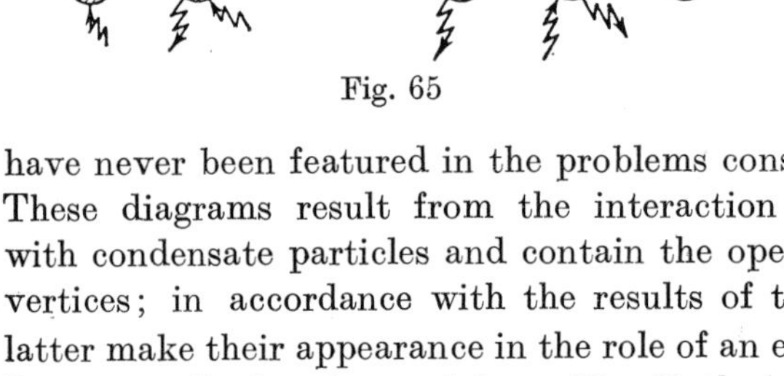

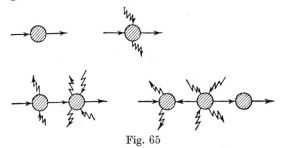

Fig. 65

These diagrams result from the interaction of uncondensed particles with condensate particles and contain the operators $\tilde{\xi}_0$ and $\tilde{\xi}_0^+$ at several vertices; in accordance with the results of the previous section, these latter make their appearance in the role of an external field: $\tilde{\xi}_0, \tilde{\xi}_0^+ \to \sqrt{n_0}$. It may easily be observed from Fig. 65 that the total number of lines entering any irreducible self-energy diagram is always the same as the total number of lines leaving it (the total number includes all the jagged lines, corresponding to the operators of creation and annihilation of particles of condensate). Since all the self-energy parts are connected by straight lines, i.e. contain two uncondensed particles either entering or leaving, the fact just mentioned enables us to classify all the irreducible self-energy diagrams as follows:

A. Diagrams with one straight line entering and one leaving, corresponding to uncondensed particles. The numbers of jagged lines that enter and leave these diagrams (the powers of the operators $\tilde{\xi}_0$ and $\tilde{\xi}_0^+$) must be the same. We shall denote the sum of all the matrix elements of these diagrams in the coordinate representation by $\Sigma_{11}(x-x')$ and illustrate it by a shaded circle, as in Fig. 66*a*.

B. Diagrams from which two uncondensed lines leave. In these diagrams there are two more jagged lines entering than there are leaving. We shall denote the corresponding sum of matrix elements by $\Sigma_{02}(x-x')$ and represent it by a shaded circle with two entrant jagged lines, as in Fig. 66*b*.

C. Diagrams in which there are two entrant uncondensed particle lines. In these diagrams, on the contrary, the number of jagged lines leaving is two greater than the number entering. We shall denote the sum of these matrix elements by $\Sigma_{20}(x-x')$; in Fig. 66*c*, the sum of such self-energy diagrams is represented by a circle with two jagged lines leaving it. All three types of irreducible self-energy parts can be combined in any order into diagrams for the Green functions $G'(x-x')$. The only obvious restriction is that the number of times that the matrix elements Σ_{02} enter into a diagram be equal to the corresponding number of matrix elements of type Σ_{20}. Fig. 67 illustrates some examples of diagrams for the Green functions for uncondensed particles.

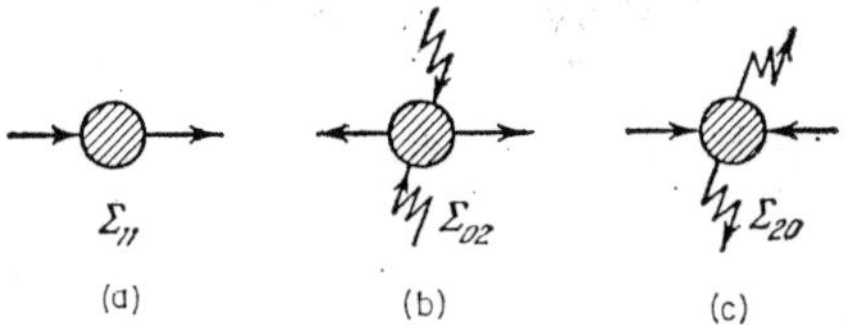

Fig. 66

We can now write an analogue of Dyson's equation for the Green function for an uncondensed particle. We shall first derive it diagrammatically. We separate out the first irreducible self-energy part from the diagram, reading from left to right. As distinct from the cases considered in previous chapters, the irreducible part can be of two types: Σ_{11} or Σ_{20}. The vertical dotted lines in Fig. 67 illustrate schematically the division of the diagram into two parts. To the right of the dotted line in Fig. 67*a* we have a chain of lines and self-energy parts, the sum of which is again a complete Green function $G'(x-x')$. On the other hand, the right-hand sides of the dotted lines in Figs. 67*b*, *c*, *d*, following the self-energy part Σ_{20}, represent, when summed over all the diagrams, a new function, which we shall denote by $\hat{G}(x-x')$. From the diagrammatic

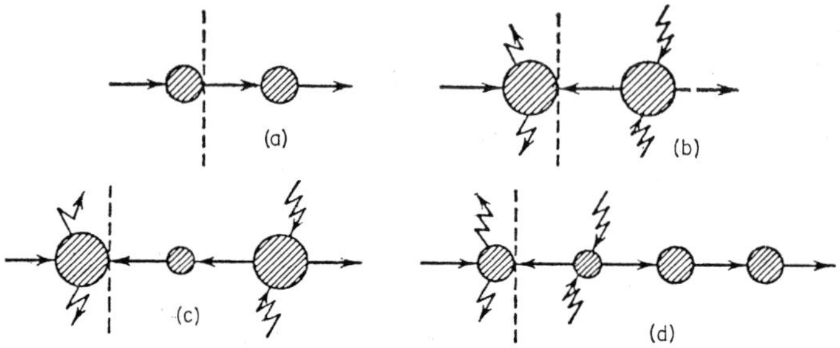

Fig. 67

point of view, it is distinguished by the fact that there are two departing uncondensed lines in the diagrams representing it. For the sake of convenience, we now introduce into the diagram arrows for each line corresponding to the two points x and x', indicating whether the line is entering or leaving at each of these points. The Green function for a

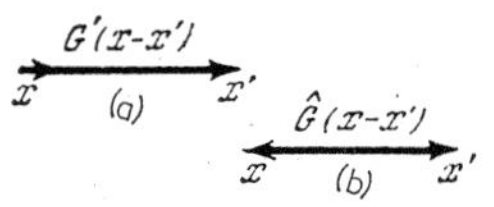

Fig. 68

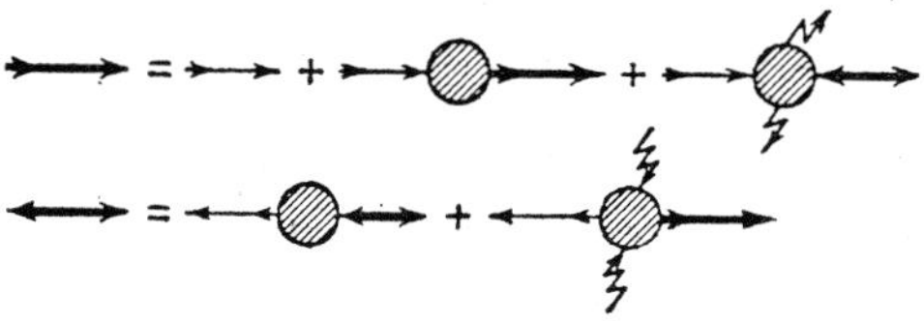

Fig. 69

non-interacting particle $G^{(0)}(x-x')$ is by definition the average of the T-product of operators $\psi'(x)\psi'^{+}(x')$ in the interaction representation. We shall put an arrow along the line at the point x in the direction from x (the operator $\psi'(x)$), and at the point x' an arrow in the direction to x' (the operator $\psi'^{+}(x')$). The Green function $G'(x-x')$ is obviously a heavy line with two arrows such as in the zero order Green function (Green function without interaction, Fig. 68*a*). As regards the function $\hat{G}(x-x')$, it is clear from Figs. 67*b*, *c*, *d* that it will be a heavy line on the diagram with two ends leaving it (Fig. 68*b*). The equations connecting the Green functions $G'(x-x')$ and $\hat{G}(x-x')$ are illustrated in Fig. 69. The structure of these equations is clear without further explanations; we shall remark here once again that the function $\hat{G}(x-x')$ appears in the theory as a result of the interaction of uncondensed particles with condensate particles and therefore has no analogue for non-interacting particles. As regards the self-energy parts Σ_{11}, Σ_{20} and Σ_{02}, as usual, the last cannot be written in the closed form in terms of the functions G and $\hat{G}$. The Feynman diagram method yields expansions for them into series, each term of which can be associated with a definite diagram. Several lower order diagrams for Σ_{11} and Σ_{20} are illustrated in Fig. 70 for the interaction Hamiltonian (23.12).

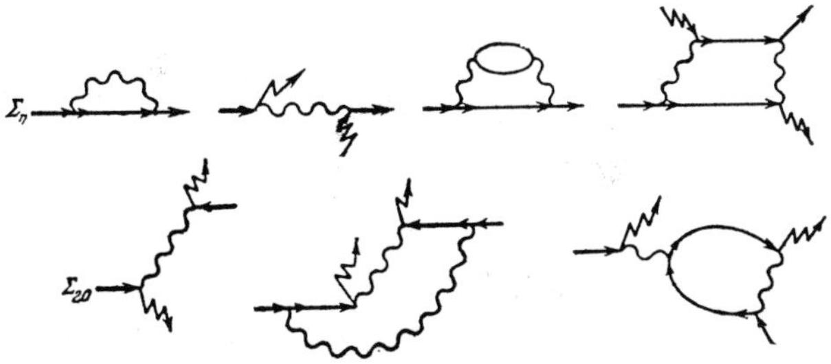

Fig. 70

We write down the equations of Fig. 69(†):

$$G'(x-x') = G^{(0)}(x-x') + \int\int G^{(0)}(x-y)\,[\Sigma_{11}(y-z)G'(z-x') + \Sigma_{20}(y-z)\hat{G}(z-x')]\,\mathrm{d}^4z\,\mathrm{d}^4y, \tag{24.1}$$

$$\hat{G}(x-x') = \int\int G^{(0)}(y-x)[\Sigma_{11}(z-y)\hat{G}(z-x') + \Sigma_{02}(y-z)G'(z-x')]\,\mathrm{d}^4z\,\mathrm{d}^4y.$$

Fourier transforming these equations, we get

$$\begin{aligned} G'(p) &= G^{(0)}(p) + G^{(0)}(p)\Sigma_{11}(p)G'(p) + G^{(0)}(p)\Sigma_{20}(p)\hat{G}(p), \\ \hat{G}(p) &= G^{(0)}(-p)\Sigma_{11}(-p)\hat{G}(p) + G^{(0)}(-p)\Sigma_{02}(p)G'(p). \end{aligned} \tag{24.2}$$

Using expression (23.9) for the Green function $G^{(0)}(p)$ for non-interacting particles, we can write (24.2) in the more convenient form

$$\begin{aligned} (\omega - \varepsilon_0(\boldsymbol{p}) + \mu - \Sigma_{11}(p))G'(p) - \Sigma_{20}(p)\hat{G}(p) &= 1, \\ (-\omega - \varepsilon_0(\boldsymbol{p}) + \mu - \Sigma_{11}(-p))\hat{G}(p) - \Sigma_{02}(p)G'(p) &= 0 \end{aligned} \tag{24.3}$$

(here $\varepsilon_0(\boldsymbol{p}) = \boldsymbol{p}^2/2m$). On introducing the notation

$$S(p) = \frac{\Sigma_{11}(p) + \Sigma_{11}(-p)}{2}, \qquad A(p) = \frac{\Sigma_{11}(p) - \Sigma_{11}(-p)}{2}$$

and expressing $G'(p)$ and $\hat{G}(p)$ in terms of $\Sigma_{11}, \Sigma_{02}, \Sigma_{20}$ via equations (24.3) we get for $G'(p)$ and $\hat{G}(p)$:

$$G'(p) = \frac{\omega + \varepsilon_0(\boldsymbol{p}) + S(p) + A(p) - \mu}{(\omega - A(p))^2 - (\varepsilon_0(\boldsymbol{p}) + S(p) - \mu)^2 + \Sigma_{20}(p)\Sigma_{02}(p)}, \tag{24.4}$$

$$\hat{G}(p) = -\frac{\Sigma_{02}(p)}{(\omega - A(p))^2 - (\varepsilon_0(\boldsymbol{p}) + S(p) - \mu)^2 + \Sigma_{20}(p)\Sigma_{02}(p)}, \tag{24.5}$$

(†) The choice of coefficients in the equations implies a suitable definition of the self-energy parts Σ_{ik} (see § 25).

These formulae generalise the usual expression for the single-particle function in terms of its self-energy part.

2. *Analytic properties of the Green functions*

The function $\hat{G}(p)$ has appeared up till now as the result of summation of definite diagrams. We shall next give a definition of it in terms of the operators ψ'^{+}. To do this, we consider

$$-i\langle T(\tilde{\xi}_0\tilde{\xi}_0\tilde{\psi}'^{+}(x)\tilde{\psi}'^{+}(x'))\rangle$$

and verify that its expansion into a perturbation theory series is the same as the expansion of $\hat{G}(x-x')$. We shall assume, as we did at the end of the previous section, that all the operators are defined with factors $e^{i\mu t}$ or $e^{-i\mu t}$ as in (23.21), and change to the interaction representation

$$\frac{-i\langle T(\xi_0\xi_0\psi'^{+}(x)\psi'^{+}(x'))S\rangle}{\langle S\rangle}.$$

We split the operations T and $\langle\cdots\rangle$ into $T=T^0T'$ and $\langle\langle\cdots\rangle'\rangle^0$. On regarding ξ_0 and ξ_0^+ as external parameters, i.e. carrying out the averaging over the uncondensed particles, we find that the diagrams for this quantity are the same as the diagrams for the function $\hat{G}(x-x')$, whereas the matrix elements differ by the presence of two extra operators ξ_0. It has been shown, that averaging over the condensate particles amounts to replacing the operators ξ_0 and ξ_0^+ in the interaction representation by the Heisenberg operators $\tilde{\xi}_0$ and $\tilde{\xi}_0^+$, and in turn replacing these latter by numbers: $\tilde{\xi}_0\to\sqrt{n_0}$ and $\tilde{\xi}_0^+\to\sqrt{n_0}$. Two equivalent definitions can therefore be used for $\hat{G}(x-x')$:

$$\hat{G}(x-x')=\frac{-i}{n_0}\langle T(\tilde{\xi}_0\tilde{\xi}_0\tilde{\psi}'^{+}(x)\tilde{\psi}'^{+}(x'))\rangle \tag{24.6}$$

or

$$\hat{G}(x-x')=-i\langle N+2\,|T(\tilde{\psi}'^{+}(x)\tilde{\psi}'^{+}(x'))|\,N\rangle, \tag{24.7}$$

where $\hat{G}(x-x')$ in the latter formula is expressed in terms of the matrix element of $T(\psi'^{+}(x)\psi'^{+}(x'))$ between the ground states of the systems with $N+2$ and N particles.

Let us investigate the properties of the Green functions $G'(x-x')$ and $\hat{G}(x-x')$. On using the definition (23.4) of $G'(x-x')$, we can represent it, precisely as we did in Chap. II, as the sum of matrix elements over the intermediate states when $t>t'$ and $t<t'$:

when $t>t'$,

$$G'(x-x')=-i\sum_m\langle N\,|\tilde{\psi}'(x)|\,m\rangle\langle m\,|\tilde{\psi}'^{+}(x')|\,N\rangle,$$

when $t<t'$,

$$G'(x-x')=-i\sum_n\langle N\,|\tilde{\psi}'^{+}(x')|\,n\rangle\langle n\,|\tilde{\psi}'(x)|\,N\rangle.$$

On separating the positional and time dependences in the matrix elements in the usual way, we get

$$G'(x-x') = \begin{cases} -i\sum_m |\psi_{Nm}|^2 \exp\{i(\boldsymbol{p}_m\cdot \boldsymbol{r}-\boldsymbol{r}') - i\omega_{mN}(t-t') + i\mu(t-t')\}, \\ \qquad t>t', \\ -i\sum_n |\psi_{Nn}|^2 \exp\{i\boldsymbol{p}_n(\boldsymbol{r}-\boldsymbol{r}') - i\omega_{nN}(t'-t) - i\mu(t-t')\}, \\ \qquad t<t'; \end{cases} \tag{24.8}$$

here, $\boldsymbol{p}_m$ and $\boldsymbol{p}_n$ are the momenta of the system in the intermediate states, $\omega_{mN} = E_m - E_{N0}$, $\omega_{nN} = E_n - E_{N0}$, where E_n, E_m are the energies of the system in states n and m, E_{N0} is the ground state energy of the system with N particles. By the properties of the operators $\tilde{\psi}'$ and $\tilde{\psi}'^{+}$, the system has $N+1$ particles in state m, and $N-1$ particles in state n. The appearance of the factors $e^{\pm i\mu t}$ in (24.8) is connected with this. On using the definition $\mu \simeq E_{N+1,0} - E_{N0}$, we can write (24.8) in the form

$$G'(x-x') = \begin{cases} -i\sum_m |\psi_{Nm}|^2 \exp\{i(\boldsymbol{p}_m\cdot \boldsymbol{r}-\boldsymbol{r}') - i(E_m - E_{N+1,0})(t-t')\} \\ \qquad (t>t'), \\ -i\sum_n |\psi_{Nn}|^2 \exp\{-i\boldsymbol{p}_n(\boldsymbol{r}-\boldsymbol{r}') + i(E_n - E_{N-1,0})(t-t')\} \\ \qquad (t<t'). \end{cases} \tag{24.9}$$

The energy differences $E_m - E_{N+1,0}$ and $E_n - E_{N-1,0}$ are the spectra, or excitation energies, of systems with $N+1$ and $N-1$ particles. Given a large number of particles, the spectra of these systems are the same up to terms of order $1/N$. Fourier transforming (24.9) both with respect to the coordinate differences and to the time, we get in the momentum representation for the Green function:

$$G'(p) = (2\pi)^3 \left[\sum_m \frac{\delta(\boldsymbol{p}-\boldsymbol{p}_m)\,|\psi_{Nm}|^2}{\omega - (E_m - E_{N+1,0}) + i\delta} - \sum_n \frac{\delta(\boldsymbol{p}+\boldsymbol{p}_n)\,|\psi_{Nn}|^2}{\omega + (E_n - E_{N-1,0}) - i\delta}\right]. \tag{24.10}$$

The poles of the function $G'(p)$ correspond to the values $\omega = \pm(E_m - E_0)$, i.e. they determine, as usual, apart from the sign, the spectrum of the system; their position relative to the real ω axis is clear from (24.10).

We now carry out a similar expansion with respect to the intermediate states for the function $\hat{G}(x-x')$, making use of its representation in the form (24.7):

$$\hat{G}(x-x') = \begin{cases} -i\sum_m \langle N+2\,|\tilde{\psi}'^{+}(x)|\,m\rangle\langle m\,|\tilde{\psi}'^{+}(x')|\,N\rangle \quad (t>t'), \\ -i\sum_m \langle N+2\,|\tilde{\psi}'^{+}(x')|\,m\rangle\langle m\,|\tilde{\psi}'^{+}(x)|\,N\rangle \quad (t<t') \end{cases}$$

or

$$G(x-x') = \begin{cases} -i \sum_m \psi^+_{N+2,m} \psi^+_{mN} \exp[i(\mathbf{p}_m \cdot \mathbf{r} - \mathbf{r}') - i(E_m - E_{N+2,0} + \mu)t \\ \qquad + i(E_m - E_{N0} - \mu)t'] \quad (t > t'), \\ -i \sum_m \psi^+_{N+2,m} \psi^+_{mN} \exp[i(\mathbf{p}_m \cdot \mathbf{r}' - \mathbf{r}) - i(E_m - E_{N+2,0} + \mu)t' \\ \qquad + i(E_m - E_{N0} - \mu)t] \quad (t < t'). \end{cases} \quad (24.11)$$

The states distinguished by the index m correspond to states of the system with $N+1$ particles. On introducing $E_{N+1,0}$, the ground state of the system with $N+1$ particles, into (24.11) and again using the definition of the chemical potential, we can transform (24.11) to the form

$$\hat{G}(x-x') = \begin{cases} -i \sum_m \psi^+_{N+2,m} \psi^+_{mN} \exp[i(\mathbf{p}_m \cdot \mathbf{r} - \mathbf{r}') - i(E_m - E_{N+1,0})(t-t')] \\ \qquad (t > t'), \\ -i \sum_m \psi^+_{N+2,m} \psi^+_{mN} \exp[i(\mathbf{p}_m \cdot \mathbf{r} - \mathbf{r}') - i(E_m - E_{N+1,0})(t'-t)] \\ \qquad (t < t'). \end{cases}$$

The Fourier component of $\hat{G}(x-x')$ is equal to

$$\hat{G}(p) = (2\pi)^3 \sum_m \psi^+_{N+2,m} \psi^+_{m,N} \times \left[\frac{\delta(\mathbf{p} - \mathbf{p}_m)}{\omega - (E_m - E_{N+1,0}) + i\delta} - \frac{\delta(\mathbf{p} + \mathbf{p}_m)}{\omega + (E_m - E_{N+1,0}) - i\delta} \right]. \quad (24.12)$$

Comparing expressions (24.10) and (24.12), we can conclude that the poles of the Green functions $G'(x-x')$ and $\hat{G}(x-x')$ are the same. In particular, on returning to the representation of $G'(p)$ and $\hat{G}(p)$ in terms of the irreducible self-energy parts (24.4) and (24.5), we see that the spectrum of the system $\omega = \varepsilon(\mathbf{p})$ is given by the equation

$$\left(\varepsilon(\mathbf{p}) - A(p)\right)^2 - \left(\varepsilon_0(\mathbf{p}) + S(p) - \mu\right)^2 + \Sigma_{20}(p)\,\Sigma_{02}(p) = 0$$
$$(p \equiv \{\varepsilon(\mathbf{p}), \mathbf{p}\}).$$

Along with the functions $\hat{G}(x-x')$ of (24.6) and (24.7), it is useful to introduce the function

$$\check{G}(x-x') = -\frac{i}{n_0} \langle T(\tilde{\psi}'(x)\tilde{\psi}'(x')\tilde{\xi}_0^+\tilde{\xi}_0^+)\rangle \equiv \langle N\,|\,T(\tilde{\psi}'(x)\tilde{\psi}'(x'))\,|\,N+2\rangle. \quad (24.13)$$

On expanding this last expression over the intermediate states, along the same lines as above for $\hat{G}(x-x')$, we can find expressions analogous to (24.12) for the Fourier components of $\check{G}(p)$:

$$\check{G}(p) = (2\pi)^3 \sum_m \psi_{Nm} \psi_{m,N+2} \times \left[\frac{\delta(\mathbf{p} - \mathbf{p}_m)}{\omega - (E_m - E_{N+1,0}) + i\delta} - \frac{\delta(\mathbf{p} + \mathbf{p}_m)}{\omega + (E_m - E_{N+1,0}) - i\delta} \right]. \quad (24.14)$$

The function $\check{G}(p)$ therefore has poles (and contours round the poles), which are the same as the poles of $\hat{G}(p)$ and $G'(p)$. As regards the coefficients, i.e. the residues at the poles, they are real in the case of $G'(p)$, whereas they are the complex conjugate of each other at the identical poles of $\hat{G}(p)$, $\check{G}(p)$.

The function $\check{G}(x - x')$ is represented pictorially by a line, with two arrows directed towards each other. The equations connecting $\check{G}(x - x')$ with the ordinary Green function are represented schematically in Fig. 71. These equations contain the Green function $G'(x' - x)$, represented with an opposite direction of the arrows. The equations of Fig. 71 may be written as follows in Fourier components of all the functions:

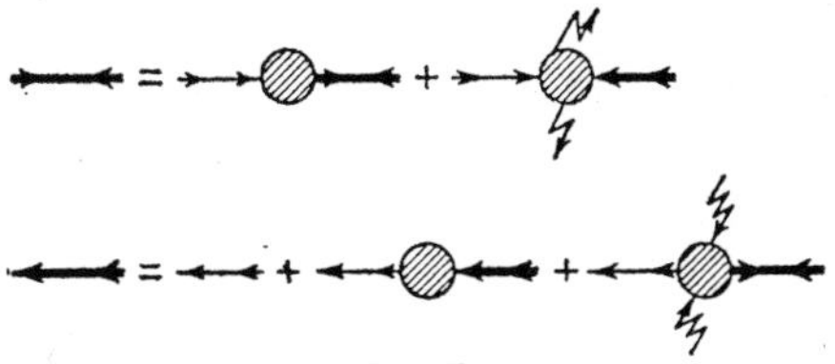

Fig. 71

$$\check{G}(p) = G^{(0)}(p)\Sigma_{11}(p)\,\check{G}(p) + G^{(0)}(p)\Sigma_{20}(p)\,G'(-p),$$
$$G'(-p) = G^{(0)}(-p) + G_0(-p)\,[\Sigma_{11}(-p)G'(-p) + \Sigma_{02}(p)\,\check{G}(p)].$$

We find on solving these equations for $\check{G}(p)$:

$$\check{G}(p) = \frac{\Sigma_{20}(p)}{(\omega - A(p))^2 - (\varepsilon_0(\mathbf{p}) + S(p) - \mu)^2 + \Sigma_{20}(p)\Sigma_{02}(p)}\,. \quad (24.15)$$

Expressions (24.5) and (24.15) for $\hat{G}(p)$ and $\check{G}(p)$ differ from one another in replacing the $\Sigma_{20}(p)$ in the numerator by $\Sigma_{02}(p)$.

3. *Behaviour of the Green functions for small momenta*

We shall conclude the present section by making some general remarks concerning the results obtained. Because of spatial uniformity, all the functions depend on the absolute value of the vector $\mathbf{p}$. It is clear from (24.12) and (24.14), that $\hat{G}(p)$ and $\check{G}(p)$ are even functions of the frequency ω. It is easily seen from this that $\Sigma_{20}(p) = \Sigma_{02}(p)$. For, inasmuch as the interaction Hamiltonian preserves the total number of particles, it is symmetric with respect to the operators ψ and ψ^+. Hence we can associate with any diagram for Σ_{20} precisely the same diagram for Σ_{02}, obtained by replacing all the lines that are entering in Σ_{20} by lines that are leaving, and vice versa; and correspondingly, the direction of the circuit round all the interior lines is reversed. But the circuit direction round all the interior lines can be changed by changing p to $-p$ in the matrix element for a given diagram $\Sigma_{20}(p)$. Since Σ_{20} is an even function, by (24.15), we have

$$\Sigma_{20}(p) = \Sigma_{02}(p) \text{ and } \hat{G}(p) = \check{G}(p).$$

We consider the equation for the poles of the Green functions:

$$\big(\omega - A(p)\big)^2 - \big(\varepsilon_0(\mathbf{p}) + S(p) - \mu\big)^2 + \Sigma_{02}^2(p) = 0. \quad (24.16)$$

It is clear from physical considerations that this equation must have a solution, no matter how small ω and $\boldsymbol{p}$. In fact, the possible solutions for the excitation energy spectrum with small $\boldsymbol{p}$ must include the acoustic spectrum $\omega = c\,|\boldsymbol{p}|$, i.e. the spectrum corresponding to long wavelength density oscillations. We therefore put ω and $\boldsymbol{p}$ equal to zero in (24.16). As a result, we get conditions connecting the chemical potential μ with $\Sigma_{11}(0)$, $\Sigma_{20}(0)$ and $\Sigma_{02}(0)$:

$$(\mu - \Sigma_{11}(0))^2 = \Sigma_{02}^2(0).$$

It will be clear from the results of the next section, that of the two roots of this equation we have to choose

$$\mu = \Sigma_{11}(0) - \Sigma_{02}(0). \tag{24.17}$$

In order to find the form of the Green functions in the neighbourhood of small ω and $\boldsymbol{p}$, we carry out expansions in the denominators of (24.4), (24.5) and (24.15), confining ourselves everywhere to second order terms in ω and $\boldsymbol{p}$. We find then using (24.17):

$$\begin{aligned} G'(p) &= \frac{\Sigma_{11}(0) - \mu}{B(\omega^2 - c^2\,|\boldsymbol{p}|^2)} = \frac{\Sigma_{20}(0)}{B(\omega^2 - c^2\,|\boldsymbol{p}|^2)}, \\ \hat{G}(p) &= \check{G}(p) = -\frac{\Sigma_{20}(0)}{B(\omega^2 - c^2\,|\boldsymbol{p}|^2)}, \end{aligned} \tag{24.18}$$

where

$$B = \left(1 - \frac{\partial \Sigma_{11}(0)}{\partial \omega}\right)^2 - \frac{\partial^2 \Sigma_{11}(0)}{\partial \omega^2}\Sigma_{20}(0) + \frac{1}{2}\frac{\partial^2}{\partial \omega^2}\Sigma_{02}^2(0),$$

$$Bc^2 = 2\Sigma_{20}(0)\left\{\frac{1}{2m} + \frac{\partial \Sigma_{11}(0)}{\partial\,|\boldsymbol{p}|^2} - \frac{\partial \Sigma_{20}}{\partial\,|\boldsymbol{p}|^2}\right\}.$$

Obviously, c is the velocity of sound. As must be the case, it vanishes if $\Sigma_{20}(0)$ is zero, since the velocity of sound is zero for an ideal Bose gas.

On comparing the results (24.18) with the general expansions of the Green functions (24.10), (24.12) and (24.14), we find that the ratio $\Sigma_{20}(0)/B$ is real and positive. Thus the form of all the Green functions $G'(p)$, $\hat{G}(p)$, $\check{G}(p)$ is the same for small ω and $\boldsymbol{p}$ ($\omega \sim c\,p$):

$$G(p) = \frac{\text{const}}{\omega^2 - c^2 p^2}. \tag{24.19}$$

§ 25. DILUTE NON-IDEAL BOSE GAS

1. Diagram technique

We shall now illustrate the methods described by considering in more detail the particular case when the interactions between the particles are binary forces (Belyaev [42]). The interaction Hamiltonian is

$$H_{int} = \frac{1}{2}\int\int \psi^+(\boldsymbol{r})\psi^+(\boldsymbol{r}')\,U(\boldsymbol{r} - \boldsymbol{r}')\,\psi(\boldsymbol{r}')\,\psi(\boldsymbol{r})\,d^3\boldsymbol{r}\,d^3\boldsymbol{r}'. \tag{25.1}$$

We separate out in H_{int} in the explicit form the operators of the condensate particles ξ_0 and ξ_0^+ in accordance with (23.1). As a result, we get eight distinct terms in a sum that can be used to represent H_{int}:

$$H_a = \frac{1}{2} \int \int \psi'^+(\boldsymbol{r})\psi'^+(\boldsymbol{r}')U(\boldsymbol{r}-\boldsymbol{r}')\psi'(\boldsymbol{r}')\psi'(\boldsymbol{r})\mathrm{d}^3\boldsymbol{r}\mathrm{d}^3\boldsymbol{r}',$$

$$H_b = \frac{1}{2} V(\xi_0^+)^2\xi_0^2 \int U(\boldsymbol{R})\mathrm{d}^3\boldsymbol{R},$$

$$H_c = \frac{1}{2} \int \int [\xi_0^+\psi'^+(\boldsymbol{r}') + \psi'^+(\boldsymbol{r})\xi_0^+]\psi'(\boldsymbol{r})\psi'(\boldsymbol{r}')U(\boldsymbol{r}-\boldsymbol{r}')\mathrm{d}^3\boldsymbol{r}\mathrm{d}^3\boldsymbol{r}',$$

$$H_d = \frac{1}{2} \int \int \psi'^+(\boldsymbol{r})\psi'^+(\boldsymbol{r}')[\psi'(\boldsymbol{r}')\xi_0 + \xi_0\psi'(\boldsymbol{r})]U(\boldsymbol{r}-\boldsymbol{r}')\mathrm{d}^3\boldsymbol{r}\mathrm{d}^3\boldsymbol{r},$$

$$H_e = \frac{1}{2} \int \int [\xi_0^+\psi'^+(\boldsymbol{r}')\xi_0\psi'(\boldsymbol{r}) + \xi_0^+\psi'^+(\boldsymbol{r})\xi_0\psi'(\boldsymbol{r}')]U(\boldsymbol{r}-\boldsymbol{r}')\mathrm{d}^3\boldsymbol{r}\mathrm{d}^3\boldsymbol{r}',$$

$$H_f = \frac{1}{2} \int \int [\xi_0^+\psi'^+(\boldsymbol{r})\xi_0\psi'(\boldsymbol{r}) + \xi_0^+\psi'^+(\boldsymbol{r}')\xi_0\psi'(\boldsymbol{r}')]U(\boldsymbol{r}-\boldsymbol{r}')\mathrm{d}^3\boldsymbol{r}\mathrm{d}^3\boldsymbol{r}',$$

$$H_g = \frac{1}{2} \int \int \xi_0^+\xi_0^+\psi'(\boldsymbol{r})\psi'(\boldsymbol{r}')U(\boldsymbol{r}-\boldsymbol{r}')\mathrm{d}^3\boldsymbol{r}\mathrm{d}^3\boldsymbol{r}',$$

$$H_h = \frac{1}{2} \int\int \xi_0\xi_0\psi'^+(\boldsymbol{r})\psi'^+(\boldsymbol{r}')U(\boldsymbol{r}-\boldsymbol{r}')\mathrm{d}^3\boldsymbol{r}\mathrm{d}^3\boldsymbol{r}'. \quad (25.2)$$

The elementary processes corresponding to each of these terms are shown in Fig. 72. Any matrix element can be formed by the usual method, Wick's theorem being applied to the operators of the uncondensed particles. In accordance with the results of the previous sections,

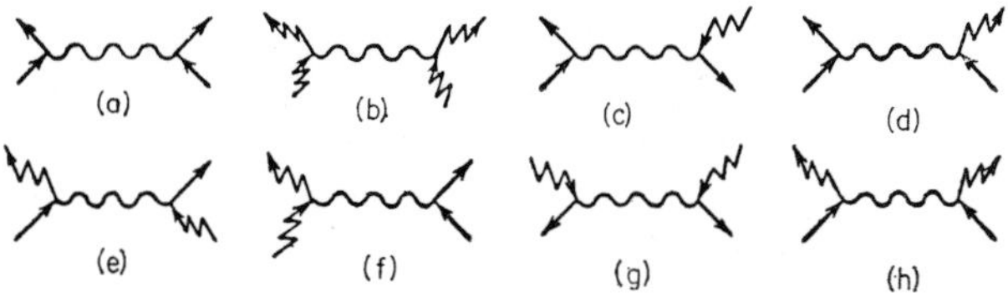

Fig. 72

it is only necessary to take the connected diagrams for the process, the operators ξ_0, ξ_0^+ being everywhere regarded as external parameters, which have to be replaced in accordance with $\xi_0, \xi_0^+ \to \sqrt{n_0}$ (if the frequencies of all the particles concerned in the process are measured from the value of the chemical potential). We shall confine ourselves to formulating the correspondence rules in the momentum representation between the matrix elements and the diagrams for the single-particle Green functions. Let us take any diagram of the mth order of perturbation theory for one of the Green functions, say $G'(x - x')$, containing s condensate lines entering and s leaving (as we have remarked several times, the total numbers of the lines entering and leaving any diagram must be the same). The diagram in question is contained in the expression

$$(-i)\frac{(-i)^m}{m!}\langle T(\psi'(x)\int H_{int}(t_1)\cdots H_{int}(t_m)\psi'^{+}(x')\mathrm{d}t_1\cdots\mathrm{d}t_m)\rangle.$$

The number of possible permutations of the m Hamiltonians $H_{int}(t_i)$, that do not destroy the order of pairings determined by the given diagram, is equal to $m!$ The number of operators ψ' (equal to the number of operators ψ'^{+}) is obviously equal to $2m-s+1$.

In accordance with the definition of the Green function $G^{(0)}$, a factor $-i$ appears with each pairing $\widehat{\psi'\psi'^{+}}$. We shall associate $G^{(0)}$ with each straight line and introduce a wavy line corresponding to the potential

$$V(x-x') = U(\mathbf{r}-\mathbf{r}')\delta(t-t').$$

The total number of ternary vertices (i.e. vertices from which a wavy line departs) is, by (25.1) and Fig. 72, twice the order of perturbation theory. It may easily be shown that, if we introduce Fourier components of all the quantities:

$$G'(x-x') = \frac{1}{(2\pi)^4}\int G'(p)e^{ip(x-x')}\mathrm{d}^4p,$$

$$V(x-x') = \frac{1}{(2\pi)^4}\int U(\mathbf{q})e^{iq(x-x')}\mathrm{d}^4q, \text{ etc.},$$

the matrix element of any diagram of the mth order of perturbation theory for a Green function can be formed according to the following rules:

(1) every straight line proceeding from left to right corresponds in $G'(p)$ to the function $G^{(0)}(p) = [\omega - \varepsilon_0(\mathbf{p}) + \mu + i\delta]^{-1}$ (a line in the opposite direction corresponds to $G^{(0)}(-p)$);

(2) each wavy line with momentum $\mathbf{q}$ corresponds to a Fourier component of the interaction potential $U(\mathbf{q})$;

(3) an entering or departing condensate line corresponds to the factor $\sqrt{n_0}$;

(4) at every ternary vertex the momentum $\mathbf{q}$ of the wavy line is equal to the difference between the momenta of the particle lines. Integrations are performed over the momenta not determined by the laws of conservation; each integration implies a corresponding factor $(2\pi)^{-4}$;

(5) the entire matrix element must be multiplied by $A_{s,m}(-i)^{s-m}$, where $A_{s,m}$ depends on what sort of terms of (25.2) figure in the diagram.

These rules remain unchanged for the Green functions $\hat{G}$ and $\check{G}$, provided we understand by s the power of the factors n_0 figuring in the given diagram of order m. If, for instance, the number of entering condensate lines in one of the diagrams for $\hat{G}$ is equal to l, the number of departing lines is $l + 2$ because of the definition (24.7) of $\hat{G}$. The number of pairings of the operators ψ' and ψ'^{+} (the number of functions $G^{(0)}$) is obviously equal to

$$2m - l.$$

Since the factor $-i$ appears in each Green function, in accordance with its definition, the factor by which the matrix element must be multiplied is equal to

$$(-i)^{l+1-m}.$$

However, $l + 1$ is just the power s of the factors n_0 arising from the condensate lines.

2. *Connection between the chemical potential and the self-energy parts of the single-particle Green functions*

We now turn to the proof of (24.17) for the chemical potential μ. Let us consider the operator $\xi_0(t)$ in the Heisenberg representation (we assume that the term $-\mu N$ is included in the complete Hamiltonian). As regards its dependence on time, $\xi_0(t)$ satisfies the usual operator equation of quantum mechanics:

$$\frac{i\,\partial\tilde{\xi}_0(t)}{\partial t} = [\tilde{\xi}_0(t), H] = -\mu\tilde{\xi}_0(t) - [H_{int}, \tilde{\xi}_0(t)].$$

We find with the aid of this equation, for the Green function $G_0(t - t')$ for condensate particles:

$$\frac{\partial G_0(t-t')}{\partial t} = i\,\mu G_0(t - t') - \langle T\left([\tilde{\xi}_0(t), H_{int}]\tilde{\xi}_0^{+}(t')\right)\rangle.$$

However, it follows from the results of the previous sections, that $G_0(t - t')$ does not depend on time and is simply equal to n_0. Hence it follows that

$$\mu n_0 = -\langle T\left([H_{int}, \tilde{\xi}_0(t)]\tilde{\xi}_0^{+}(t')\right)\rangle. \tag{25.3}$$

Let us evaluate the average on the right-hand side of this equation. Changing to the interaction representation

$$\langle T\left([H_{int}, \tilde{\xi}_0(t)]\tilde{\xi}_0^{+}(t')\right)\rangle = \frac{\langle T\left([H_{int}, \xi_0(t)]\xi_0^{+}(t')S\right)\rangle}{\langle S\rangle},$$

we can briefly reproduce the arguments of § 23. We first carry out the averaging and time-ordering over the uncondensed particles in the operations T and $\langle\cdots\rangle$. In accordance with the general method, we only need to take here the connected diagrams into account, whilst regarding the operators ξ_0 and ξ_0^{+} as external parameters. In the present instance the connected diagrams are different vacuum loops; we denote the result of the averaging, which only affects the operators of the uncondensed particles in H_{int}, by $\bar{H}_{int}^{con}$. The function $\bar{H}_{int}^{con}$ depends on ξ_0, ξ_0^+ as parameters. To obtain the final result, we have to replace all the operators ξ_0, ξ_0^+ appearing in $\bar{H}_{int}^{con}$ by the exact Heisenberg operators $\tilde{\xi}_0$ and $\tilde{\xi}_0^+$, after which we get for (25.3):

$$\mu n_0 = -\frac{\langle T^0([\bar{H}_{int}^{con}, \xi_0]\xi_0^+ \langle S\rangle'))^0}{\langle S\rangle} = -\langle T([\tilde{\bar{H}}_{int}^{con}, \tilde{\xi}_0]\tilde{\xi}_0^+)\rangle. \quad (25.4)$$

The commutator $[\bar{H}_{int}^{con}, \xi_0]$ contains the commutations of ξ_0 with different products of operators ξ_0, ξ_0^+ in the vacuum averages $\bar{H}_{int}^{con}$. When evaluating $\bar{H}_{int}^{con}$, the usual averaging in accordance with Wick's method of the uncondensed particle operators has been carried out. Since this averaging consists in pairwise averaging of the operators ψ' and ψ'^+, the numbers of operators ξ_0, ξ_0^+ are also equal.

Let us consider, for example, the vacuum loop of the mth order of perturbation theory, containing s operators ξ_0 and s operators ξ_0^+. The result of the commutation of ξ_0 with one of the operators ξ_0^+ is $[\xi_0^+, \xi_0] = -1/V$; although ξ_0 can commute with all s operators ξ_0^+. Hence, if the correction to the ground state energy corresponding to the given mth order vacuum loop with n_0^s is denoted by $\langle H_{int}\rangle_{m,s}^{con}$ (this latter is obtained from $\bar{H}_{int}^{con}$ by replacing the operators ξ_0, ξ_0^+ by $\sqrt{n_0}$), it may easily be seen that (25.4) is

$$\mu n_0 = \sum_{s,m} \frac{s}{V} \langle H_{int}\rangle_{m,s}^{con}$$

or

$$\mu = \sum_{m,s} \frac{\partial}{\partial n_0} \frac{\langle H_{int}\rangle_{s,m}^{con}}{V} = \frac{\partial}{\partial n_0} \frac{\langle H_{int}\rangle}{V}. \quad (25.5)$$

(The vacuum average $\langle H_{int}\rangle$ is a function of the parameters μ and n_0. Hence the partial derivative with respect to n_0 at constant μ is seen on the right-hand side of (25.5).) The idea of the rest of the proof (Hugenholtz and Pines [43]) is based on the fact that the operators ξ_0, ξ_0^+ appear symmetrically with the operators ψ' and ψ'^+ in the interaction Hamiltonian. Each vacuum loop $\langle H_{int}\rangle$ with a number of condensate lines on the diagram can therefore be formally associated with diagrams for the irreducible self-energy parts $\Sigma_{11}(0)$, $\Sigma_{20}(0)$, the necessary number of incoming and outgoing condensate lines (operators ξ_0^+ and ξ_0) being replaced in the loop by entering and departing straight lines (operators ψ'^+ and ψ'). Fig. 73 shows some simple examples of lower order diagrams (†). Since the entering and departing condensate particle lines carry the 4-momentum $p = 0$, the correspondence in question will also hold for the matrix elements $\Sigma_{11}(p)$, $\Sigma_{20}(p) = \Sigma_{02}(p)$ when $p = 0$.

(†) Some of these diagrams are zero (see below), but this is of no importance for illustrating our arguments.

The matrix element of an arbitrary irreducible diagram $\langle H_{int}\rangle_{m,s}^{\text{con}}$ is formed in accordance with the same correspondence rules as were formulated above for the Green functions; it is easily seen, that the only difference lies in the numerical factor, by which the entire integral has to be multiplied, which is equal to

$$(-i)^{s-m-2},$$

where m is the order of perturbation theory, and s is the power of n_0 in the diagram. The corresponding factor in the matrix element for the self-energy parts has the form

$$(-i)^{s-m}.$$

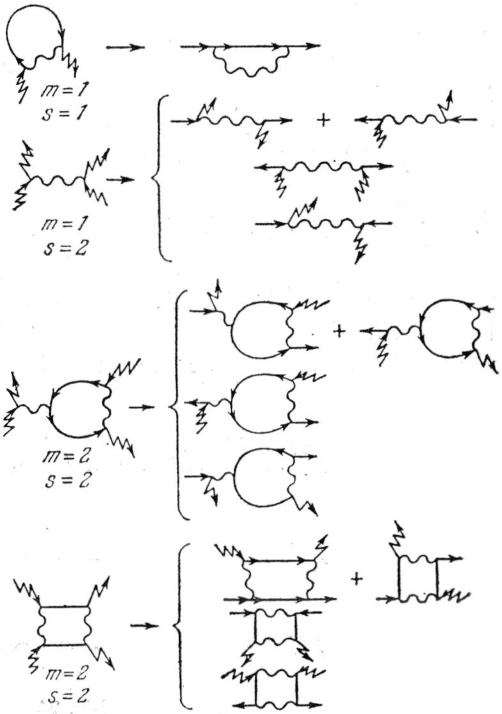

Fig. 73

The power of n_0 is reduced by one when differentiating the vacuum loop with respect to n_0. Let us consider all the possible diagrams for $\Sigma_{11}(0)$ of the mth order of perturbation theory, containing $s-1$ factors n_0 (and the same number of entering and departing condensate particle lines). All these diagrams can be got from the vacuum loop $\langle \overline{H}_{int}^{\text{con}}\rangle_{m-1,s}$ by replacing one of the s operators ξ_0^+ by an entrant, and a ξ_0 by a departing straight line, i.e. these diagrams can be obtained by s^2 methods:

$$\left(\Sigma_{11}(0)\right)_{m,s-1} = \frac{s^2}{n_0 V}\langle H_{int}\rangle_{m-1,s}^{\text{con}}.$$

As regards the diagrams for $\Sigma_{20}(0)$, these latter are obtained from $(\overline{H}_{int}^{\text{con}})_{m-1,s}$ by replacing two entrant condensate lines by two entrant straight lines. There are altogether $s(s-1)$ such diagrams, as follows from the number of methods by which this replacement can be made. Therefore,

$$\left(\Sigma_{20}(0)\right)_{m,s-1} = \frac{s^2-s}{n_0 V}\langle H_{int}\rangle_{m-1,s}^{\text{con}}.$$

On comparing the difference $\Sigma_{11}(0) - \Sigma_{20}(0)$ with expression (25.5), we obtain at once:

$$\mu = \Sigma_{11}(0) - \Sigma_{20}(0). \tag{25.6}$$

As already remarked in the previous section, the validity of this relation extends beyond the case of binary forces between the particles.

Equation (25.6), in conjunction with (23.19), connecting the chemical potential with the density of the total number of particles in the system, gives us a set of two conditions from which the values of the parameters μ and n_0 can be determined. We shall not dwell here on the proof of the equivalence of conditions (23.20) and (25.6); we shall only remark that, for perturbation theory calculations (i.e. for the case of a gas of weakly interacting particles), condition (25.6) is more convenient, since it expresses μ directly in terms of the familiar functions of an ideal Bose gas.

3. *Low density approximation*

Let us apply the method developed above to a gas of interacting bosons with the Hamiltonian (25.1). We have already considered this example in Chap. I, where it was assumed that the interaction forces between the particles were small. Thus expression (4.11) for the excitation spec-

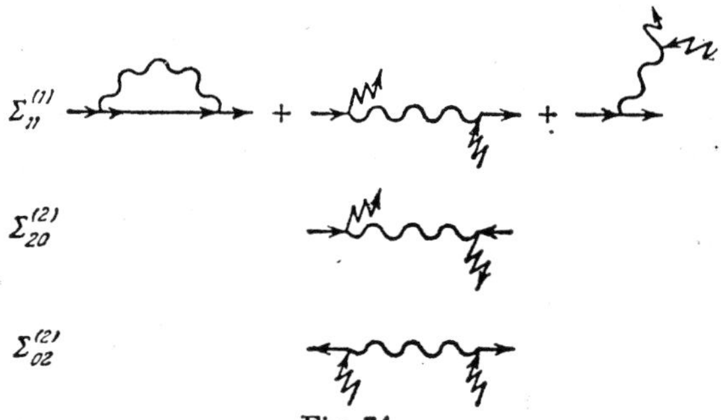

Fig. 74

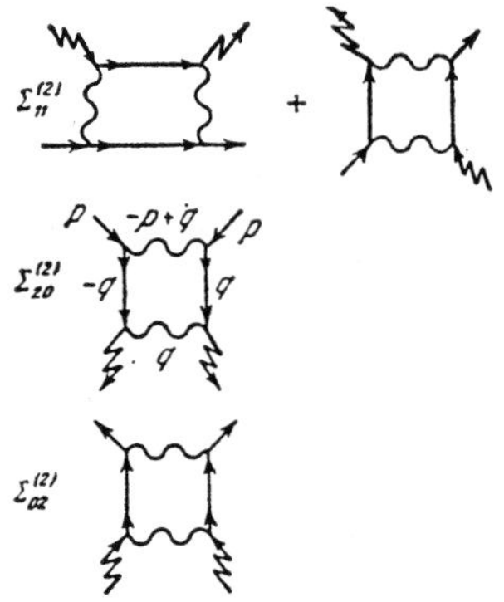

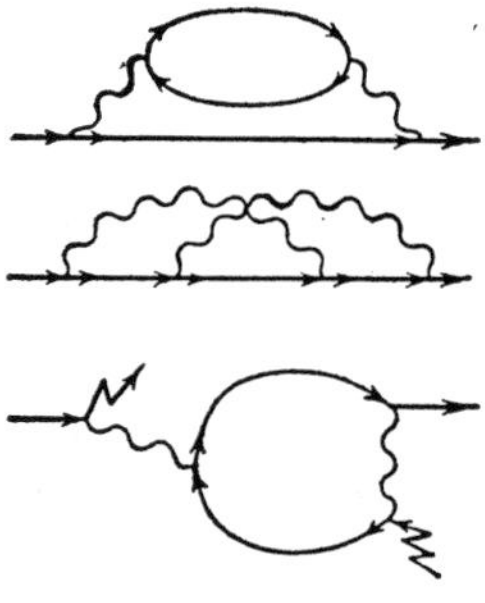

Fig. 75 Fig. 76

trum contained the Fourier components of the potential, proportional, in the Born approximation, to the scattering amplitude of the particles with one another. We shall soon show that this result is in fact valid in the more general case when the gas density, but not the interaction, is assumed small, i.e. when the dimensions of the particles are small compared with the average distance apart of the particles in the gas (if the particle dimensions are characterized by the amplitude f_0 of the S-scattering with one another, the condition is $f_0 n^{1/3} \ll 1$). Let us consider the diagrams of the first order of perturbation theory for $\Sigma_{11}(p)$, $\Sigma_{20}(p)$ (Fig. 74). Of the three Σ_{11} diagrams, the first is due to the averaging of the term H_a of (25.2) of the interaction Hamiltonian (25.1). This diagram is zero, since the interior line in it is the average $-i\langle\psi'^{+}(\boldsymbol{r}')\psi'(\boldsymbol{r})\rangle \equiv 0$ (remember that the wavy line corresponds to the interaction potential $V(x-x') = U(\boldsymbol{r}-\boldsymbol{r}')\,\delta(t-t')$, whilst the order of the operators ψ'^{+} and ψ' in H_a is given by (25.1)). The remaining terms yield:

$$\Sigma_{11}^{(1)}(p) = n_0\big(U(0) + U(\boldsymbol{p})\big), \qquad \Sigma_{20}^{(1)}(p) = \Sigma_{02}^{(1)}(p) = n_0 U(\boldsymbol{p}),$$

where the $U(\boldsymbol{p})$ are the Fourier components of the interaction potential. By (25.6),

$$\mu = n_0 U(0).$$

The only diagrams of the second order of perturbation theory that are non-zero are shown in Fig. 75. For instance, the diagrams $\Sigma_{11}^{(2)}(p)$ of Fig. 76 are zero, since each of them contains products of Green functions of the form $G^{(0)}(\boldsymbol{r}-\boldsymbol{r}', t_1-t_2)G^{(0)}(\boldsymbol{r}''-\boldsymbol{r}''', t_2-t_1)$ (in the coordinate representation). Incidentally, we know that $G^{(0)}(\boldsymbol{r}_1-\boldsymbol{r}_2, t_1-t_2) \equiv 0$ for $t_1 < t_2$.

To estimate the diagrams of Fig. 75, we assume for simplicity that the Fourier component of the interaction potential has the form:

$$U(\boldsymbol{p}) = \begin{cases} U_0 & \text{for } |\boldsymbol{p}| \ll \dfrac{1}{a}, \\[2ex] 0 & \text{for } |\boldsymbol{p}| \gg \dfrac{1}{a}, \end{cases}$$

where $a \sim f_0$ is of the order of magnitude of the particle radius. The estimate of any diagram may involve the following parameters of the problem: U_0 and a, characterising the interaction, the mass m, and the density n_0 of the number of condensate particles. Two dimensionless quantities can be formed from these parameters:

$$\zeta \sim \frac{m U_0}{a}, \qquad \beta = \sqrt{n_0 a^3}\,.$$

The quantity ζ is the perturbation theory parameter (expansion into a Born series), whilst β is the "gas" parameter. Formally, the perturbation theory series is an expansion in powers of $\zeta \ll 1$; however, we shall below only assume $\beta \ll 1$.

For greater simplicity, let us consider the diagram for $\Sigma_{20}^{(2)}(p)$ in Fig. 75. For this,

$$\Sigma_{20}^{(0)}(p) \sim n_0 \int G^{(0)}(q) G^{(0)}(-q) U(\boldsymbol{q}) U(\boldsymbol{p}-\boldsymbol{q}) \mathrm{d}^3\boldsymbol{q}\, \mathrm{d}\omega$$

or, on substituting values (23.9) for the Green functions $G^{(0)}$ and integrating over ω, we get

$$\Sigma_{20}^{(2)}(p) \sim n_0 U_0^2 \int \frac{\mathrm{d}^3\boldsymbol{q}}{\mu - \varepsilon_0(\boldsymbol{q})}.$$

Of importance in the last integral are the larger values of $|\boldsymbol{q}| \sim 1/a$, for which $\mu/\varepsilon_q^0 \sim m n_0 U_0 a^2 = \zeta\beta^2 \ll 1$, so that

$$\Sigma_{20}^{(2)} \sim \frac{m n_0 U_0^2}{a} \sim \Sigma_{20}^{(1)} \zeta.$$

A similar estimate for the $\Sigma_{11}^{(2)}$ diagrams shows that $\Sigma_{11}^{(2)} \sim \Sigma_{11}^{(1)}\zeta$.

Let us now consider the third order diagram for $\Sigma_{11}^{(3)}$ in Fig. 77*a*. We obtain for this:

$$\Sigma_{20}^{(3)} \sim n_0^2 \int G^{(0)}(-q) [G^{(0)}(q)]^2 [U(\boldsymbol{q})]^2 U(\boldsymbol{p}+\boldsymbol{q}) \mathrm{d}^3\boldsymbol{q}\, \mathrm{d}\omega$$

$$\sim n_0^2 U_0^3 \int \frac{\mathrm{d}^3\boldsymbol{q}}{[\mu - \varepsilon_0(\boldsymbol{q})]^2}.$$

As distinct from the previous case, our integral here is convergent at the upper limit, and the chief role is played in it by the region $|\boldsymbol{q}| \sim \sqrt{m\mu} \sim \sqrt{n_0 U_0 m}$, so that

$$\Sigma_{20}^{(3a)} \sim \frac{n_0^2 U_0^3 m^{3/2}}{\mu^{1/2}} \sim \Sigma_{20}^{(1)} \zeta^{3/2} \beta. \tag{25.7}$$

At the same time, the third order diagram of Fig. 77*b* has the value

$$\Sigma_{20}^{(3b)} \sim \Sigma_{20}^{(1)} \zeta^2. \tag{25.8}$$

It is clear from (25.7) and (25.8) that $\Sigma_{20}^{(3a)} \sim \beta \zeta^{-1/2} \Sigma_{20}^{(3b)}$. This result is a consequence of the fact that $\Sigma_{20}^{(3b)}$ contains two integrals of the product of two $G^{(0)}$, each formally divergent at the upper limit, whereas three $G^{(0)}$ functions are integrated in $\Sigma_{20}^{(3a)}$; the integral is convergent without cutting it off and is determined by the values of the integrand in the region of momenta $|\boldsymbol{q}| \sim \sqrt{m\mu}$. This distinction is shown on the diagrams by the number of continuous lines in a closed loop (formed by straight and wavy lines).

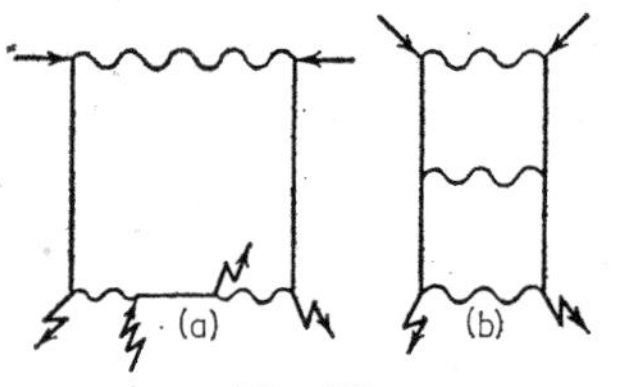

Fig. 77

Hence every loop in Σ_{ik} with more than two straight lines contributes the small parameter β, whereas the loops with two lines do not contain β. Only diagrams of the second type need be considered for lower approximations. This means formally that, of all the diagrams for Σ_{11} and Σ_{20}, we need only select those in which there are two condensate lines, entering or leaving, i.e. the diagrams of the first degree in n_0. For it is clear from dimensional considerations that all the diagrams that contain a higher power of n_0 will include an extra order of smallness in β. All the necessary diagrams are of the ladder type (Fig. 78). Let us denote by $\Gamma^{(0)}(p_1, p_2; p_3, p_4)$ the set of diagrams of Fig. 79 in the momentum representation. The first approximation in β differs, therefore, from the first approximation of perturbation theory in replacing the Fourier components of the potential $U(\boldsymbol{q})$ (the first "rung" of the ladder) by $\Gamma^{(0)}(p_1, p_2; p_1 - q, p_2 + q)$. Naturally, in all the more complicated diagrams, summation of the "ladder" loops contained in them also leads to the appearance of $\Gamma^{(0)}$ (for our purposes, however, discussion of these diagrams is superfluous, since, as already mentioned, they yield terms of higher order in β). The potential $U(\boldsymbol{q})$ is therefore eliminated from the problem; the role of effective potential is played by $\Gamma^{(0)}$.

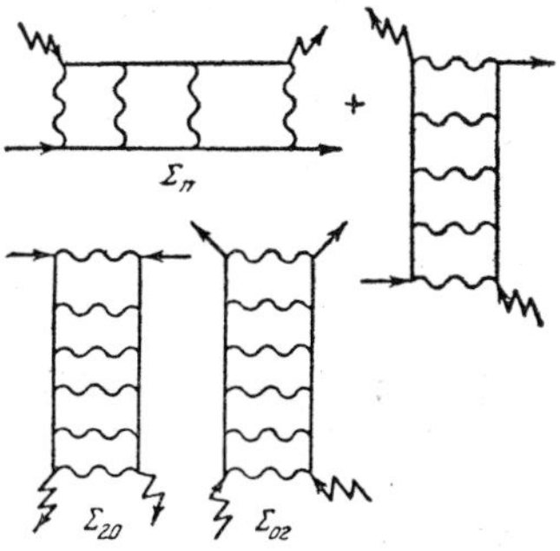

Fig. 78

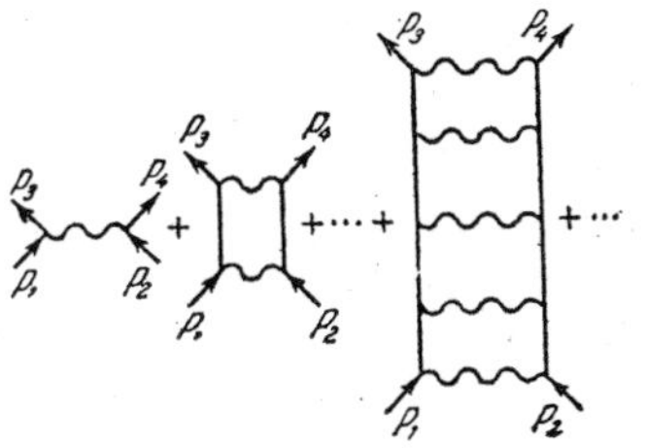

Fig. 79

The integral equation for $\Gamma^{(0)}(p_1, p_2; p_3, p_4)$ follows at once from the structure of the diagrams of Fig. 79:

$$\Gamma^{(0)}(p_1, p_2; p_3, p_4) = U(\boldsymbol{p}_3 - \boldsymbol{p}_1) + \frac{i}{(2\pi)^4} \int U(\boldsymbol{p}_1 - \boldsymbol{k}) G^{(0)}(k) \times G^{(0)}(p_1 + p_2 - k)\, \Gamma^{(0)}(k, p_1 + p_2 - k; p_3, p_4)\, \mathrm{d}^4 k. \tag{25.9}$$

4. *Effective interaction potential*

Let us now stop to examine (25.9). We introduce the total and relative momenta:

$$p_1 + p_2 = p_3 + p_4 = P;\ p_1 - p_2 = 2k;\ p_3 - p_4 = 2k'.$$

Equation (25.9) for $\Gamma^{(0)}(p_1, p_2; p_3, p_4) \equiv \Gamma^{(0)}(k, k'; P)$ now transforms to

$$\Gamma^{(0)}(k, k'; P) = U(\boldsymbol{k} - \boldsymbol{k}') + \frac{i}{(2\pi)^4} \int U(\boldsymbol{k} - \boldsymbol{p}) G^{(0)}\left(\frac{P}{2} + p\right) G^{(0)}\left(\frac{P}{2} - p\right) \Gamma^{(0)}(p, k'; P)\, \mathrm{d}^4 p. \tag{25.10}$$

The interaction potential $V(x - x')$ does not contain retardation effects: $V(x - x') = U(\boldsymbol{r} - \boldsymbol{r}')\,\delta(t - t')$. We have thus for the Fourier components: $V(q) \equiv U(\boldsymbol{q})$, i.e. it does not depend on the fourth component of the 4-vector q. As a result, $\Gamma^{(0)}(p_1, p_2; p_3, p_4)$ only depends on one combination of the fourth components $\omega_1 + \omega_2 = \omega_3 + \omega_4 = \Omega$, where $P = \{\boldsymbol{P}, \Omega\}$. Thus $\Gamma^{(0)}(k, k'; P)$ does not depend on the fourth components of the first two arguments, and this enables us to perform the integration over $d\omega$ in the integral of equation (25.10):

$$\int d\omega G^{(0)}\left(\frac{P}{2} + p\right) G^{(0)}\left(\frac{P}{2} - p\right) = -\frac{2\pi i}{\Omega - \frac{\boldsymbol{P}^2}{4m} + 2\mu - \frac{\boldsymbol{p}^2}{m} + i\delta}.$$

After substituting in (25.10), the equation for $\Gamma^{(0)}(k, k'; P)$ becomes

$$\Gamma^{(0)}(k, k'; P) = U(\boldsymbol{k} - \boldsymbol{k}') + \frac{1}{(2\pi)^3}\int d^3\boldsymbol{p} \frac{U(\boldsymbol{k} - \boldsymbol{p})\Gamma^{(0)}(p, k'; P)}{\frac{\varkappa^2}{m} - \frac{\boldsymbol{p}^2}{m} + i\delta}, \quad (25.11)$$

where

$$\frac{\varkappa^2}{m} = \Omega - \frac{\boldsymbol{P}^2}{4m} + 2\mu .$$

This equation cannot be solved in a general form for any arbitrary type of interaction; however, its solution can be expressed in terms of the scattering amplitude of two particles with one another in vacuo.

Let us remind the reader of the statement of the problem of the scattering of particles in a potential $U(\boldsymbol{r})$. The Schrödinger equation of a particle in a field $U(\boldsymbol{r})$ can be written as

$$(\nabla^2 + \boldsymbol{k}^2)\psi_{\boldsymbol{k}}(\boldsymbol{r}) = 2m U(\boldsymbol{r})\psi_{\boldsymbol{k}}(\boldsymbol{r}),$$

where $\boldsymbol{k}^2/2m$ is the eigenvalue of the particle energy and $\psi_{\boldsymbol{k}}(r)$ is its wave function. This equation may be conveniently written in terms of the solution of the Poisson equation:

$$\psi_{\boldsymbol{k}}(\boldsymbol{r}) = -\frac{m}{2\pi}\int \frac{e^{i|\boldsymbol{k}||\boldsymbol{r}-\boldsymbol{r}'|}}{|\boldsymbol{r} - \boldsymbol{r}'|} U(\boldsymbol{r}')\psi_{\boldsymbol{k}}(\boldsymbol{r}')d^3\boldsymbol{r}' + \psi_{0\boldsymbol{k}}(\boldsymbol{r}), \quad (25.12)$$

where $\psi_{0\boldsymbol{k}}$ is the wave function of a free particle with the same energy. The scattering amplitude is determined from the condition that, at large distances from the scattering centre, the wave function is the sum of a plane wave (free particle) and an outgoing wave(†):

$$\psi_{\boldsymbol{k}}(\boldsymbol{r}) = e^{i(\boldsymbol{k}\cdot\boldsymbol{r})} - f(\theta)\frac{e^{i|\boldsymbol{k}||\boldsymbol{r}|}}{|\boldsymbol{r}|},$$

where θ is the scattering angle relative to the direction of the vector $\boldsymbol{k}$. On comparing the behaviour of (25.12) at large $|\boldsymbol{r}|$ with this definition, we get

$$f(\theta) = \frac{m}{2\pi}\int e^{-i(\boldsymbol{k}'\cdot\boldsymbol{r}')} U(\boldsymbol{r}')\psi_{\boldsymbol{k}}(\boldsymbol{r}')d^3\boldsymbol{r}',$$

(†) Our definition of the scattering amplitude differs in sign from the generally accepted one (see e.g. [16]).

where the vector $\boldsymbol{k}'$ is directed along $\boldsymbol{r}$. On changing to the momentum representation for the wave function:

$$\psi_{\boldsymbol{k}}(\boldsymbol{r}) = (2\pi)^{-3} \int \psi_{\boldsymbol{k}}(\boldsymbol{p}) e^{i(\boldsymbol{p}\cdot\boldsymbol{r})} \mathrm{d}^3\boldsymbol{p},$$

we get

$$f(\theta) \equiv f(\boldsymbol{k}, \boldsymbol{k}') = \frac{m}{(2\pi)^4} \int U(\boldsymbol{k}' - \boldsymbol{p}) \psi_{\boldsymbol{k}}(\boldsymbol{p}) \mathrm{d}^3\boldsymbol{p} \tag{25.13}$$

(the particle is incident along the direction of $\boldsymbol{k}$). By the scattering amplitude, we usually understand (25.13) with $|\boldsymbol{k}| = |\boldsymbol{k}'|$; we shall make use of a generalised scattering amplitude $f(\boldsymbol{k}, \boldsymbol{k}')$, defined in accordance with (25.13) with any vectors $\boldsymbol{k}, \boldsymbol{k}'$. Equation (25.12) becomes in the momentum representation:

$$\psi_{\boldsymbol{k}}(\boldsymbol{p}) = (2\pi)^3\, \delta(\boldsymbol{k} - \boldsymbol{p}) + \frac{4\pi f(\boldsymbol{k}, \boldsymbol{p})}{\boldsymbol{k}^2 - \boldsymbol{p}^2 + i\delta}. \tag{25.14}$$

After substituting (25.14) in (25.13), we get

$$\frac{2\pi}{m} f(\boldsymbol{k}, \boldsymbol{k}') = U(\boldsymbol{k}' - \boldsymbol{k}) + \frac{1}{(2\pi)^3} \int \frac{U(\boldsymbol{k}' - \boldsymbol{p}) \left[\dfrac{2\pi}{m} f(\boldsymbol{k}, \boldsymbol{p})\right]}{\dfrac{\boldsymbol{k}^2}{2m} - \dfrac{\boldsymbol{p}^2}{2m} + i\delta} \mathrm{d}^3\boldsymbol{p}. \tag{25.15}$$

We now return to equation (25.11). We know that the scattering by one another of particles interacting with the potential energy $U(\boldsymbol{r} - \boldsymbol{r}')$ reduces to the scattering of one particle with a reduced mass $m^* = m_1 m_2/(m_1 + m_2)$ in the potential $U(\boldsymbol{r})$. On carrying out the substitution $m \to m^* = 1/2m$ everywhere in (25.15), the latter can be written as

$$U(\boldsymbol{k} - \boldsymbol{k}') = \left[\frac{4\pi}{m} f(\boldsymbol{k}', \boldsymbol{k})\right] - \frac{1}{(2\pi)^3} \int \frac{U(\boldsymbol{k} - \boldsymbol{p}) \left[\dfrac{4\pi}{m} f(\boldsymbol{k}', \boldsymbol{p})\right] \mathrm{d}^3\boldsymbol{p}}{\dfrac{\boldsymbol{k}'^2}{m} - \dfrac{\boldsymbol{p}^2}{m} + i\delta} \equiv \hat{L}\left(\frac{4\pi}{m} f\right), \tag{25.16}$$

where $\hat{L}$ denotes the operator on the right-hand side of (25.16). By subtracting the same expression from both sides of (25.11), we can reduce this to the following:

$$\Gamma^{(0)}(k, k'; P) - \frac{1}{(2\pi)^3} \int \frac{U(\boldsymbol{k} - \boldsymbol{p})\, \Gamma^{(0)}(p, k'; P)}{\dfrac{\boldsymbol{k}'^2}{m} - \dfrac{\boldsymbol{p}^2}{m} + i\delta} \mathrm{d}^3\boldsymbol{p}$$

$$= U(\boldsymbol{k} - \boldsymbol{k}') + \frac{1}{(2\pi)^3} \int U(\boldsymbol{k} - \boldsymbol{p})$$

$$\times \left\{\frac{1}{\dfrac{\varkappa^2}{m} - \dfrac{\boldsymbol{p}^2}{m} + i\delta} - \frac{1}{\dfrac{\boldsymbol{k}'^2}{m} - \dfrac{\boldsymbol{p}^2}{m} + i\delta}\right\} \Gamma^{(0)}(p, k'; P) \mathrm{d}^3\boldsymbol{p}.$$

The left-hand side here is $\hat{L}(\Gamma^{(0)})$. On acting on it by the operator $\hat{L}^{-1}$, we finally get the equation for $\Gamma^{(0)}$:

$$\Gamma^{(0)}(k, k'; P) = \frac{4\pi}{m} f(\boldsymbol{k}', \boldsymbol{k}) + \frac{1}{(2\pi)^3}\int \left[\frac{4\pi}{m} f(\boldsymbol{p}, \boldsymbol{k})\right] \times \left\{\frac{1}{\dfrac{\varkappa^2}{m} - \dfrac{\boldsymbol{p}^2}{m} + i\delta} - \frac{1}{\dfrac{\boldsymbol{k}'^2}{m} - \dfrac{\boldsymbol{p}^2}{m} + i\delta}\right\} \Gamma^{(0)}(p, k'; P)\,\mathrm{d}^3\boldsymbol{p}. \tag{25.17}$$

It is clear from this that, to a first approximation, $\Gamma^{(0)}(k, k'; P)$ is equal to $(4\pi/m) f(\boldsymbol{k}', \boldsymbol{k})$. The integral on the right-hand side of (25.17) is convergent, even if we assume f and $\Gamma^{(0)}$ constant, so that it is of order $\sim (|\boldsymbol{k}|/m) f^2$. It will be clear from what follows that the important region of momenta $|\boldsymbol{k}|$ is $|\boldsymbol{k}| \sim \sqrt{m\mu} \sim \sqrt{n_0 f}$, i.e. $|\boldsymbol{k}| f \ll 1$, and it is sufficient to confine ourselves to the first term for $\Gamma^{(0)}(k, k'; P)$. Notice, furthermore, that we can now neglect the dependence on $\boldsymbol{k}$, $\boldsymbol{k}'$ in the expression for $f(\boldsymbol{k}, \boldsymbol{k}')$. At small energies this dependence is an expansion in powers of the ratio of the particle dimensions a to the wavelength $\lambda \sim 1/|\boldsymbol{k}|$. Since a is of the order of the scattering amplitude f, and $|\boldsymbol{k}| f \ll 1$, we can finally write

$$\Gamma^{(0)}(k, k'; P) \simeq \frac{4\pi}{m} f(0, 0) \equiv \frac{4\pi}{m} f_0. \tag{25.18}$$

5. *Green functions of a Bose gas in the low density approximation. Spectrum*

We have on the basis of the above:

$$\Sigma_{11}(p) = \frac{8\pi}{m} f_0 n_0; \quad \Sigma_{20}(p) = \Sigma_{02}(p) = \frac{4\pi}{m} f_0 n_0; \quad \mu = \frac{4\pi}{m} f_0 n_0. \tag{25.19}$$

Substitution of these expressions in (24.4) and (24.5) gives

$$G(p) = \frac{\omega + \dfrac{\boldsymbol{p}^2}{2m} + \dfrac{4\pi n_0 f_0}{m}}{\omega^2 - \varepsilon^2(\boldsymbol{p}) + i\delta}, \qquad \hat{G}(p) = -\frac{4\pi n_0 f_0}{m} \frac{1}{\omega^2 - \varepsilon^2(\boldsymbol{p}) + i\delta},$$

where

$$\varepsilon(\boldsymbol{p}) = \sqrt{\left(\frac{\boldsymbol{p}^2}{2m} + \frac{4\pi n_0 f_0}{m}\right)^2 - \frac{16\pi^2 n_0^2 f_0^2}{m^2}} = \sqrt{\frac{\boldsymbol{p}^4}{4m^2} + \frac{4\pi n_0 f_0}{m^2}\boldsymbol{p}^2} \tag{25.20}$$

is the spectrum of the system for small momenta. The difference between (25.20) and expression (4.11) of Chap. I for the spectrum lies in replacing the Born amplitude by the exact S-scattering amplitude. It follows from (25.20) that the quasi-particles with $|\boldsymbol{p}| \ll \sqrt{n_0 f_0}$ have the acoustic dispersion $\varepsilon(\boldsymbol{p}) \simeq |\boldsymbol{p}| \sqrt{4\pi n_0 f_0/m^2}$, whilst for $|\boldsymbol{p}| \gg \sqrt{n_0 f_0}$ they become "almost free" particles $\varepsilon(\boldsymbol{p}) \simeq \varepsilon_0(\boldsymbol{p}) + (4\pi n_0 f_0/m)$ (this form of the spectrum corresponds to a particle moving in a continuous medium that has an index of refraction). The transition in the formula for the dispersion from the phonon region to the "free particle" region occurs when $|\boldsymbol{p}| \sim \sqrt{n_0 f_0} \ll 1/f_0$, so that both regions can be validly considered in the constant amplitude approximation.

Notice in conclusion that the model investigated can never be associated with the properties of real helium. Apart from the fact that the low density approximation does not correspond to liquid He II, it has to be emphasized that the form (25.20) of the spectrum for small $\boldsymbol{p}$ is actually unstable. For, when $\boldsymbol{p} \neq 0$, $\partial\varepsilon/\partial|\boldsymbol{p}|$ — the excitation velocity — is greater than the velocity of sound $\sqrt{4\pi n_0 f_0/m^2}$, i.e. the excitation can create phonons (see the next section). This leads to the appearance of damping in the spectrum with an excitation life inversely proportional to $|\boldsymbol{p}|^5$ at small $\boldsymbol{p}$. The spectrum of helium at small $\boldsymbol{p}$ does not possess such an instability.

§ 26. PROPERTIES OF THE SINGLE-PARTICLE EXCITATION SPECTRUM CLOSE TO ITS END-POINT

1. Statement of the problem

The spectrum of the single-particle excitations in a real Bose liquid, i.e. in helium, can evidently not be calculated theoretically. The dependence of the energy on the momentum (the phonon part of the spectrum) is only linear for very small momenta; it ceases to be linear at larger momenta and its form becomes dependent on the actual properties of the interaction between the particles of liquid.

The characteristics of the excitation spectrum in a Bose as compared with a Fermi liquid is that undamped Bose excitations can exist. This means from the mathematical point of view that the solutions of equation (24.16) are rea At finite temperatures, the damping of the excitations is due to the possibility of their colliding with one another. There are no actual excitations at absolute zero. Hence the only possible mechanism leading to a finite life of the excitation is its breakdown into excitations of lower energy, provided such a process is admissible by virtue of the laws of conservation of momentum and energy. In a Fermi liquid, breakdown accompanied by the formation of particles and holes is always possible; this leads to a finite life of the quasi-particles, inversely proportional to $(|\boldsymbol{p}| - p_0)^2$. Given sufficiently small momenta, the excitations in a Bose liquid can be undamped. It is only when the momentum increases that the excitation energy finally reaches a threshold value, above which the excitations are unstable from the point of view of breakdown into two or more excitations with lower energies. We shall call this threshold the end-point of the spectrum. It is a singular point of the spectral curve. We shall try below to explain the nature of this singularity; as will be clear from what follows, the entire investigation can be carried out in a general way, without any assumptions whatever regarding the weakness of the interaction (Pitaevskii [44]). Our only restriction (which still leaves plenty of physical generality, one would think) will be to assume that the end-point of the spectrum corresponds to the threshold of a breakdown into two (and not more than two) excitations.

The energy and momentum conservation laws must be satisfied during the decay of the excitation into two. This fact can be expressed by the equation

$$\varepsilon(\boldsymbol{p}) = \varepsilon(\boldsymbol{q}) + \varepsilon(\boldsymbol{p} - \boldsymbol{q}). \tag{26.1}$$

Here, $\boldsymbol{p}$ and $\varepsilon(\boldsymbol{p})$ are the momentum and energy of the decaying excitation, $\boldsymbol{q}$ and $\varepsilon(\boldsymbol{q})$ the momentum and energy of one of the resultant excitations, and $\boldsymbol{p} - \boldsymbol{q}$, $\varepsilon(\boldsymbol{p} - \boldsymbol{q})$ the momentum and energy of the other. If, given $\boldsymbol{p}$, (26.1) has no solutions for $\boldsymbol{q}$, this implies that decay is impossible. The decay threshold (we denote the excitation momentum at the threshold point by p_c, and the energy by $\varepsilon_c = \varepsilon(p_c)$) is characterised by the fact that (26.1) has no solutions for $\boldsymbol{q}$ when $\varepsilon < \varepsilon_c$ and has solutions when $\varepsilon = \varepsilon_c$. It is necessary, for this, that the right-hand side of (26.1), regarded as a function of the vector $\boldsymbol{q}$, have a minimum for certain values of $\boldsymbol{q}$ when $|\boldsymbol{p}| = p_c$. When $|\boldsymbol{p}| = p_c$, the right-hand side of (26.1) depends on two variables; on the absolute value of $\boldsymbol{q}$, and on $\cos\theta$, where θ is the angle between the vectors $\boldsymbol{p}$ and $\boldsymbol{q}$. The expression in question can have a minimum for both zero and finite θ.

Let the right-hand side of (26.1) have a minimum at some momentum $\boldsymbol{q}$. We write an expansion up to second order terms in the increment $\Delta\boldsymbol{q}$:

$$\varepsilon(\boldsymbol{q} + \Delta\boldsymbol{q}) + \varepsilon(\boldsymbol{p} - \boldsymbol{q} - \Delta\boldsymbol{q}) \simeq \varepsilon(\boldsymbol{q}) + \varepsilon(\boldsymbol{p} - \boldsymbol{q}) + \frac{\partial\varepsilon(\boldsymbol{q})}{\partial \boldsymbol{q}_i}\Delta \boldsymbol{q}_i$$

$$- \frac{\partial\varepsilon(\boldsymbol{p} - \boldsymbol{q})}{\partial \boldsymbol{p}_i}\Delta\boldsymbol{q}_i + \frac{1}{2}\frac{\partial^2\varepsilon(\boldsymbol{q})}{\partial\boldsymbol{q}_i\partial\boldsymbol{q}_k}\Delta\boldsymbol{q}_i\Delta\boldsymbol{q}_k + \frac{1}{2}\frac{\partial^2\varepsilon(\boldsymbol{p} - \boldsymbol{q})}{\partial\boldsymbol{p}_i\partial\boldsymbol{p}_k}\Delta\boldsymbol{q}_i\Delta\boldsymbol{q}_k.$$

The linear terms must fall out at the minimum. There are obviously two possibilities:

(1) $\partial\varepsilon(\boldsymbol{q})/\partial\boldsymbol{q} = \partial\varepsilon(\boldsymbol{p} - \boldsymbol{q})/\partial\boldsymbol{p} \neq 0$. This case corresponds to decay into two excitations moving in the direction of the vector $\boldsymbol{p}$ with the same velocity $\boldsymbol{v} = \partial\varepsilon/\partial\boldsymbol{q}$. Two cases are possible here. Firstly, one of the excitations can have a momentum arbitrarily close to zero. This corresponds to the case when the excitation velocity is equal, at the point p_c, to the

sound velocity c and the excitation can create a phonon (case a). Secondly, both the excitations can have a finite momentum (case b).

(2) $\partial\varepsilon(\boldsymbol{q})/\partial\boldsymbol{q} = 0$; $\partial\varepsilon(\boldsymbol{p}-\boldsymbol{q})/\partial\boldsymbol{p} = 0$. It is necessary for this that each of the excitations be created with momentum equal to p_0, at which the excitation energy $\varepsilon(\boldsymbol{p})$ is a minimum. For liquid helium, such a point on the spectrum corresponds to $p_0 = 2\times10^{-19}$ g cm/sec. The spectrum $\varepsilon(\boldsymbol{p})$ has a so-called roton form in the neighbourhood of this point:

$$\varepsilon(\boldsymbol{p}) = \Delta + \frac{(|\boldsymbol{p}|-p_0)^2}{2m^*}\,(|\boldsymbol{p}|-p_0| \ll p_0). \tag{26.2}$$

If $\varepsilon_c = 2\Delta$, the excitation decays into two rotons with momenta $\boldsymbol{q}$ and $\boldsymbol{q}_1$, where $|\boldsymbol{q}|$, $|\boldsymbol{q}_1| = p_0$ and $\varepsilon(\boldsymbol{q})$, $\varepsilon(\boldsymbol{q}_1) = \Delta$. The angle θ at which the two rotons depart is determined by the condition that the sum of their momenta be equal to p_c (case c). The three cases described cover all the types of threshold of decay into two excitations.

2. *System of equations*

For our investigation of the form of the spectrum close to the threshold point, we make use of the methods of quantum field theory described above, i.e. we seek the form of the Green function close to the end-point of the spectrum, since the spectrum is itself determined by the poles of the Green function. It is obvious physically that the singularities of the Green function are connected with those diagrams in which one line forks into two, which represents pictorially the decay of an excitation into two. Let us consider say the diagrams of Fig. 80. The different Green functions G', $\hat{G}$ and $\check{G}$ figure in these diagrams. Each of the loops is a self-energy part, characterised by the fact that it consists of two ternary vertices (counting only the number of uncondensed ends), joined by two continuous lines. The corresponding integral for such a loop is

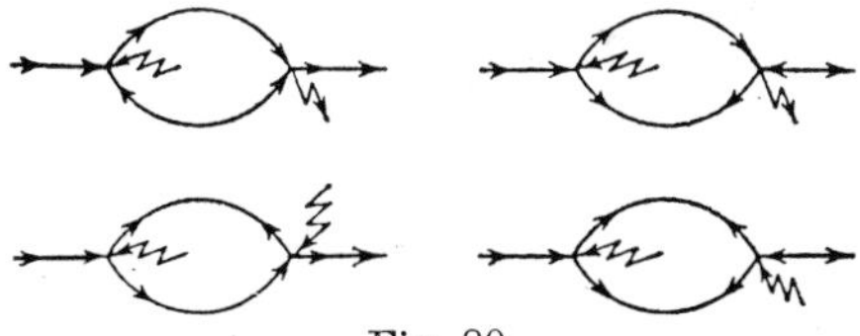

Fig. 80

$$\int \mathrm{d}\omega'\,\mathrm{d}^3\boldsymbol{q}\,G(q)G(p-q)\Gamma_1\Gamma_2, \tag{26.3}$$

where G may denote any of the three Green functions G', $\hat{G}$ and $\check{G}$, whilst Γ_1, Γ_2 are the vertices on the right and left-hand sides of the diagrams. Suppose that the values of ω and $\boldsymbol{p}$ for the outer ends lie close to the pole $\omega = \varepsilon(\boldsymbol{p})$ (we have shown above that the poles of all three Green functions are the same). The singularity of the integral (26.3), if it exists, is connected with the domain of integration over ω' and $\boldsymbol{q}$, in which the functions $G(q)$ and $G(p-q)$ are close to their pole. In accordance with

(24.10), (24.12) and (24.14), the two functions have the following forms close to their pole:

$$\left.\begin{aligned} G(q) &= \frac{A_1}{\omega' - \varepsilon(\mathbf{q}) + i\delta} \quad \text{or} \quad \frac{A_2}{\omega' + \varepsilon(\mathbf{q}) - i\delta}, \\ G(p-q) &= \frac{B_1}{\omega - \omega' - \varepsilon(\mathbf{p}-\mathbf{q}) + i\delta} \quad \text{or} \quad \frac{B_2}{\omega - \omega' + \varepsilon(\mathbf{p}-\mathbf{q}) - i\delta}, \end{aligned}\right\} \tag{26.4}$$

depending on whether the function is being considered near the positive or the negative pole. On substituting these expressions into (26.3), it will be seen that our interest is in the terms such as $A_1 B_1$. The integration over ω' in these terms can be performed between the limits $-\infty$ to $+\infty$, after which the remaining integral over $\mathbf{q}$ has the form, in a certain domain of values of $\mathbf{q}$:

$$\int \frac{\Gamma_1 \Gamma_2 A B \, d^3\mathbf{q}}{\varepsilon(\mathbf{q}) + \varepsilon(\mathbf{p}-\mathbf{q}) - \omega}. \tag{26.5}$$

The singularities of the last integral are determined by the possible vanishing of the denominator of the integrand at certain values of $\mathbf{q}$. By our previous analysis, the denominator is always greater than zero when $\omega < \varepsilon(p_c)$; when $\omega = \varepsilon(p_c)$ the integrand becomes infinite, i.e. $\omega = \varepsilon(p_c)$ is a singular point (in the mathematical sense of the word) of (26.5). The nature of the singularity is therefore determined purely by the analytic properties of the Green functions and does not depend on which of the diagrams of Fig. 80 we have actually chosen for the self-energy part. This last fact enables our future discussion to be considerably simplified. For, as we have just shown, to determine the nature of the singularity, we only need the expressions for the Green functions close to the pole. All three functions have the same form close to the pole. Provided, therefore, we are not interested in the actual size of the regular terms and in various unimportant coefficients, and inasmuch as the diagrams for all three functions have the same structure, we need not make any distinction between G', $\hat{G}$ and $\check{G}$ close to the poles. We form, for instance, equation (24.2) and introduce a new function $G_1(p) = G'(p) + \hat{G}(p)$. We get the following equation for $G_1(p)$:

$$G_1(p) = G^{(0)}(p) + G^{(0)}(p)\,[\Sigma_{11}(p) + \Sigma_{20}(p)]\,G_1(p).$$

We split the set of all self-energy parts $\Sigma = \Sigma_{11} + \Sigma_{20}$ into diagrams that have no singularities at the point $\omega = \varepsilon_c(\Sigma_0)$, and diagrams Σ_1 that have singularities, which take the pictorial form shown in Fig. 80. We introduce a function $G^{(0)}(p)$ as follows:

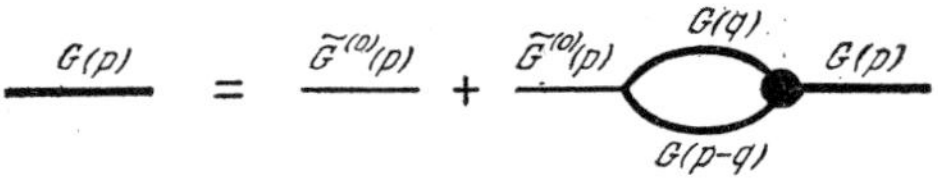

Fig. 81

$$\tilde{G}^{(0)}(p) = \frac{1}{G^{(0)-1}(p) - \Sigma_0(p)}.$$

The remaining equation can now be written as Dyson's equation, illustrated schematically in Fig. 81. Since the nature of the singularity is determined by the form of all the Green functions at the pole, where they only differ from each other as regards their coefficients, we can replace all the interior lines G', $\hat{G}$ or $\check{G}$ by G_1. To the left of the loop there is a "bare" vertex $\Gamma^{(0)}$, which, from the point of view of the general method, is the result of the interaction of three uncondensed particles with condensate particles (as for instance in the diagrams of Fig. 80); to the right we have the ternary vertex Γ, which is obtained from $\Gamma^{(0)}$ as a result of the interaction of the lines leaving it.

The diagrams expressing $\Gamma(p, p-q, q)$ in terms of $\Gamma^{(0)}(p, p-q, q)$ are illustrated in Fig. 82*a*. We use a square here to denote the complete irreducible vertex part for the scattering of two uncondensed particles by one another $\Gamma(p_1, p_2, p_3, p_4)$. Summation of these diagrams is carried out by the simple equation illustrated in Fig. 82*b*. This is the set of all

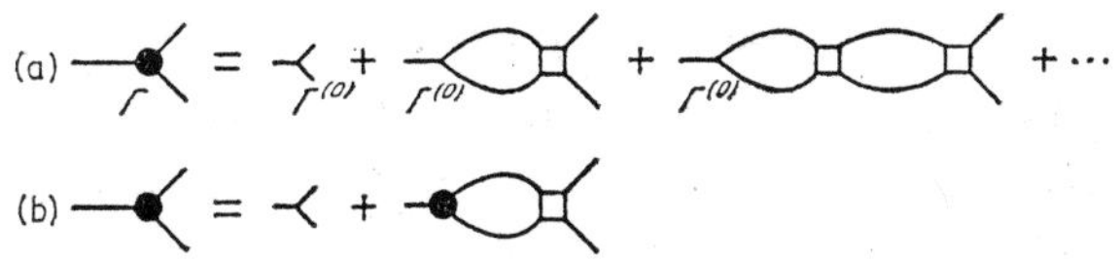

Fig. 82

the four-particle diagrams that cannot be split up between the ends p_1, p_2 and p_3, p_4 into two parts, joined only by one or two lines. On now omitting the index of G_1 everywhere, we can write the equations in the analytic form:

$$G^{-1}(p) - \tilde{G}^{(0)-1}(p) = \frac{i}{(2\pi)^4} \int \Gamma^{(0)}(p, p-q, q) G(q) \times G(p-q)\, \Gamma(p, p-q, q)\, d^4q, \quad (26.6)$$

$$\Gamma(p, p-q, q) = \Gamma^{(0)}(p, p-q, q) + \frac{i}{(2\pi)^4} \int \Gamma(p, p-k, k) \times G(k) G(p-k) \Gamma(k, p-k; p-q, q)\, d^4k. \quad (26.7)$$

The properties of equations (26.6) and (26.7) are entirely different close to the three types of threshold described at the start of this section, so that we have to consider the three cases separately.

3. *Properties of the spectrum close to the phonon creation threshold*

We consider the properties of the excitation spectrum close to the point where the excitation velocity becomes equal to the velocity of sound. As from this point, the excitation can create a phonon. The laws of conservation (26.1) becomes in this case

$$\varepsilon(\mathbf{p}) = \varepsilon(\mathbf{p} - \mathbf{q}) + \omega(\mathbf{q}), \quad (26.1')$$

where $\omega(\boldsymbol{q})$ is the phonon frequency, $\boldsymbol{q}$ is its wave vector. At small $\boldsymbol{q}$, the frequency $\omega(\boldsymbol{q})$ has the form

$$\omega(\boldsymbol{q}) = c|\boldsymbol{q}| - \alpha|\boldsymbol{q}|^3. \tag{26.8}$$

We shall assume that $\alpha > 0$, i.e. the phonon spectrum is stable, although the third order terms in $\omega(\boldsymbol{q})$ will not be required below. The function $\varepsilon(\boldsymbol{p})$ has a singularity at $|\boldsymbol{p}| = p_c$. We shall assume (this will be supported by the final result) that the singularity makes itself felt in terms of a higher order of smallness than the second in powers of $\Delta p = |\boldsymbol{p}| - p_c$, i.e. that close to p_c,

$$\varepsilon(\boldsymbol{p}) \simeq \varepsilon_c + c\Delta p + \beta(\Delta p)^2. \tag{26.9}$$

(By hypothesis, the excitation velocity $v = \partial\varepsilon/\partial|\boldsymbol{p}|$ is equal to the sound velocity at $|\boldsymbol{p}| = p_c$.)

At $|\boldsymbol{p}| = p_c$ and $\cos\theta = 1$ (θ is the angle between $\boldsymbol{q}$ and $\boldsymbol{p}$), the right-hand side of (26.1′) becomes, taking (26.8) and (26.9) into account:

$$\varepsilon_c + \beta|\boldsymbol{q}|^2. \tag{26.10}$$

The point $|\boldsymbol{p}| = p_c$ is actually the threshold only on condition that expression (26.10) has a minimum at $\boldsymbol{q} = 0$ for which it is necessary that the condition

$$\beta > 0$$

be fulfilled.

Since, at $|\boldsymbol{p}| = p_c$, the excitation can create in our present case a phonon with q arbitrarily close to zero, the important region as regards finding the singularity in integral (26.6) will be that of small values of the argument of one of the Green functions, say $G(q)$. The Green function is given by (24.19) for small ω and $\boldsymbol{q}$:

$$G(q) = \frac{a}{\omega^2 - \omega^2(\boldsymbol{q}) + i\delta} \tag{26.11}$$

and is proportional to the phonon propagation function. (It is impossible to use form (26.4) for the function $G(q)$, since both poles are almost the same at small $\boldsymbol{q}$.)

Close to $|\boldsymbol{p}| = p_c$ and $\varepsilon = \varepsilon_c$, the Green function has a singularity. We assume, however, that, in accordance with (26.9), $G^{-1}(p)$ has the following form close to zero (i.e. close to the pole of $G(p)$):

$$G^{-1}(p) = A^{-1}[\Delta\varepsilon - c\Delta p - \beta(\Delta p)^2 + i\delta] \tag{26.12}$$

$$(\Delta p = |\boldsymbol{p}| - p_c,\ \Delta\varepsilon = \omega - \varepsilon_c).$$

We also have to determine the higher order terms in $G^{-1}(p)$ that contain the singularity.

We consider the properties of the vertex part $\Gamma(p, p - q, q)$. At small $\boldsymbol{q}$, this vertex represents a process in which a particle with momentum $\boldsymbol{p}$ emits a long-wave excitation or phonon. Such a vertex must necessarily be proportional to the magnitude of the momentum $|\boldsymbol{q}|$ of the emitted phonon, since, from the macroscopic point of view, the process is the scattering of the excitation by density oscillations (sound). In the limit of infinite wavelengths of the acoustic oscillations, this interaction must vanish, since the excitation is not scattered in a homogeneous medium. We shall therefore use the expression

$$\Gamma(p, p-q, q) = g\,|\boldsymbol{q}| \tag{26.13}$$

for $\Gamma(p, p-q, q)$ in the region of small $|\boldsymbol{q}|$.

We now consider the integral on the right-hand side of (26.6). Because of the definition of Γ and Γ_0 (Fig. 82), this integral is in every order of Γ a chain consisting of loops joined by four-particle vertex functions (Fig. 83). Each of these loops yields a contribution to the singularity of the Green function, the contribution of each loop being only countable once because of our assumption that the non-regular terms are small.

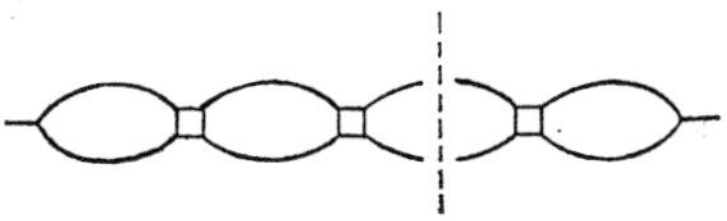
Fig. 83

If we take a given loop, the set of diagrams to the left and right of it can be summed independently, and, by the definition of the exact three-vertex function $\Gamma(p, p-q, q)$, form at both vertices of the loop a three-particle function $\Gamma(p, p-q, q)$. The small non-regular additive correction to the inverse Green function $G^{-1}(p)$ can therefore be found from a consideration of the non-regular part of the expression

$$\int \Gamma^2(p, p-q, q)G(q)G(p-q)\,\mathrm{d}^4q .$$

We consider the regions of small $\boldsymbol{q}$ in the expression and substitute in it the values (26.11)–(26.13) for $\Gamma(p, p-q, q)$, $G(q)$ and $G(p-q)$. We get

$$Aag^2\int\frac{|\boldsymbol{q}|^4\,\mathrm{d}|\boldsymbol{q}|\,\mathrm{d}\Omega\,\mathrm{d}\omega'}{(\omega'^2-c^2\boldsymbol{q}^2+i\delta)\,[\omega'-\omega+\varepsilon(\boldsymbol{p}-\boldsymbol{q})-i\delta]}\,.$$

The integration over ω' can be performed from $-\infty$ to $+\infty$ and amounts to taking the residues at the point $\omega' = c\,|\boldsymbol{q}|$. On omitting from now on the coefficients of no interest to us, we can write our expression, after integrating over ω', as

$$\int\frac{|\boldsymbol{q}|^3\,\mathrm{d}|\boldsymbol{q}|\,\mathrm{d}\cos\theta}{c\,|\boldsymbol{q}|+\varepsilon(\boldsymbol{p}-\boldsymbol{q})-\omega}\,.$$

Although the integral itself is convergent at the upper limit, its singularity is determined by the behaviour of the integrand in the region of small $|\boldsymbol{q}|$. We use (26.10) to expand the denominator. Small angles $\theta \ll 1$ are important for finding the singularity, so that we can put

$\cos\theta = 1$ to the required accuracy in the quadratic terms. We have as a result:

$$\int \frac{|\boldsymbol{q}|^3 \, \mathrm{d}|\boldsymbol{q}| \, \mathrm{d}\cos\theta}{x + c\,|\boldsymbol{q}|\,(1-\cos\theta) - 2\beta\Delta p\,|\boldsymbol{q}| + \beta\,|\boldsymbol{q}|^2} \sim \int |\boldsymbol{q}|^2 \ln\,(x - 2\beta\,|\boldsymbol{q}|\,\Delta p + \beta|\boldsymbol{q}|^2)\,\mathrm{d}|\boldsymbol{q}|.$$

We have put here $x = c\Delta p - \Delta\varepsilon + \beta(\Delta p)^2$. Factorising the expression under the logarithm and integrating, we get

$$a_1\left(\frac{k_1}{2}\right)^3 \ln k_1 + a_2\left(\frac{k_2}{2}\right)^3 \ln k_2, \tag{26.14}$$

where

$$k_{1,2} = \beta\Delta p \pm \sqrt{(\beta\,\Delta p)^2 - \beta x}\,.$$

It is clear from (26.14) that $G^{-1}(p)$ has in fact a singularity in terms of a higher order than those which were used in deriving the last expression. This fact justified all the assumptions that have been made regarding the smallness of the non-regular terms.

We determine the latter terms in the immediate neighbourhood of the pole of $G(p)$, i.e. for x satsifying

$$|x| \ll \beta(\Delta p)^2.$$

In this case we can neglect the term containing k_2. We now get from (26.14):

$$(\Delta p)^3 \ln\,(-\Delta p). \tag{26.15}$$

By (26.12) and (26.15), $G(p)$ has the form

$$G(p) = \frac{A}{\omega - \varepsilon_c - c\Delta p - \beta(\Delta p)^2 - a(\Delta p)^3 \ln\,(-\Delta p)}$$

in the neighbourhood of its pole. This function determines the energy of an elementary excitation close to the threshold. Damping is absent below the threshold:

$$\varepsilon(\boldsymbol{p}) = \varepsilon_c + c(|\boldsymbol{p}| - p_c) + \beta(|\boldsymbol{p}| - p_c)^2 + a(|\boldsymbol{p}| - p_c)^3 \ln(|\boldsymbol{p}| - p_c).$$

Above the threshold $|\boldsymbol{p}| > p_c$ the excitation energy has a negative imaginary part, equal to $-a\pi(\Delta p)^3$:

$$\varepsilon(\boldsymbol{p}) = \varepsilon_c + c(|\boldsymbol{p}| - p_c) + \beta(|\boldsymbol{p}| - p_c)^2 + a(|\boldsymbol{p}| - p_c)^3 \ln\,|p_c - |\boldsymbol{p}|| - a\pi i(|p| - p_c)^3.$$

Hence it follows, in particular, that we must have $a > 0$. There are thus no undamped excitations for $|\boldsymbol{p}| > p_c$: the life of an excitation is inversely proportional to $(|\boldsymbol{p}| - p_c)^3$. The smallness of the damping close to the threshold is connected with the fact that the interaction with long-wave phonons is always weak, due to the presence of the factor $|\boldsymbol{q}|$ in Γ.

4. Properties of the spectrum close to the threshold of break-up into two excitations with parallel non-vanishing momenta

When integrating over $\boldsymbol{q}$ in (26.6), it follows from physical considerations that the important values of $\boldsymbol{q}$ and the frequency ω' are those for which excitations are created close to the threshold. But these values of the momentum and energy are not singular for the Green functions of the created excitations. The only singular feature of such a point is that in its neighbourhood, the given excitation could "stick" to another — a process which is impossible at absolute zero because of the absence of actual excitations. The Green functions under the integral sign in (26.6) therefore have the simple form (26.4) close to the pole:

$$G(q) = \frac{A}{\omega - \varepsilon(\boldsymbol{q}) + i\delta},$$

where $\varepsilon(\boldsymbol{q})$ is real and has no singularities in the vicinity of those values of the vector $\boldsymbol{q}$ which we consider. This fact greatly simplifies an investigation of the subject.

Let us consider one of the loops in the set of chains corresponding, in accordance with Fig. 83 or equation (26.7), to the right-hand side of equation (26.6). The functions $\Gamma(p_1 p_2; p_3 p_4)$ or $\Gamma^{(0)}(p, p-q, q)$ appearing at the vertices of the loop obviously have no singularities. We shall in future always omit them from our calculations, as they lead only to inessential coefficients or to regular additive corrections to the Green functions. Let us confine ourselves in our loop to the domain of integration over $\boldsymbol{q}$, close to the values of the momentum $\boldsymbol{q}_0$ and energy ε_0 at which excitations are created. On substituting expressions (26.4) for the Green functions and integrating over ω', we find that the part of the integral in the loop, which contains the singularity, can be written in the form

$$\int \frac{\mathrm{d}^3\boldsymbol{q}}{\varepsilon(\boldsymbol{q}) + \varepsilon(\boldsymbol{p}-\boldsymbol{q}) - \omega}.$$

Since $\varepsilon(\boldsymbol{q}) + \varepsilon(\boldsymbol{p}-\boldsymbol{q})$ must have a minimum at $|\boldsymbol{p}| = p_c$, for $|\boldsymbol{p}|$ close to p_c it will have the form

$$\varepsilon(\boldsymbol{q}) + \varepsilon(\boldsymbol{p}-\boldsymbol{q}) \simeq \varepsilon_c + v_c \Delta p + \alpha(\boldsymbol{q}-\boldsymbol{q}_0)^2 + \frac{\beta(\boldsymbol{q}-\boldsymbol{q}_0 \cdot \boldsymbol{p}_c)^2}{p_c^2},$$

where v_c is the velocity of each of the excitations forming at the threshold point, $\boldsymbol{q}_0$ is the momentum of one of the excitations (remember that excitations leaving after the decay have their momentum directed along the vector $\boldsymbol{p}_c$). The coefficients α and β in this expansion are determined by the form of the functions $\varepsilon(\boldsymbol{p}-\boldsymbol{q})$ and $\varepsilon(\boldsymbol{q})$:

$$\alpha = \frac{v_c p_c}{2q_0(p_c - q_0)}, \quad \beta = \frac{1}{2}\left\{\left(\frac{\partial^2\varepsilon}{\partial \boldsymbol{q}^2}\right)_{q=q_0} + \left(\frac{\partial^2\varepsilon}{\partial \boldsymbol{q}^2}\right)_{q=p_c-q_0} - \frac{v_c p_c}{q_0(p_c-q_0)}\right\}.$$

On introducing the new variable $\boldsymbol{u} = \boldsymbol{q} - \boldsymbol{q}_0$, $(\boldsymbol{u} \cdot \boldsymbol{p}_c) = u\, p_c \cos\psi$, we get:

$$\int \frac{u^2 du\, d\cos\psi}{v_c \Delta p - \Delta\varepsilon + \alpha u^2 + \beta u^2 \cos^2\psi} \sim \sqrt{v_c \Delta p - \Delta\varepsilon}\,.$$

Summation of all the loops does not alter the nature of the singularity, since, as distinct from the phonon case, the complete three-particle function $\Gamma(p, p-q, q)$ must neither vanish nor become infinite at the values $|\boldsymbol{q}| \sim q_0$ of interest to us. The non-regular part of the inverse Green function therefore has the form, close to p_c and ε_c:

$$a\sqrt{v_c \Delta p - \Delta\varepsilon}\,.$$

By hypothesis, $|\boldsymbol{p}| = p_c$ and $\varepsilon = \varepsilon_c$ is a point of the spectrum, so that $G^{-1}(p)$ must vanish for $\Delta p = 0$ and $\Delta\varepsilon = 0$, i.e. the regular part of $G^{-1}(p)$ must be of the form $a_1 \Delta p + b_1 \Delta\varepsilon$ for small $\Delta\varepsilon$ and Δp. Finally,

$$G^{-1}(p) = A^{-1}\left[a\Delta p + \Delta\varepsilon + b\sqrt{v_c \Delta p - \Delta\varepsilon}\right].$$

The excitation energy is determined by the equation

$$G^{-1}(p) = 0. \tag{26.16}$$

The formal solution of this equation yields two roots:

$$\Delta\varepsilon_{1,2} = -a\Delta p - \frac{b^2}{2} \pm \sqrt{ab^2 \Delta p + \frac{b^4}{4} + b^2 v_c \Delta p}\,,$$

where we have to choose the $\Delta\varepsilon_1$ with the "plus" sign in front of the radical, in order for $\Delta\varepsilon \to 0$ as $\Delta p \to 0$. On expanding the expression under the radical close to the threshold at small Δp, we get

$$\varepsilon \simeq \varepsilon_c + v_c(|\boldsymbol{p}| - p_c) - \left(\frac{a + v_c}{b}\right)^2 (\Delta p)^2.$$

On substituting this expression in (26.16), we see that the necessary condition for (26.16) to have a solution for small negative Δp (before the threshold) is

$$\frac{a + v_c}{b} > 0.$$

Given $|\boldsymbol{p}| > p_c$, if $a\Delta p + \Delta\varepsilon$ and $b\sqrt{v_c \Delta p - \Delta\varepsilon}$ are both positive, the equation has no roots at all, either real or complex. In this case, therefore, the curve of the energy spectrum cannot be continued beyond the threshold point and terminates at it with a slope equal to v_c.

5. *Break-up into two excitations at an angle to each other*

In this case also, the important domain for the integrations includes the $|\boldsymbol{q}|$ at which excitations are created close to the threshold point. The Green functions have their usual form (26.4) in this domain, for the same reasons as in the previous section. We cannot now assert, however, that the vertex Γ is terminated at $\varepsilon = \varepsilon_c$.

For a start, let us consider, as above, one of the loops in Fig. 83. By definition the functions at the vertices of the loop do not contain harmful integrations. It is therefore natural to assume with regard to them that they remain finite at the threshold point. We consider, as usual, integration over the domain of ω' and $\boldsymbol{q}$ close to the values at which excitations are created near the threshold. The singularities of the loop amount to singularities of the integral

$$\int \frac{\mathrm{d}^3 \boldsymbol{q}}{\varepsilon(\boldsymbol{q}) + \varepsilon(\boldsymbol{p} - \boldsymbol{q}) - \omega},$$

in which we can use for $\varepsilon(\boldsymbol{q})$ and $\varepsilon(\boldsymbol{p} - \boldsymbol{q})$ the expansion (26.2) of $\varepsilon(\boldsymbol{q})$ close to the roton part of the spectrum (remember that case b corresponds to decay into two rotons with momenta equal in magnitude to p_0 and directed at some finite angle θ_0 to each other, $\cos \theta_0/2 = p_c/2p_0$). After substituting (26.2), the above expression transforms to

$$\int \frac{\mathrm{d}^3 \boldsymbol{q}}{2\Delta - \omega + \frac{(|\boldsymbol{q}| - p_0)^2}{2m^*} + \frac{(|\boldsymbol{q} - \boldsymbol{p}| - p_0)^2}{2m^*}}. \tag{26.17}$$

We change to cylindrical polars q_z', q_ϱ', φ in accordance with the formula (the z axis is along the vector $\boldsymbol{p}$):

$$q_z = p_0 \cos\frac{\theta_0}{2} + q_z', \qquad q_x = \left(p_0 \sin\frac{\theta_0}{2} + q_\varrho'\right)\cos\varphi,$$

$$q_y = \left(p_0 \sin\frac{\theta_0}{2} + q_\varrho'\right)\sin\varphi. \tag{26.18}$$

On substituting (26.18) in (26.17) and neglecting higher powers of q_z' and q_ϱ', we get

$$\int \frac{\mathrm{d}q_\varrho' \,\mathrm{d}q_z'}{2\Delta - \omega + \frac{1}{m^*}\left(\sin^2\frac{\theta_0}{2}\, q_\varrho'^2 + \cos^2\frac{\theta_0}{2}\, q_z'^2\right)}.$$

It is convenient to introduce polar coordinates r, ψ into this last expression:

$$\frac{1}{\sqrt{m^*}}\, q_\varrho' \sin\frac{\theta_0}{2} = r\cos\psi, \qquad \frac{1}{\sqrt{m^*}}\, q_z' \cos\frac{\theta_0}{2} = r\sin\psi.$$

We find as a result that

$$\int \frac{r\,\mathrm{d}r}{2\Delta - \omega + r^2} \sim \ln(2\Delta - \omega).$$

Each loop thus reduces to a large term $\ln(2\Delta - \omega)$, dependent only on the frequency of the outer end ω. Let us take a given loop; the set of all the loops to the left and right of it represents, in accordance with

Fig. 80*a* and equation (26.7), an exact three-particle vertex Γ. The principal term in the right-hand side of (26.6) at small $2\Delta - \omega$ is therefore of the form

$$\Gamma^2(p, p - q_0, q_0) \ln(2\Delta - \omega), \tag{26.19}$$

where $\boldsymbol{q}_0$ is the critical value of the vector $\boldsymbol{q}$ — the momentum of a roton formed on decay at the threshold point.

We can determine $\Gamma(p, p - q, q)$ in the neighbourhood $\boldsymbol{q} \sim \boldsymbol{q}_0$ by solving equation (26.7). It is simpler, however, to sum directly the principal terms of the series to which this equation corresponds, using the above-mentioned fact that the principal term in each loop depends only on the frequency ω of the outer end and is the same for each loop. Formally, the series in question is a geometric progression, the sum of which is

$$\Gamma(p, p - q_0, q_0) \sim \frac{P}{1 + Q \ln\left(\frac{2\Delta - \omega}{2\Delta}\right)}.$$

Substituting this expression for $\Gamma(p, p - q_0, q_0)$ in (26.19), we see that, by (26.6), the main non-regular term in $G^{-1}(p)$ close to the threshold is

$$\frac{a}{\ln\left(\frac{2\Delta - \omega}{\alpha}\right)}.$$

We have finally, on recalling that $G^{-1}(p_c) = 0$ by hypothesis:

$$G^{-1}(p) = A^{-1}\left[|\boldsymbol{p}| - p_c - \frac{a}{\ln\left(\frac{2\Delta - \omega}{\alpha}\right)}\right]. \tag{26.20}$$

The equation

$$G^{-1}(p) = 0$$

yields in this case the following expression for the curve of the energy spectrum when $|\boldsymbol{p}| < p_c$:

$$\varepsilon(\boldsymbol{p}) = 2\Delta - \alpha e^{-a/(p_c - |\boldsymbol{p}|)}$$

(the exponential smallness of $\varepsilon(\boldsymbol{p}) - 2\Delta$ has enabled us to neglect powers of $\Delta\varepsilon$ in expansion (26.20) of the regular part of $G^{-1}(p)$). In this case also, therefore, the curve $\varepsilon(\boldsymbol{p})$ ends at the point $|\boldsymbol{p}| = p_c$, where it has a horizontal tangent of infinite order.

Notice that, in all the cases considered, the Green function has a branch-point at $\omega = \varepsilon_c$, $|\boldsymbol{p}| = p_c$.

We must again emphasise the fact that the foregoing analysis is not based on the actual form of the particle interactions with one another, nor on the weakness of this interaction. We have made use only of general fundamental relations between the exact functions determined by the diagram technique(†).

(†) Recent experimental data [45, 46] would seem to indicate that decay into two rotons actually occurs in He^4.

§ 27. APPLICATION OF FIELD THEORY METHODS TO A SYSTEM OF INTERACTING BOSONS AT FINITE TEMPERATURES

We shall conclude this chapter by considering the possibility of generalising the above approach to the case of a system of bosons, interacting at a finite temperature. It seems natural to try to construct the generalisation on the basis of the general method. With this aim, we shall return straight away to the thermodynamic description of the system, in which the role of independent variable is played by the chemical potential μ instead of by the total number of particles N in the system. We have already indicated above that, in an ideal Bose gas, at temperatures below the point of Bose condensation, such a description is impossible, because the chemical potential of the gas, determined from the condition that the mean number of particles in the system be constant, proves to be identically zero in this temperature region.

As we know, in an ideal gas the distribution of the Bose particles over the states with momentum $\boldsymbol{p}$ has the form

$$n_{\boldsymbol{p}} = \frac{1}{\exp\{[\varepsilon_0(\boldsymbol{p}) - \mu]/T\} - 1}.$$

This distribution only has a meaning for negative μ. Given sufficiently high temperatures, $\mu < 0$; the point where the chemical potential vanishes in fact determines the temperature of the Bose condensation. At lower temperatures we have to put μ identically zero. The condition that the mean number of particles in the system be constant is fulfilled by virtue of the particles "condensing" into the lowest level. As we have mentioned more than once, the number of the latter is comparable with the total number of particles in the system, i.e. is proportional to the volume of the system.

The thermodynamic method formally features the Green function for the non-interacting particles, the Fourier component of which is equal to

$$\mathfrak{G}^{(0)}(p) = \frac{1}{i\omega - \varepsilon_0(\boldsymbol{p}) + \mu}.$$

The temperature at which Bose condensation of the system occurs is characterized by the fact that a pole first appears for

$$\frac{1}{\mu - \varepsilon_0(\boldsymbol{p})}$$

when the momentum $\boldsymbol{p} = 0$. If we were to try to continue this quantity into the region $\mu > 0$, we should have to deal with a function which changes sign and becomes infinite at entirely arbitrary values of $\boldsymbol{p}$. At the same time, if $\mu < 0$, the system of non-interacting particles finds itself above the temperature of Bose condensation, as is clear from our above discussions.

These difficulties are actually imaginary. This is because, for a gas of interacting Bose particles, there is no region below the temperature of Bose condensation in which perturbation theory is applicable. For instance, we saw in § 25 that, in the region of small momenta, the individual terms of the perturbation theory lead to divergent expressions. To obtain a physical result, we had to carry out a summation of the entire series of principal terms of perturbation theory. In all similar cases, the general equations and relations between the different quantities of the theory have a wider meaning and are applicable beyond the limits within which perturbation theory is applicable. The exact quantities, such as the Green functions when the interactions are taken into account, have sensible properties even at temperatures below the point of Bose condensation.

As regards the chemical potential, it is determined from the condition that the mean number of particles in the system be equal to a given number. In the presence of interactions between the particles, μ is never identically zero and can therefore be chosen as the independent variable right from the start, the value of this variable being determined for a given system from the condition mentioned that the number of particles in the system be constant. It is not possible to establish from general considerations what sign the chemical potential will have for interacting Bose particles below the point of condensation (†).

We have already seen in § 23 that the exact number of condensate particles appears in the perturbation theory formula at $T = 0$; the number of particles of condensate in an ideal gas, from which the derivation starts, never actually appears in the theory. The same situation exists, as we shall shortly see, at finite temperatures. This fact is extremely important. In particular, it enables us to make precise what is actually meant by the temperature of the Bose condensation of a system of interacting particles. It is quite clear that this temperature, which might be called the λ-transition temperature, does not necessarily coincide with the Bose condensation temperature of an ideal gas. A physical definition of the λ-transition temperature is provided by the condition that the density of the number of particles in the condensate vanishes. We cannot predict from general considerations to which side the transition temperature will shift with interaction. There is thus the possibility in principle both of a situation such that the physical conden-

(†) In the model considered in § 25, μ is positive by (25.19) ($f > 0$; a choice of $f < 0$ would correspond to attraction between particles, and such a system would be unstable in this approximation). For real helium, $\mu < 0$ at temperatures below the λ-point, since otherwise there would be no temperature at which helium could be in equilibrium with its vapour. As we know, the chemical potential of both phases must be equal at equilibrium. Helium vapour is a dilute Boltzmann gas, the chemical potential of which is negative.

sate exists at a temperature above the condensation temperature of an ideal gas, and conversely, a situation when the appearance of interaction at lower temperatures leads to the disappearance of the condensate that would exist in the ideal Bose gas.

It is pertinent to emphasize here that, in essence, the treatment of § 23 in the statement of the diagram technique at $T = 0$ nowhere makes use of the fact that the number of particles with zero momentum is equal to infinity in an ideal gas. The special feature of the perturbation theory treated there consists in the fact that, since the particles with zero momentum play a special role, we tried to carry out the derivation in such a way that the operators ξ_0, ξ_0^+ were taken into account exactly. In other words, we did not make the usual statistical assumptions in this case, that the contribution of these particles is relatively small. As regards the remaining particles, they could be treated in the usual way.

The approach indicated is also possible at finite temperatures. In our construction of perturbation theory, we shall start out from the representations of the Green functions in the form

$$\frac{\langle T(\psi_1', \psi_2', \ldots, \psi_1^+, \ldots, \mathfrak{S})\rangle}{\langle \mathfrak{S}\rangle}.$$

The averaging sign $\langle \cdots \rangle$ denotes in this representation the operation of taking the trace of the averaged expression over the states of the Hamiltonian of the non-interacting particles $H_0 - \mu N$:

$$\langle \cdots \rangle = \frac{\mathrm{Tr}\,[\mathrm{e}^{(\mu N - H_0)/T}\cdots]}{\mathrm{Tr}\,\mathrm{e}^{(\mu N - H_0)/T}}. \tag{27.1}$$

Since the total number of particles is not preserved in the variables μ, the thermodynamic averaging in (27.1) occurs independently for all the particles, including those with zero momentum. Hence, precisely as in § 23, the operations T and $\langle \cdots \rangle$ can each be written in the form of two successive operations: $T = T^0 T'$ and $\langle \cdots \rangle = \langle\langle \cdots \rangle'\rangle^0$, relating to particles with zero momentum and to other particles, respectively. The meaning of this subdivision is the same: the particles in the states with zero momentum play a special role, so that we consider them exactly. The possibility of subdividing the operations T and $\langle \cdots \rangle$ into two has played an essential role in our arguments at absolute zero.

On repeating almost word for word the arguments that led to (23.14), we find that, in the Matsubara method also, an evaluation of any Green function for uncondensed particles requires a knowledge of all the "exact" n-particle Green functions for particles in zero momentum states. As in (23.14), the Green functions for condensate particles are determined by the relations

$$\mathfrak{G}_{0n}(\tau_1 \cdots \tau_n; \tau_1' \cdots \tau_n') = \frac{\langle T(\xi_0(\tau_1) \cdots \xi_0(\tau_n); \xi_0^+(\tau_1') \cdots \xi_0^+(\tau_n')\mathfrak{S})\rangle}{\langle \mathfrak{S}\rangle},$$

where the τ are the time parameters of the method for $T \neq 0$. The last expression can be written in a form analogous to (23.16), i.e. in terms of the "Heisenberg" operators:

$$\mathfrak{G}_{0n}(\tau_1 \cdots \tau_n; \tau_1' \cdots \tau_n') \tag{27.2}$$
$$= \frac{\mathrm{Tr}\,\{\exp[(\mu N - H)/T]\, T(\tilde{\xi}_0(\tau_1) \cdots \tilde{\xi}_0(\tau_n); \tilde{\xi}_0^+(\tau_1') \cdots \tilde{\xi}_0^+(\tau_n'))\}}{\mathrm{Tr}\,\{\exp(\mu N - H)/T\}},$$

where the "Heisenberg" operators $\tilde{\xi}_0(\tau)$ and $\tilde{\xi}_0^+(\tau)$ satisfy the equations

$$\frac{\partial}{\partial \tau} \tilde{\xi}_0(\tau) = [H - \mu N, \tilde{\xi}_0(\tau)],$$

$$\frac{\partial}{\partial \tau} \tilde{\xi}_0^+(\tau) = [H - \mu N, \tilde{\xi}(\tau)]$$

and are connected with the usual Schrödinger operators by the relations

$$\tilde{\xi}_0(\tau) = e^{(H-\mu N)\tau} \xi_0 e^{-(H-\mu N)\tau}, \qquad \tilde{\xi}_0^+(\tau) = e^{(H-\mu N)\tau} \xi_0^+ e^{-(H-\mu N)\tau}.$$

The Green functions $\mathfrak{G}$ are therefore ensemble averages of the time-ordered product of the operators $\tilde{\xi}_0$, $\tilde{\xi}_0^+$.

It may be recalled in this connection that, in quantum statistics, the averaging of quantities can be carried out in two equivalent ways: on the one hand, the averaging can be regarded as a quantum mechanical averaging over the actual state in which the system finds itself. This state is characterised by the values of the energy and the number of particles. On the other hand, the averaging can be performed with the aid of a grand ensemble, for which the system is regarded as open; this enables it, at a given temperature, to find itself with a definite probability in different quantum mechanical states with different values of energy and number of particles. The equivalence of the two methods has its basis in the fact that a grand ensemble has an extraordinarily narrow maximum about the mean values of the energy and number of particles, so that, for instance, the relative energy fluctuation $\sqrt{(E - \bar{E})^2}/\bar{E} \sim 1/\sqrt{N}$ and tends to zero when the dimensions of the system tend to infinity. Given quantum mechanical averaging, the state energy and number of particles in a closed system are obviously the same as the corresponding averages in a grand ensemble. From the thermodynamic point of view, the difference between the two averaging methods amounts to the fact that, in the first case, the value of the averaged quantity is expressed in terms of the energy as the thermodynamic variable, whereas in the case of an ensemble average the same value will be expressed as a function of temperature. The introduction of the chemical potential has the same meaning in statistics. Noting what has been said, we shall regard (27.2) as the average over the quantum mechanical state of the system:

$$\mathfrak{G}_{0n}(\tau_1 \cdots \tau_n; \tau_1' \cdots \tau_n') = \langle \Phi^* \,|\, T(\tilde{\xi}_0(\tau_1) \cdots \tilde{\xi}_0(\tau_n);\ \tilde{\xi}_0^+(\tau_1') \cdots \tilde{\xi}_0^+(\tau_n'))\,|\Phi\rangle. \tag{27.3}$$

Hence, in accordance with (27.2), the perturbation theory series for any Green function for uncondensed particles contains the exact Green functions for particles in states with zero momentum, which, on being written in the form (27.3), depend only on the properties of the operators ξ_0 and ξ_0^+ with respect to the state Φ for the interacting particles. If, therefore, a condensate exists in this state, the operators ξ_0, ξ_0^+ can be regarded as numbers, and, in accordance with this, all the Green functions of form (27.2) must be replaced by the products of factors $\sqrt{n_0(T)}$:

$$\mathfrak{G}_{0m}(\tau_1 \cdots \tau_n; \tau_1' \cdots \tau_m') = [n_0(T)]^m. \tag{27.4}$$

This last can be verified in precisely the same way as when deriving the analogous formula for absolute zero in § 23.

In (27.4) $n_0(T)$ is the density of the number of particles at a given temperature. The condition $n_0(T_\lambda) = 0$ determines the transition temperature T_λ. As already mentioned, this temperature can be either higher or lower than the "Bose condensation" temperature T_0 of an ideal gas. In the latter case the perturbation theory expansion for $T_0 > T > T_\lambda$ has just the usual form, in spite of the fact that, in the interaction representation, the condensate exists and the operators ξ_0, ξ_0^+ are extremely large. Let us emphasise once more that this fact is connected with the appearance of the exact Green functions for condensate particles in the perturbation theory expansion.

CHAPTER VI

ELECTROMAGNETIC RADIATION IN AN ABSORBING MEDIUM

§ 28. RADIATION GREEN FUNCTIONS IN AN ABSORBING MEDIUM

THE electromagnetic field plays a fundamental role in the range of phenomena with which statistical physics is concerned. In essence, all the forces acting between the particles of condensed media — solids and liquids — are of an electromagnetic type. The distinctive feature of these forces is their short-range nature; they fall off at distances of interatomic order and determine the adhesion between the particles.

We shall not touch on short-range forces in this chapter: we confine ourselves to the range of problems connected with electromagnetic radiation whose wavelength exceeds the interatomic spacings. The problems cover both phenomena relating to the passage of electromagnetic waves through a substance and to the various effects connected with long-range electromagnetic forces (so-called van der Waals forces).

The interaction of long-wave electromagnetic radiation with a substance is well known to be describable in a purely macroscopic way by the introduction of a complex dielectric constant $\varepsilon(\omega) = \varepsilon'(\omega) + i\varepsilon''(\omega)$ (see [47]), dependent on the frequency ω of the radiation (†). We shall find in this section expressions for the Green functions of the electromagnetic radiation in an absorbing medium in terms of the dielectric constant of the latter.

An electromagnetic field is usually described in quantum mechanics by the Schrödinger operators of a vector potential $\boldsymbol{A}(\boldsymbol{r})$ and a scalar potential $\varphi(\boldsymbol{r})$. We shall use a "four-dimensional" notation (‡) $A_\alpha = (\boldsymbol{A}, \varphi)$;

(†) We shall assume that the magnetic permeability $\mu(\omega)$ is equal to unity, since it only differs from unity in narrow frequency bands that, as a rule, will be of no interest to us.

(‡) The Greek indices in this chapter number the components of the four-dimensional vector potential. Its spatial components will occasionally be denoted by Latin indices $i, k, \ldots$ Summation is assumed over all twice repeated indices, whether Latin or Greek.

$\alpha = 1, 2, 3, 0$, for these operators. Together with the operators in the Schrödinger representation, we shall use the Heisenberg operators $A_\alpha(\boldsymbol{r}, t)$, defined as usual by

$$A_\alpha(\boldsymbol{r}, t) = e^{it\hat{H}} A_\alpha(\boldsymbol{r}) e^{-it\hat{H}}. \tag{28.1}$$

The operators $A_\alpha(\boldsymbol{r}, t)$ are connected with the operators of the electric and magnetic field-strengths $\boldsymbol{E}(\boldsymbol{r}, t)$, $\boldsymbol{H}(\boldsymbol{r}, t)$ by the usual relations (we have put here $c = 1$):

$$\boldsymbol{E}(\boldsymbol{r}, t) = -\frac{\partial}{\partial t}\boldsymbol{A}(\boldsymbol{r}, t) - \operatorname{grad}\varphi(\boldsymbol{r}, t),$$

$$\boldsymbol{H}(\boldsymbol{r}, t) = \operatorname{curl}\boldsymbol{A}(\boldsymbol{r}, t). \tag{28.2}$$

The vector potential of the electromagnetic field (as also the operators representing it in the second quantisation representation) is not uniquely defined. An arbitrariness always remains, connected with the so-called gauge invariance of the theory, amounting to the fact that $A(\boldsymbol{r}, t)$ can be subjected to the transformation:

$$\boldsymbol{A}(\boldsymbol{r}, t) \to \boldsymbol{A}(\boldsymbol{r}, t) + \operatorname{grad}\chi(\boldsymbol{r}, t),$$

$$\varphi(\boldsymbol{r}, t) \to \varphi(\boldsymbol{r}, t) - \frac{\partial}{\partial t}\chi(\boldsymbol{r}, t),$$

where χ is an arbitrary operator. It is easily verified that $\boldsymbol{E}$, $\boldsymbol{H}$, which have a direct physical meaning, are unchanged under this so-called gauge transformation.

The large wavelength of the electromagnetic field is indicated by the fact that a closed system of equations — the Maxwell equations — exists for the mean (†) values of the electric and magnetic fields $\langle \boldsymbol{E}\rangle$, $\langle \boldsymbol{H}\rangle$:

$$\operatorname{curl}\langle \boldsymbol{H}(\boldsymbol{r}, t)\rangle = \frac{\partial}{\partial t}\left(\hat{\varepsilon}\langle \boldsymbol{E}(\boldsymbol{r}, t)\rangle\right),$$

$$\operatorname{curl}\langle \boldsymbol{E}(\boldsymbol{r}, t)\rangle = -\frac{\partial}{\partial t}\langle \boldsymbol{H}(\boldsymbol{r}, t)\rangle. \tag{28.3}$$

The dielectric constant $\hat{\varepsilon}$ occurring in (28.3) depends only on the properties of the medium and, in the case of absorbing media, is an operator acting on functions of the time in accordance with the rule

$$\hat{\varepsilon}\langle \boldsymbol{E}(\boldsymbol{r}, t)\rangle = \langle \boldsymbol{E}(\boldsymbol{r}, t)\rangle + \int_{-\infty}^{t} f(\boldsymbol{r}, t - t')\langle \boldsymbol{E}(\boldsymbol{r}, t')\rangle \, dt'. \tag{28.4}$$

In Fourier components, the action of the operator $\hat{\varepsilon}$ amounts simply to multiplying $\langle \boldsymbol{E}(\boldsymbol{r}, \omega)\rangle$ by the dielectric constant $\varepsilon(\omega)$ of the medium:

$$\varepsilon(\boldsymbol{r}, \omega) = 1 + \int_{0}^{\infty} e^{i\omega t} f(\boldsymbol{r}, t)\, dt, \tag{28.5}$$

(†) We understand the averaging to mean the statistical averaging

$$\langle \boldsymbol{E}(\boldsymbol{r}, t)\rangle = \operatorname{Tr}\{\exp[(F - \hat{H})/T]\, \boldsymbol{E}(\boldsymbol{r}, t)\}.$$

whilst the system (28.3) transforms into

$$\operatorname{curl}\langle \boldsymbol{H}(\boldsymbol{r},\omega)\rangle = -i\omega\varepsilon(\boldsymbol{r},\omega)\langle \boldsymbol{E}(\boldsymbol{r},\omega)\rangle,$$
$$\operatorname{curl}\langle \boldsymbol{E}(\boldsymbol{r},\omega)\rangle = i\omega\langle \boldsymbol{H}(\boldsymbol{r},\omega)\rangle. \tag{28.6}$$

It can be shown (see Landau and Lifshitz [47]) that the $\varepsilon(\omega)$ defined by means of (28.5) is an analytic function in the upper half-plane of the complex variable ω, with no zeros in this half-plane.

The properties of the electromagnetic radiation at finite temperatures are determined by the temperature-dependent Green functions (†) $\mathfrak{D}_{\alpha\beta}(\boldsymbol{r}_1, \tau_1;\ \boldsymbol{r}_2, \tau_2)$:

$$\mathfrak{D}_{\alpha\beta}(\boldsymbol{r}_1, \boldsymbol{r}_2; \tau_1-\tau_2) = \begin{cases} -\operatorname{Tr}\{e^{(F-\hat{H})/T} e^{\hat{H}(\tau_1-\tau_2)} A_\alpha(\boldsymbol{r}_1) e^{-\hat{H}(\tau_1-\tau_2)} A_\beta(\boldsymbol{r}_2)\} & \text{for } \tau_1 > \tau_2, \\ -\operatorname{Tr}\{e^{(F-\hat{H})/T} e^{-\hat{H}(\tau_1-\tau_2)} A_\beta(\boldsymbol{r}_2) e^{\hat{H}(\tau_1-\tau_2)} A_\alpha(\boldsymbol{r}_1)\} & \text{for } \tau_1 < \tau_2. \end{cases} \tag{28.7}$$

To express $\mathfrak{D}_{\alpha\beta}$ in terms of the dielectric constant $\varepsilon(\omega)$, we use the connexion established in Chap. III between the temperature-dependent Green function and the retarded function, defined in our case as

$$D^R_{\alpha\beta}(\boldsymbol{r}_1, \boldsymbol{r}_2; t_1-t_2) = \begin{cases} -i\operatorname{Tr}\{e^{(F-\hat{H})/T}[A_\alpha(\boldsymbol{r}_1, t_1)A_\beta(\boldsymbol{r}_2, t_2) - A_\beta(\boldsymbol{r}_2, t_2)A_\alpha(\boldsymbol{r}_1, t_1)]\} & \text{for } t_1 > t_2, \\ 0 & \text{for } t_1 < t_2. \end{cases} \tag{28.8}$$

Since we are concerned with a later application of the results obtained to non-uniform bodies, we shall no longer make the assumption here that $\mathfrak{D}$ and D^R are functions of the differences of the positional coordinates. Similarly, we assume that the dielectric constant varies at different points of the body: $\varepsilon = \varepsilon(\boldsymbol{r}, \omega)$.

If we repeat all the arguments of § 17, except for omitting the Fourier transformation in the spatial coordinates, we easily arrive at the following Lehmann-type representations for $\mathfrak{D}$ and D^R:

$$\mathfrak{D}_{\alpha\beta}(\boldsymbol{r}_1, \boldsymbol{r}_2; \omega_n) = \int_{-\infty}^{\infty} \frac{\varrho_{\alpha\beta}(\boldsymbol{r}_1, \boldsymbol{r}_2; x)}{x - i\omega_n}\, dx, \tag{28.9}$$

$$D^R_{\alpha\beta}(\boldsymbol{r}_1, \boldsymbol{r}_2; \omega) = \int_{-\infty}^{\infty} \frac{\varrho_{\alpha\beta}(\boldsymbol{r}_1, \boldsymbol{r}_2; x)}{x - \omega - i\delta}\, dx, \tag{28.10}$$

where

$$\varrho_{\alpha\beta}(\boldsymbol{r}_1, \boldsymbol{r}_2; \omega) = -(2\pi)^3 \sum_{n,m} \exp[(F-E_n)/T]\big(A_\alpha(\boldsymbol{r}_1)\big)_{nm}\big(A_\beta(\boldsymbol{r}_2)\big)_{mn}(1-e^{-\omega_{mn}/T})\,\delta(\omega-\omega_{mn}).$$

(†) Instead of the thermodynamic potential Ω, the free energy F is introduced into (28.7) (cf. (11.1), (11.2)). This replacement can be made because the chemical potential of an electromagnetic field is identically zero.

It follows from (28.10) that D^R is analytic in ω in the upper half-plane. Comparison of (28.10), (28.9) shows that, for $\omega_n > 0$,

$$\mathfrak{D}_{\alpha\beta}(\boldsymbol{r}_1, \boldsymbol{r}_2; \omega_n) = D^R_{\alpha\beta}(\boldsymbol{r}_1, \boldsymbol{r}_2; i\omega_n). \tag{28.11}$$

To find $\mathfrak{D}$ for $\omega_n < 0$, we observe that $\mathfrak{D}(\tau)$ is an even function of τ (see § 11), because the electromagnetic field operators are real ($A^+_\alpha = A_\alpha$). Its Fourier component $\mathfrak{D}(\omega_n)$ is therefore even in ω_n, whence follows the relation, valid for all ω_n(†):

$$\mathfrak{D}_{\alpha\beta}(\boldsymbol{r}_1, \boldsymbol{r}_2; \omega_n) = D^R_{\alpha\beta}(\boldsymbol{r}_1, \boldsymbol{r}_2; i\,|\omega_n|). \tag{28.12}$$

We now turn to finding the retarded function. An important point here is the gauge of the vector potential. The tensor $D^R_{\alpha\beta}$ has altogether ten independent components (like every symmetric tensor of the second rank). There remains a substantial degree of arbitrariness in our arrangement, however, connected with the gauge invariance. In fact, a physical meaning attaches, not to the $D^R_{\alpha\beta}$ themselves, which are formed from the components of the vector potential, but only to the six quantities, formed from the operators $E_i(\boldsymbol{r}, t)$ in accordance with the same rules as those by which D^R_α is formed from $A_\alpha(\boldsymbol{r}, t)$ (formula (28.8)). Hence only six physical conditions are imposed on the ten functions $D^R_{\alpha\beta}$, i.e. there are four arbitrary functions in our arrangement. We can use this arbitrariness in order to make the components D^R_{00} and D^R_{i0} vanish. This choice evidently corresponds to a zero scalar potential. The operators $\boldsymbol{E}$, $\boldsymbol{H}$ are connected with $\boldsymbol{A}$ in this case by the formulae

$$\boldsymbol{E} = -\frac{\partial \boldsymbol{A}}{\partial t}, \quad \boldsymbol{H} = \operatorname{curl} \boldsymbol{A}. \tag{28.2′}$$

In order to express D^R_{ik} in terms of $\varepsilon(\omega)$, we proceed as follows. Imagine that our system, consisting of a body and equilibrium electromagnetic radiation, is situated in an external field, produced by imposed currents $\boldsymbol{j}^{\text{imp}}(\boldsymbol{r}, t)$. The Maxwell equations for the average fields become in this case

$$\begin{aligned} \operatorname{curl}\langle \boldsymbol{H}(\boldsymbol{r}, \omega)\rangle &= 4\pi \boldsymbol{j}^{\text{imp}}(\boldsymbol{r}, \omega) - i\omega\varepsilon(\boldsymbol{r}, \omega)\langle \boldsymbol{E}(\boldsymbol{r}, \omega)\rangle, \\ \operatorname{curl}\langle \boldsymbol{E}(\boldsymbol{r}, \omega)\rangle &= i\omega\langle \boldsymbol{H}(\boldsymbol{r}, \omega)\rangle. \end{aligned} \tag{28.6′}$$

In the case of the gauge (28.2′), the average potential $\langle A^{\text{imp}}(\boldsymbol{r}, t)\rangle$ will satisfy the equation

$$[\varepsilon(\boldsymbol{r}, \omega)\omega^2\delta_{il} - \operatorname{curl}_{im}\operatorname{curl}_{ml}]\langle A_l(\boldsymbol{r}, \omega)\rangle = -4\pi j^{\text{imp}}_i(\boldsymbol{r}, \omega). \tag{28.13}$$

The solution of equation (28.13) is

$$\langle A^{\text{imp}}_i(\boldsymbol{r}, \omega)\rangle = -\int \bar{D}_{il}(\boldsymbol{r}, \boldsymbol{r}'; \omega) j^{\text{imp}}_l(\boldsymbol{r}', \omega)\, d^3\boldsymbol{r}', \tag{28.14}$$

(†) Remember that, like every boson Green function, $\mathfrak{D}(\omega_n)$ only has non-zero components for "even" frequencies $\omega_n = 2n\pi T$.

where $\bar{D}$ is the so-called Green function of equation (28.13). It is a solution of the equation

$$[\varepsilon(\boldsymbol{r}, \omega)\omega^2 \delta_{il} - \mathrm{curl}_{im}\,\mathrm{curl}_{ml}]\bar{D}_{lk}(\boldsymbol{r}, \boldsymbol{r}'; \omega) = 4\pi\delta_{ik}\delta(\boldsymbol{r} - \boldsymbol{r}'). \quad (28.15)$$

In view of the analyticity of $\varepsilon(\omega)$ in the upper half-plane, $\bar{D}$ is also analytic in the upper half-plane.

On the other hand, $\langle A^{\mathrm{imp}}\rangle$ can be evaluated, in the presence of imposed currents, directly from the definition (28.1). The Hamiltonian of the system has the form $\hat{H} + \hat{H}^{\mathrm{imp}}$ in this case, where $\hat{H}$ is the Hamiltonian of the body and radiation, whilst

$$\hat{H}^{\mathrm{imp}} = -\int \boldsymbol{j}^{\mathrm{imp}}(\boldsymbol{r}, t)\boldsymbol{A}(\boldsymbol{r})\,\mathrm{d}^3\boldsymbol{r}.$$

On denoting the operators in the presence of an imposed field by the index "imp", we have (retaining the indexless notation for the operators without an imposed field)

$$\boldsymbol{A}^{\mathrm{imp}}(\boldsymbol{r}, t) = e^{it(\hat{H}+\hat{H}^{\mathrm{imp}})}\boldsymbol{A}(\boldsymbol{r})e^{-it(\hat{H}+\hat{H}^{\mathrm{imp}})}.$$

Next, as in Chap. II, we write $\exp\{-it(\hat{H} + \hat{H}^{\mathrm{imp}})\}$ in the form

$$e^{-it(\hat{H}+\hat{H}^{\mathrm{imp}})} = e^{-it\hat{H}}S_{\mathrm{imp}}(t).$$

If the imposed currents satisfy the condition $\boldsymbol{j}^{\mathrm{imp}}(t \to -\infty) \to 0$, $S_{\mathrm{imp}}(t)$ has the form (cf. Chap. II)

$$S_{\mathrm{imp}}(t) = T_t \exp\left\{-i \int_{-\infty}^{t} \hat{H}^{\mathrm{imp}}(t')\,\mathrm{d}t'\right\}.$$

The mean value of the vector potential operator in the presence of imposed currents now becomes

$$\langle \boldsymbol{A}^{\mathrm{imp}}(\boldsymbol{r}, t)\rangle = \langle S^{-1}_{\mathrm{imp}}(t)\boldsymbol{A}(\boldsymbol{r}, t)S_{\mathrm{imp}}(t)\rangle.$$

On expanding S_{imp} into a series in $\hat{H}^{\mathrm{imp}}$ and retaining terms of the first order in $\boldsymbol{j}^{\mathrm{imp}}$, we get

$$\langle A_i^{\mathrm{imp}}(\boldsymbol{r}, t)\rangle = -i \int_{-\infty}^{t} \mathrm{d}t' \int \mathrm{d}^3\boldsymbol{r}'\, j_k^{\mathrm{imp}}(\boldsymbol{r}', t')$$
$$\times \langle\{A_k(\boldsymbol{r}', t')A_i(\boldsymbol{r}, t) - A_i(\boldsymbol{r}, t)A_k(\boldsymbol{r}', t')\}\rangle. \quad (28.16)$$

This function can be expressed in terms of the retarded electromagnetic field function D^R_{ik} which we have introduced. We have, from the definition (28.8):

$$\langle A_i^{\mathrm{imp}}(\boldsymbol{r}, t)\rangle = -\int_{-\infty}^{\infty} \mathrm{d}t' \int \mathrm{d}^3\boldsymbol{r}\, D^R_{ik}(\boldsymbol{r}, \boldsymbol{r}'; t - t')j_k^{\mathrm{imp}}(\boldsymbol{r}', t').$$

Fourier transforming in this relation, we finally get

$$\langle A_i^{\mathrm{imp}}(\boldsymbol{r}, \omega)\rangle = -\int \mathrm{d}^3\boldsymbol{r}'\, D^R_{ik}(\boldsymbol{r}, \boldsymbol{r}'; \omega)j_k^{\mathrm{imp}}(\boldsymbol{r}', \omega). \quad (28.17)$$

Comparison of (28.14), (28.17) shows that, by virtue of the arbitrariness of $\boldsymbol{j}^{\text{imp}}$, D^R_{ik} is the same as the Green function $\bar{D}_{ik}$ of equation (28.13) introduced above. We thus arrive at the conclusion that D^R_{ik} also satisfies equation (28.15) (Dzyaloshinskii and Pitaevskii [48]). The analyticity of D^R_{ik} is, of course, fully in accord with the analytic properties of $\varepsilon(\omega)$.

The equation for $\mathfrak{D}_{ik}(\omega_n)$ is obtained from (28.13) by replacing ω in the latter by $i\,|\omega_n|$:

$$[\varepsilon(\boldsymbol{r}, i\,|\omega_n|)\,\omega_n^2\,\delta_{il} + \operatorname{curl}_{im}\operatorname{curl}_{ml}]\,\mathfrak{D}_{lk}(\boldsymbol{r}, \boldsymbol{r}'; \omega_n) = -4\pi\,\delta(\boldsymbol{r}-\boldsymbol{r}')\,\delta_{ik}. \tag{28.18}$$

The dielectric constant, with an imaginary frequency, that appears here is simply connected with the imaginary part $\varepsilon''(\omega)$ of $\varepsilon(\omega)$ at real frequencies (see e.g. [47], § 58):

$$\varepsilon(i\,|\omega_n|) = 1 + \frac{2}{\pi}\int\limits_0^\infty \frac{\omega\,\varepsilon''(\omega)}{\omega^2+\omega_n^2}\,d\omega. \tag{28.19}$$

Since always $\varepsilon'' > 0$, it is clear from (28.19) that $\varepsilon(i\,|\omega_n|)$ is a real, positive, monotonically decreasing function.

By solving (28.14) or (28.18), we can thus express the Green functions of the electromagnetic field in terms of the imaginary part of the dielectric constant. This is usually a very difficult problem in the case of non-uniform media. We consider in later sections the particular case of layer-type media, for which a complete solution of the problem can be given.

Let us return to the case of a uniform medium, in which ε is independent of the coordinates. Since D^R and $\mathfrak{D}$ now depend only on the difference $\boldsymbol{r}-\boldsymbol{r}'$, we have, on Fourier transforming in (28.15):

$$[(\omega^2\varepsilon(\omega) - k^2)\,\delta_{il} + k_i k_l]\,D^R_{lk}(\boldsymbol{k}, \omega) = 4\pi\,\delta_{ik}. \tag{28.20}$$

This equation determines D^R in the gauge where the scalar potential is zero. In order to find D^R for any gauge, we use (28.20) to find the function $D^E_{ik} = \omega^2 D^R_{ik}$, where D^R_{ik} is the retarded function for the gauge $\varphi = 0$. The function thus defined is already gauge invariant, since it differs by the constant term from the retarded function formed from the components of the electric field strength operators. The function D^E_{ik} satisfies the equation

$$[(\omega^2\varepsilon(\omega) - k^2)\,\delta_{il} + k_i k_l]\,D^E_{lk}(\boldsymbol{k}, \omega) = 4\pi\,\omega^2\,\delta_{ik} \tag{28.21}$$

and is connected with the function $D^R_{\alpha\beta}$ in an arbitrary gauge by the obvious relation (cf. (28.2))

$$D^E_{ik} = \omega^2 D^R_{ik} - \omega\,k_i D^R_{0k} - \omega\,k_k D^R_{i0} + k_i k_k D^R_{00} \tag{28.22}$$

(here and in what follows, D^R_{00}, D^R_{i0}, D^R_{ik} denote the time, displacement and positional components of $D^R_{\alpha\beta}$ in any given gauge). By symmetry considerations, the vector D^R_{i0} must be directed along $\boldsymbol{k}$ — the only vector appearing in (28.21) and (28.22):

$$D^R_{i0} = D^R_{0i} = k_i d; \tag{28.23}$$

for the same reason,

$$D^R_{ik} = a\,\delta_{ik} + b\,k_i k_k. \tag{28.24}$$

On substituting (28.22)—(28.24) in (28.21), we get two equations for D^R_{00}, a, b, and d:

$$\begin{aligned} a\left(\varepsilon(\omega)\,\omega^2 - k^2\right) &= 4\pi, \\ a + \varepsilon(\omega)\,(\omega^2 b + D^R_{00} - 2\,\omega\,d) &= 0. \end{aligned} \tag{28.25}$$

We thus see that, in the uniform case, $D^R_{\alpha\beta}$ is defined up to only two arbitrary functions, instead of four as in the case of a non-uniform medium.

Let us now obtain the formulae for some particular cases. We put $d = b = 0$. Then

$$D^R_{ik} = \frac{4\pi\,\delta_{ik}}{\varepsilon(\omega)\,\omega^2 - k^2}, \quad D^R_{00} = -\frac{4\pi}{\varepsilon(\omega)\,(\varepsilon(\omega)\,\omega^2 - k^2)}, \quad D^R_{i0} = 0. \tag{28.26a}$$

The case $\varphi = 0$ corresponds to

$$D^R_{ik} = \frac{4\pi}{\varepsilon(\omega)\,\omega^2 - k^2}\left(\delta_{ik} - \frac{k_i k_k}{\varepsilon(\omega)\,\omega^2}\right), \quad D^R_{00} = D^R_{i0} = 0. \tag{28.26b}$$

Finally, we have in the case of the so-called transverse (or radiation) gauge (div $A = 0$):

$$D^R_{ik} = \frac{4\pi}{\varepsilon(\omega)\,\omega^2 - k^2}\left(\delta_{ik} - \frac{k_i k_k}{k^2}\right), \quad D^R_{00} = \frac{4\pi}{\varepsilon(\omega)\,k^2}, \quad D^R_{i0} = 0. \tag{28.26c}$$

The formulae for the temperature-dependent Green function $\mathfrak{D}$ follow from (28.26) by the substitution $\omega \to i\,|\omega_n|$:

$$\begin{aligned} \mathfrak{D}_{ik} &= -\frac{4\pi\,\delta_{ik}}{\varepsilon(i|\omega_n|)\,\omega_n^2 + k^2}, \\ \mathfrak{D}_{00} &= \frac{4\pi}{\varepsilon(i|\omega_n|)\,(\varepsilon(i\,|\omega_n|)\,\omega_n^2 + k^2)}, \quad \mathfrak{D}_{i0} = 0, \end{aligned} \tag{28.27a}$$

$$\mathfrak{D}_{ik} = -\frac{4\pi\,\delta_{ik}}{\varepsilon(i|\omega_n|)\,\omega_n^2 + k^2}\left(\delta_{ik} + \frac{k_i k_k}{\varepsilon(i\,|\omega_n|)\,\omega_n^2}\right), \quad \mathfrak{D}_{00} = \mathfrak{D}_{i0} = 0, \tag{28.27b}$$

$$\begin{aligned} \mathfrak{D}_{ik} &= -\frac{4\pi\,\delta_{ik}}{\varepsilon(i|\omega_n|)\,\omega_n^2 + k^2}\left(\delta_{ik} - \frac{k_i k_k}{k^2}\right), \\ \mathfrak{D}_{00} &= \frac{4\pi}{\varepsilon(i|\omega_n|)\,k^2}, \quad \mathfrak{D}_{i0} = 0. \end{aligned} \tag{28.27c}$$

The ordinary time-dependent Green function of the electromagnetic field, defined as

$$D_{\alpha\beta}(\boldsymbol{r}_1 - \boldsymbol{r}_2, t_1 - t_2) = -i \operatorname{Tr} \{\exp[(F - \hat{H})/T]\, T_t[A_\alpha(\boldsymbol{r}_1, t_1), A_\beta(\boldsymbol{r}_2, t_2)]\}, \tag{28.28}$$

can prove useful in a number of problems. As we saw in § 17, its Fourier components are connected with $D^R(\boldsymbol{k}, \omega)$ by the relations

$$\begin{aligned} \operatorname{Re} D(\boldsymbol{k}, \omega) &= \operatorname{Re} D^R(\boldsymbol{k}, \omega), \\ \operatorname{Im} D(\boldsymbol{k}, \omega) &= \coth \frac{\omega}{2T} \operatorname{Im} D^R(\boldsymbol{k}, \omega). \end{aligned} \tag{28.29}$$

We shall not derive the resulting unwieldy formulae for D.

The function D is of special interest at absolute zero, where it can be evaluated by using the ordinary method of quantum field theory. On passing to the limit $T = 0$ in (28.29), we get:

$$\begin{aligned} \operatorname{Re} D(\boldsymbol{k}, \omega) &= \operatorname{Re} D^R(\boldsymbol{k}, \omega), \\ \operatorname{Im} D(\boldsymbol{k}, \omega) &= \operatorname{sign} \omega \operatorname{Im} D^R(\boldsymbol{k}, \omega). \end{aligned} \tag{28.30}$$

Using the fact that the real part of $\varepsilon(\omega)$ is an even function of ω, and that the imaginary part an odd function (see e.g. [47]), we can easily show that we get D at $T = 0$ from D^R by replacing ω everywhere in (28.26) by $|\omega|$. In particular, we obtain for the gauge (28.26a):

$$\begin{aligned} D_{ik} &= \frac{4\pi\,\delta_{ik}}{\varepsilon(|\omega|)\,\omega^2 - k^2}, \\ D_{00} &= -\frac{4\pi}{\varepsilon(|\omega|)\,(\varepsilon(|\omega|)\,\omega^2 - k^2)}, \quad D_{0i} = 0. \end{aligned} \tag{28.31}$$

Formulae (28.31) are a generalisation of the ordinary formulae for the photon Green function in quantum electrodynamics (see e.g. [25]). They were obtained by another method by Ryazanov [49] for the particular case of a transparent medium ($\varepsilon''(\omega) = 0$).

§ 29. CALCULATION OF THE DIELECTRIC CONSTANT

A different approach can be used for finding the temperature-dependent Green function of the electromagnetic field $\mathfrak{D}$ in absorbing media, namely the application of the diagram technique developed in Chap. III. Since we are only interested in the electromagnetic field with long wavelengths, substantially greater than the interatomic spacings, we write the interaction Hamiltonian of the particles and field as a sum of two terms:

$$\hat{H}_{int} = \hat{H}_{int}^{(1)} + \hat{H}_{int}^{(2)},$$

the energy of the non-interacting particles and free photons being referred to as the zero-order Hamiltonian $\hat{H}_0$. We include in $\hat{H}_{int}^{(1)}$ that part of the interaction which leads to the short-range forces mentioned at the beginning of the previous section, whilst $\hat{H}^{(2)}_{int}$ is the Hamiltonian of the interaction of the long-wave electromagnetic field and the particles.

In the case of the gauge with zero scalar potential,

$$\hat{H}^{(2)}_{int} = -\int \left(\boldsymbol{A}(\boldsymbol{r}) \cdot \boldsymbol{j}(\boldsymbol{r})\right) \mathrm{d}^3\boldsymbol{r}, \tag{29.1}$$

where $\boldsymbol{j}(\boldsymbol{r})$ is the operator of the particle current density. The large value of the wavelength in (29.1) means that the Fourier expansion of $\boldsymbol{A}(\boldsymbol{r})$ only contains momenta $\boldsymbol{k}$ which do not exceed some limiting momentum k_0, much less than the reciprocal of the interatomic spacing $1/a$. In view of this, all the integrals arising in the diagram technique over $\boldsymbol{k}$ need to be cut off for $k_0 \ll 1/a$.

At non-relativistic particle velocities (a condition which is fulfilled in all macroscopic systems), the current density operator has the form (see e.g. [16])

$$\boldsymbol{j}(\boldsymbol{r}) = \sum_a \left\{ -i \frac{e_a}{2m_a} \left(\psi_a^+(\boldsymbol{r}) \nabla \psi_a(\boldsymbol{r}) - \nabla \psi_a^+(\boldsymbol{r}) \psi_a(\boldsymbol{r})\right) - \frac{e_a^2}{2m_a} \boldsymbol{A}(\boldsymbol{r}) \psi_a(\boldsymbol{r}) \psi_a(\boldsymbol{r}) \right\}.$$

The summation is over the different types of particle.

We shall distinguish the diagrams for corrections to the long-wavelength radiation the Green function according to the number of long-wavelength photon lines. The parts of the diagrams not containing such lines will be denoted for brevity by shaded polygons. Obviously, we can understand by such polygons the sum of all possible parts possessing the property in question, the perturbation theory series in the charge e being thus reduced to a series in the number of long-wavelength photon lines. The different types of diagram of this series are illustrated in Fig. 84. The functions corresponding to the polygons are wholly determined by the properties of the condensed body, which has been formed as a result of the action of the short-range forces.

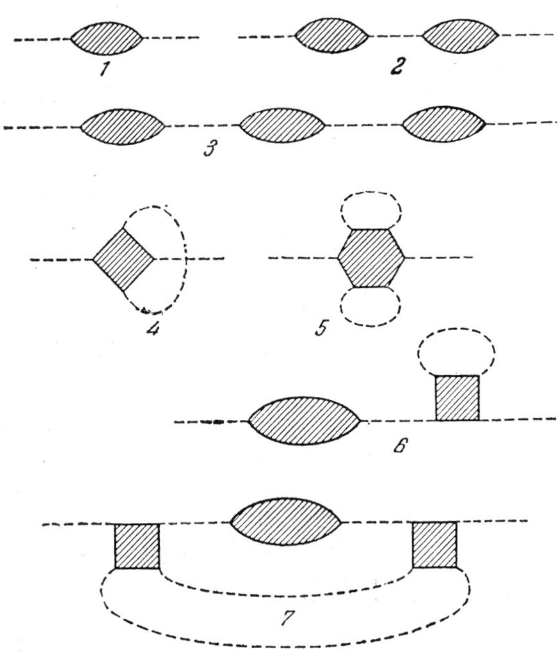

Fig. 84

It is fairly obvious, at once, from physical considerations, that diagrams 4—7 of Fig. 84 yield a negligibly small contribution, since they correspond to different non-linear processes such as the scattering of light by light. We can also prove this assertion as follows. As already remarked, all the integrals over the momenta of long-wave photon lines have to be cut off at some $k_0 \ll 1/a$. It follows from dimensional considerations that a small factor of order $k_0 a$ corresponds to every photon line, over the momentum of which an integration is performed. The only diagrams in which no integrations over momenta of long-wavelength phonon lines make an appearance are diagrams such as 1, 2 or 3 of Fig. 84.

We carried out the summation of a sequence of diagrams of this kind in § 10 when deriving Dyson's equation. We can therefore write down the equation for $\mathfrak{D}_{ik}$ at once(†):

$$\mathfrak{D}_{ik}(\boldsymbol{r}_1, \boldsymbol{r}_2; \omega_n) = \mathfrak{D}^{(0)}_{ik}(\boldsymbol{r}_1 - \boldsymbol{r}_2; \omega_n) + \int d^3\boldsymbol{r}_3 d^3\boldsymbol{r}_4 \mathfrak{D}^{(0)}_{il}(\boldsymbol{r}_1 - \boldsymbol{r}_3; \omega_n) \mathcal{H}_{lm}(\boldsymbol{r}_3, \boldsymbol{r}_4; \omega_n) \mathfrak{D}_{mk}(\boldsymbol{r}_4, \boldsymbol{r}_2; \omega_n). \tag{29.2}$$

The function $\mathcal{H}$, representing the contribution of the shaded loop in Fig. 82, is called the polarisation operator. It is clear from the foregoing arguments, that it is completely determined by the properties of the medium.

Let us express the polarisation operator $\mathcal{H}$ in terms of the dielectric constant of the system. We observe that the long-wavelength radiation Green function $\mathfrak{D}_{ik}$ (in the gauge where $\varphi = 0$) satisfies equation (28.18). On acting on equation (29.2) from the left by the operator

$$\omega_n^2 \delta_{ik} + \mathrm{curl}_{il}\, \mathrm{curl}_{lk}$$

and remarking that $\mathfrak{D}^{(0)}$ satisfies equation (28.18) with $\varepsilon = 1$, we get easily:

$$\int d^3\boldsymbol{r}\, \mathcal{H}_{il}(\boldsymbol{r}_1, \boldsymbol{r}'; \omega_n) \mathfrak{D}_{lk}(\boldsymbol{r}', \boldsymbol{r}_2; \omega_n) = \frac{\varepsilon(\boldsymbol{r}_1, i\,|\omega_n|) - 1}{4\pi} \omega_n^2 \mathfrak{D}_{ik}(\boldsymbol{r}_1, \boldsymbol{r}_2; \omega_n),$$

whence the required formula follows:

$$\mathcal{H}_{ik}(\boldsymbol{r}_1, \boldsymbol{r}_2; \omega_n) = \frac{1}{4\pi} \left(\varepsilon(\boldsymbol{r}_1, i\,|\omega_n|) - 1\right) \omega_n^2 \delta_{ik} \delta(\boldsymbol{r}_1 - \boldsymbol{r}_2). \tag{29.3}$$

The fact that the polarisation operator has turned out to be proportional to $\delta(\boldsymbol{r}_1 - \boldsymbol{r}_2)$ is connected with our neglect in the macroscopic theory of the effects of spatial correlations. The latter only become important in the frequency range where the so-called anomalous skin-effect occurs. The properties of $\mathcal{H}$ in this frequency range will be discussed in Chap. VII.

(†) Since we have in mind the case of non-uniform media, we have written equation (29.2) in the coordinate representation, retaining only the Fourier transform in τ.

For the rest of the present chapter we shall be concerned with the much greater frequencies at which the anomalous skin-effect is absent.

If we take the case of a uniform body, $\mathcal{H}$ is a function of $\boldsymbol{r}_1 - \boldsymbol{r}_2$ only, and ε is independent of $\boldsymbol{r}$. On carrying out the Fourier transformation in $\boldsymbol{r}_1 - \boldsymbol{r}_2$ in (29.4), we arrive at a simple equation connecting the polarisation operator of the system, $\mathcal{H}(\boldsymbol{k}, \omega_n)$, with its dielectric constant:

$$\mathcal{H}_{ik}(\boldsymbol{k}, \omega_n) = \frac{1}{4\pi}\left(\varepsilon(i\,|\omega_n|) - 1\right)\omega_n^2\delta_{ik}. \tag{29.4}$$

Formula (29.4) enables us to find the dielectric constant of a medium at $T \neq 0$ using the methods of quantum field theory. For, if we can find the polarisation operator of the system, we automatically find the value of $\varepsilon(\omega)$ at a discrete set of points of the imaginary axis: $\omega_n = 2n\pi Ti$. On recalling that $\varepsilon(\omega)$ is an analytic function with no singularities in the upper half-plane of ω, and repeating word for word as applied to ε all the arguments of § 17 relating to $\mathfrak{G}$ and G_R, we arrive at the conclusion that $\varepsilon(\omega)$ can be found simply by the analytic continuation of

$$\frac{4\pi}{3\,\omega_n^2}\mathcal{H}_{ii}(\omega_n)$$

from a discrete set of points on the positive imaginary semi-axis to the whole of the upper half-plane. Although this problem has no solution in a general form, the analytic continuation in question can be performed in a number of particular cases.

The problem of finding $\varepsilon(\omega)$ is greatly simplified at absolute zero. In this case, the polarisation operator can be found by using the time-dependent method of field theory described in Chap. II. On repeating all the calculations carried out in this section for the case of the time-dependent diagram technique, and noting what was said in § 28 regarding the functions $D_{ik}(\boldsymbol{k}, \omega)$, we get the formula (for the gauge in which $\varphi = 0$):

$$\Pi_{ik}(\omega) = \frac{1}{4\pi}\left(\varepsilon\left(\varepsilon(|\omega|\right) - 1\right)\omega^2\delta_{ik}. \tag{29.5}$$

On further recalling that $\varepsilon'(\omega) = \varepsilon'(-\omega)$, and $\varepsilon''(\omega) = -\varepsilon''(-\omega)$, we can express ε in terms of Π:

$$\begin{aligned} \varepsilon'(\omega) &= 1 + \frac{4\pi}{3\,\omega^2}\,\mathrm{Re}\,\Pi_{ii}(\omega), \\ \varepsilon''(\omega) &= \frac{4\pi}{3\,\omega\,|\omega|}\,\mathrm{Im}\,\Pi_{ii}(\omega). \end{aligned} \tag{29.6}$$

The evaluation of $\varepsilon(\omega)$ at $T = 0$ therefore reduces to finding the polarisation operator of the system.

§ 30. VAN DER WAALS FORCES IN A NON-UNIFORM DIELECTRIC

The long-wavelength electromagnetic field is a source of specific long-range forces, which may be called van der Waals forces, since their nature is similar to that of the attraction between molecules at large distances. Although the contribution of these forces to the free energy of the system is extremely small by comparison with the contribution of the short-range adhesive forces, they lead to a qualitatively new effect — non-additiveness of the free energy. It is this non-additiveness, connected with the long-range nature of the van der Waals forces, that enables their contribution to be distinguished in the thermodynamic functions.

This non-additiveness is easily understood if we return to the connexion between the van der Waals forces and the long-wavelength electromagnetic field. Indeed, every change in the density, and with this, in the electrical properties of the medium, in a certain region leads, by virtue of Maxwell's equations, to a change in the field outside this region as well. Hence, that part of the free energy which is connected with the long-wavelength radiation is not determined by the properties of the substance at a given point only, i.e. it is non-additive.

This leads to the fact that the chemical potential of a thin layer of liquid on the surface of a solid depends on the thickness of the layer. On the other hand, the van der Waals forces are a source of interaction forces between solids, i.e. the free energy depends on the distances between them. It is obvious that an important role in these phenomena is played by the electromagnetic field with wavelengths of the order of the thickness of the layer or the distance between the solids, which enables us to express the quantities of interest in terms of the dielectric constants $\varepsilon(\omega)$ of the bodies.

To find the correction to the ground state energy due to the long-wavelength electromagnetic field, we use the diagram method developed in § 15 for the thermodynamic potential Ω (it is the same as the ground state energy F in the case of photons). On repeating the relevant arguments of the previous section, we can show that only the sequence of diagrams of Fig. 85 will contribute to F. One might have thought that according to what was said in § 15 about the coefficients in front of the diagrams for F, we can no longer understand, by the shaded blocks in Fig. 85, the sum of all possible parts of diagrams not containing long-wavelength photon lines. It is easy to show that this is not the case. It turns out that the coefficient in front of the

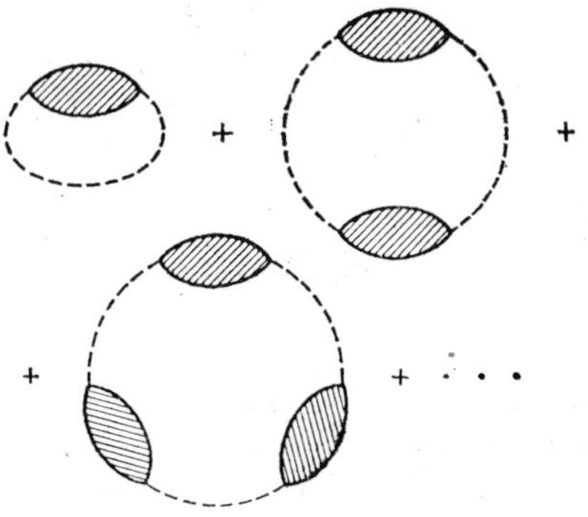

Fig. 85

diagrams is equal to $1/m$, where m is the number of photon lines (†). This circumstance enables us to carry out a summation of the shaded blocks and to associate the shaded loops in Fig. 85 with the polarisation operator $\mathcal{H}$ calculated in § 29.

The sequence of diagrams of Fig. 85 corresponds to the following series for the ground state energy F:

$$F = F_0 - \frac{T}{2} \sum_{n=-\infty}^{\infty} \Big\{ \int \mathcal{H}_{ik}(\boldsymbol{r}_1, \boldsymbol{r}_2; \omega_n) \mathfrak{D}_{ki}^{(0)}(\boldsymbol{r}_2 - \boldsymbol{r}_1; \omega_n)\, d^3\boldsymbol{r}_1 d^3\boldsymbol{r}_2$$

$$+ \frac{1}{2} \int \mathcal{H}_{ik}(\boldsymbol{r}_1, \boldsymbol{r}_2; \omega_n) \mathfrak{D}_{kl}^{(0)}(\boldsymbol{r}_2 - \boldsymbol{r}_3; \omega_n) \mathcal{H}_{lp}(\boldsymbol{r}_3, \boldsymbol{r}_4; \omega_n) \mathfrak{D}_{pi}^{(0)}(\boldsymbol{r}_4 - \boldsymbol{r}_1)$$

$$\times\, d^3\boldsymbol{r}_1 d^3\boldsymbol{r}_2 d^3\boldsymbol{r}_3 d^3\boldsymbol{r}_4 + \frac{1}{m} \int \mathcal{H}_{ik}(\boldsymbol{r}_1, \boldsymbol{r}_2; \omega_n) \mathfrak{D}_{kl}^{(0)}(\boldsymbol{r}_2 - \boldsymbol{r}_3; \omega_n)$$

$$\cdots \mathcal{H}_{qr}(\boldsymbol{r}_{2m-1}, \boldsymbol{r}_{2m}; \omega_n) \mathfrak{D}_{ri}^{(0)}(\boldsymbol{r}_{2m} - \boldsymbol{r}_1)\, d^3\boldsymbol{r}_1 \cdots d^3\boldsymbol{r}_{2m} + \cdots \Big\}, \quad (30.1)$$

where F_0 is the ground state energy of the medium (F_0 includes all the corrections due to the short-range forces).

Series (30.1) is not summed directly. Instead of the ground state energy we determine the extra pressure (more precisely, the extra stress tensor), arising as a result of the interaction of the substance with the long-wavelength electromagnetic field. For this, we imagine that the body is subjected to a small deformation with the displacement vector $\boldsymbol{u}(\boldsymbol{r})$. The change in the ground state energy δF is now equal to $-\int (\boldsymbol{f} \cdot \boldsymbol{u}) dV$, where $\boldsymbol{f}$ is the force acting per unit volume of the body during deformation. The corresponding change in F_0 is

$$\delta F_0 = \int (\boldsymbol{u} \cdot \operatorname{grad} p_0) dV,$$

where $p_0(\varrho, T)$ is the pressure when no account is taken of the corrections at the given density ϱ and temperature.

Given this displacement, only the polarisation operator will change in series (30.1), since it alone depends on the properties of the medium. In fact (see (29.3)),

$$\delta \mathcal{H}_{ik}(\boldsymbol{r}_1, \boldsymbol{r}_2; \omega_n) = \frac{1}{4\pi} \omega_n^2 \delta_{ik}(\boldsymbol{r}_1 - \boldsymbol{r}_2) \delta\varepsilon(\boldsymbol{r}_1, i\, |\omega_n|).$$

(†) We saw in § 15 that the number of topologically equivalent diagrams of the nth order is equal to $(n-1)!$, which led to the appearance of the factor $1/n$ ($1/n = (n-1)!/n!$). In our case, when $\hat{H}_{int} = \hat{H}_{int}^{(1)} + \hat{H}_{int}^{(2)}$ (see § 29), the coefficient produced by the expansion of the exponent is equal to $1/l!(2m)!$, where m is the number of photon lines, and l the number of vertices connected with $\hat{H}_{int}^{(1)}$. A calculation analogous to that carried out in § 15 shows that the number of topologically equivalent diagrams is equal to $l!(2m-1)!$, whence the assertion of the text follows.

The coefficients $1/m$ cancel on variation of (30.1) and we get

$$\delta F = \delta F_0 - \frac{T}{8\pi} \sum_{n=-\infty}^{\infty} \omega_n^2 \int d^3\boldsymbol{r}\, \delta\varepsilon(\boldsymbol{r}, i\,|\omega_n|) \{\mathfrak{D}_{ii}^{(0)}(\boldsymbol{r}-\boldsymbol{r};\omega_n)$$
$$+ \int \mathfrak{D}_{ik}^{(0)}(\boldsymbol{r}-\boldsymbol{r}_1;\omega_n)\, \mathcal{H}_{kl}(\boldsymbol{r}_1, \boldsymbol{r}_2;\omega_n)\, \mathfrak{D}_{li}^{(0)}(\boldsymbol{r}_2-\boldsymbol{r};\omega_n)\, d^3\boldsymbol{r}_1 d^3\boldsymbol{r}_2$$
$$+ \int \mathfrak{D}_{ik}^{(0)}(\boldsymbol{r}-\boldsymbol{r}_1;\omega_n)\, \mathcal{H}_{kl}(\boldsymbol{r}_1, \boldsymbol{r}_2;\omega_n)\, \mathfrak{D}_{lp}^{(0)}(\boldsymbol{r}_2-\boldsymbol{r}_3;\omega_n)\, \mathcal{H}_{pq}(\boldsymbol{r}_3, \boldsymbol{r}_4;\omega_n)$$
$$\times \mathfrak{D}_{qi}^{(0)}(\boldsymbol{r}_4-\boldsymbol{r};\omega_n)\, d^3\boldsymbol{r}_1 \cdots d^3\boldsymbol{r}_4 + \cdots\}.$$

The series in the curly brackets is none other than the series for the long-wavelength photon Green function $\mathfrak{D}$, corresponding to the sequence of diagrams 1, 2, 3 of Fig. 82. Hence

$$\delta F = \delta F_0 - \frac{T}{8\pi} \sum_{n=-\infty}^{\infty} \omega_n^2 \int \mathfrak{D}_{ii}(\boldsymbol{r}, \boldsymbol{r};\omega_n)\, \delta\varepsilon(\boldsymbol{r}, i\,|\omega_n|)\, d^3\boldsymbol{r}.$$

Recalling that $\mathfrak{D}$ is an even function of ω_n, we finally get

$$\delta F = \delta F_0 - \frac{T}{4\pi} \sum_{n=0}^{\infty}{}' \omega_n^2 \int \mathfrak{D}_{ii}(\boldsymbol{r}, \boldsymbol{r};\omega_n)\, \delta\varepsilon(\boldsymbol{r}, i\omega_n)\, d^3\boldsymbol{r}. \tag{30.2}$$

The prime on the summation sign indicates that the term with $n = 0$ is taken with half its weight. Remember that $\omega_n = 2n\pi T$.

The variation $\delta\varepsilon$ is connected with the displacement $\boldsymbol{u}$ by the formula (†)

$$\delta\varepsilon = -\boldsymbol{u}\,\mathrm{grad}\,\varepsilon - \varrho \frac{\partial\varepsilon}{\partial\varrho}\,\mathrm{div}\,\boldsymbol{u}.$$

On substituting this in (30.2) and integrating by parts, we get the following expression for $\boldsymbol{f}$:

$$\boldsymbol{f} = -\mathrm{grad}\, p_0 - \frac{T}{4\pi} \sum_{n=0}^{\infty}{}' \omega_n^2 \mathfrak{D}_{ii}(\boldsymbol{r}, \boldsymbol{r};\omega_n)\, \mathrm{grad}\, \varepsilon(\boldsymbol{r}, i\omega_n)$$
$$+ \frac{T}{4\pi} \sum_{n=0}^{\infty}{}' \omega_n^2\, \mathrm{grad} \left[\mathfrak{D}_{ii}(\boldsymbol{r}, \boldsymbol{r};\omega_n)\, \varrho \frac{\partial\varepsilon(\boldsymbol{r}, i\omega_n)}{\partial\varrho}\right]. \tag{30.3}$$

This formula enables us to find the correction to the chemical potential of the body. We notice first that, in mechanical equilibrium, $\boldsymbol{f} = 0$. On equating (30.3) to zero and noting that the relations

$$\mathrm{grad}\,\varepsilon(\varrho, T) = \frac{\partial\varepsilon}{\partial\varrho}\,\mathrm{grad}\,\varrho, \qquad dp_0(\varrho, T) = \varrho\, d\mu_0(\varrho, T)$$

(†) The variation of ε at a given point consists of two parts. The first part is connected with the change in ε due to the fact that different parts of the material are at a given point when displacements occurs:

$$\delta_1 \varepsilon = \varepsilon(\boldsymbol{r}-\boldsymbol{u}) - \varepsilon(\boldsymbol{r}) = -\boldsymbol{u}\,\mathrm{grad}\,\varepsilon,$$

and the second with the change in the density on deformation:

$$\delta_2 \varepsilon = \frac{\partial\varepsilon}{\partial\varrho}\,\delta\varrho = -\frac{\partial\varepsilon}{\partial\varrho}\,\varrho\,\mathrm{div}\,\boldsymbol{u}.$$

hold at constant temperature, where μ_0 is the undisturbed chemical potential per unit mass, we obtain after a simple transformation

$$-\varrho \operatorname{grad}\left\{\mu_0(\varrho, T) - \frac{T}{4\pi}\sum_n{}' \omega_n^2 \mathfrak{D}_{ii}(\mathbf{r}, \mathbf{r}; \omega_n)\frac{\partial \varepsilon}{\partial \varrho}\right\} = 0. \qquad (30.4)$$

We know that the condition for mechanical equilibrium of any non-uniform body is that the chemical potential be constant in the body. Hence it is immediately clear from (30.4) that

$$\mu(\varrho, T) = \mu_0(\varrho, T) - \frac{T}{4\pi}\sum_n{}' \omega_n^2 \mathfrak{D}_{ii}(\mathbf{r}, \mathbf{r}; \omega_n)\frac{\partial \varepsilon}{\partial \varrho}. \qquad (30.5)$$

We now evaluate the stress tensor. We first need to reduce the expression (30.3) for the force $\boldsymbol{f}$ to the form

$$f_i = \frac{\partial \sigma_{ik}}{\partial x_k}. \qquad (30.6)$$

As a preliminary, we introduce, in addition to the Green function $\mathfrak{D}_{ik}(\mathbf{r}, \mathbf{r}'; \omega_n)$, two further functions:

$$\begin{aligned} \mathfrak{D}^E_{ik}(\mathbf{r}, \mathbf{r}'; \omega_n) &= -\omega_n^2 \mathfrak{D}_{ik}(\mathbf{r}, \mathbf{r}'; \omega_n), \\ \mathfrak{D}^H_{ik}(\mathbf{r}, \mathbf{r}'; \omega_n) &= \operatorname{curl}_{il} \operatorname{curl}'_{km} \mathfrak{D}_{lm}(\mathbf{r}, \mathbf{r}'; \omega_n). \end{aligned} \qquad (30.7)$$

We rewrite the expression for the force as follows:

$$\begin{aligned} f_i = -\frac{\partial p_0}{\partial x_i} + \frac{T}{4\pi}\sum_{n=0}^{\infty}{}' \frac{\partial}{\partial x_i}\Big[\varepsilon(\mathbf{r}, i\omega_n)\mathfrak{D}^E_{kk}(\mathbf{r}, \mathbf{r}; \omega_n) \\ - \varrho(\mathbf{r})\frac{\partial \varepsilon(\mathbf{r}, i\omega_n)}{\partial \varrho}\mathfrak{D}^E_{kk}(\mathbf{r}, \mathbf{r}; i\omega_n)\Big] - \frac{T}{4\pi}\sum_{n=0}^{\infty}{}' \varepsilon(\mathbf{r}, i\omega_n)\frac{\partial}{\partial x_i}\mathfrak{D}^E_{kk}(\mathbf{r}, \mathbf{r}; \omega_n). \end{aligned} \qquad (30.8)$$

It now only remains for us to transform the last term in (30.8). We write it (omitting the summation and the factor $T/4\pi$) as

$$\varepsilon(\mathbf{r}')\frac{\partial}{\partial x_i}\mathfrak{D}^E_{kk}(\mathbf{r}, \mathbf{r}') + \varepsilon(\mathbf{r})\frac{\partial}{\partial x'_i}\mathfrak{D}^E_{kk}(\mathbf{r}, \mathbf{r}'), \qquad (30.9)$$

where we put $\mathbf{r} = \mathbf{r}'$ after the differentiations.

On further carrying out obvious transformations, we obtain for (30.9):

$$\begin{aligned} 2\frac{\partial}{\partial x_k}\varepsilon(\mathbf{r})\mathfrak{D}^E_{ik}(\mathbf{r}, \mathbf{r}) - \frac{\partial}{\partial x_k}\varepsilon(\mathbf{r})\mathfrak{D}^E_{ki}(\mathbf{r}, \mathbf{r}') - \frac{\partial}{\partial x'_k}\varepsilon(\mathbf{r}')\mathfrak{D}^E_{ik}(\mathbf{r}, \mathbf{r}') \\ + \varepsilon(\mathbf{r}')\left(\frac{\partial}{\partial x_i}\mathfrak{D}^E_{kk}(\mathbf{r}, \mathbf{r}') - \frac{\partial}{\partial x_k}\mathfrak{D}^E_{ik}(\mathbf{r}, \mathbf{r}')\right) \\ + \varepsilon(\mathbf{r})\left(\frac{\partial}{\partial x'_i}\mathfrak{D}^E_{kk}(\mathbf{r}, \mathbf{r}') - \frac{\partial}{\partial x'_k}\mathfrak{D}^E_{ki}(\mathbf{r}, \mathbf{r}')\right). \end{aligned} \qquad (30.10)$$

We can obtain from equation (28.18) for the Green function $\mathfrak{D}$ the identities

$$\frac{\partial}{\partial x_k} \varepsilon(\boldsymbol{r}) \mathfrak{D}^E_{ki}(\boldsymbol{r}, \boldsymbol{r}') = -4\pi \frac{\partial}{\partial x_i} \delta(\boldsymbol{r} - \boldsymbol{r}'),$$

$$\frac{\partial}{\partial x'_k} \varepsilon(\boldsymbol{r}') \mathfrak{D}^E_{ik}(\boldsymbol{r}, \boldsymbol{r}') = 4\pi \frac{\partial}{\partial x_i} \delta(\boldsymbol{r} - \boldsymbol{r}'),$$

$$\frac{\partial}{\partial x_k} \mathfrak{D}^H_{ik}(\boldsymbol{r}, \boldsymbol{r}') = -\frac{\partial}{\partial x'_k} \mathfrak{D}^H_{ik}(\boldsymbol{r}, \boldsymbol{r}') = -4\pi \frac{\partial}{\partial x_i} \delta(\boldsymbol{r} - \boldsymbol{r}'),$$

$$\varepsilon(\boldsymbol{r}') \left(\frac{\partial}{\partial x_k} \mathfrak{D}^E_{ik}(\boldsymbol{r}, \boldsymbol{r}') - \frac{\partial}{\partial x_i} \mathfrak{D}^E_{kk}(\boldsymbol{r}, \boldsymbol{r}') \right)$$
$$= -\frac{\partial}{\partial x'_k} \mathfrak{D}^H_{ki}(\boldsymbol{r}, \boldsymbol{r}') + \frac{\partial}{\partial x'_i} \mathfrak{D}^H_{kk}(\boldsymbol{r}, \boldsymbol{r}') + 8\pi \frac{\partial \delta(\boldsymbol{r} - \boldsymbol{r}')}{\partial x_i},$$

$$\varepsilon(\boldsymbol{r}) \left(\frac{\partial}{\partial x'_k} \mathfrak{D}^E_{ki}(\boldsymbol{r}, \boldsymbol{r}') - \frac{\partial}{\partial x'_k} \mathfrak{D}^E_{kk}(\boldsymbol{r}, \boldsymbol{r}') \right)$$
$$= -\frac{\partial}{\partial x_k} \mathfrak{D}^H_{ik}(\boldsymbol{r}, \boldsymbol{r}') + \frac{\partial}{\partial x_i} \mathfrak{D}^H_{kk}(\boldsymbol{r}, \boldsymbol{r}') - 8\pi \frac{\partial \delta(\boldsymbol{r} - \boldsymbol{r}')}{\partial x_i}.$$

On substituting these in (30.10) and putting $\boldsymbol{r} + \boldsymbol{r}'$, we get

$$\varepsilon(\boldsymbol{r}) \frac{\partial}{\partial x_i} \mathfrak{D}^E_{kk}(\boldsymbol{r}, \boldsymbol{r}) = 2 \frac{\partial}{\partial x_k} \varepsilon(\boldsymbol{r}) \mathfrak{D}^E_{ik}(\boldsymbol{r}, \boldsymbol{r}) + 2 \frac{\partial}{\partial x_k} \mathfrak{D}^H_{ik}(\boldsymbol{r}, \boldsymbol{r}) - \frac{\partial}{\partial x_i} \mathfrak{D}^H_{kk}(\boldsymbol{r}, \boldsymbol{r}).$$

On substituting this expression in turn in (30.8), we find finally that the force $\boldsymbol{f}$ can be written in the form (30.6) with the stress tensor given by

$$\sigma_{ik} = -\delta_{ik} p_0(\varrho, T) - \frac{T}{2\pi} \sum_{n=0}^{\infty}{}' \left\{ -\frac{1}{2} \delta_{ik} \left[\varepsilon(\boldsymbol{r}, i\omega)_n - \varrho \frac{\partial \varepsilon(\boldsymbol{r}, i\omega_n)}{\partial \varrho} \right] \right.$$
$$\times \mathfrak{D}^E_{ll}(\boldsymbol{r}, \boldsymbol{r}; \omega_n) - \frac{1}{2} \delta_{ik} \mathfrak{D}^H_{ll}(\boldsymbol{r}, \boldsymbol{r}; \omega_n)$$
$$\left. + \varepsilon(\boldsymbol{r}, i\omega_n) \mathfrak{D}^E_{ik}(\boldsymbol{r}, \boldsymbol{r}; \omega_n) + \mathfrak{D}^H_{ik}(\boldsymbol{r}, \boldsymbol{r}; \omega_n). \right\} \quad (30.11)$$

Formula (30.11) has no direct physical meaning, since the $\mathfrak{D}^E(\boldsymbol{r}, \boldsymbol{r}')$ and $\mathfrak{D}^H(\boldsymbol{r}, \boldsymbol{r}')$ appearing in it become infinite at $\boldsymbol{r} = \boldsymbol{r}'$. This is connected with the fact that the short wavelength electromagnetic oscillations yield an infinite contribution in σ_{ik}, unless a suitable cut-off is introduced. However, the short-wavelength oscillations have no relation with the effects due to the non-uniformities of the body, since their contribution is the same in a uniform body as in non-uniform bodies having the same value of ε at the point considered.

The long-wavelength contribution of interest, which does not depend in fact on the nature of the cut-off, is obtained from (30.11) by means of a suitable subtraction. In fact, we have to understand by the Green function $\mathfrak{D}^E_{ik}(\boldsymbol{r}, \boldsymbol{r})$ (and similarly by $\mathfrak{D}^H_{ik}(\boldsymbol{r}, \boldsymbol{r})$) in this formula the limit of the difference:

$$\lim_{\boldsymbol{r}\to\boldsymbol{r}'}[\mathfrak{D}^E_{ik}(\boldsymbol{r}, \boldsymbol{r}') - \overline{\mathfrak{D}}^E_{ik}(\boldsymbol{r}, \boldsymbol{r}')],$$

where $\overline{\mathfrak{D}}^E$ is the Green function for a uniform unbounded medium, the dielectric constant of which is the same as that of a non-uniform body at the point at which the stress tensor is calculated. To avoid unnecessary unwieldiness, we shall in future write (30.11) in the earlier form, on the assumption that the subtraction in question has already been made.

The same remark applies to (30.5) for the chemical potential, which can be written, in view of (30.7), as

$$\mu(\varrho, T) = \mu_0(\varrho, T) + \frac{T}{4\pi}\sum_{n=0}^{\infty}{}' \frac{\partial\varepsilon(\boldsymbol{r}, i\omega_n)}{\partial\varrho}\mathfrak{D}^E_{ii}(\boldsymbol{r}, \boldsymbol{r}; \omega_n). \qquad (30.12)$$

Notice that we also include in the category of non-uniform media systems consisting of several bodies, each of which is uniform. In this case, when solving (28.18), the components of $\mathfrak{D}_{ik}$ must satisfy certain conditions on the boundaries between the bodies. The independent variables in (28.18) are the coordinates $\boldsymbol{r}$, whilst the coordinates $\boldsymbol{r}'$ play the role of parameters. We are therefore talking about the boundary conditions relative to the variables $\boldsymbol{r}$. These conditions amount to the continuity of the tangential components of the electric and magnetic fields. Since a point $\boldsymbol{r}$ corresponds to one of the indices (i) of the tensor $\mathfrak{D}_{ik}$, the tangential components of tensors $\mathfrak{D}^E_{ik}$, $\mathfrak{D}^H_{ik}$ in this index must be continuous.

Formulae (30.11) and (30.12), which were obtained by Dzyaloshinkii and Pitaevskii [48], solve in principle the problem of finding the van der Waals part of the thermodynamic functions of a body. This problem amounts to the solution of equations (28.18) for the Green function $\mathfrak{D}_{ik}$.

§ 31. MOLECULAR INTERACTION FORCES BETWEEN SOLIDS

1. *Interaction forces between solids*

Let us apply the general theory developed above to the calculation of the van der Waals forces acting between solids, the surfaces of which are a very small distance apart. The gap separating the bodies may be filled with a liquid. We shall use the indices 1, 2 below to distinguish functions relating to the two bodies, and the index 3 for functions relating to the medium filling the gap.

Although we shall assume that the gap is plane-parallel, it must be borne in mind that in reality a correct statement of the problem of the interaction force between two bodies requires that we regard at least one of them as possessing finite dimensions and being surrounded on all sides by the medium 3, and that we find the total forces acting on it; in view of the very rapid decrease in the molecular forces with distance, this resultant force can in fact be entirely referred to the forces acting through the narrow gap separating the bodies.

The total force acting on body 2 can be calculated as the total flux of momentum flowing into the body from the surrounding medium 3, which is equal to the integral of the flux density over an arbitrary surface enclosing the body. It has to be borne in mind here that medium 3 is in thermodynamic equilibrium, one of the conditions for which is that its chemical potential be constant: $\mu = \text{const}$, where μ is given by (30.12).

Since the corrections to the density of the medium, connected with the long-wavelength fluctuations of the field, are small, the density ϱ can be regarded as constant in medium 3; the change in the chemical potential $\mu_0(\varrho, T)$ is here the same (by virtue of (30.4)) as the change in $p_0(\varrho, T)/\varrho$. The condition $\mu = \text{const}$ can therefore be rewritten as

$$p_0(\varrho, T) + \frac{T}{4\pi} \sum_{n=0}^{\infty}{}' \varrho \frac{\partial \varepsilon_3(i\omega_n)}{\partial \varrho} \mathfrak{D}_{ll}^E(\mathbf{r}, \mathbf{r}; \omega_n) = \text{const}. \qquad (31.1)$$

By virtue of this condition, part of the complete stress tensor (30.11) turns out to be a uniform pressure which is constant over the liquid and gives no contribution to the total force acting on the body; to determine the force, we in fact only need to write the stress tensor in medium 3 as

$$\sigma'_{ik} = -\frac{T}{2\pi} \sum_{n=0}^{\infty}{}' \left\{ \varepsilon_3(i\omega_n) \left[\mathfrak{D}_{ik}^E(\mathbf{r}, \mathbf{r}; \omega_n) - \frac{1}{2} \delta_{ik} \mathfrak{D}_{ll}^E(\mathbf{r}, \mathbf{r}; \omega_n) \right] \right.$$
$$\left. + \mathfrak{D}_{ik}^H(\mathbf{r}, \mathbf{r}; \omega_n) - \frac{1}{2} \delta_{ik} \mathfrak{D}_{ll}^H(\mathbf{r}, \mathbf{r}; \omega_n) \right\}. \qquad (31.2)$$

We choose the x-axis to be perpendicular to the plane of the gap, the width of which we denote by l (so that the surfaces of bodies 1, 2 are the $x = 0$, $x = l$ planes). The force F per unit area of surface 2 is now equal to

$$F(l) = \sigma'_{xx}(l) = \frac{T}{4\pi} \sum_{n=0}^{\infty}{}' \{ \varepsilon_3(i\omega_n) [\mathfrak{D}_{yy}^E(l, l; \omega_n) + \mathfrak{D}_{zz}^E(l, l; \omega_n)$$
$$- \mathfrak{D}_{xx}^E(l, l; \omega_n)] + \mathfrak{D}_{yy}^H(l, l; \omega_n) + \mathfrak{D}_{zz}^H(l, l; \omega_n) - \mathfrak{D}_{xx}^H(l, l; \omega_n) \}; \qquad (31.3)$$

a positive force corresponds to an attraction between the bodies, and a negative one to a repulsion.

Since the problem is uniform in the y, z directions, the Green function $\mathfrak{D}_{ik}(\mathbf{r}, \mathbf{r}'; \omega_n)$ depends only on the differences $y - y'$, $z - z'$. We carry out a Fourier transformation in these variables:

$$\mathfrak{D}_{ik}(x, x'; \mathbf{q}; \omega_n) = \int e^{-iq_y(y-y') - iq_z(z-z')} \mathfrak{D}_{ik}(\mathbf{r}, \mathbf{r}'; \omega_n) \, d(y - y') \, d(z - z')$$

and take the y axis along the vector $\boldsymbol{q}$. Equations (28.18) for the Green function become

$$\left(w^2 - \frac{\mathrm{d}^2}{\mathrm{d}x^2}\right)\mathfrak{D}_{zz}(x, x') = -4\pi\delta(x - x'),$$

$$\left(\varepsilon\omega_n^2 - \frac{\mathrm{d}^2}{\mathrm{d}x^2}\right)\mathfrak{D}_{yy}(x, x') + iq\frac{\mathrm{d}}{\mathrm{d}x}\mathfrak{D}_{xy}(x, x') = -4\pi\delta(x - x'),$$

$$w^2\mathfrak{D}_{xy}(x, x') + iq\frac{\mathrm{d}}{\mathrm{d}x}\mathfrak{D}_{yy}(x, x') = 0,$$

$$w^2\mathfrak{D}_{xx}(x, x') + iq\frac{\mathrm{d}}{\mathrm{d}x}\mathfrak{D}_{xy}(x, x') = -4\pi\delta(x - y),$$

$$\left(\varepsilon\omega_n^2 - \frac{\mathrm{d}^2}{\mathrm{d}x^2}\right)\mathfrak{D}_{xy}(x, x') + iq\frac{\mathrm{d}}{\mathrm{d}x}\mathfrak{D}_{xx}(x, x') = 0,$$

where $w = \sqrt{\varepsilon\omega_n^2 + q^2}$, whilst x' plays the role of a parameter (the components $\mathfrak{D}_{xz}$, $\mathfrak{D}_{yz}$ of the Green function vanish, since the equations for them prove to be homogeneous).

Solving this system amounts to solving altogether two equations:

$$\begin{aligned}\left(w^2 - \frac{\mathrm{d}^2}{\mathrm{d}x^2}\right)\mathfrak{D}_{zz}(x, x') &= -4\pi\delta(x - x'),\\ \left(w^2 - \frac{\mathrm{d}^2}{\mathrm{d}x^2}\right)\mathfrak{D}_{yy}(x, x') &= -\frac{4\pi w^2}{\varepsilon\omega_n^2}\delta(x - x'),\end{aligned} \tag{31.4}$$

after which $\mathfrak{D}_{xy}$ and $\mathfrak{D}_{xx}$ are obtained as

$$\mathfrak{D}_{xy} = -\frac{iq}{w^2}\frac{\mathrm{d}}{\mathrm{d}x}\mathfrak{D}_{yy}, \qquad \mathfrak{D}_{xx} = -\frac{iq}{w^2}\frac{\mathrm{d}}{\mathrm{d}x}\mathfrak{D}_{xy} - \frac{4\pi}{w^2}\delta(x - x'). \tag{31.5}$$

The boundary conditions, corresponding to the continuity of the tangential components of the electric and magnetic field strengths, amount to the requirement of continuity of $\mathfrak{D}^E_{yk}$, $\mathfrak{D}^H_{yk}$, $\mathfrak{D}^E_{zk}$, $\mathfrak{D}^H_{zk}$, or what amounts to the same thing, to continuity of $\mathfrak{D}_{yk}$, $\mathfrak{D}_{zk}$, $\mathrm{curl}_{yl}\,\mathfrak{D}_{lk}$, $\mathrm{curl}_{zl}\,\mathfrak{D}_{lk}$. On using the first of equations (31.5), we find that

$$\mathfrak{D}_{zz}, \quad \frac{\mathrm{d}\mathfrak{D}_{zz}}{\mathrm{d}x}, \quad \mathfrak{D}_{yy}, \quad \frac{\varepsilon}{w^2}\frac{\mathrm{d}\mathfrak{D}_{yy}}{dx} \tag{31.6}$$

must be continuous on the dividing boundary.

Since we are only interested in the Green function in the gap region, we can confine ourselves at once to the case $0 < x' < l$. In domain 3 $(0 < x < l)$ the functions $\mathfrak{D}_{yy}$, $\mathfrak{D}_{zz}$ are given by equations (31.4) with $\varepsilon = \varepsilon_3$, $w = w_3 = \sqrt{\varepsilon_3\omega_n^2 + q^2}$. In domains 1 $(x < 0)$ and 2 $(x > l)$ they satisfy the same equations with vanishing right-hand sides (since we always have $x \neq x'$ here), and with ε_1, w_1 and ε_2, w_2 respectively instead of ε, w.

The subtraction mentioned at the end of § 30 amounts to subtracting from all the $\mathfrak{D}$ functions in the gap region their values for $\varepsilon_1 = \varepsilon_2 = \varepsilon_3$, $w_1 = w_2 = w_3$. A particular consequence of this is that we can at once omit the term containing the δ-functions in the second of relations (31.5), so that $\mathfrak{D}_{xy}$, $\mathfrak{D}_{xx}$ are given in the gap region by

$$\mathfrak{D}_{xy} = -\frac{iq}{w_3^2}\frac{\mathrm{d}}{\mathrm{d}x}\mathfrak{D}_{yy}, \quad \mathfrak{D}_{xx} = -\frac{iq}{w_3^2}\frac{\mathrm{d}}{\mathrm{d}x}\mathfrak{D}_{xy}. \tag{31.7}$$

Before proceeding to the solution of the equations, one remark should be made. The general solution of equations (31.4) has the form $f_1(x - x') + f_2(x + x')$. On using equations (31.4), (31.7) and the definition of $\mathfrak{D}^E$, $\mathfrak{D}^H$, we can show that the parts of the Green functions that depend on $x + x'$ provide no contribution to expression (31.3) for the force. We shall not dwell here on this, since the result is obvious *a priori* from physical considerations: if we were to put $x = x'$ in a solution of the form $f_2(x + x')$, we should get the flux of momentum in the gap, which would depend on the coordinates, in contradiction to its law of conservation. In future, therefore, we shall as a rule only give the expressions for the parts of the Green functions $\mathfrak{D}^+$ which depend only on $x - x'$.

Let us evaluate $\mathfrak{D}_{zz}$. It satisfies the equations

$$\left(w_3^2 - \frac{\mathrm{d}^2}{\mathrm{d}x^2}\right)\mathfrak{D}_{zz} = -4\pi\delta(x - x') \quad \text{for} \quad 0 < x < l,$$

$$\left(w_1^2 - \frac{\mathrm{d}^2}{\mathrm{d}x^2}\right)\mathfrak{D}_{zz} = 0 \quad \text{for} \quad x < 0; \quad \left(w_2^2 - \frac{\mathrm{d}^2}{\mathrm{d}x^2}\right)\mathfrak{D}_{zz} = 0 \quad \text{for} \quad x > l.$$

We obtain from this:

$$\mathfrak{D}_{zz} = Ae^{w_1 x} \text{ for } x < 0, \; \mathfrak{D}_{zz} = Be^{-w_2 x} \text{ for } x > l,$$

$$\mathfrak{D}_{zz} = C_1 e^{w_3 x} + C_2 e^{-w_3 x} - \frac{2\pi}{w_3} e^{-w_3|x - x'|} \text{ for } 0 < x < l.$$

Having determined the constants A, B, C_1, C_2 from the boundary conditions for the continuity of $\mathfrak{D}_{zz}$ and $\mathrm{d}\mathfrak{D}_{zz}/\mathrm{d}x$, we get for $\mathfrak{D}_{zz}^+$:

$$\mathfrak{D}_{zz}^+ = \frac{4\pi}{w_3 \Delta}\cosh w_3(x - x') - \frac{2\pi}{w_3} e^{-w_3|x - x'|} \text{ for } 0 < x < l,$$

where

$$\Delta = 1 - e^{2w_3 l}\frac{(w_1 + w_3)(w_2 + w_3)}{(w_1 - w_3)(w_2 - w_3)}. \tag{31.8}$$

On subtracting the value of $\mathfrak{D}_{zz}^+$ for $w_1 = w_2 = w_3$ (Δ becomes infinite here), we finally get

$$\mathfrak{D}_{zz}^+ = \frac{4\pi}{w_3 \Delta}\cosh w_3(x - x'). \tag{31.9}$$

Similarly, the solution of the equation for $\mathfrak{D}_{yy}$ gives us (after subtraction):

$$\mathfrak{D}_{yy}^{+} = \frac{4\pi w_3}{\omega_n^2 \varepsilon_3 \bar{\Delta}} \cosh w_3(x - x'), \tag{31.10}$$

$$\bar{\Delta} = 1 - e^{2w_3 l} \frac{(\varepsilon_1 w_3 + \varepsilon_3 w_1)(\varepsilon_2 w_3 + \varepsilon_3 w_2)}{(\varepsilon_1 w_3 - \varepsilon_3 w_1)(\varepsilon_2 w_3 - \varepsilon_3 w_2)} \tag{31.11}$$

and, on using (31.7), we get

$$\begin{aligned} \mathfrak{D}_{xy}^{+} &= -\frac{4\pi i q}{\omega_n^2 \varepsilon_3 \bar{\Delta}} \sinh w_3(x - x'), \\ \mathfrak{D}_{xx}^{+} &= -\frac{4\pi q^2}{\omega_n^2 \varepsilon_3 w_3 \bar{\Delta}} \cosh w_3(x - x'). \end{aligned} \tag{31.12}$$

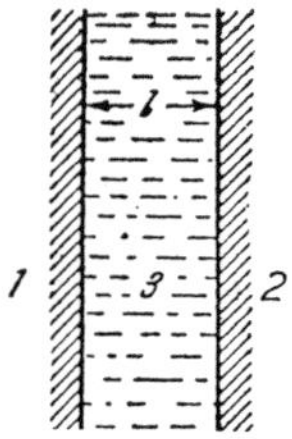
Fig. 86

If we now work out $\mathfrak{D}_{ik}^{E}(x, x'; q, \omega_n)$ and $\mathfrak{D}_{ik}^{H}(x, x'; q, \omega_n)$ and substitute them in (31.3), we get

$$F(l) = -\frac{T}{2\pi} \sum_{n=0}^{\infty}{}' \int_0^{\infty} q\, dq\, w_3 \left(\frac{1}{\Delta} + \frac{1}{\bar{\Delta}}\right).$$

On changing to a new integration variable, $q = \sqrt{\varepsilon_3}\, \omega_n \sqrt{p^2 - 1}$, and returning to the usual system of units, we arrive at the final expression (Dzyaloshinskii, Lifshitz, Pitaevskii [50]) for the force F, per unit area of each of two bodies, separated by a gap of width l, filled with liquid (Fig. 86):

$$F(l) = \frac{T}{\pi c^3} \sum_{n=0}^{\infty}{}' \varepsilon_3^{3/2} \omega_n^3 \int_1^{\infty} p^2 dp \left\{ \left[\frac{(s_1 + p)(s_2 + p)}{(s_1 - p)(s_2 - p)} \exp\left(\frac{2p\omega_n}{c} l \sqrt{\varepsilon_3}\right) - 1 \right]^{-1} \right.$$
$$\left. + \left[\frac{(s_1 + p\varepsilon_1/\varepsilon_3)(s_2 + p\varepsilon_2/\varepsilon_3)}{(s_1 - p\varepsilon_1/\varepsilon_3)(s_2 - p\varepsilon_2/\varepsilon_3)} \exp\left(\frac{2p\omega_n}{c} l \sqrt{\varepsilon_3}\right) - 1 \right]^{-1} \right\}, \tag{31.13}$$

where

$$s_1 = \sqrt{\varepsilon_1/\varepsilon_3 - 1 + p^2}, \quad s_2 = \sqrt{\varepsilon_2/\varepsilon_3 - 1 + p^2}, \quad \omega_n = \frac{2\pi n T}{\hbar},$$

and ε_1, ε_2, ε_3 are functions of the imaginary frequency $i\omega_n$ ($\varepsilon = \varepsilon(i\omega_n)$). This formula was first obtained by Lifshitz [51] for the case $\varepsilon_3 = 1$, i.e. for bodies separated by a vacuum gap, by means of a method that made no use of quantum field theory.

The general formula (31.13) is extremely complicated. However, it can be considerably simplified because of the fact that the effect of temperature on the interaction force between the bodies is usually quite insignificant(†).

The fact is that, owing to the presence of the exponential functions in the integrands in (31.13), the main role in the sum is played by those

(†) When speaking of temperature effects, we are not thinking of the result of the direct dependence of the dielectric constant on temperature.

terms for which $\omega_n \sim c/l$, or $n \sim c\hbar/lT$. Thus, in the case $lT/c\hbar \ll 1$, large values of n will be important, and we can change in (4.13) the summation to an integration over $\mathrm{d}n = (\hbar/2\pi T)\mathrm{d}\omega$. The temperature does not now appear explicitly in the formula, and we get the following result ($\varepsilon = \varepsilon(i\omega)$):

$$F(l) = \frac{\hbar}{2\pi^2 c^3} \int_0^\infty \mathrm{d}\omega \int_1^\infty \mathrm{d}p\, p^2 \omega^3 \varepsilon_3^{3/2} \left\{ \left[\frac{(s_1+p)(s_2+p)}{(s_1-p)(s_2-p)} \times \exp\left(\frac{2p\omega}{c} l\sqrt{\varepsilon_3}\right) - 1 \right]^{-1} + \left[\frac{(s_1+p\varepsilon_1/\varepsilon_3)(s_2+p\varepsilon_2/\varepsilon_3)}{(s_1-p\varepsilon_1/\varepsilon_3)(s_2-p\varepsilon_2/\varepsilon_3)} \times \exp\left(\frac{2p\omega}{c} l\sqrt{\varepsilon_3}\right) - 1 \right]^{-1} \right\}. \quad (31.14)$$

Formula (31.14) is still complicated. It admits of considerable further simplicifation in two important limiting cases.

We shall dwell first on the limiting case of "small" distances, by which we understand distances which are small compared with the wavelengths λ_0 that are characteristic for the absorption spectra of certain bodies. The temperatures which may be under discussion here for condensed solids are always small compared to the $\hbar\omega$ that play a role here (for instance, in the visual part of the spectrum), so that the inequality $Tl/\hbar c \ll 1$ is most certainly satisfied.

Owing to the presence of the exponential factor

$$\exp\left(2p\omega l\sqrt{\varepsilon_3}/c\right)$$

in the denominators of the integrand, the main role in the integration over p is played by the values of p such that $p\omega l/c \sim 1$. In this case, $p \gg 1$, so that we can put $s_1 \approx s_2 \approx p$ when finding the principal terms. The first term in the curly brackets in (31.14) vanishes in this approximation. We get from the second term, after using as integration variable $x = 2lp\,\omega\sqrt{\varepsilon_3}/c$:

$$F = \frac{\hbar}{16\pi^2 l^3} \int_0^\infty \int_0^\infty x^2 \mathrm{d}x\, \mathrm{d}\omega \left[\frac{(\varepsilon_1+\varepsilon_3)(\varepsilon_2+\varepsilon_3)}{(\varepsilon_1-\varepsilon_3)(\varepsilon_2-\varepsilon_3)} e^x - 1 \right]^{-1} \quad (31.15)$$

(the lower limit of the integration over x is replaced by zero in this approximation). The force in this case proves to be inversely proportional to the cube of the distance, which is what we might in fact expect from the usual laws of van der Waals forces between two atoms. The functions $\varepsilon(i\omega) - 1$ are monotonically decreasing as ω increases, and tend to zero. Consequently, as from some value $\omega \sim \omega_0$, the values of ω cease to provide a significant contribution to the integral; the condition that l is small implies that we must have $l \ll c/\omega_0$.

Let us turn to the opposite limiting case of "large" distances $l \gg \lambda_0$. Here, however, we shall assume that the distances are not so great that the condition $lT/\hbar c \ll 1$ is destroyed. We introduce a new variable of integration $x = 2pl\omega/c$ into the general formula (31.14); whilst we take p instead of ω (as before) as the second variable:

$$F = \frac{\hbar}{32\pi^2 l^4} \int_0^\infty dx \int_1^\infty dp \frac{x^3}{p^2} \varepsilon_3^{3/2} \left\{ \left[\frac{(s_1+p)(s_2+p)}{(s_1-p)(s_2-p)} e^{x\sqrt{\varepsilon_3}} - 1 \right]^{-1} \right.$$

$$\left. + \left[\frac{(s_1+p\varepsilon_1/\varepsilon_3)(s_2+p\varepsilon_2/\varepsilon_3)}{(s_1-p\varepsilon_1/\varepsilon_3)(s_2-p\varepsilon_2/\varepsilon_3)} e^{x\sqrt{\varepsilon_3}} - 1 \right]^{-1} \right\}, \qquad \varepsilon = \varepsilon\left(i\frac{xc}{2pl}\right).$$

Owing to the presence of $\exp(x\sqrt{\varepsilon_3})$, in the denominators in the integral over x, the important region is the one where $x \approx 1/\sqrt{\varepsilon_3} \leq 1$, and since $p \geq 1$, the argument of the functions ε is close to zero at large l throughout the important range of values of the variables. We can accordingly replace ε_1, ε_2, ε_3 simply by their values at $\omega = 0$, i.e. by the electrostatic dielectric constants. On substituting after this $x \to x/\sqrt{\varepsilon_{30}}$, we get the final result:

$$F = \frac{\hbar c}{32\pi^2 l^4 \sqrt{\varepsilon_{30}}} \int_0^\infty dx \int_1^\infty dp \frac{x^3}{p^2} \left\{ \left[\frac{(s_{10}+p)(s_{20}+p)}{(s_{10}-p)(s_{20}-p)} e^x - 1 \right]^{-1} \right.$$

$$\left. + \left[\frac{(s_{10}+p\varepsilon_{10}/\varepsilon_{30})(s_{20}+p\varepsilon_{20}/\varepsilon_{30})}{(s_{10}-p\varepsilon_{10}/\varepsilon_{30})(s_{20}-p\varepsilon_{20}/\varepsilon_{30})} e^x - 1 \right]^{-1} \right\}, \tag{31.16}$$

$$s_{10} = \sqrt{\frac{\varepsilon_{10}}{\varepsilon_{30}} - 1 + p^2}, \qquad s_{20} = \sqrt{\frac{\varepsilon_{20}}{\varepsilon_{30}} - 1 + p^2},$$

where ε_{10}, ε_{20}, ε_{30} are the electrostatic values of the dielectric constant.

Let us now dwell on the high temperature case. When $lT/\hbar c \gg 1$, we only need retain the first term in the sum (31.13). However, we cannot put $n = 0$ directly in it, because of the resulting indeterminacy (the factor ω_n^3 vanishes, but the integral over p is divergent). We can get round this difficulty by first replacing p by the new variable of integration $x = 2p\omega_n l\sqrt{\varepsilon_{30}}/c$ (as a result of which the factor ω_n^3 goes out). On then putting $\omega_n = 0$, we get

$$F = \frac{T}{16\pi l^3} \int_0^\infty x^2 \left[\frac{(\varepsilon_{10}+\varepsilon_{30})(\varepsilon_{20}+\varepsilon_{30})}{(\varepsilon_{10}-\varepsilon_{30})(\varepsilon_{20}-\varepsilon_{30})} e^x - 1 \right]^{-1} dx. \tag{31.17}$$

Thus, at sufficiently large distances the decrease of the interaction force slows down and once more proceeds in accordance with an l^{-3} law with a coefficient dependent on the temperature and the electrostatic dielectric constant.

2. *Interaction forces between atoms in solutions*

We shall now show how we can pass from the macroscopic formula (31.14) to the interaction of individual atoms in a vacuum. To do this, we shall make a formal assumption that both bodies are sufficiently "rarefied".

This means, from the point of view of macroscopic electrodynamics, that their dielectric constants are close to unity, i.e. the differences $\varepsilon_1 - 1$ and $\varepsilon_2 - 1$ are small.

We start with the case of "small" distances. Formula (31.15) with $\varepsilon_3 = 1$ gives us to the requisite accuracy:

$$F = \frac{\hbar}{64\pi^2 l^3} \int_0^\infty \int_0^\infty x^2 e^{-x}(\varepsilon_1 - 1)(\varepsilon_2 - 1)\,dx\,d\omega$$

$$= \frac{\hbar}{32\pi^2 l^3} \int_0^\infty [\varepsilon_1(i\omega) - 1][\varepsilon_2(i\omega) - 1]\,d\omega. \tag{31.18}$$

On expressing $\varepsilon(i\omega)$ in terms of $\varepsilon''(\omega)$ on the real axis in accordance with (28.19), we get

$$\int_0^\infty [\varepsilon_1(i\omega) - 1][\varepsilon_2(i\omega) - 1]\,d\omega$$

$$= \frac{4}{\pi^2} \int_0^\infty \int_0^\infty \int_0^\infty \frac{\omega_1 \omega_2 \varepsilon_1''(\omega_1)\varepsilon_2''(\omega_2)}{(\omega_1^2 + \omega^2)(\omega_2^2 + \omega^2)}\,d\omega_1\,d\omega_2\,d\omega$$

$$= \frac{2}{\pi} \int_0^\infty \int_0^\infty \frac{\varepsilon_1''(\omega_1)\varepsilon_2''(\omega_2)}{\omega_1 + \omega_2}\,d\omega_1\,d\omega_2,$$

whence we find for the force F:

$$F = \frac{\hbar}{16\pi^3 l^3} \int_0^\infty \int_0^\infty \frac{\varepsilon_1''(\omega_1)\varepsilon_2''(\omega_2)}{\omega_1 + \omega_2}\,d\omega_1\,d\omega_2. \tag{31.19}$$

This force corresponds to an inter-atomic interaction with an energy(†)

$$U(R) = -\frac{3\hbar}{8\pi^4 R^6 N_1 N_2} \int_0^\infty \int_0^\infty \frac{\varepsilon_1''(\omega_1)\varepsilon_2''(\omega_2)}{\omega_1 + \omega_2}\,d\omega_1\,d\omega_2, \tag{31.20}$$

where R is the inter-atomic distance, N_1, N_2 are the numbers of atoms per unit volume in the first and second bodies respectively. The imaginary part of the dielectric constant is connected with the spectral density of the "oscillator strengths" $f(\omega)$, familiar from spectroscopy by the relationship

$$\omega\varepsilon''(\omega) = \frac{2\pi^2 e^2}{m} N f(\omega)$$

(†) If the potential energy of the interaction of molecules 1 and 2 is $U = -a/R^6$, the total energy of the binary interactions of all the molecules in two half-spaces separated by a gap l is

$$U = -\frac{a\pi N_1 N_2}{12\, l^2}.$$

The force F, on the other hand, is

$$F = -\frac{dU}{dl} = \frac{a\pi N_1 N_2}{6 l^3}.$$

This gives us the correspondence of (31.19) and (31.20).

(see e.g. [47], § 62). On substituting it in (31.20), we get

$$U(R) = -\frac{3\hbar e^4}{2m^2 R^6} \int_0^\infty \int_0^\infty \frac{f_1(\omega_1) f_2(\omega_2)}{\omega_1 + \omega_2} \, d\omega_1 \, d\omega_2. \tag{31.21}$$

This expression is exactly the same as the familiar formula of London [52], obtained by means of ordinary perturbation theory, applied to the dipole interaction of two atoms. Suppose, for example, that we are discussing the interaction of two hydrogen atoms. On using the familiar expression

$$f_{0n} = \frac{2m}{\hbar^2} (E_n - E_0) \, |x_{0n}|^2$$

for the oscillator strength for a transition between states E_0 and E_n (x_{0n} is the corresponding matrix element of the coordinate of an electron in the atom) and passing in (31.19) from integration over the frequency to a summation over the energy levels of the atom, we get the London formula for hydrogen atoms:

$$U(R) = -\frac{6e^4}{R^6} \sum_{n,m} \frac{|x_{0n}|^2 \, |x_{0m}|^2}{E_n - E_0 + E_m - E_0}.$$

At "great" distances the formula for the attraction between two "rarefied" bodies has the form

$$F = \frac{\hbar c}{32\pi^2 l^4} (\varepsilon_{10} - 1)(\varepsilon_{20} - 1) \int_0^\infty x^3 e^{-x} dx \int_1^\infty \frac{1 - 2p^2 + 2p^4}{8p^6} dp$$

$$= \frac{\hbar c}{l^4} \frac{23}{640\pi^2} (\varepsilon_{10} - 1)(\varepsilon_{20} - 1). \tag{31.22}$$

This force corresponds to the interaction of two atoms with an energy

$$U(R) = -\frac{23\hbar c}{4\pi R^7} \alpha_1 \alpha_2, \tag{31.23}$$

where α_1, α_2 are the static polarisabilities of the two atoms ($\varepsilon_0 = 1 - 4\pi N\alpha$). Formula (31.23) is the same as the result obtained by quantum mechanical methods by Casimir and Polder [53] for the attraction between two atoms at a fairly large distance, when the retardation effects become important.

We now consider the interaction of two atoms located in a liquid (Pitaevskii [54]). Suppose we have weak solutions of atoms of different kinds with concentrations (the number of particles per cm³) N_1, N_2 respectively in the same solvent. Suppose further that the gap is filled with pure solvent. The dielectric constants $\varepsilon_1, \varepsilon_2$ of solutions in which the concentrations of the dissolved atoms are small only differ slightly from the dielectric constant of the pure solvent, which we denote by $\varepsilon_3 = \varepsilon$. Up to the first order in the concentrations,

$$\varepsilon_1 = \varepsilon + N_1 \left(\frac{\partial \varepsilon_1}{\partial N_1}\right)_{N_1=0}, \qquad \varepsilon_2 = \varepsilon + N_2 \left(\frac{\partial \varepsilon_2}{\partial N_2}\right)_{N_2=0}.$$

If we retain only terms of the same order in (31.15) for the force at "small" distances, we get (in the same way as when deriving (31.18)):

$$F(l) = \frac{\hbar}{32\pi^2 l^3} N_1 N_2 \int\limits_0^\infty \left(\frac{\partial \varepsilon_1(i\omega)}{\partial N_1}\right)_{N_1=0} \left(\frac{\partial \varepsilon_2(i\omega)}{\partial N_2}\right)_{N_2=0} \frac{d\omega}{\varepsilon^2(i\omega)}.$$

This force corresponds to an interaction energy between the dissolved atoms equal to

$$U(R) = -\frac{3\hbar}{16\pi^3 R^6} \int\limits_0^\infty \left(\frac{\partial \varepsilon_1(i\omega)}{\partial N_1}\right)_{N_1=0} \left(\frac{\partial \varepsilon_2(i\omega)}{\partial N_2}\right)_{N_2=0} \frac{d\omega}{\varepsilon^2(i\omega)}. \tag{31.24}$$

We similarly find for the energy at "great" distances:

$$U(R) = -\frac{23\hbar c}{64\pi^3 \varepsilon_0^{5/2} R^7} \left(\frac{\partial \varepsilon_{10}}{\partial N_1}\right)_{N_1=0} \left(\frac{\partial \varepsilon_{20}}{\partial N_2}\right)_{N_2=0}. \tag{31.25}$$

We see that, when the molecules of the dissolved substance interact strongly with the solvent, the forces of interaction between them are no longer determined by their polarisabilities.

3. *Thin films on a solid surface*

The general theory of van der Waals forces described above can also be used for finding the thermodynamic functions of a thin liquid film on the surface of a solid; the thickness l of the film, is of course assumed large compared with the inter-atomic distances.

We obtained formula (30.12) for the chemical potential of the liquid per unit mass in terms of the Green functions of the long-wavelength electromagnetic field existing in it. This formula is inconvenient for two reasons, however: firstly, it contains $\partial\varepsilon/\partial\varrho$, which has never been investigated experimentally throughout the whole frequency interval; secondly, it yields the chemical potential μ as a function of the density ϱ, whereas we generally want to know μ as a function of the pressure p.

We consider a layer *3*, on the surface of a solid *1* and in equilibrium with its vapour *2* (Fig. 87). We shall regard the vapour as a vacuum as far as its electromagnetic properties are concerned, i.e. we shall put its dielectric constant equal to unity everywhere: $\varepsilon_2 = 1$.

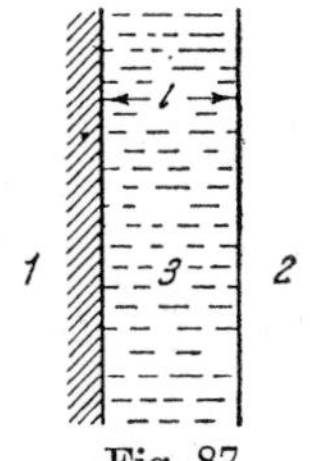

Fig. 87

By the condition for mechanical equilibrium, the normal component σ_{xx} of the stress tensor must be continuous on the surface of the layer. This gives us the equation

$$p = p_0(\varrho, T) - \bar{\sigma}_{xx},$$

where p is the vapour pressure, $p_0(\varrho, T)$ is the liquid pressure for the given density and temperature, and $\bar{\sigma}_{xx}$ denotes the set of all terms except the first in expression (30.11) for the stress tensor in the layer. Solving this equation for ϱ gives us the density as a function of the pressure (†)

$$\varrho = \varrho_0(p + \bar{\sigma}_{xx}, T).$$

On substituting this expression in (30.12) for the chemical potential, we get

$$\mu = \mu_0(p + \bar{\sigma}_{xx}, T) + \frac{T}{4\pi} \sum_{n=0}^{\infty}{}' \frac{\partial \varepsilon(i\omega_n)}{\partial \varrho} \mathfrak{D}_{ii}^E(\mathbf{r}, \mathbf{r}; \omega_n),$$

where $\mu_0(p, T)$ is the chemical potential of the liquid. On expanding μ_0 in powers of the small quantity $\bar{\sigma}_{xx}$ and taking into account the thermodynamic equation $(\partial\mu/\partial p)_T = 1/\varrho$, the last equation reduces to

$$\mu(p, T) = \mu_0(p, T) + \frac{1}{\varrho}\bar{\sigma}_{xx} + \frac{T}{4\pi} \sum_{n=0}^{\infty}{}' \frac{\partial \varepsilon(i\omega_n)}{\partial \varrho} \mathfrak{D}_{ii}^E(\mathbf{r}, \mathbf{r}; \omega_n).$$

Finally, on substituting in this the expression for $\bar{\sigma}_{xx}$ from (30.11), we find that the term with $\partial\varepsilon/\partial\varrho$ falls out and there remains

$$\mu(p, T) = \mu_0(p, T) + \frac{1}{\varrho}\sigma'_{xx}.$$

Here σ'_{xx} is the component of the "contracted" stress tensor (31.2). This quantity is constant along the thickness of the layer (because the flux of momentum is constant), and it in fact determines the force $F(l)$, in accordance with (31.3).

We introduce the notation ζ for the "van der Waals part" of the chemical potential of the film, per unit volume of the liquid,

$$\mu = \mu_0 + \frac{\zeta}{\varrho}. \tag{31.26}$$

By what has been said above,

$$\zeta = \sigma'_{xx} = F(l). \tag{31.27}$$

As l tends to ∞, i.e. when the liquid stretches to infinity, ζ tends to zero.

Thus there is no need for new calculations to find the quantity ζ which is of interest to us. It is determined by the formulae for $F(l)$ obtained above (general formula (31.13) and the subsequent limiting formulae), in which we only have to put $\varepsilon_2 = 1$.

The reader specially interested in the problems outlined in §§ 3, 4, can turn to the more detailed articles by Lifshitz [51] and by Dzyaloshinskii, Lifshitz and Pitaevskii [50].

(†) $\bar{\sigma}_{xx}$ is also a function of ϱ, but since it represents a small correction to the pressure, we can put $\varrho = \varrho_0(p, T)$.

CHAPTER VII

THEORY OF SUPERCONDUCTIVITY

§ 32. GENERAL INTRODUCTION. CHOICE OF MODEL

1. Superconductivity

The problem of superconductivity is one of the most important and difficult presented to quantum statistics. It is well known that many metals undergo a phase change to a new "superconducting" state at sufficiently low temperatures. In this state the metal has thermodynamic and electromagnetic properties that are quite different from those in the normal state. Perhaps the most striking way in which the transition manifests itself experimentally is that the metal, on cooling to a critical temperature, suddenly ceases to present a resistance to electric current. In other words, there is no dissipation of energy when a current flows in a superconductor.

Experimental investigations have shown that the properties of a superconductor in a magnetic field are very different from the relatively simple properties of the normal metal. The magnetic field does not penetrate into the body of the superconductor (the Meissner–Ochsenfeld effect). The effective depth, measured from the surface of a superconductor located in a constant magnetic field, at which the field is still different from zero (the so-called penetration depth) is extremely small, of the order of 10^{-5} to 10^{-6} cm. The thermodynamic transition from the normal to the superconducting state is a phase transition of the second kind and is characterized by a discontinuity in the specific heat at the transition temperature.

A substantial advance in the understanding of this phenomenon has been achieved in recent years. It has been found that the development of a theory of superconductivity requires a wide use of the methods of quantum field theory. These methods will be treated in subsequent sections; in the present section, we shall dwell mainly on the physical side of the subject.

It has been clear for a long time that superconductivity is somewhat akin to superfluidity. This is evident, first of all, from the fact that the maintainance of an electric current in a superconductor does not require an external potential difference, i.e. does not require work from external sources. The current carriers in a metal are electrons; we are therefore concerned with none other than superfluidity of the electron liquid.

In Chap. I, in reference to the superfluidity of helium, we dwelt in detail on the properties of the energy spectrum of the excitations required for producing superfluidity. It must be remarked right away, however, that, at small momenta, the spectrum for a superconductor cannot be of the same type as that which we have to associate with liquid helium. For helium has a phonon acoustic branch as the initial part of its spectrum. As is well known, the propagation of sound is connected with long-wavelength density oscillations. But, for an electron liquid in a metal, a change in density involves a fairly substantial amount of energy, since it is hindered by the Coulomb forces acting between the electrons and the lattice, and between the electrons themselves. A change in the density of the electron liquid destroys the condition of electric neutrality, so that the corresponding spectrum of the long-wavelength oscillations starts at some finite frequency, just as in the case of a plasma. This frequency is in fact extremely high in a metal (~ 1 eV $\simeq 10^4$ °K). Our remarks naturally do not apply to the short-wavelength excitations, with a wave vector of the order of the reciprocal of an inter-atomic distance. We know that it is precisely these electron excitations that play the main role in a normal metal. In accordance with the results of Chap. I, a sufficient condition for the existence of superfluidity is that such excitations be separated by a gap from the ground state, i.e. that the spectrum has the shape illustrated in Fig. 88. It may be remarked that, apart from the arguments adduced above, the presence of a spectrum of this type in superconductors has been indicated by experimental data on the electronic specific heat at low temperatures; these data lead to a temperature dependence of the specific heat of the form $e^{-\Delta/T}$.

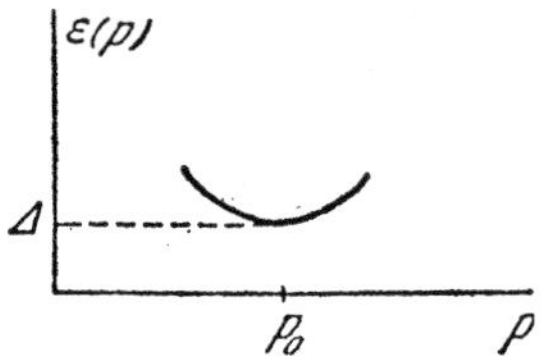

Fig. 88

We shall not dwell here on a treatment of various phenomenological theories; though they often give an adequate description of experimental data, they cannot provide an explanation of the microscopic mechanism involved.

The isotopic effect discovered in 1950 [55] provided a key to an understanding of the relative roles of the different interactions in a metal when it becomes superconducting. The critical temperature T_c (the temperature at which the transition from the normal to the superconducting state occurs) was found to be a function of the mass of an isotope, of the form $T_c \sim (M)^{-1/2}$. Fröhlich [56] proposed independently that the main interaction corresponding to superconductivity must be of electrons with phonons. This interaction involves a strong dependence on the mass of the ions.

2. *Model. Interaction Hamiltonian*

The interaction Hamiltonian has already been discussed in Chap. II, and we know that its form is

$$H_{int}(x) = g\psi^+(x)\psi(x)\varphi(x). \tag{32.1}$$

Let us find the matrix element of the scattering of two electrons by one another, in the process of which the electrons exchange one phonon. This process is illustrated schematically in Fig. 89. The dotted line represents the exchange of a phonon, which corresponds in the matrix element to the phonon D-function,

$$g^2 D(\varepsilon_3 - \varepsilon_1; \boldsymbol{p}_3 - \boldsymbol{p}_1) = g^2 \frac{u^2(\boldsymbol{p}_3 - \boldsymbol{p}_1)^2}{(\varepsilon_3 - \varepsilon_1)^2 - u^2(\boldsymbol{p}_3 - \boldsymbol{p}_1)^2},$$

where $\varepsilon_3 - \varepsilon_1$, $\boldsymbol{p}_3 - \boldsymbol{p}_1$ are respectively the changes in the energy and momentum of one of the electrons on collision. Close to the Fermi surface the momentum change on collision is in general of the order p_0 (i.e. $u\,|\boldsymbol{p}_3 - \boldsymbol{p}_1|$ is of the order of the Debye frequency ω_D, since $p_0 \sim a^{-1}$), so that the change in the energy of the electrons may be fairly small. In this region, i.e. when $|\varepsilon_3 - \varepsilon_1| \ll \omega_D$, the effective interaction, which is determined by the foregoing expression, simply reduces to the constant $-g^2$, i.e. there is an attraction.

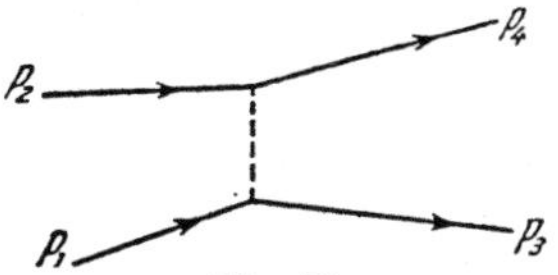

Fig. 89

L. Cooper [57] discovered in 1957 that the effective attraction between electrons close to the Fermi surface, resulting from an arbitrarily weak electron-phonon interaction, necessarily leads to the formation of bound electron pairs. Since binding involves an increase in energy, a readjustment of the ground state of the system must result when interaction takes place. Excitation of such a system requires the expenditure of a finite energy, equal to the binding energy of the pair, which will in fact play the role of a gap in the excitation spectrum. It proved possible, on the basis of this idea, to construct a complete theory of superconductivity, explaining the vast mass of facts accumulated in several decades of intensive study of the phenomenon.

The basis chosen for our statement of the theory is not the same as in the original statements (Bardeen, Cooper and Schrieffer [58], Bogolyubov [59](†)), since it seems to us that the methods of quantum field theory offer substantial advantages. Apart from its simplicity and harmony, the approach described below enables a number of important new results to be obtained.

Before proceeding further, let us remark that the electron-phonon interaction is not the only one for electrons in a metal. Repulsive Coulomb forces also operate between the electrons. The effective interaction between them will therefore be either an attraction or a repulsion, depending on

(†) See also [60].

the ratio of the magnitudes of the electron-phonon attraction and the Coulomb repulsion of the electrons. In general, the problem of taking into account both interactions for actual metals is extremely difficult. In addition to this, actual superconductors are anisotropic. A proviso must therefore be made regarding the present-day theory of superconductivity, that in essence it considers a simple model with a quadratic dispersion law for the electrons, in which it is postulated in advance that the interaction of the electrons has the nature of an attraction in a narrow band of their energies close to the Fermi surface. This energy region is obviously of the order of the maximum energy of the emitted phonons, i.e. $\sim\omega_D$, where ω_D is the Debye frequency. In addition, we shall assume below, for simplicity, that the interaction is constant in this region and is fairly small.

To date, no theory of superconductivity has been developed on the basis of the Fermi liquid concept and which also takes the anisotropic properties into account. Nevertheless, it is interesting to note that, in spite of the crudity of the model, the theory not only explains the phenomena qualitatively, but also leads to good quantitative agreement with the available experimental data.

We write down the effective interaction Hamiltonian of the electrons with one another in the second quantization representation as follows:

$$H_{int} = \frac{\lambda}{2(2\pi)^3} \sum_{p_1+p_2=p_3+p_4} a^+_{p_1\sigma_1} a^+_{p_2\sigma_2} a_{p_3\sigma_2} a_{p_4\sigma_1} \theta_{p_1} \theta_{p_2} \theta_{p_3} \theta_{p_4}, \tag{32.2}$$

where $\lambda < 0$, and θ_p are the cut-off factors:

$$\theta_p = \begin{cases} 1, & |\varepsilon(p) - \varepsilon_F| < \omega_D, \\ 0, & |\varepsilon(p) - \varepsilon_F| > \omega_D. \end{cases}$$

The presence of these factors denotes that the only electrons taking part in the interaction are those with energies in a narrow band of width $2\omega_D$ close to the Fermi surface ($\omega_D \ll \varepsilon_F$). This Hamiltonian will often be written below in terms of the operators $\psi_\alpha(r)$ and $\psi^+_\beta(r)$ in the coordinate representation

$$H_{int} = \frac{\lambda}{2} \int \psi^+_\alpha(r)\psi^+_\beta(r)\psi_\beta(r)\psi_\alpha(r)\, d^3r. \tag{32.3}$$

It should be understood, of course, that the values of the four arguments of the ψ-operators in (32.3) are in fact somewhat different. This last is connected with the presence of the factors θ_p in expression (32.2) for the Hamiltonian. It would be more exact to write, instead of (32.3):

$$H_{int} = \frac{\lambda}{2} \int\int\int\int\int \theta(r-\xi_1)\theta(r-\xi_2)\,\theta(r-\xi_3)\,\theta(r-\xi_4)$$
$$\times \psi^+_\alpha(\xi_1)\psi^+_\beta(\xi_2)\psi_\beta(\xi_3)\psi_\alpha(\xi_4)\, d^3r\, d^3\xi_1 \cdots d^3\xi_4, \tag{32.4}$$

where $\theta(\boldsymbol{x})$ is the Fourier transform of θ_p:

$$\theta(\boldsymbol{x}) = \frac{1}{(2\pi)^3} \int e^{i(\boldsymbol{p}\cdot\boldsymbol{x})} \theta_{\boldsymbol{p}} \mathrm{d}^3\boldsymbol{p}. \tag{32.5}$$

It is easily verified by Fourier transforming that the functions $\theta(\boldsymbol{x})$ have δ-function characteristics:

$$\int \theta(\boldsymbol{x}-\boldsymbol{y}) f(\boldsymbol{y}) \mathrm{d}^3\boldsymbol{y} = f(\boldsymbol{x}),$$

if the function $f(\boldsymbol{x})$ has non-zero Fourier components $f_{\boldsymbol{p}}$ only for momenta $\boldsymbol{p}$ close to the Fermi surface. It is precisely with these functions that we shall be concerned in the theory described below. Expression (32.3) is to be understood in this sense.

§ 33. COOPER PHENOMENON. INSTABILITY OF THE GROUND STATE OF A SYSTEM OF NON-INTERACTING FERMIONS WITH RESPECT TO ARBITRARILY WEAK ATTRACTIONS BETWEEN THE PARTICLES

1. Equation for the vertex part

We consider the properties of the system with the interaction (32.3). For this, we return to a study of the vertex part $\Gamma_{\alpha\beta,\gamma\delta}(p_1, p_2; p_3, p_4)$ at absolute zero. Let us write down the perturbation theory series for this quantity. The vertex part is, to a first approximation:

$$\lambda(\delta_{\alpha\gamma}\delta_{\beta\delta} - \delta_{\alpha\delta}\delta_{\gamma\beta}). \tag{33.1}$$

The diagrams corresponding to the first terms of the perturbation theory series are illustrated in Fig. 90. As we know, singularities in the vertex part of "zero sound" type are connected with diagrams (*a*) and (*c*), i.e. these singularities are important for small momentum transfer. The diagrams of type (*b*) are connected with singularities in $\Gamma_{\alpha\beta,\gamma\delta}(p_1, p_2; p_3, p_4)$ at small values of the total 4-momentum $q = p_1 + p_2$. Let us investigate the last case in more detail. By using the concrete properties of the model in question, we can obtain more detailed information on the vertex part in the region of small q in comparison with the general results of (20.8).

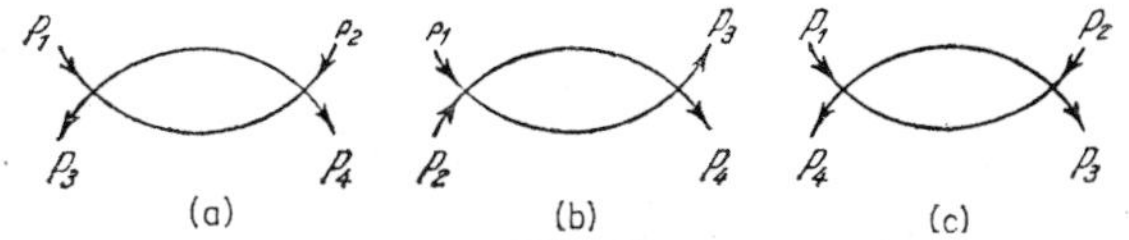

Fig. 90

The matrix element for the diagram of Fig. 90*b* is equal to

$$\lambda^2 \frac{i}{(2\pi)^4} (\delta_{\alpha\gamma}\delta_{\beta\delta} - \delta_{\alpha\delta}\delta_{\gamma\beta}) \int \mathrm{d}^4k G(k) G(q-k),$$

where $q = \{\omega_0, \boldsymbol{q}\} = \{\omega_1 + \omega_2, \boldsymbol{p}_1 + \boldsymbol{p}_2\}$. On substituting in this the expressions for the Green functions and integrating over the frequencies, we get

$$\begin{gathered}\frac{\lambda^2}{(2\pi)^3}(\delta_{\alpha\gamma}\delta_{\beta\delta} - \delta_{\alpha\delta}\delta_{\beta\gamma}) \int \frac{\mathrm{d}^3k}{\omega_0 - \varepsilon_0(\boldsymbol{k}) - \varepsilon_0(\boldsymbol{q}-\boldsymbol{k}) + 2\mu + i\delta} \\ (\varepsilon_0(\boldsymbol{k}) > \mu,\ \varepsilon_0(\boldsymbol{q}-\boldsymbol{k}) > \mu), \\ -\frac{\lambda^2}{(2\pi)^3}(\delta_{\alpha\gamma}\delta_{\beta\delta} - \delta_{\alpha\delta}\delta_{\beta\gamma}) \int \frac{\mathrm{d}^3k}{\omega_0 - \varepsilon_0(\boldsymbol{k}) - \varepsilon_0(\boldsymbol{q}-\boldsymbol{k}) + 2\mu - i\delta} \\ (\varepsilon_0(\boldsymbol{k}) < \mu,\ \varepsilon_0(\boldsymbol{q}-\boldsymbol{k}) < \mu).\end{gathered} \tag{33.2}$$

In the model which we are discussing, the only electrons that interact are those in a narrow region of energies close to the Fermi energy $\varepsilon_F \simeq \mu$. The integration over $\boldsymbol{k}$ in integrals (33.2) is therefore limited by the conditions $|\varepsilon_0(\boldsymbol{k}) - \mu|$, $|\varepsilon_0(\boldsymbol{q}-\boldsymbol{k}) - \mu| < \omega_D$. On putting ω_0, $|\boldsymbol{q}|v \ll \omega_D$, we change in the ordinary way to an integration over $\xi = v(|\boldsymbol{k}| - p_0)$. If we also neglect in the integrals a change in the upper limit of the order ω_0, or $|\boldsymbol{q}|v$, we can transform (33.2) as follows:

$$-\frac{\lambda^2 m p_0}{2\pi^2}(\delta_{\alpha\gamma}\delta_{\beta\delta} - \delta_{\alpha\delta}\delta_{\beta\gamma}) \int\limits_0^{\omega_D} \mathrm{d}\xi \times \int\limits_0^1 \left[\frac{1}{\omega_0 + 2\xi + v|\boldsymbol{q}|x - i\delta} + \frac{1}{2\xi + v|\boldsymbol{q}|x - \omega_0 - i\delta}\right] \mathrm{d}x$$

(where $x = \cos\theta$, θ is the angle between the directions of the vectors $\boldsymbol{q}$ and $\boldsymbol{k}$). The remaining integrations are performed by elementary methods. Having chosen the branches of the logarithms from the condition that the integral of the first term in the square brackets be positive for $\omega_0 > 0$, and of the second, for $\omega_0 < 0$, we get the following expression for the diagram:

$$\begin{aligned}-\frac{\lambda^2 m p_0}{2\pi^2}(\delta_{\alpha\gamma}\delta_{\beta\delta} - \delta_{\alpha\delta}\delta_{\beta\gamma}) \Bigg[1 &+ \frac{1}{2}\ln\frac{2\omega_D - i\delta}{\omega_0 + v|\boldsymbol{q}| - i\delta} \\ &+ \frac{1}{2}\ln\frac{2\omega_D - i\delta}{-\omega_0 + v|\boldsymbol{q}| - i\delta} \\ &+ \frac{\omega_0}{2v|\boldsymbol{q}|}\left(\ln\frac{\omega_0 - i\delta}{\omega_0 + v|\boldsymbol{q}| - i\delta} + \ln\frac{v|\boldsymbol{q}| - \omega_0 - i\delta}{-\omega_0 - i\delta}\right)\Bigg].\end{aligned} \tag{33.3}$$

The principal term in this expression becomes, for small ω_0 and $v|\boldsymbol{q}|$:

$$-\lambda^2\frac{m p_0}{2\pi^2}(\delta_{\alpha\gamma}\delta_{\beta\delta} - \delta_{\alpha\delta}\delta_{\beta\gamma}) \ln\frac{\omega_D}{\max\{\omega_0, v|\boldsymbol{q}|\}}.$$

Thus, when $\omega_D \gg \omega_0$, $v|\boldsymbol{q}|$, the smallness of the interaction constant λ can be compensated by the large value of the logarithm, as a result of which this term becomes of the same order as the first term of the perturbation theory (33.1). In order to find the vertex part in the neighbourhood of small ω_0 and $v|\boldsymbol{q}|$, when $\lambda \ln \omega_D/\max\{\omega_0, v|\boldsymbol{q}|\} \sim 1$, we must therefore sum the set of principal terms of the perturbation theory series, as in Chap. IV.

For this purpose, we write the equation for the vertex part in a form in which the terms leading to singularities of $\Gamma_{\alpha\beta,\gamma\delta}(p_1, p_2; p_3, p_4)$ for small $q = p_1 + p_2$ are distinguished:

$$\Gamma_{\alpha\beta,\gamma\delta}(p_1, p_2; p_3, p_4) = \tilde{\Gamma}_{\alpha\beta,\gamma\delta}(p_1, p_2; p_3, p_4) \tag{33.4}$$
$$+ \frac{i}{2(2\pi)^4} \int \tilde{\Gamma}_{\alpha\beta,\xi\eta}(p_1, p_2; k, q-k) G(k) G(q-k) \Gamma_{\xi\eta,\gamma\delta}(k, q-k; p_3, p_4) \mathrm{d}^4k.$$

In this equation $\tilde{\Gamma}_{\alpha\beta,\gamma\delta}(p_1, p_2; p_3, p_4)$ is the sum of all the matrix elements whose diagrams are irreducible in the sense that interests us, i.e. cannot be split into two parts, one containing only incoming and the other only outgoing exterior ends and connected by two electron lines. The kernel of integral equation (33.4) contains a large logarithmic term from the integration of the two Green functions. In view of the smallness of the interaction constant, it is sufficient for us to take for $\tilde{\Gamma}$ its expression in terms of the first terms of the perturbation theory, since the corresponding expressions for $\tilde{\Gamma}$ do not contain large quantities. We have made it our aim, however, to calculate the kernel of equation (33.4) without confining ourselves to terms of order $\lambda \ln \omega_D/\omega_0 \sim 1$; we try to find an expression for it up to and including terms of order λ. It would therefore seem at first sight that we need to know $\tilde{\Gamma}$ up to terms of order λ^2 in the perturbation theory, since the logarithmic integration in (33.4) can compensate one of the orders of λ. Let us consider terms of the second order of perturbation theory for $\tilde{\Gamma}$. The relevant diagrams are illustrated in Fig. 91. Let us find the order of the matrix element corresponding to, say, diagram (*a*). Omitting numerical coefficients, we get

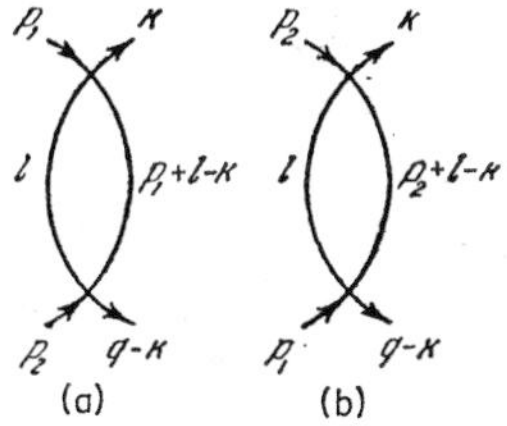

Fig. 91

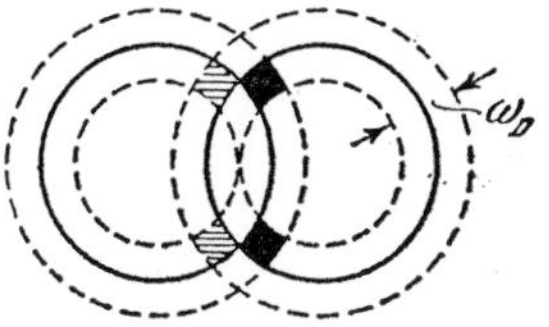

Fig. 92

$$\lambda^2 \int G(l) G(l - k + p_1) \mathrm{d}^4 l.$$

On substituting the expressions for the Green functions, we can integrate over the frequency:

$$\lambda^2 \int \frac{\mathrm{d}^3 \boldsymbol{l}}{\omega_k - \omega_1 + \varepsilon_0(\boldsymbol{l} - \boldsymbol{k} + \boldsymbol{p}_1) - \varepsilon_0(\boldsymbol{l})} \quad (\varepsilon_0(\boldsymbol{l}) > \mu;\ \varepsilon_0(\boldsymbol{l} - \boldsymbol{k} + \boldsymbol{p}_1) < \mu),$$
$$-\lambda^2 \int \frac{\mathrm{d}^3 \boldsymbol{l}}{\omega_k - \omega_1 + \varepsilon_0(\boldsymbol{l} - \boldsymbol{k} + \boldsymbol{p}_1) - \varepsilon_0(\boldsymbol{l})} \quad (\varepsilon_0(\boldsymbol{l}) < \mu;\ \varepsilon_0(\boldsymbol{l} - \boldsymbol{k} + \boldsymbol{p}_1) > \mu). \tag{33.5}$$

The domain of integration over $\boldsymbol{l}$ is in fact much narrower than if it were determined only by the last conditions. This is connected with the properties of the model, in which only electrons with momenta in the neighbourhood of the Fermi momentum $|v(|\boldsymbol{p}|-p_0)|<\omega_D$ can interact. The actual domain of integration is shaded in Fig. 92 for the first condition in (33.5) and shown in black for the second In both cases the expression $\varepsilon(\boldsymbol{l}-\boldsymbol{k}+\boldsymbol{p}_1)-\varepsilon(\boldsymbol{l})$ in the denominator of the integrand is equal to ω_D in order of magnitude in the domain in question, whereas the volume of the domain over which the integration is carried out is $\sim m^2\omega_D^2/p_0$. The matrix element for the diagrams of Fig. 91 is therefore of order $\lambda^2 m^2\omega_D/p_0$, i.e. their relative order compared with the simple vertex is $(\lambda m p_0)\omega_D/\varepsilon_F$. (As is clear from (33.3), the quantity $\lambda m p_0 \ll 1$ is a small dimensionless parameter in our model.) Since, by its physical meaning, $\omega_D \ll \varepsilon_F$, this extra order of smallness cannot be compensated in the domain considered by the large value of the logarithm. In view of this, we can confine ourselves in equation (33.4) for $\tilde{\Gamma}$ to the simple first order vertex of perturbation theory (33.1).

The equation obtained for the vertex part can now be solved easily. We notice that, as is clear from (33.3), $\Gamma_{\alpha\beta,\gamma\delta}(p_1, p_2; p_3, p_4)$ depends only on the sum of the variables $q=p_1+p_2$. Hence the integral on the right-hand side of (33.4) reduces to the integral already calculated by us for the matrix element of the second order of perturbation theory (33.3). We obtain as a result $(\omega_0 > v\,|\boldsymbol{q}|)$:

$$\Gamma_{\alpha\beta,\gamma\delta}(p_1, p_2; p_3, p_4) \equiv \Gamma(q)\,(\delta_{\alpha\gamma}\delta_{\beta\delta}-\delta_{\alpha\delta}\delta_{\beta\gamma}),$$

$$\Gamma(q)=\lambda\left\{1+\left(\frac{\lambda m p_0}{2\pi^2}\right)\left[1+\ln\left|\frac{2\omega_D}{\omega_0}\right|+\frac{\pi i}{2}+\frac{1}{2}\ln\left|\frac{\omega_0^2}{\omega_0^2-v^2|\boldsymbol{q}|^2}\right|\right.\right.$$
$$\left.\left.+\frac{\omega_0}{2v|\boldsymbol{q}|}\ln\left|\frac{\omega_0-v|\boldsymbol{q}|}{\omega_0+v|\boldsymbol{q}|}\right|\right]\right\}^{-1}. \tag{33.6}$$

2. *Properties of the vertex part*

Let us first consider for simplicity (33.6) with $\boldsymbol{q}=0$. We have for real and positive ω_0:

$$\Gamma(\omega_0)=\frac{\lambda}{1+\left(\dfrac{\lambda m p_0}{2\pi^2}\right)\left[\ln\left|\dfrac{2\omega_D}{\omega_0}\right|+\dfrac{\pi i}{2}\right]}. \tag{33.7}$$

We shall now regard $\Gamma(\omega_0)$ as a function of the complex variable ω_0, defining it as the analytic continuation of (33.7) in the upper half-plane $\operatorname{Im}\omega_0>0$. We now get

$$\Gamma(\omega_0)=\frac{\lambda}{1+\left(\dfrac{\lambda m p_0}{2\pi^2}\right)\left[\ln\left|\dfrac{2\omega_D}{\omega_0}\right|+\dfrac{\pi i}{2}-i\varphi\right]}.$$

Hence, if the interaction is an attraction ($\lambda < 0$), $\Gamma(\omega_0)$ has a pole at the point $\omega_0 = i\Omega$, where

$$\Omega = 2\omega_D \exp[-2\pi^2/|\lambda| m p_0]. \tag{33.8}$$

In the neighbourhood of the pole $\Gamma(\omega_0)$ has the form

$$\Gamma(\omega_0) = -\frac{2\pi^2}{m p_0} \frac{i\Omega}{\omega_0 - i\Omega}.$$

This result has to be connected with the above-mentioned idea of Cooper about the formation of bound-electron pairs. The vertex part $\Gamma_{\alpha\beta,\gamma\delta}(p_1, p_2; p_3, p_4)$ is defined in terms of the Fourier components of the two-particle Green function by the relation (10.17). Thus the fact that Γ has a pole means that the two-particle Green function has the same pole. The formation of bound pairs implies the instability of the ground state of the gas of interacting fermions from which we started out. The imposition of arbitrarily weak forces of attraction between the particles implies a readjustment of the entire system. The existence of an instability finds its reflexion in the appearance of poles in the vertex function with respect to the variable $\omega_0 = \omega_1 + \omega_2$ in the upper half-plane. This pole, being pure imaginary, determines the relaxation time of the unstable ground state. By the principle of indeterminacy, this time corresponds to the binding energy of the actual pair. In the new ground state the pairs behave as Bose formations and, as is the case for bosons, are capable of any degree of accumulation at the level of least energy. In the superconducting state these pairs are at the zero momentum level of the motion of the pair as a whole, in complete analogy with what occurs on the "Bose condensation" of ordinary bosons.

When $v|\boldsymbol{q}|$ is non-zero, (33.6) can be written as ($\omega_0 > v|\boldsymbol{q}|$):

$$\Gamma(\boldsymbol{q}, \omega_0) = \lambda \left\{1 + \left(\frac{\lambda m p_0}{2\pi^2}\right)\left[1 + \ln\left|\frac{2\omega_D}{\omega_0}\right| + \frac{i\pi}{2} - \frac{1}{2}\ln\left(1 - \frac{v^2|\boldsymbol{q}|^2}{\omega_0^2}\right) + \frac{\omega_0}{2v|\boldsymbol{q}|}\ln\left(\frac{\omega_0 - v|\boldsymbol{q}|}{\omega_0 + v|\boldsymbol{q}|}\right)\right]\right\}^{-1}.$$

After continuation into the half-plane $\operatorname{Im}\omega_0 > 0$ and use of definition (33.8) for Ω, we get

$$\Gamma(\boldsymbol{q}, \omega_0) = -\frac{2\pi^2}{m p_0}\left\{\ln\frac{\omega_0}{i\Omega} - 1 + \frac{1}{2}\ln\left(1 - \frac{v^2|\boldsymbol{q}|^2}{\omega_0^2}\right) - \frac{\omega_0}{2v|\boldsymbol{q}|}\ln\left(\frac{\omega_0 - v|\boldsymbol{q}|}{\omega_0 + v|\boldsymbol{q}|}\right)\right\}^{-1}. \tag{33.9}$$

With small $v|\boldsymbol{q}| \ll \Omega$,

$$\Gamma(\boldsymbol{q}, \omega_0) = -\frac{2\pi^2}{m p_0} \frac{i\Omega}{\omega_0 - i\Omega + i(v^2|\boldsymbol{q}|^2/6\Omega)}.$$

Equating to zero the denominator of (33.9) determines the pole of $\Gamma(\boldsymbol{q}, \omega_0)$ as a functions of $|\boldsymbol{q}|$. At small $|\boldsymbol{q}|$,

$$\omega_0 = i\Omega\left(1 - \frac{v^2|\boldsymbol{q}|^2}{6\Omega^2}\right),$$

i.e. the absolute value of ω_0 has decreased. At some value, $v|\boldsymbol{q}|_{\max}$, the pole ω_0 becomes zero, after which for greater $v|\boldsymbol{q}|$ there is no pole of Γ. We can easily find the value of $v|\boldsymbol{q}|_{\max}$ for which $\omega_0 = 0$:

$$v|\boldsymbol{q}|_{\max} = e\Omega. \tag{33.10}$$

Since $\boldsymbol{q}$ is the momentum of the system of two particles as a whole, this result means that only those electrons which are moving almost towards one another reveal a tendency to the formation of bound pairs.

3. *Determination of the transition temperature*

Let us remark once more that the arguments adduced throw light on the instability of the ordinary ground state of a system of "attracting" particles at low temperatures. This instability amounts to an ability of two particles whose centre of mass is almost at rest to form bound pairs, i.e. a type of bosons, which "condense" in the lowest level.

The temperature at which such an instability first appears will be the transition temperature of the metal from the normal to the superconducting state.

To determine this, we can make use of the above-mentioned analogy with a boson gas. In the approximation in which we neglect the scattering of particles by one another (the weak interaction model), the bound pairs form an ideal gas. We know that the temperature-dependent Green function of an ideal boson gas is

$$\mathfrak{G}(\boldsymbol{q}, i\omega_n) = \left[i\omega_n - \frac{q^2}{2m} + \mu\right]^{-1}$$

and represents the values at the points $\omega = i\omega_n = i2n\pi T$ of a function $G^R(\boldsymbol{q}, \omega)$, analytic in the upper half-plane of ω. This function is equal to $[\mu - (q^2/2m)]^{-1}$ at $\omega_n = 0$. The latter first becomes infinite at the point $\boldsymbol{q} = 0$ at some temperature $T = T_0$, called the temperature of the "Bose condensation". The temperature T_0 is determined by the condition $\mu = 0$.

The analogue of the boson Green function is, for a bound pair, the two-particle fermion Green function (16.5). At the transition point, this latter must have analogous properties in the sense of its dependence on the variables $\omega_{0n} = (\omega_1 + \omega_2)_n$ and $\boldsymbol{q} = \boldsymbol{p}_1 + \boldsymbol{p}_2$, corresponding to the centre of mass of the pair. The fermion Green functions in (16.5) have no singularities with respect to these variables. We therefore consider the vertex part $\Gamma_R(\boldsymbol{q}, \omega_0)$ (we shall omit the spin indices everywhere; the meaning of $\boldsymbol{q}$ and ω_0 is indicated above) and define it as an analytic function in the upper half-plane of ω_0, coinciding at the points $\omega_0 = i(\omega_1 + \omega_2)_n$ with the vertex part in the thermodynamic method. In other words, the function $\Gamma_R(\boldsymbol{q}, \omega_0)$ is the analytic continuation of the thermodynamic vertex part $\mathcal{T}_{\alpha\beta,\gamma\delta}(p_1\omega_1, p_2\omega_2;\ p_3\omega_3, p_4\omega_4) \equiv \mathcal{T}(\boldsymbol{q}, \omega_0)(\delta_{\alpha\gamma}\delta_{\beta\delta} - \delta_{\alpha\delta}\delta_{\beta\gamma})$ (we shall show below that, in the approximation of interest to us, this function, like (33.9), actually depends only on the variables $\boldsymbol{q}$ and ω_0). We assume at the basis of the above that, at temperatures below the transition temperature, $\Gamma_R(\boldsymbol{q}, \omega_0)$ has poles $\operatorname{Im}\omega_0 > 0$. At the transition temperature, a pole $\omega_0 = 0$ first makes its appearance in the function $\Gamma_R(\boldsymbol{q}, \omega_0)$.

The necessary equation for the thermodynamic vertex part has the same structure as (33.4):

$$\begin{aligned}\mathcal{T}_{\alpha\beta,\gamma\delta}(\boldsymbol{p}_1\omega_1, \boldsymbol{p}_2\omega_2; \boldsymbol{p}_3\omega_3, \boldsymbol{p}_4\omega_4) &= \tilde{\mathcal{T}}_{\alpha\beta,\gamma\delta}(\boldsymbol{p}_1\omega_1, \boldsymbol{p}_2\omega_2; \boldsymbol{p}_3\omega_3, \boldsymbol{p}_4\omega_4)\\ &- \frac{T}{2(2\pi)^3}\sum_{\omega'}\int \tilde{\mathcal{T}}_{\alpha\beta,\xi\eta}(\boldsymbol{p}_1\omega_1, \boldsymbol{p}_2\omega_2; \boldsymbol{k}\omega'; \boldsymbol{q}-\boldsymbol{k}, \omega_0-\omega')\\ &\times \mathfrak{G}(k)\mathfrak{G}(q-k)\mathcal{T}_{\xi\eta,\gamma\delta}(\boldsymbol{k}\omega', \boldsymbol{q}-\boldsymbol{k}, \omega_0-\omega'; \boldsymbol{p}_3\omega_3, \boldsymbol{p}_4\omega_4)\,\mathrm{d}^3\boldsymbol{k},\end{aligned} \tag{33.11}$$

where $\tilde{\mathcal{T}}$ again is the sum of the matrix elements for all the diagrams which cannot be split by a vertical line into two parts, joined by two lines in the same direction. For the same reasons as above, we can confine ourselves to the first approximation of perturbation theory for $\tilde{\mathcal{T}}_{\alpha\beta,\gamma\delta}(\boldsymbol{p}_1\omega_1, \boldsymbol{p}_2\omega_2; \boldsymbol{p}_3\omega_3, \boldsymbol{p}_4\omega_4)$. The problem of finding the vertex part now reduces to finding the sum and the integral in the matrix element:

$$-\frac{\lambda^2}{(2\pi)^3}T\sum_{\omega'}\int \mathfrak{G}(k)\mathfrak{G}(q-k)\,\mathrm{d}^3\boldsymbol{k}. \tag{33.12}$$

On substituting (14.6) for the Green functions in this, we can easily perform the elementary summation over the frequencies.

We shall not evaluate (33.12) for arbitrary values of ω_0 and $|\boldsymbol{q}|$. It is clear from uniformity considerations that, as in a Bose gas, a pole first makes an appearance in $\Gamma_R(\boldsymbol{q}, \omega_0)$ for values $\omega_0 = |\boldsymbol{q}| = 0$. It is therefore sufficient to find the solution of equation (33.11) for $|\boldsymbol{q}| = \omega_0 = 0$. The instant at which this quantity becomes infinite determines the transition temperature from the normal to the superconducting state. When $|\boldsymbol{q}| = \omega_0 = 0$, integral (33.12) can be transformed to

$$-\frac{\lambda^2}{2\pi^2}mp_0\int_0^{\omega_D}\tanh\left(\frac{\xi}{2T}\right)\frac{\mathrm{d}\xi}{\xi} = -\lambda^2\frac{mp_0}{2\pi^2}\left(\ln\frac{\omega_D}{2T} - \int_0^\infty\frac{\ln x\,\mathrm{d}x}{\cosh^2 x}\right) \tag{33.13}$$

(after integration by parts, since the remaining integral is convergent, the limit $x = \omega_D/2T$ can be replaced by infinity. The integral is equal to $-\ln(2\gamma/\pi)$, where $\ln\gamma = C = 0{\cdot}577$). Hence

$$\Gamma_R(0, 0) = \mathcal{T}(0, 0) = \frac{\lambda}{1 + \dfrac{\lambda m p_0}{2\pi^2}\ln\left(\dfrac{2\omega_D\gamma}{\pi T}\right)}.$$

Close to the transition temperature, this expression can be written as

$$\mathcal{T}(0,0) = -\frac{2\pi^2}{mp_0}\frac{T_c}{T-T_c}, \tag{33.14}$$

where the transition temperature T_c is equal to

$$T_c = \frac{\gamma}{\pi} 2\omega_D \exp\left[-(2\pi^2/|\lambda| mp_0)\right]. \tag{33.15}$$

The value of the frequency Ω, introduced above and characterising the instability of the system at absolute zero, is equal to

$$\Omega = \frac{\pi}{\gamma} T_c. \tag{33.16}$$

§ 34. SYSTEM OF FUNDAMENTAL EQUATIONS FOR A SUPERCONDUCTOR

1. *Superconductor at absolute zero*

We now proceed to the derivation of a system of equations for the Green functions describing the properties of a metal in the superconducting state (Gor'kov [61]). We shall confine ourselves for a start to the case of absolute zero. In our model, the complete Hamiltonian of the system of electrons in the second quantisation representation has the form

$$\hat{H} = \int \left\{ -\left(\psi^+ \frac{\nabla^2}{2m}\psi\right) + \frac{\lambda}{2}\left(\psi^+(\psi^+\psi)\psi\right)\right\} d^3\boldsymbol{r},$$

where $(\psi^+\psi) \equiv \psi^+_\alpha \psi_\alpha$ and the operators $\psi(\boldsymbol{r})$, $\psi^+(\boldsymbol{r})$ in the Schrödinger representation satisfy the usual commutation relations

$$\begin{aligned} \{\psi_\alpha(\boldsymbol{r}), \psi^+_\beta(\boldsymbol{r}')\} &= \delta_{\alpha\beta}\delta(\boldsymbol{r}-\boldsymbol{r}'), \\ \{\psi_\alpha(\boldsymbol{r}), \psi_\beta(\boldsymbol{r}')\} &= \{\psi^+_\alpha(\boldsymbol{r}), \psi^+_\beta(\boldsymbol{r}')\} = 0. \end{aligned} \tag{34.1}$$

We change to the Heisenberg representation, in which the operators $\tilde{\psi}$, $\tilde{\psi}^+$ depend on the time and obey the following operator equations:

$$\begin{aligned} \left\{i\frac{\partial}{\partial t} + \frac{\nabla^2}{2m}\right\}\tilde{\psi}_\alpha(x) - \lambda\left(\tilde{\psi}^+(x)\tilde{\psi}(x)\right)\tilde{\psi}_\alpha(x) &= 0, \\ \left\{i\frac{\partial}{\partial t} - \frac{\nabla^2}{2m}\right\}\tilde{\psi}^+_\alpha(x) + \lambda\tilde{\psi}^+_\alpha(x)\left(\tilde{\psi}^+(x)\tilde{\psi}(x)\right) &= 0. \end{aligned} \tag{34.2}$$

The equation for the Green function of the system

$$G_{\alpha\beta}(x,x') = -i\langle T\left(\tilde{\psi}_\alpha(x)\tilde{\psi}^+_\beta(x')\right)\rangle$$

can be obtained in an obvious way from equations (34.2):

$$\left\{i\frac{\partial}{\partial t} + \frac{\nabla^2}{2m}\right\}G_{\alpha\beta}(x,x') + i\lambda\langle T([\tilde{\psi}^+(x)\tilde{\psi}(x)]\tilde{\psi}_\alpha(x), \tilde{\psi}^+_\beta(x'))\rangle = \delta(x-x'). \tag{34.3}$$

This equation contains the mean of the product of four $\tilde{\psi}$-operators for a system of non-interacting electrons, which can be expanded by Wick's theorem into averages of pairs of operators $\tilde{\psi}$, $\tilde{\psi}^+$. The product of four $\tilde{\psi}$-operators for interacting particles is already expressible in terms of the vertex part, i.e. includes the contribution of the different scattering processes. In our weak interaction model, the scattering of different particles by one another can be neglected. At the same time we have to take into account that the ground state of the system differs from the ordinary state with a filled Fermi sphere by the presence of bound electron pairs. As already mentioned in the previous section, such pairs are Bose formations and therefore possess the ability to accumulate in any amount at the level with the lowest energy. In the absence of an external field, and neglecting scattering processes, the pairs obviously "condense" in the state in which they are at rest as a whole. Let us consider the product of operators $\tilde{\psi}\tilde{\psi}$ or $\tilde{\psi}^+\tilde{\psi}^+$. The first annihilates, whilst the second creates, two electrons. In particular, these two electrons may find themselves in a bound state, in other words, the operators $\tilde{\psi}\tilde{\psi}$ and $\tilde{\psi}^+\tilde{\psi}^+$ contain terms corresponding to the annihilation and creation of bound pairs, including pairs at the lowest level. Since there are many such pairs (the number of them is proportional to the total number of particles), the corresponding contribution to the operators $\tilde{\psi}\tilde{\psi}$ and $\tilde{\psi}^+\tilde{\psi}^+$ can be regarded as a c-number, just as in the case of a system of bosons. Notice that there are special reasons, in a metal, why we do not consider pairs which are not at the lowest level: a bound pair of electrons, when the motion as a whole has finite momentum, is a Bose excitation with zero spin. As we have already remarked, the condition for electric neutrality in a metal actually implies that a substantial energy ($\sim$1 eV) is required for the excitation of such a "Bose condensate" pair, this being much greater than the characteristic energies which we encounter when developing the theory of superconductivity.

Returning to equation (34.3) for the Green function, in the light of what has been said, the average of the product of four ψ-operators can be written as follows, e.g.

$$\langle T(\tilde{\psi}_\alpha(x_1)\,\tilde{\psi}_\beta(x_2)\,\tilde{\psi}_\gamma^+(x_3)\,\tilde{\psi}_\delta^+(x_4))\rangle = -\langle T(\tilde{\psi}_\alpha(x_1)\,\tilde{\psi}_\gamma^+(x_3))\rangle$$
$$\times \langle T(\tilde{\psi}_\beta(x_2)\,\tilde{\psi}_\delta^+(x_4))\rangle + \langle T(\tilde{\psi}_\alpha(x_1)\,\tilde{\psi}_\delta^+(x_4))\rangle\,\langle T(\tilde{\psi}_\beta(x_2)\,\tilde{\psi}_\gamma^+(x_3))\rangle$$
$$+ \langle N\,|\,T(\tilde{\psi}_\alpha(x_1)\tilde{\psi}_\beta(x_2))\,|\,N+2\rangle\,\langle N+2\,|\,T(\tilde{\psi}_\gamma^+(x_3)\tilde{\psi}_\delta^+(x_4))\,|\,N\rangle, \tag{34.4}$$

where $|N\rangle$ and $|N+2\rangle$ are the ground states of the systems with N and $N+2$ particles. This way of writing the average implies that we have neglected all the effects of scattering of particles by one another. The existence of interactions is only taken into account in so far as it leads to the formation of bound pairs. The third term on the right-hand side of (34.4) is written in complete analogy with the case of a Bose gas, in accordance with the fact that a large number of bound pairs has "condensed" into the lowest level. The quantity

$$\langle N \,|\, T(\tilde{\psi}\tilde{\psi}) \,|\, N+2\rangle \langle N+2 \,|\, T(\tilde{\psi}^+\tilde{\psi}^+) \,|\, N\rangle$$

is obviously of the order of the density of the number of pairs.

It is easily verified that the functions thus introduced can be written as

$$\begin{aligned} \langle N \,|\, T\big(\tilde{\psi}_\alpha(x)\,\tilde{\psi}_\beta(x')\big) \,|\, N+2\rangle &= e^{-2i\mu t} F_{\alpha\beta}(x-x'), \\ \langle N+2 \,|\, T\big(\tilde{\psi}_\alpha^+(x)\tilde{\psi}_\beta^+(x')\big) \,|\, N\rangle &= e^{2i\mu t} F_{\alpha\beta}^+(x-x'). \end{aligned} \tag{34.5}$$

For the uniform problem (in the absence of an external field), the Green function $G(x-x')$ depends only on the coordinate difference $x-x'$. The source of the extra dependence on t in expressions (34.5) is clear from the general formulae of quantum mechanics for the time derivative of an arbitrary operator $\tilde{A}(t)$:

$$\frac{\partial}{\partial t}\langle N \,|\, \tilde{A}(t) \,|\, N+2\rangle = i(E_N - E_{N+2})\langle N \,|\, \tilde{A}(t) \,|\, N+2\rangle.$$

By definition, the chemical potential $\mu = \partial E/\partial N$, so that the energy difference $E_{N+2} - E_N$ is equal to 2μ.

When substituting expression (34.4) into equation (34.3) for the Green function, we shall always omit the first two terms on the right-hand side of (34.4), since they may easily be shown to lead, in the equations for functions G, F, F^+, to an additive correction to the chemical potential, and are of no interest. As a result, we get the following equation connecting G and F^+:

$$\left\{i\frac{\partial}{\partial t} + \frac{\nabla^2}{2m}\right\}\hat{G}(x-x') - i\lambda\hat{F}(0+)\hat{F}^+(x-x') = \delta(x-x'). \tag{34.6}$$

(Here, $\hat{G}$, $\hat{F}$, $\hat{F}^+$ denote the matrix forms of the functions $G_{\alpha\beta}$, $F_{\alpha\beta}$, $F_{\alpha\beta}^+$ in the spin indices, their products being the matrix products.)

The function $\hat{F}(0+)$ is defined as follows:

$$F_{\alpha\beta}(0+) = e^{2i\mu t}\langle N \,|\, \tilde{\psi}_\alpha(x)\tilde{\psi}_\beta(x) \,|\, N+2\rangle \equiv \lim_{r\to r',\, t\to t'+0} F_{\alpha\beta}(x-x'). \tag{34.7}$$

The equation for $\hat{F}^+(x-x')$ can be obtained in the same way by using the second of equations (34.2):

$$\left\{i\frac{\partial}{\partial t} - \frac{\nabla^2}{2m} - 2\mu\right\}\hat{F}^+(x-x') + i\lambda\hat{F}^+(0+)\,\hat{G}(x-x') = 0. \tag{34.8}$$

In accordance with (34.7),

$$F_{\alpha\beta}^+(0+) = e^{-2i\mu t}\langle N+2 \,|\, \tilde{\psi}_\alpha^+(x)\tilde{\psi}_\beta^+(x) \,|\, N\rangle. \tag{34.9}$$

In the absence of spin-dependent interactions, the Green function $\hat{G}(x-x')$ is proportional to the unit matrix $\delta_{\alpha\beta}$ in the spin variables:

$$\hat{G}_{\alpha\beta}(x-x') = \delta_{\alpha\beta} G(x-x').$$

The functions $\hat{F}$, $\hat{F}^+$ are proportional to the matrix $\hat{I}$, anti-symmetric with respect to its indices. For, since the operators $\tilde{\psi}_\alpha(x)$ and $\tilde{\psi}_\beta(x')$ anticommute at the same instant, $F_{\alpha\beta}(\boldsymbol{r}-\boldsymbol{r}', 0) = -F_{\beta\alpha}(\boldsymbol{r}'-\boldsymbol{r}, 0)$. It follows from this that

$$\{F^+_{\alpha\beta}(\boldsymbol{r}-\boldsymbol{r}', 0)\}^* = -F_{\alpha\beta}(\boldsymbol{r}-\boldsymbol{r}', 0). \tag{34.10}$$

In particular,

$$\{F^+_{\alpha\beta}(0+)\}^* = -F(0+). \tag{34.11}$$

It is convenient to write $\hat{F}$, $\hat{F}^+$ as

$$\begin{aligned} \hat{F}^+(x-x') &= \hat{I}F^+(x-x'), \\ \hat{F}(x-x') &= -\hat{I}F(x-x'), \end{aligned} \tag{34.12}$$

where $(\hat{I}^2)_{\alpha\beta} = -\delta_{\alpha\beta}$.

It is clear from (34.10) that $F^+(x-x')$ and $F(x-x')$ satisfy the relationship

$$\{F^+(\boldsymbol{r}-\boldsymbol{r}', 0)\}^* = F(\boldsymbol{r}-\boldsymbol{r}', 0).$$

The anti-symmetry of $\hat{F}$, $\hat{F}^+$ with respect to the spin indices corresponds to the fact that the bound pairs are at a displaced level. The function $F_{\alpha\beta}(\boldsymbol{r}-\boldsymbol{r}', 0)$ can obviously, apart from a factor, be regarded as the wave function of a pair of particles in a bound state (the centre of mass of the pair is at rest).

Let us write down the system of equations for these functions, omitting everywhere the dependence on the spin variables:

$$\begin{aligned} \left\{i\frac{\partial}{\partial t} + \frac{\nabla^2}{2m}\right\} G(x-x') - i\lambda F(0+)F^+(x-x') &= \delta(x-x'), \\ \left\{i\frac{\partial}{\partial t} - \frac{\nabla^2}{2m} - 2\mu\right\} F^+(x-x') + i\lambda F^+(0+)G(x-x') &= 0, \end{aligned} \tag{34.13}$$

where $(F(0+))^* = F^+(0+)$.

Fourier transforming everywhere(†) we obtain:

$$\begin{aligned} \left(\omega - \frac{\boldsymbol{p}^2}{2m}\right) G(p) - i\lambda F(0+)F^+(p) &= 1, \\ \left(\omega + \frac{\boldsymbol{p}^2}{2m} - 2\mu\right) F^+(p) + i\lambda F^+(0+)G(p) &= 0. \end{aligned} \tag{34.14}$$

(†) This system of equations has a great similarity to the system for the functions G' and $\hat{G}$ in the Bose system. It must be borne in mind, however, that the analogues of the functions F and F^+ are in this case the operators ξ_0, ξ_0^+ of the condensate particles. We therefore use the notations G and F^+ as distinct from G' and $\hat{G}$ for bosons.

The foregoing discussion has been carried through in the thermodynamic variables, where the number of particles is given. It is much more convenient to use the chemical potential μ as the independent variable. As usual, the transformation to this variable can be accomplished formally by substituting $\omega = \omega' + \mu$. Omitting the primes from the frequency, (34.14) can be rewritten as

$$\begin{aligned}(\omega - \xi)G(p) - i\lambda F(0+)F^+(p) &= 1,\\ (\omega + \xi)F^+(p) + i\lambda F^+(0+)G(p) &= 0,\end{aligned} \tag{34.15}$$

where $\xi = v(|\boldsymbol{p}| - p_0)$, $p_0 \simeq \sqrt{2m\mu}$ is the Fermi limiting momentum, and $v = p_0/m$.

The solution is

$$G(p) = \frac{\omega + \xi}{\omega^2 - \xi^2 - \Delta^2}, \qquad F^+(p) = -i\lambda \frac{F^+(0+)}{\omega^2 - \xi^2 - \Delta^2}$$

where

$$\Delta^2 = \lambda^2 |F^+(0+)|^2. \tag{34.16}$$

The determinant of the left-hand side of system (34.15) vanishes at the points $\omega = \pm\varepsilon(\boldsymbol{p})$, where $\varepsilon(\boldsymbol{p}) = \sqrt{\xi^2 + \Delta^2}$. The solution of (34.15) is therefore obtained up to arbitrary terms of the form

$$A_1(\boldsymbol{p})\delta(\omega - \varepsilon(\boldsymbol{p})) + A_2(\boldsymbol{p})\delta(\omega + \varepsilon(\boldsymbol{p})).$$

The boundary condition, determining the choice of the arbitrary A_1 and A_2 in the functions G and F^+, is supplied by Landau's theorem (Chap. II), in accordance with which the sign of the imaginary part of the Green function G is opposite to the sign of ω, whilst the function

$$G_R(\omega, \boldsymbol{p}) = \mathrm{Re}\,G(\omega, \boldsymbol{p}) + i\,\mathrm{sign}(\omega)\,\mathrm{Im}\,G(\omega, \boldsymbol{p})$$

must be analytic and cannot have singularities in the upper half-plane. It may easily be verified that the solution satisfying these requirements is (†)

$$G(p) = \frac{u_{\boldsymbol{p}}^2}{\omega - \varepsilon(\boldsymbol{p}) + i\delta} + \frac{v_{\boldsymbol{p}}^2}{\omega + \varepsilon(\boldsymbol{p}) - i\delta}, \tag{34.17}$$

$$F^+(p) = -i\lambda \frac{F^+(0+)}{(\omega - \varepsilon(\boldsymbol{p}) + i\delta)(\omega + \varepsilon(\boldsymbol{p}) - i\delta)}, \tag{34.18}$$

where the functions $u_{\boldsymbol{p}}^2$ and $v_{\boldsymbol{p}}^2$ are equal to

$$u_{\boldsymbol{p}}^2 = \frac{1}{2}\left(1 + \frac{\xi}{\varepsilon(\boldsymbol{p})}\right); \qquad v_{\boldsymbol{p}}^2 = \frac{1}{2}\left(1 - \frac{\xi}{\varepsilon(\boldsymbol{p})}\right). \tag{34.19}$$

(†) We have chosen $F^+(0+)$ to be real. This is always possible in the absence of an external field, since equations (34.13) admit of the transformations $\{F(x-x'),\ F(0+)\} \to \{F(x-x')\,e^{2i\varphi},\ F(0+)\,e^{2i\varphi}\}$ and $\{F^+(x-x'),\ F^+(0+)\} \to \{F^+(x-x')\,e^{-2i\varphi},\ F^+(0+)\,e^{-2i\varphi}\}$ with constant phase. For more details about this, see subsection 2 below.

The positive pole $\omega = \varepsilon(\boldsymbol{p})$ in Green function (34.17) determines the excitation spectrum $\varepsilon(\boldsymbol{p}) = \sqrt{\xi^2 + \Delta^2}$. This spectrum has a gap Δ, which we find by starting from the relation

$$F^+(0) = (2\pi)^{-4} \int F^+(p)\, \mathrm{d}\omega\, \mathrm{d}^3\boldsymbol{p}. \tag{34.20}$$

On substituting (34.18) in this, we get the equation

$$1 = -\frac{\lambda}{2(2\pi)^3} \int \frac{\mathrm{d}^3\boldsymbol{p}}{\sqrt{\xi^2 + \Delta^2}}. \tag{34.21}$$

The divergence is cut out of this integral due to the condition that, in our model, the only electrons taking part in the interaction are those with energies in a layer of width $2\omega_D$ about the Fermi surface. We find on performing the integration that

$$1 = -\frac{\lambda}{2\pi^2} m p_0 \ln \frac{2\omega_D}{\Delta}.$$

We get as a result:

$$\Delta = 2\omega_D e^{-1/\eta}, \tag{34.22}$$

where $\eta = |\lambda|\, m p_0 / 2\pi^2$. On comparing this expression with the result of the previous subsection, we find that the size of the gap in the energy spectrum at absolute zero has the following relation with the transition temperature:

$$\Delta = \frac{\pi}{\gamma} T_c. \tag{34.23}$$

2. The equations in the presence of an external electromagnetic field. Gauge-invariance

If a superconductor is located in an external field, say in an electromagnetic field, system (34.13) becomes rather more complicated. Notice, first of all, that in an external field the functions are no longer functions of the coordinate difference only. The electromagnetic field can be brought into (34.13) in the usual way by the following substitution:

$$\nabla \to \nabla - ie\boldsymbol{A} \text{ or } \nabla \to \nabla + ie\boldsymbol{A}, \tag{34.24}$$

depending on whether the differentiation refers to the operator $\tilde{\psi}$ or $\tilde{\psi}^+$. (It is generally convenient to use the gauge in which the scalar potential φ is zero.) The equations for G and F^+ in the case of a field are

$$\left\{i\frac{\partial}{\partial t} + \frac{1}{2m}\left(\frac{\partial}{\partial \boldsymbol{r}} - ie\boldsymbol{A}(\boldsymbol{r})\right)^2 + \mu\right\} G(x, x') - i\lambda F(x, x) F^+(x, x') = \delta(x - x'),$$
$$\left\{i\frac{\partial}{\partial t} - \frac{1}{2m}\left(\frac{\partial}{\partial \boldsymbol{r}} + ie\boldsymbol{A}(\boldsymbol{r})\right)^2 - \mu\right\} F^+(x, x') + i\lambda F^+(x, x) G(x, x') = 0. \tag{34.25}$$

Notice the obvious gauge-invariance of these equations. A gauge transformation of the vector potential,

$$\boldsymbol{A} \to \boldsymbol{A} + \frac{\partial \varphi}{\partial \boldsymbol{r}}, \tag{34.26}$$

leads to different transformations of the functions G, F and F^+:

$$\begin{aligned} G(x, x') &\to G(x, x')\, e^{ie[\varphi(\mathbf{r})-\varphi(\mathbf{r}')]}, \\ F(x, x') &\to F(x, x')\, e^{ie[\varphi(\mathbf{r})+\varphi(\mathbf{r}')]}, \\ F^+(x, x') &\to F^+(x, x')\, e^{-ie[\varphi(\mathbf{r})+\varphi(\mathbf{r}')]}, \end{aligned} \tag{34.27}$$

whilst the "gap" $|\lambda|\, F(x, x)$ or $|\lambda|\, F^+(x, x)$, which, in an external field, is generally speaking a function of x, transforms in accordance with

$$F(x, x) \to F(x, x)\, e^{2ie\varphi(\mathbf{r})}, \qquad F^+(x, x) \to F^+(x, x)\, e^{-2ie\varphi(\mathbf{r})}. \tag{34.28}$$

The gauge-invariance of the equations enables us to study now the properties of a superconductor in a magnetic field. It must be emphasized, as regards the gauge-invariance of (34.25), that it is connected with the use of the form (32.3) for writing the interaction Hamiltonian. Strictly speaking, Hamiltonian (32.2) is not gauge-invariant, this being, of course, a property of the model. It may easily be shown that, in this model, equation (34.13) contains, instead of $F(0+)$ and $F^+(0+)$ — the values of the functions F and F^+ at coincident points — the functions

$$\overline{F(x, x)} = \int \theta(\mathbf{r}-\mathbf{y})\,\theta(\mathbf{r}-\mathbf{z})\, F(y, z)\, d^3\mathbf{z}\, d^3\mathbf{y}$$

and $\overline{F^+(x, x)}$ respectively. As the wave function of a pair, $F(\mathbf{y}, \mathbf{z}; t = t')$ has a correlation radius of the order of the pair dimensions $\xi_0 \sim hv/T_c$, and falls off rapidly when $|\mathbf{y}-\mathbf{z}| \gg \xi_0$. The functions θ are of the δ-function type, as mentioned in § 32, with a half-width of the order v/ω_D. Hence replacing of $\overline{F(x, x)}$ by $F(x, x)$ involves an error of the order T_c/ω_D, which is always small in real superconductors.

3. *Superconductor at finite temperatures*

We shall end the present section by considering the question of extending the above approach to the case of non-zero temperatures. This extension is obviously possible on the basis of the method described in Chap. III for $T \neq 0$. A system is characterised in the superconducting state by the non-zero averages

$$\mathfrak{F}(x, x') = \frac{\langle T_\tau(\psi(x)\psi(x')\mathfrak{S})\rangle}{\langle\mathfrak{S}\rangle}, \qquad \mathfrak{F}^+(x, x') = \frac{\langle T_\tau(\overline{\psi}(x)\overline{\psi}(x')\mathfrak{S})\rangle}{\langle\mathfrak{S}\rangle}$$

(the meaning of the averaging and the definition of operators $\psi(x)$, $\overline{\psi}(x)$ being the same as in Chap. III. Remember that the chemical potential is chosen as the independent thermodynamic variable). If we regard the ensemble averages in the definitions of $\mathfrak{F}(x, x')$ and $\mathfrak{F}^+(x, x')$ as quantum mechanical averages over the state with energy equal to the mean energy $\bar{E}$, and with a number of particles equal to the average number of particles, this means that the creation or annihilation of a pair of electrons leaves the state virtually unchanged. It is necessary for this that the pair of electrons in question belong to those bound pairs that find themselves in the "Bose condensate" state. Since the number of these pairs is very large (proportional to the total number of particles in the total volume of the system), the addition or annihilation of one such pair makes no difference to the total state of the system. In other words, as in the case of absolute zero, in the system the superconducting state possesses terms in the operator products $\psi\psi$ and $\psi^+\psi^+$ that can be regarded as c-numbers. We assume that the (thermodynamic) average of the product of four ψ-operators can be written in terms of the Green function $\mathfrak{G}(x, x') = -\langle T_\tau(\psi(x)\bar{\psi}(x')\mathfrak{S})\rangle/\langle\mathfrak{S}\rangle$ and the functions $\mathfrak{F}(x, x')$, $\mathfrak{F}^+(x, x')$, just as was done in expression (34.4) at absolute zero. As above, this implies neglecting the effects of scattering of particles by one another. We have:

$$\frac{\langle T_\tau(\psi_\alpha(x_1)\psi_\beta(x_2)\bar{\psi}_\gamma(x_3)\bar{\psi}_\delta(x_4)\mathfrak{S})\rangle}{\langle\mathfrak{S}\rangle} = -\mathfrak{G}_{\alpha\gamma}(x_1, x_3)\,\mathfrak{G}_{\beta\delta}(x_2, x_4)$$
$$+\mathfrak{G}_{\alpha\delta}(x_1, x_4)\,\mathfrak{G}_{\beta\gamma}(x_2, x_3) + \mathfrak{F}_{\alpha\beta}(x_1, x_2)\,\mathfrak{F}^+_{\gamma\delta}(x_3, x_4). \quad (34.29)$$

We shall not repeat here the derivation of the equations for the functions $\mathfrak{G}$ and $\mathfrak{F}^+$. It is very similar to the derivation of (34.13); we shall just quote the final form of these equations:

$$\begin{aligned} &\left\{-\frac{\partial}{\partial\tau} + \frac{\nabla^2}{2m} + \mu\right\}\mathfrak{G}(x - x') + \Delta\mathfrak{F}^+(x - x') = \delta(x - x'), \\ &\left\{\frac{\partial}{\partial\tau} + \frac{\nabla^2}{2m} + \mu\right\}\mathfrak{F}^+(x - x') - \Delta^*\mathfrak{G}(x - x') = 0, \end{aligned} \quad (34.30)$$

where

$$\Delta = |\lambda|\,\mathfrak{F}(0+), \qquad \Delta^* = |\lambda|\,\mathfrak{F}^+(0+). \quad (34.31)$$

It sometimes proves necessary to determine the function $\mathfrak{F}(x - x')$. The corresponding equation is easily obtained:

$$\left\{-\frac{\partial}{\partial\tau} + \frac{\nabla^2}{2m} + \mu\right\}\mathfrak{F}(x - x') - \Delta\mathfrak{G}(x' - x) = 0.$$

This contains the function $\mathfrak{G}$ with arguments interchanged.

It is easily shown that the set of four equations for $\mathfrak{G}(x - x')$, $\mathfrak{F}^+(x - x')$, $\mathfrak{F}(x - x')$ and $\mathfrak{G}(x' - x)$ can be written symbolically in the matrix form

$$\begin{pmatrix} \left\{-\frac{\partial}{\partial\tau} + \frac{\nabla^2}{2m} + \mu\right\} & \Delta \\ -\Delta^* & \left\{\frac{\partial}{\partial\tau} + \frac{\nabla^2}{2m} + \mu\right\} \end{pmatrix} \times \begin{pmatrix} \mathfrak{G}(x - x') & \mathfrak{F}(x - x') \\ \mathfrak{F}^+(x - x') & -\mathfrak{G}(x' - x) \end{pmatrix} = \hat{1}. \quad (34.32)$$

In other words, the four functions form one Green function matrix for the operator on the left-hand side of (34.30).

In the temperature-dependent method, as we know, all the functions are expanded into Fourier series, rather than into a Fourier integral over the frequency. It is clear from these equations that, if we Fourier transform the functions $\mathfrak{F}$ and $\mathfrak{F}^+$, in the same way as we Fourier transformed the Green function in Chap. III:

$$\begin{aligned} \mathfrak{F}^+(x-x') &= (2\pi)^{-3}\, T \sum_n e^{-i\omega_n \tau} \int d^3p\, e^{i(p\cdot r)} \mathfrak{F}^+_\omega(p), \\ \mathfrak{F}(x-x') &= (2\pi)^{-3}\, T \sum_n e^{-i\omega_n \tau} \int d^3p\, e^{i(p\cdot r)} \mathfrak{F}_\omega(p), \end{aligned} \tag{34.33}$$

where $\omega = (2n+1)\pi T$, equations (34.15) now correspond to the set

$$\begin{aligned} (i\omega - \xi)\,\mathfrak{G}_\omega(p) + \Delta \mathfrak{F}^+_\omega(p) &= 1, \\ (i\omega + \xi)\,\mathfrak{F}^+_\omega(p) + \Delta^* \mathfrak{G}_\omega(p) &= 0, \end{aligned} \tag{34.34}$$

which is satisfied by

$$\mathfrak{G}_\omega(p) = -\frac{i\omega + \xi}{\omega^2 + \xi^2 + \Delta^2}, \qquad \mathfrak{F}^+_\omega(p) = \frac{\Delta^*}{\omega^2 + \xi^2 + \Delta^2}. \tag{34.35}$$

Notice also that, in the absence of a field, $\mathfrak{F}$ and $\mathfrak{F}^+$ are equal to one another, and Δ is real. This solution is single-valued, as distinct from the situation holding for the system (34.15). This is connected with the fact that the analytic properties of the thermodynamic functions are uniquely defined. The size of the gap is determined from the condition

$$1 = \frac{|\lambda| T}{(2\pi)^3} \sum_n \int \frac{d^3p}{\omega + \xi^2 + \Delta^2}. \tag{34.36}$$

The series over the frequencies is easily summed. Instead of condition (34.21) at $T = 0$, a new relation is obtained, from which we can find the size of the gap at finite temperatures:

$$1 = \frac{|\lambda|\, m p_0}{2\pi^2} \int_0^{\omega_D} d\xi \frac{\tanh \dfrac{\sqrt{\xi^2 + \Delta^2(T)}}{2T}}{\sqrt{\xi^2 + \Delta^2(T)}}. \tag{34.37}$$

At the point of the phase transition, i.e. at the temperature $T = T_c$, the gap $\Delta(T)$ vanishes and, as must be the case, condition (34.37) turns into (33.15), from which the transition temperature T_c can be determined.

§ 35. DEDUCTION OF THE SUPERCONDUCTIVITY EQUATIONS IN THE PHONON MODEL

Let us dwell on the derivation of the superconductivity equations in the model in which the electrons interact with one another through the electron-phonon interaction. Of course, such a model suffers from the same defect as the approach described above, inasmuch as no account is taken in it of the Coulomb forces acting in the metal. Nevertheless it evidently has a more direct physical significance than the model with a four-fermion interaction, even though the latter is more useful when it comes to obtaining practical results. The basic advantage of the phonon model lies, first of all, in the electron-phonon interaction Hamiltonian (32.1) being gauge-invariant right from the start, as distinct from the method with the four-fermion interaction Hamiltonian (32.2), which is only approximately gauge-invariant, by virtue of the relation $T_c \ll \omega_D$. As regards this relation, it is in general fulfilled only in the weak binding approximation(†). We shall show below that the restriction to weak binding is not essential in the theory of superconductors and that only the ratio $\omega_D/\varepsilon_F \ll 1$ is actually a small parameter of the theory ($\omega_D/\varepsilon_F \sim u/v \sim 10^{-2} - 10^{-3}$, where u is the sound velocity in the body, and v is the electron velocity on the Fermi surface(‡)). We shall confine ourselves to deriving the equations at absolute zero.

Suppose, then, that the interaction Hamiltonian of the system of electrons and phonons is

$$H_{int}(x) = g\left(\tilde{\psi}^+(x)\tilde{\psi}(x)\right)\tilde{\varphi}(x).$$

If the system is in the superconducting state, its properties are characterised by two functions F, F^+, in addition to the Green function G. Hence, instead of the usual Dyson equation (§ 21), it is generally necessary to investigate three equations, connecting the three functions

$$\begin{aligned} \hat{G}_{\alpha\beta}(x, x') &= -i\langle T(\tilde{\psi}_\alpha(x), \tilde{\psi}^+_\beta(x'))\rangle \equiv \delta_{\alpha\beta}G(x-x'), \\ \hat{F}^+_{\alpha\beta}(x, x') &= \langle T(\tilde{\psi}^+_\alpha(x), \tilde{\psi}^+_\beta(x'))\rangle = I_{\alpha\beta}F^+(x-x'), \\ \hat{F}_{\alpha\beta}(x, x') &= \langle T(\tilde{\psi}_\alpha(x), \tilde{\psi}_\beta(x'))\rangle = -I_{\alpha\beta}F(x-x'). \end{aligned} \tag{35.1}$$

As regards the equation for the phonon Green function

$$D(x_1 - x_2) = -i\langle T(\varphi(x_1), \varphi(x_2))\rangle,$$

it remains almost unchanged, as we shall see.

The equation for the Green function can be obtained from the diagram technique of perturbation theory. Just as in the case of a boson system below the "Bose condensation" point, the set of possible diagrams of perturbation theory is widened due to the appearance in them of lmes corre-

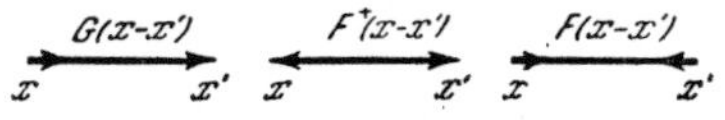

Fig. 93

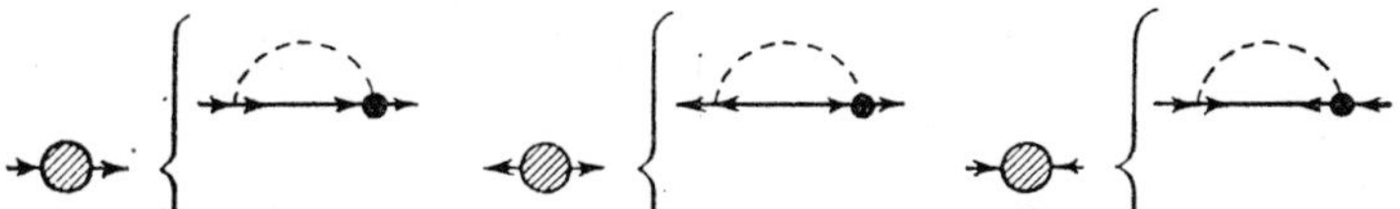

Fig. 94

(†) We always have, however, $T_c \ll \omega_D$ for actual superconductors.

(‡) Electron-phonon interactions in the theory of superconductors were investigated in [59], [60].

sponding to the functions F and F^+. Let us agree to represent on the diagram the functions G, F^+ and F by heavy lines with two arrows, the directions of which at the points x, x' are chosen in accordance with (35.1) in such a way that the operator $\tilde{\psi}$ at the point in question corresponds to an arrow directed away from the point, whereas $\tilde{\psi}^+$ corresponds to an arrow directed towards the point. All three lines are illustrated in Fig. 93. It may easily be seen now that, in complete analogy with a boson gas, there are three types of irreducible self-energy parts, which we denote by $\Sigma_{11}(x, x')$, $\Sigma_{20}(x, x')$ and $\Sigma_{02}(x, x')$. In Fig. 94, illustrating the Σ_{ik} diagrams, the dotted line corresponds to the phonon D-functions, the point to a simple vertex, or factor g, and the heavy point or rectangle denotes the modification to the simple vertex as a result of the different electron-phonon interactions.

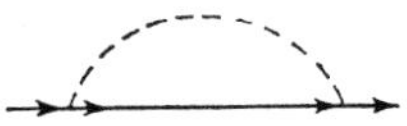

Fig. 95

Let us Fourier transform all the functions and consider one of the self-energy diagrams, say $\Sigma_{11}(p)$. It will easily be seen that, up to terms of order ω_D/ε_F, we can neglect all the phonon corrections to the electron-phonon vertex part in the simplest diagram for $\Sigma_{11}(p)$ of Fig. 95. In fact, as we proved in § 21, the values of the D-function (and phonon vertex) of importance in the relevant integrand are those for phonons with momenta of the order of the Fermi momentum of the electrons. For this reason, the estimate obtained in § 21 for the corrections to the phonon vertex resulting from the electron-phonon interaction still remains valid in the present case, since the size of these corrections is determined by the values of the Green functions in the energy and momentum region remote from the Fermi surface. It is quite clear that the Fourier components of the electron Green function for a metal in the superconducting state differ from their values in the normal metal only in a narrow region close to the Fermi surface, with excitation energies not greater than the order of the maximum phonon energies, i.e. of order ω_D. To the same degree, the functions $F^+(p)$, $F(p)$, which are characteristic of the superconducting state, also differ from zero only in the region mentioned. Fig. 94 illustrates the two possible types of self-energy diagram, for each of the parts Σ_{11}, Σ_{20} and Σ_{02}, depending on the choice of the modification to the phonon vertex. On the basis of what has just been said, we can immediately omit the diagrams of the second type, where the phonon vertex is marked by a heavy rectangle, since the diagrams of this kind can be formed only by using the superconductor Green functions F and F^+.

We can therefore confine ourselves to the zero order approximation of perturbation theory for the phonon vertex in the irreducible parts of Σ_{11}, Σ_{20} and Σ_{02}.

For the same reasons, the phonon Green function $D(x_1 - x_2)$ will remain unchanged, and we can use (21.14) directly for its Fourier component. The structure of the equations for G and F^+ is illustrated in Fig. 96 and is clear without further explanation. Let us write these equations in analytic form. In the coordinate representation,

$$\left\{i\frac{\partial}{\partial t}+\frac{\nabla^2}{2m}+\mu\right\}G(x-x')=\delta(x-x')+g^2i\int G(x-z)$$
$$\times D(x-z)G(z-x')\,d^4z+g^2i\int F(x-z)D(x-z)F^+(z-x')\,d^4z,$$
$$\left\{-i\frac{\partial}{\partial t}+\frac{\nabla^2}{2m}+\mu\right\}F^+(x-x')=g^2i\int G(z-x)D(z-x)F^+(z-x')\,d^4z$$
$$+g^2i\int F^+(x-z)D(x-z)G(z-x')\,d^4z. \qquad (35.2)$$

An electromagnetic field can be included in these equations in the usual way, precisely as was done above in § 34. Let us emphasise that the resulting system is completely gauge-invariant, as distinct from system

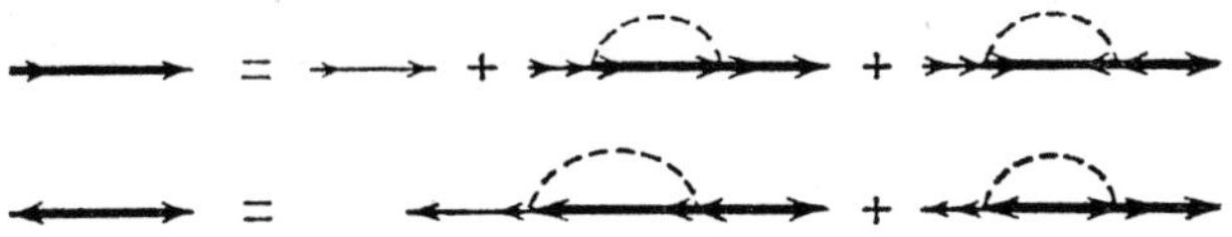

Fig. 96

(34.13), in which the gauge-invariance was only approximate, up to terms T_c/ω_D. Unfortunately, as is clear from (35.2), this system has a much more complicated form, with a non-linear integral term, which makes it less suitable for solution in the coordinate representation, as is required in a number of problems in which a non-uniform magnetic field exists. On the other hand, the practical results obtained are as a rule equivalent for the two models.

In the absence of a magnetic field, on transforming to the momentum representation in equations (35.2), we easily get equations for the Fourier components of all the functions:

$$(\omega-\xi-g^2i\overline{G}_\omega)G(p)-g^2i\overline{F}_\omega F^+(p)=1,$$
$$(-\omega-\xi-g^2i\overline{G}_{-\omega})F^+(p)-g^2i\overline{F^+_\omega}G(p)=0. \qquad (35.3)$$

Here, we have written

$$\overline{G}_\omega=\frac{1}{(2\pi)^4}\int G(p-k)D(k)\,d^4k, \qquad \overline{F}_\omega=\frac{1}{(2\pi)^4}\int F(p-k)D(k)\,d^4k,$$
$$\overline{F^+_\omega}=\frac{1}{(2\pi)^4}\int F^+(p-k)D(k)\,d^4k, \qquad \overline{G}_{-\omega}=\overline{G}_\omega(-p). \qquad (35.4)$$

This system is completely analogous to the system (34.15) obtained above. The only difference is this: whereas, in equations (34.15),

$\overline{G}_\omega$, $\overline{F}_\omega$ and $\overline{F_\omega^+}$ are constant in the region $|v(|\mathbf{p}|-p_0)| < \omega_D$ about the Fermi surface and vanish outside this region, the functions (35.4) are in general functions of ω and p, smoothly decreasing to zero when ω, $|v(|\mathbf{p}|-p_0)| \gg \omega_D$.

Let us express G and F^+ in terms of $\overline{G}_\omega$, $\overline{F}_\omega$ and $\overline{F_\omega^+}$:

$$G(p) = \frac{\omega - g^2 i\overline{G}_\omega + \xi}{[\omega - \xi - g^2 i\overline{G}_\omega]\,[\omega + \xi + g^2 i\overline{G}_{-\omega}] - g^4\,|\overline{F_\omega^+}|^2},$$

$$F^+(p) = \frac{-ig^2\overline{F_\omega^+}}{[\omega - \xi - g^2 i\overline{G}_\omega]\,[\omega + \xi + g^2 i\overline{G}_{-\omega}] - g^4\,|\overline{F_\omega^+}|^2}.$$

After substituting these expressions into the definitions (35.4) of $\overline{G}_\omega$ and $\overline{F_\omega^+}$, we get two integral equations for these functions instead of the one equation (21.16) of § 21. The solution of these equations was obtained by Eliashberg [62]. We shall give the final result, without dwelling on the details. It turns out that, at small energies, the excitation spectrum has the form

$$\varepsilon(\mathbf{p}) = \sqrt{\xi^2 + \Delta^2},$$

where, however, $\xi = v_1(p - p_0)$ contains, in accordance with § 21, the renormalised velocity on the Fermi surface v_1. The gap at absolute zero is connected with $\overline{F_{\omega=0}^+}$ by the relation

$$\Delta = g^2\frac{v_1}{v_0}\,|\overline{F_{\omega=0}^+}|.$$

In the weak interaction limit, $g^2 \ll 1$, these formulae are the same as the results of the previous section.

§ 36. THERMODYNAMICS OF SUPERCONDUCTORS

1. *Temperature-dependence of the gap*

Let us investigate in greater detail the temperature-dependence of the gap. We first take the case of low temperatures $T \ll T_c$ and carry out a suitable expansion of condition (34.37). We have identically:

$$\frac{1}{\eta} = \int\limits_0^{\omega_D} \frac{d\xi}{\sqrt{\xi^2 + \Delta^2}} - 2\int\limits_0^{\infty} \frac{d\xi}{\sqrt{\xi^2 + \Delta^2}} \cdot \frac{1}{\exp\left(\sqrt{\xi^2 + \Delta^2}/T\right) + 1}, \tag{36.1}$$

where $\eta = |\lambda|\, mp_0/2\pi^2$ (the second integral is convergent, so that its upper limit can be put equal to infinity). On expanding with respect to the exponent in the second integrand, changing to an integration over ε, and using the definitions of the corresponding Bessel functions, we can write (36.1) as a series in zero order Bessel functions:

$$\ln\frac{\Delta_0}{\Delta} = 2\sum_{n=1}^{\infty}(-1)^{n+1}K_0\left(\frac{n\Delta}{T}\right) \tag{36.2}$$

(here $\Delta_0 \equiv \Delta(T=0)$).

At low temperatures $\Delta \gg T$, we obtain on using the asymptotic expansions of the Bessel functions:

$$\Delta = \Delta_0 - \sqrt{2\pi T\Delta_0}\left(1-\frac{T}{8\Delta_0}\right)e^{-\Delta_0/T}. \tag{36.3}$$

The behaviour of the gap at temperatures close to the transition temperature T_c is most conveniently determined by starting from (34.36). The gap is small close to T_c, so that we can expand in powers of Δ^2/T^2 in (34.36):

$$\frac{1}{\eta} = T\sum_n \int_{-\omega_D}^{\omega_D} d\xi\left\{\frac{1}{\omega^2+\xi^2} - \frac{\Delta^2}{(\omega^2+\xi^2)^2} + \frac{\Delta^4}{(\omega^2+\xi^2)^3} + \cdots\right\}.$$

On interchanging the order of summation over the frequency and integration over ξ in the convergent terms on the right-hand side we get

$$\frac{1}{\eta} = \int_0^{\omega_D}\frac{d\xi}{\xi}\tanh\frac{\xi}{2T} - \frac{\Delta^2}{(\pi T)^2}\sum_0^{\infty}\frac{1}{(2n+1)^3}$$

$$+\frac{3}{4}\frac{\Delta^4}{(\pi T)^4}\sum_0^{\infty}\frac{1}{(2n+1)^5}+\cdots. \tag{36.4}$$

The series that appear here can be expressed in terms of the Riemann ζ-functions:

$$\sum_0^{\infty}\frac{1}{(2n+1)^z} = \frac{2^z-1}{2^z}\zeta(z). \tag{36.5}$$

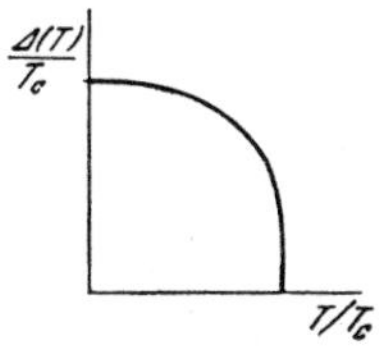

Fig. 97

We find on substituting this into the previous expression:

$$\ln\frac{T}{T_c} = -\frac{7\zeta(3)}{8}\frac{\Delta^2}{(\pi T)^2} + \frac{93\zeta(5)}{128}\frac{\Delta^4}{(\pi T)^4} + \cdots.$$

We obtain to a first approximation for the size of the gap close to T_c:

$$\Delta = \pi T_c\sqrt{\frac{8}{7\zeta(3)}}\sqrt{1-\frac{T}{T_c}} \simeq 3{\cdot}06T_c\sqrt{1-\frac{T}{T_c}}. \tag{36.6}$$

Fig. 97 illustrates the graph of the temperature-dependence of the gap throughout the temperature interval.

2. *Specific heat*

To find the different thermodynamic functions, we use the relation derived above for the derivative of the thermodynamic potential with respect to the interaction constant:

$$\frac{\delta \Omega}{\delta \lambda} = \frac{1}{\lambda} \langle H_{int} \rangle .$$

In our case H_{int} is given by (32.3). If we retain in this average terms which differ from zero only in the superconducting phase, we obtain, since $\lambda < 0$,

$$\frac{\delta \Omega}{\delta |\lambda|} = -\frac{1}{\lambda^2} |\Delta|^2.$$

The connection between $1/|\lambda|$ and Δ at a given temperature is determined by the relation (36.1). Hence the difference between the potentials Ω for the metal in the superconducting and normal phases is

$$\Omega_s - \Omega_n = \int_0^{\Delta} \left(\frac{\mathrm{d}\dfrac{1}{|\lambda|}}{\mathrm{d}\Delta} \right) \Delta^2 \mathrm{d}\Delta .$$

By a general rule of statistical physics, this added correction, when expressed in suitable variables, is the same for all the thermodynamic potentials. Using (36.2), according to which

$$\frac{1}{|\lambda|} = \frac{m p_0}{2\pi^2} \left\{ \ln \frac{2\omega_D}{\Delta} - 2 \sum_{n=1}^{\infty} (-1)^{n+1} K_0 \left(\frac{n\Delta}{T} \right) \right\},$$

and using the familiar formula of Bessel function theory: $K_0'(x) = -K_1(x)$ we get

$$F_s - F_n = -\left(\frac{m p_0}{2\pi^2} \right) \left\{ \frac{\Delta^2}{2} - 2 \sum_{n=1}^{\infty} (-1)^{n+1} \frac{T^2}{n^2} \int_0^{n\Delta/T} K_1(x)\, x^2 \mathrm{d}x \right\}.$$

At low temperatures $\Delta/T \gg 1$. In this case $\int_0^{n\Delta/T} K_1(x)\, x^2 \mathrm{d}x = 2 - \int_{n\Delta/T}^{\infty} K_1(x) x^2 \mathrm{d}x$. The latter integral only needs to be found for $n = 1$. the asymptotic expansion of $K_1(x)$ being used for this purpose.

The remaining series over n is easily summed. We get as a result:

$$F_s - F_n = \frac{m p_0 T^2}{6} - \frac{m p_0}{2\pi^2} \left[\frac{\Delta^2}{2} + \sqrt{2\pi \Delta_0^3 T} \left(1 + \frac{15}{8} \frac{T}{\Delta_0} \right) e^{-\Delta_0/T} \right]. \qquad (36.7)$$

The first term on the right-hand side is equal to the main term (with reversed sign) in the expansion of the free energy of the normal metal in powers of T. We know that this latter term leads to a linear law for the electron part of the specific heat in the normal phase:

$$C_n = \frac{m p_0 T}{3}.$$

On substituting (36.3), we find that the entropy in the superconducting phase at low temperatures is equal to

$$S_s = \frac{m p_0}{\pi^2}\sqrt{\frac{2\pi \Delta_0^3}{T}}\, e^{-\Delta_0/T},$$

whilst the specific heat is

$$C_s = \frac{m p_0}{\pi^2}\sqrt{\frac{2\pi \Delta_0^5}{T^3}}\, e^{-\Delta_0/T}.$$

This gives us, with the aid of (34.23), for the ratio of the specific heats of the superconducting phase in the region $T \ll T_c$ and of the normal phase at $T = T_c$:

$$\frac{C_s(T)}{C_n(T_c)} = \frac{3}{\gamma}\sqrt{\frac{2}{\pi}}\left(\frac{\Delta_0}{T}\right)^{3/2} \exp\left(-\frac{\Delta_0}{T}\right). \tag{36.8}$$

In order to obtain asymptotic formulae in the temperature region close to T_c, we start from expansion (36.4):

$$\delta \frac{1}{|\lambda|} = -\left(\frac{m p_0}{2\pi^2}\right)\frac{7\zeta(3)}{(2\pi T)^2}\Delta \delta \Delta.$$

We find with the aid of (36.6) for the difference between the free energies:

$$F_s - F_n = -\left(\frac{m p_0}{2\pi^2}\right)\frac{7\zeta(3)}{16(\pi T)^2}\Delta^4 = -\frac{2 m p_0 T_c^2}{7\zeta(3)}\left(1-\frac{T}{T_c}\right)^2. \tag{36.9}$$

Hence the entropy in the superconducting phase is

$$S_s = -\frac{4 m p_0 T_c}{7\zeta(3)}\left(1-\frac{T}{T_c}\right) + S_n.$$

On differentiating a second time and retaining the main terms, we get the following expression for the specific heat of the superconductor at the transition point:

$$C_s(T_c) = C_n(T_c) + \frac{4 m p_0}{7\zeta(3)} T_c.$$

Thus the specific heat of a metal has a jump of $4 m p_0 T_c / 7\zeta(3)$ at the point of the phase transition to the superconducting state. If we take into account higher order terms in $T_c - T$ in expansion (36.9), we can find the ratio of the specific heat $C_s(T)$ to the specific heat $C_n(T_c)$ in the normal phase close to T_c:

$$\frac{C_s(T)}{C_n(T_c)} = 2{\cdot}43 + 3{\cdot}77\left(\frac{T}{T_c} - 1\right).$$

This behaviour of the electron part of the specific heat of a metal close to T_c is illustrated in Fig. 98 on p. 304.

3. *Critical field*

The so-called critical magnetic field H_c is an important thermodynamic quantity in the theory of superconductivity. At a given temperature $T < T_c$, a metal in a magnetic field may find itself both in the superconducting and in the normal state.

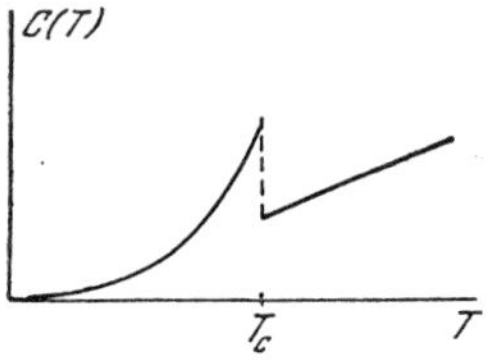

Fig. 98

If a superconductor is situated in a magnetic field, the surface currents screening the field produce a magnetic moment, which interacts with the external field. The additional energy arising from this interaction is $-(\boldsymbol{H}.\boldsymbol{M})/2$ (per unit volume). Let us consider a superconducting cylinder located in a field parallel to the cylinder axis. Having found the surface current from the condition that the magnetic field be equal to zero in the body of the superconductor, and having determined the magnetic moment produced by these currents, we discover that the extra magnetic energy is $H^2/8\pi$, i.e. the free energy of the superconductor in the magnetic field is $F_{sH} = F_s + (H^2/8\pi)$. Thus a transition from the superconducting to the normal phase occurs when the magnetic field is increased at a given temperature; this transition is a first order transition. The critical field is

$$\frac{H_c^2}{8\pi} = F_n - F_s.$$

We shall again confine ourselves to limiting cases. At low temperatures $(T \ll T_c)$, we obtain from (36.3) and (36.7), neglecting exponentially small terms:

$$H_c(0) = \sqrt{\frac{2mp_0}{\pi}}\Delta_0 = T_c\frac{\pi}{\gamma}\sqrt{\frac{2mp_0}{\pi}} \tag{36.10}$$

and

$$H_c(T) = H_c(0)\left(1 - \frac{\gamma^2}{3}\frac{T^2}{T_c^2}\right). \tag{36.11}$$

On using (36.9) and expressing $H_c(T)$ close to the transition point in terms of $H_c(0)$ with the aid of (36.10), the temperature dependence $H_c(T)$ in the region close to T_c is found to be

$$H_c(T) = H_c(0)\gamma\sqrt{\frac{8}{7\zeta(3)}}\left(1 - \frac{T}{T_c}\right) \simeq 1{\cdot}73\,H_c(0)\left(1 - \frac{T}{T_c}\right). \tag{36.12}$$

Notice that the experimental data usually correspond to the function

$$H_c(T) = H_c(0)\left(1 - \frac{T^2}{T_c^2}\right). \tag{36.13}$$

The theoretical formulae (36.11) and (36.12) are in fairly close agreement with the experimental relationship (36.13) in both the limiting cases (see [58], [63]).

§ 37. SUPERCONDUCTORS IN A WEAK ELECTROMAGNETIC FIELD

1. Constant weak magnetic field

Let us return to the question of the electromagnetic properties of superconductors. We shall confine ourselves in this section to the behaviour of superconductors in fairly weak fields, which are small compared with the critical magnetic field. Suppose that a superconductor with a plane surface occupies the half-space $z < 0$ as illustrated in Fig. 99, and is located in a constant magnetic field, directed parallel to its surface. We introduce the vector potential $\boldsymbol{A}$:

$$\boldsymbol{H} = \operatorname{curl} \boldsymbol{A}.$$

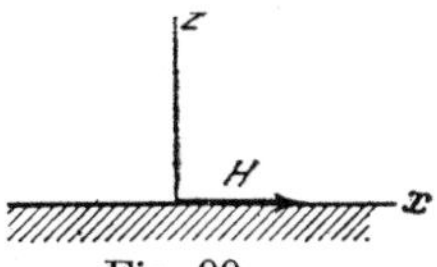

Fig. 99

In vacuo $\boldsymbol{H} = \text{const}$ and we can take the vector potential as, say,

$$A_y = -Hz, \qquad A_x = A_z = 0. \tag{37.1}$$

Current is produced in the superconductor under the action of the magnetic field; the field distribution in the superconductor is subject to Maxwell's equation

$$\nabla^2 \boldsymbol{A} = -4\pi \boldsymbol{j}. \tag{37.2}$$

Since the current density is in turn due to the presence of the field, it will be proportional to $\boldsymbol{A}$ in a linear approximation with respect to the field. It follows from uniformity considerations that, in an infinite superconductor, the connection of the current density with the field must in general be of the form

$$\boldsymbol{j}(x) = -\int Q(x-y)\,\boldsymbol{A}(y)\,\mathrm{d}^4 y \tag{37.3}$$

or in Fourier components:

$$\boldsymbol{j}(k) = -Q(k)\,\boldsymbol{A}(k). \tag{37.3'}$$

We shall not dwell in detail below on the solution of the actual electromagnetic problem as defined by (37.2), (37.3) for the half-space, but confine ourselves to deriving an expression for the kernel $Q(x-y)$, with a view to demonstrating the application of quantum field theory methods to this case.

As usual, the current density $\boldsymbol{j}$ at a given point is the thermodynamic average of the familiar quantum mechanical expression for the current operator $\boldsymbol{j}(x)$ in second quantisation:

$$\boldsymbol{j}(x) = \frac{ie}{2m}(\nabla_{r'} - \nabla_r)_{r' \to r}\, \overset{\approx}{\psi}(x')\,\tilde{\psi}(x) - \frac{e^2 \boldsymbol{A}(x)}{m}\, \overset{\approx}{\psi}(x)\,\tilde{\psi}(x). \tag{37.4}$$

The current density $\boldsymbol{j}(x)$ can thus be written directly in terms of the Green function of the system (we have put $c = 1$):

$$\boldsymbol{j}(x) = 2\left\{\frac{ie}{2m}(\nabla_{r'} - \nabla_r)\,\mathfrak{G}(x, x') - \frac{e^2 \boldsymbol{A}(x)}{m}\,\mathfrak{G}(x, x')\right\}_{r' \to r,\, \tau' \to \tau + 0} \tag{37.5}$$

We proceed to finding the Green function, or more precisely, the correction to the Green function of first order in the field. In a constant magnetic field all the Green functions $\mathfrak{G}$, $\mathfrak{F}$, and $\mathfrak{F}^+$ depend only on the difference of the "time" coordinates $\tau = \tau_1 - \tau_2$. We change to the Fourier components $\mathfrak{G}_\omega$ and $\mathfrak{F}_\omega$. The system of equations for these latter reads as follows in a constant magnetic field:

$$\left\{i\omega + \frac{1}{2m}\left(\frac{\partial}{\partial \boldsymbol{r}} - ie\boldsymbol{A}(\boldsymbol{r})\right)^2 + \mu\right\}\mathfrak{G}_\omega(\boldsymbol{r}, \boldsymbol{r}') + \Delta(\boldsymbol{r})\mathfrak{F}^+_\omega(\boldsymbol{r}, \boldsymbol{r}') = \delta(\boldsymbol{r} - \boldsymbol{r}'),$$

$$\left\{-i\omega + \frac{1}{2m}\left(\frac{\partial}{\partial \boldsymbol{r}} + ie\boldsymbol{A}(\boldsymbol{r})\right)^2 + \mu\right\}\mathfrak{F}^+_\omega(\boldsymbol{r}, \boldsymbol{r}') - \Delta^*(\boldsymbol{r})\mathfrak{G}_\omega(\boldsymbol{r}, \boldsymbol{r}') = 0. \tag{37.6}$$

We write the Green functions $\mathfrak{G}$ and $\mathfrak{F}^+$ as

$$\mathfrak{G} = \mathfrak{G}_0 + \mathfrak{G}^{(1)}; \quad \mathfrak{F} = \mathfrak{F}_0 + \mathfrak{F}^{(1)}; \quad \mathfrak{F}^+ = \mathfrak{F}^+_0 + \mathfrak{F}^{+(1)},$$

where $\mathfrak{G}_0$, $\mathfrak{F}_0$, $\mathfrak{F}^+_0$ are the Green functions in the absence of a field, and $\mathfrak{G}^{(1)}$, $\mathfrak{F}^{(1)}$, $\mathfrak{F}^{+(1)}$ are the added terms, linear in the field. We obtain on linearising equations (37.6):

$$\left\{i\omega + \frac{\nabla^2}{2m} + \mu\right\}\mathfrak{G}^{(1)}_\omega(\boldsymbol{r}, \boldsymbol{r}') + \Delta_0\mathfrak{F}^{+(1)}_\omega(\boldsymbol{r}, \boldsymbol{r}')$$
$$= -\Delta^{(1)}(\boldsymbol{r})\mathfrak{F}^+_{0\omega}(\boldsymbol{r} - \boldsymbol{r}') + \frac{ie}{2m}[(\nabla\cdot\boldsymbol{A}) + (\boldsymbol{A}\cdot\nabla)]\mathfrak{G}_{0\omega}(\boldsymbol{r} - \boldsymbol{r}'),$$
$$\left\{-i\omega + \frac{\nabla^2}{2m} + \mu\right\}\mathfrak{F}^{+(1)}_\omega(\boldsymbol{r}, \boldsymbol{r}') - \Delta_0\mathfrak{G}^{(1)}_\omega(\boldsymbol{r}, \boldsymbol{r}') \tag{37.7}$$
$$= \Delta^{*(1)}(\boldsymbol{r})\mathfrak{G}_{0\omega}(\boldsymbol{r} - \boldsymbol{r}') - \frac{ie}{2m}[(\nabla\cdot\boldsymbol{A}) + (\boldsymbol{A}\cdot\nabla)]\mathfrak{F}^+_{0\omega}(\boldsymbol{r} - \boldsymbol{r}').$$

Using these equations, we can very easily express $\mathfrak{G}^{(1)}_\omega(\boldsymbol{r}, \boldsymbol{r}')$ and $\mathfrak{F}^{+(1)}_\omega(\boldsymbol{r}, \boldsymbol{r}')$ in terms of the quantities on the right-hand sides of (37.7). This is best done by using expression (34.32) for the inverse operator to the left-hand sides of (37.7). Before writing down to result for $\mathfrak{G}^{(1)}_\omega(\boldsymbol{r}, \boldsymbol{r}')$, however, let us consider in more detail the structure of equations (37.7).

System (37.6) and hence (37.7) also are gauge-invariant, i.e. are invariant with respect to the transformations (34.26) and (34.27). Thus when finding the current $\boldsymbol{j}(\boldsymbol{r})$, which is equal in the linear approximation to

$$\boldsymbol{j}(\boldsymbol{r}) = \frac{ie}{m}T\sum_\omega(\nabla_{r'} - \nabla_r)_{r'\to r}\mathfrak{G}^{(1)}_\omega(\boldsymbol{r}, \boldsymbol{r}') - \frac{e^2\boldsymbol{A}(\boldsymbol{r})N}{m}, \tag{37.8}$$

the final result can depend only on the transverse part of the vector potential $\boldsymbol{A}$. In other words, the addition to $\boldsymbol{A}$ of the gradient of any scalar φ: $\boldsymbol{A} \to \boldsymbol{A} + \partial\varphi/\partial\boldsymbol{r}$ cannot change the current $\boldsymbol{j}(\boldsymbol{r})$.

As regards $\mathfrak{G}^{(1)}_\omega(\boldsymbol{r}, \boldsymbol{r}')$, $\mathfrak{F}^{+(1)}_\omega(\boldsymbol{r}, \boldsymbol{r}')$, by (34.27), they are by no means invariant with respect to a change of the vector potential. The same applies to $\Delta^{(1)}(\boldsymbol{r})$, $\Delta^{*(1)}(\boldsymbol{r})$, which appear on the right-hand sides of (37.7) and must themselves be determined from the integral equation

$$\Delta^{*(1)}(\boldsymbol{r}) = |\lambda|\, T \sum_{\omega} \mathfrak{F}_{\omega}^{+(1)}(\boldsymbol{r}, \boldsymbol{r}').$$

Given any $\boldsymbol{A}(\boldsymbol{r})$, $\Delta(\boldsymbol{r})$ is in general an unknown function of the potential $\boldsymbol{A}$. It can nevertheless be asserted that, in the linear approximation with respect to the field, by virtue of the uniformity of the problem, the function $\Delta^{*(1)}(\boldsymbol{r})$, being a scalar, depends only on div $\boldsymbol{A}$. Owing to this fact, if we choose our gauge such that the vector potential $\boldsymbol{A}(\boldsymbol{r})$ satisfies the equation

$$\operatorname{div} \boldsymbol{A} = 0,$$

it proves possible to simplify the problem considerably, inasmuch as, with this choice of $\boldsymbol{A}(\boldsymbol{r})$, the function $\Delta^{*(1)}(\boldsymbol{r})$ vanishes identically. The result obtained below refers only to a purely "transverse" vector potential $\boldsymbol{A}(\boldsymbol{r})$.

The method described for making $\Delta^{*(1)}$, $\Delta^{(1)}$ vanish can be extended in such a way as to make it usable in problems that do not have spatial uniformity in the absence of a field (say superconductors of finite dimensions in a magnetic field). In these cases it is possible to form a scalar from the field $\boldsymbol{A}(\boldsymbol{r})$ and the vector $\boldsymbol{r}$ or any other vector characterising the problem. In the general case of the non-uniform problem, therefore, the $\Delta^{(1)}(\boldsymbol{r})$, $\Delta^{*(1)}(\boldsymbol{r})$ in equations (37.7) (where we now have to understand by $\mathfrak{G}_0$ and $\mathfrak{F}_0$ the Green functions of the body in question with the relevant boundary conditions) depend both on the longitudinal and on the transverse components of the vector $\boldsymbol{A}(\boldsymbol{r})$. It is always possible, however, to choose the longitudinal part $\boldsymbol{A}_{\text{long}} = \operatorname{grad} \varphi$ in such a way that $\Delta^{(1)}$, $\Delta^{*(1)}$ vanish. As regards the function φ, for which this condition is fulfilled, it can be found from the condition $\operatorname{div} \boldsymbol{j} = 0$, i.e. from the condition for charge conservation.

Let us now return to equations (37.7) for an infinite superconductor. Putting $\Delta^{(1)}(\boldsymbol{r}) = 0$ and using (34.32), we obtain the following expression for the correction, linear in the field, to the Green function:

$$\mathfrak{G}_{\omega}^{(1)}(\boldsymbol{r}, \boldsymbol{r}') = \frac{ie}{m} \int \{\mathfrak{G}_{0\omega}(\boldsymbol{r}-\boldsymbol{l}) \left(\boldsymbol{A}(\boldsymbol{l}) \cdot \nabla_l\right) \mathfrak{G}_{0\omega}(\boldsymbol{l}-\boldsymbol{r}') + \mathfrak{F}_{0\omega}(\boldsymbol{l}-\boldsymbol{r}) \left(\boldsymbol{A}(\boldsymbol{l}) \cdot \nabla_l\right) \mathfrak{F}_{0\omega}^{+}(\boldsymbol{r}'-\boldsymbol{l})\} \mathrm{d}^3\boldsymbol{l}. \quad (37.9)$$

We have already made use of the condition div $\boldsymbol{A} = 0$ in this formula, so that it can be assumed that the differentiation in brackets $(\boldsymbol{A}(\boldsymbol{l}) \cdot \nabla)$ refers only to the functions $\mathfrak{G}_{0\omega}(\boldsymbol{l}-\boldsymbol{r}')$ and $\mathfrak{F}_{0\omega}^{+}(\boldsymbol{r}'-\boldsymbol{l})$. By (37.8), the current density $\boldsymbol{j}(\boldsymbol{r})$ is equal to

$$\boldsymbol{j}(\boldsymbol{r}) = \frac{e^2}{m^2} T \sum_{\omega} (\nabla_r - \nabla_{r'}) \int \{\mathfrak{G}_{0\omega}(\boldsymbol{r}-\boldsymbol{l}) \left(\boldsymbol{A}(\boldsymbol{l}) \cdot \nabla_l\right) \mathfrak{G}_{0\omega}(\boldsymbol{l}-\boldsymbol{r}') + \mathfrak{F}_{0\omega}(\boldsymbol{l}-\boldsymbol{r}) \left(\boldsymbol{A}(\boldsymbol{l}) \cdot \nabla_l\right) \mathfrak{F}_{0\omega}^{+}(\boldsymbol{r}'-\boldsymbol{l})\}_{\boldsymbol{r}' \to \boldsymbol{r}} \mathrm{d}^3\boldsymbol{l} - \frac{Ne^2}{m} \boldsymbol{A}(\boldsymbol{r}). \quad (37.10)$$

It is convenient here to change to the Fourier transforms, introducing the Fourier components for the current density $\boldsymbol{j}(\boldsymbol{r})$ and the potential $\boldsymbol{A}(\boldsymbol{r})$ in the usual way:

$$\boldsymbol{j}(\boldsymbol{r}) = \frac{1}{(2\pi)^3}\int \boldsymbol{j}(\boldsymbol{k})\, e^{i(\boldsymbol{k}\cdot\boldsymbol{r})} \mathrm{d}^3\boldsymbol{k}; \quad \boldsymbol{A}(\boldsymbol{r}) = \frac{1}{(2\pi)^3}\int \boldsymbol{A}(\boldsymbol{k})\, e^{i(\boldsymbol{k}\cdot\boldsymbol{r})} \mathrm{d}^3\boldsymbol{k}.$$

The equation for the connection between the components $\boldsymbol{j}(\boldsymbol{k})$ and $\boldsymbol{A}(\boldsymbol{k})$ is as follows:

$$\boldsymbol{j}(\boldsymbol{k}) = -\frac{2e^2T}{(2\pi)^3 m^2}\sum_{\omega}\int \boldsymbol{p}\,(\boldsymbol{p}\cdot\boldsymbol{A}(\boldsymbol{k}))\,\{\mathfrak{G}_{\omega}(\boldsymbol{p}_+)\,\mathfrak{G}_{\omega}(\boldsymbol{p}_-) + \mathfrak{F}_{\omega}(\boldsymbol{p}_+)\,\mathfrak{F}^+_{\omega}(\boldsymbol{p}_-)\}\mathrm{d}^3\boldsymbol{p} - \frac{Ne^2}{m}\boldsymbol{A}(\boldsymbol{k}), \tag{37.11}$$

where $\boldsymbol{p}_{\pm} = \boldsymbol{p} \pm 1/2\,\boldsymbol{k}$. Let us emphasise again that our result refers only to the purely transverse gauge. (It can be shown, however, that the formulae obtained remain in force for any gauge, i.e. the longitudinal part falls out from the final result.)

The field $\boldsymbol{A}(\boldsymbol{r})$ and current $\boldsymbol{j}(\boldsymbol{r})$ in a superconductor vary at distances of the order of the penetration depth δ, which is usually of the order $\sim 10^{-5}$ to 10^{-6} cm (i.e. at distances much greater than atomic distances). Hence only the components $\boldsymbol{j}(\boldsymbol{k})$ and $\boldsymbol{A}(\boldsymbol{k})$ in the range $k \sim 1/\delta \ll p_0$ are important in (37.11). We shall see below that the integration in the kernel (37.11) is performed essentially close to the Fermi surface over a narrow range of values of $|\boldsymbol{p}|$ of the order $||\boldsymbol{p}| - p_0| \sim |\boldsymbol{k}|$. Further, only the two vectors $\boldsymbol{k}$ and $\boldsymbol{A}(\boldsymbol{k})$ figure in (37.11), and $(\boldsymbol{k}\cdot\boldsymbol{A}) = 0$. On choosing the vector $\boldsymbol{k}$ as the polar axis of coordinates for the variable of integration $\boldsymbol{p}$ and carrying out the averaging over the angle in the azimuthal plane, we find at once that the vector $\boldsymbol{j}(\boldsymbol{k})$ is directed along the vector $\boldsymbol{A}(\boldsymbol{k})$. In view of what has been said above, on substituting expressions (34.35) for the functions $\mathfrak{G}$ and $\mathfrak{F}$, we get

$$\boldsymbol{j}(\boldsymbol{k}) = -\frac{Ne^2}{m}\bar{Q}(\boldsymbol{k})\,\boldsymbol{A}(\boldsymbol{k}),$$

where

$$\bar{Q}(\boldsymbol{k}) = 1 + \frac{3T}{4}\sum_{\omega}\int_0^{\pi}\sin^3\theta\,\mathrm{d}\theta\int_{-\infty}^{+\infty}\mathrm{d}\xi\,\frac{(i\omega+\xi_+)(i\omega+\xi_-)+\Delta^2}{[\omega^2+\xi_+^2+\Delta^2][\omega^2+\xi_-^2+\Delta^2]} \tag{37.12}$$

(we have used here the fact that $p_0^3/3\pi^2 = N$). For small k, $\xi_{\pm} = \xi \pm \frac{1}{2}(\boldsymbol{v}\cdot\boldsymbol{k})$.

It must be borne in mind for future calculations that, for large ω and ξ, the integrand on the right-hand side of (37.12) behaves like ω^{-2} when $\omega \gg \xi$ and like ξ^{-2} when $\xi \gg \omega$. Strictly speaking, therefore, the integral over ξ and sum over the frequencies ω are divergent. To understand the essence of the matter, we consider the singularities of the expression for a normal metal (i.e. when $\Delta = 0$):

$$\frac{3}{4} T \sum_{\omega} \int d\xi \int_0^{\pi} \sin^3\theta \, d\theta \frac{1}{(i\omega - \xi_+)(i\omega - \xi_-)}. \tag{37.13}$$

Notice, first of all, that an unusually important role is played in this integral by the order in which we carry out the summation over the frequencies and the integration over ξ. For, if we first integrate over ξ, the poles of the integrand, whatever the sign of ω, lie to one side of the real axis, so that the result is equal to zero. Suppose, now, that we first sum over the frequencies $\omega = (2n+1)\pi T$. It is easily shown that the following is obtained as a result of summation of this simple series:

$$\frac{3}{8} \int_0^{\pi} \sin^3\theta \, d\theta \int \frac{d\xi}{(\boldsymbol{v}\cdot\boldsymbol{k})} \left[\tanh\frac{\xi_-}{2T} - \tanh\frac{\xi_+}{2T}\right] = -1. \tag{37.14}$$

The reason why the final result changes on interchanging the order of the summation and integration is to be traced to the formal divergence of the entire expression. The essence of the matter is clearly, however, that the result when we first sum over the frequencies only differs from zero in a very narrow range of energies close to the Fermi surface (this region has a width $\sim(\boldsymbol{v}\cdot\boldsymbol{k})$, as follows from (37.14)). In this region the integral over the momentum proves to be rapidly convergent, and the expression for the excitation energy, measured from the Fermi surface, can be written approximately as $\xi = (\boldsymbol{p}^2 - p_0^2)/2m \simeq v(|\boldsymbol{p}| - p_0)$. It is for this reason that, in integrals of the type in question, we must first sum over the frequencies and then carry out the integration over ξ: otherwise, the integration over ξ embraces the domain $||\boldsymbol{p}| - p_0| \sim p_0$, where our expansion of the functions about the Fermi surface becomes unsuitable.

However, it is possible to avoid the need for carrying out the fairly complicated summation over the frequencies in (37.12). We have to proceed as follows. We add to and subtract from the integrand of (37.12) the corresponding expression (37.13) for a normal metal. The integral and sum over the frequencies of the difference of the integrands is now rapidly convergent, as a result of which we can change the order of integration and summation. The relevant expression for the normal metal was evaluated in (37.14); it cancels the unity in (37.12). Integration over ξ gives us

$$\bar{Q}(\boldsymbol{k}) = \frac{3\pi T}{4} \sum_{\omega} \int_{-1}^{+1} \frac{(1-\beta^2)\, d\beta}{\sqrt{\omega^2 + \Delta^2}} \cdot \frac{\Delta^2}{\omega^2 + \Delta^2 + \frac{1}{4} v^2 |\boldsymbol{k}|^2 \beta^2}. \tag{37.15}$$

Further transformation of this kernel is difficult unless some assumptions are made about $|\boldsymbol{k}|$. It is clear from the structure of the integrand that it is only the ratio of $v\,|\boldsymbol{k}|$ to the transition temperature T_c that is important. For, when $T \ll T_c$, the gap Δ_0 is of the order T_c; close to T_c, i.e. when $|T - T_c| \ll T_c$, the gap is small, though now $\omega = (2n+1)\pi T \sim T_c$. The quantity $\xi_0 \sim v/T_c$, with the dimension of a length, plays the part of a characteristic parameter in present-day superconductor theory, and is in fact the radius of correlation of the bound electrons. The penetration depth δ can be either greater or less than ξ_0. In the former case the important region $|\boldsymbol{k}| \sim 1/\delta$ satisfies the inequality $v\,|\boldsymbol{k}| \ll T_c$ whereas in the latter case it is described by the reverse inequality $v\,|\boldsymbol{k}| \gg T_c$.

Let us start with the first case. Let $v\,|\boldsymbol{k}| \ll T_c$. We retain in (37.15) only the first non-vanishing term of the expansion in $v\,|\boldsymbol{k}|$:

$$\bar{Q}(\boldsymbol{k}) = \frac{3\pi}{4}\Delta^2 T \sum_{\omega} \int_{-1}^{+1} \frac{(1-\beta^2)\,\mathrm{d}\beta}{(\omega^2+\Delta^2)^{3/2}} = \pi T \Delta^2 \sum_{\omega} \frac{1}{(\omega^2+\Delta^2)^{3/2}}. \qquad (37.16)$$

Thus, when $\delta \gg \xi_0$, the kernel $\bar{Q}(\boldsymbol{k})$ is independent of $\boldsymbol{k}$ and the connection between the current and field is of a local kind, in the sense that the current at a given point $\boldsymbol{r}$ is determined by the field $\boldsymbol{A}(\boldsymbol{r})$ at this point alone:

$$\boldsymbol{j}(\boldsymbol{r}) = -\frac{e^2 N_s}{m}\boldsymbol{A}(\boldsymbol{r}). \qquad (37.17)$$

An equation of this type was first proposed by H. and F. London [64]. It will therefore be natural to describe a superconductor in which $\delta \gg \xi_0$ as a London type superconductor. The function $N_s(T)$ plays the role of the number of "superconducting" electrons. Formula (37.16) expresses the ratio $N_s(T)/N$ as a function of the temperature. Let us emphasise that the gap Δ figuring here is the equilibrium gap in the absence of field at a given temperature, determined by condition (34.37). When $T = 0$, the summation over the frequencies can be replaced by integration: $2\pi T\,\delta n = \mathrm{d}\omega$. On evaluating the integral, we find that, at $T = 0$, the number of superconducting electrons is equal to their total number N.

Close to T_c, $\Delta(T)$ is small compared with T_c and ω. Neglecting Δ^2 in the denominator, we get the series

$$\frac{N_s(T)}{N} = \frac{2\Delta^2}{\pi^2 T^2}\sum_{n>0}\frac{1}{(2n+1)^3},$$

which has already been evaluated in § 36. Using expression (36.6) for the gap close to T_c, we find that in the present case

$$\frac{N_s(T)}{N} = 2\left(1 - \frac{T}{T_c}\right).$$

Let us now consider the second limiting case $v|\boldsymbol{k}| \gg T_c$. The integrand has poles at the points $v\,|\boldsymbol{k}|\,\beta/2 = \pm i\sqrt{\omega^2+\Delta^2}$. Since $v\,|\boldsymbol{k}|$ is large, this means that the integrand has a sharp maximum in the region of angles $\beta = \cos\theta \simeq T_c/v\,|\boldsymbol{k}| \ll 1$. Hence the term containing β^2 in the numerator can be neglected by comparison with unity. The remaining expression is a rapidly convergent integral over β, since the integrand decreases as $1/\beta^2$ in the domain $T_c/v\,|\boldsymbol{k}| \ll \beta \ll 1$. On making the substitution $v\,|\boldsymbol{k}|\,\beta = x$ and putting the limits of integration equal to infinity, the integral can be found with the aid of theory of residues:

$$\bar{Q}(\boldsymbol{k}) = \frac{3T\pi^2}{v\,|\boldsymbol{k}|} \sum_{\omega>0} \frac{\Delta^2}{\omega^2 + \Delta^2} = \frac{3\pi^2}{4v\,|\boldsymbol{k}|} \Delta \tanh \frac{\Delta}{2T}. \tag{37.18}$$

We see that the kernel $\bar{Q}(\boldsymbol{k})$ is in this case essentially dependent on $\boldsymbol{k}$. Hence, if the field penetration depth $\delta \ll \xi_0$, relationship (37.3) is non-local, in other words, the current density $\boldsymbol{j}(\boldsymbol{r})$ at a given point is determined by the values of the vector potential throughout a neighbourhood of the point with linear dimensions of order ξ_0. The non-local connection between the field and current for certain superconductors was first predicted by Pippard on the basis of an analysis of experimental data [65]. Throughout what follows we shall describe the case when $\delta \ll \xi_0$ as Pippard's case.

There is an important fact that must be mentioned at once. As we have already remarked, the only important factor for deciding which case actually holds is the ratio of the penetration depth δ to the parameter $\xi_0 \sim v/T_c$. If, therefore, the condition $\delta \ll \xi_0$ is fulfilled at low temperatures, the increase in δ as we approach T_c will mean that the reverse situation arises at temperatures sufficiently close to T_c, δ will become much greater than ξ_0. In other words, the London case will always hold in the immediate neighbourhood of T_c. A considerable number of the familiar superconductors belongs to the Pippard type almost throughout the whole temperature range and they only change over to the London type in an extremely narrow neighbourhood $T_c - T \ll T_c$. The remaining pure superconductors represent an intermediate case at low temperatures, i.e. they have a fairly wide London range of temperatures close to T_c. We shall not touch here on the subject of alloys (see § 39).

Having expression (37.15) for the kernel $\bar{Q}(\boldsymbol{k})$ at our disposal, we can use Maxwell's equations to solve the problem of the field penetration into a superconductor with a plane surface. The solution is particularly simple in the London case: on substituting (37.17) into (37.2) and taking all the functions as functions of z only, we find that the vector potential distribution in the superconductor is described by

$$A_y(z) = -H_0\,\delta e^{z/\delta},$$

where δ is the London penetration depth

$$\delta = \sqrt{\frac{m}{4\pi N_s e^2}} \quad \left(\text{in ordinary units } \delta = \sqrt{\frac{mc^2}{4\pi N_s e^2}}\right). \tag{37.19}$$

The solution of the problem in the Pippard case is much more difficult and requires the use of special mathematical methods. We shall not dwell on this, but refer the reader who is interested in the theory of superconductors as such to the original literature (see [58], and also [63]).

2. *Superconductor in variable fields*

We have confined ourselves so far to a discussion of the properties of superconductors in constant magnetic fields. A problem of great physical interest is also the behaviour of a superconductor in an alternating electromagnetic field, or, in more concrete terms, the nature of the absorption and reflection of electromagnetic radiation incident on the surface of a superconductor. The thermodynamic, equilibrium approach on which the previous treatment has been based, is not immediately applicable in the case of an alternating field. We shall find extremely useful in these circumstances the analytic expressions obtained in Chap. III, connecting the various time-dependent functions with the corresponding functions determined by the thermodynamic method. Suppose that an alternating field $\boldsymbol{A}$ (as above, we shall assume that $\varphi = 0$) of frequency ω exists inside an infinite superconductor. The current in the superconductor resulting from the action of the field is obviously connected, as before, with the field by a relationship of the type (37.3). The difference here is that, with an alternating field, we have to know the Fourier component $Q(\boldsymbol{k}, \omega)$ at non-zero ω. (The kernel $\overline{Q}(\boldsymbol{k})$ defined above is evidently $\overline{Q}(\boldsymbol{k}) \equiv Q(\boldsymbol{k}, 0)$.) We shall again start from the quantum mechanical expression for the current operator:

$$\hat{\tilde{\boldsymbol{j}}}(\boldsymbol{x}) = \frac{ie}{2m}(\nabla_{r'} - \nabla_r)_{r' \to r} \tilde{\psi}^+(x')\tilde{\psi}(x) - \frac{e^2}{m}\boldsymbol{A}(x)\tilde{\psi}^+(x)\tilde{\psi}(x)$$

$$\equiv \hat{\tilde{\boldsymbol{j}}}_1(x) - \frac{e^2}{m}\boldsymbol{A}(x)\tilde{\psi}^+(x)\psi(x),$$

where the operators are written in the Heisenberg representation and contain the field dependence. The connection with the corresponding operators in the interaction representations is, by (6.28), given by

$$\hat{\tilde{\boldsymbol{j}}} = S^{-1}(t)\boldsymbol{j}S(t),$$

where

$$S(t) = T\exp\left\{i\int_{-\infty}^{t}(\boldsymbol{j}(x)\boldsymbol{A}(x))\,\mathrm{d}^4x\right\}.$$

In the approximation linear in the field:

$$\hat{\tilde{\boldsymbol{j}}}_\alpha(x) = \hat{\boldsymbol{j}}_{1\alpha}(x) - \frac{e^2}{m}\boldsymbol{A}_\alpha(x)\psi^+(x)\psi(x) + i\int_{-\infty}^{t}[\hat{\boldsymbol{j}}_{1\alpha}(x), \hat{\boldsymbol{j}}_{1\beta}(y)]\boldsymbol{A}_\beta(y)\,\mathrm{d}^4y.$$

The current in the superconductor at a given point and given instant is the average

$$\boldsymbol{j}(x) = \langle \hat{\hat{\boldsymbol{j}}}(x) \rangle = \sum_m \exp\left[(\Omega + \mu N_m - E_m)/T\right] \langle m | \hat{\hat{\boldsymbol{j}}}(x) | m \rangle .$$

Since $\langle \hat{\boldsymbol{j}}_1 \rangle \equiv 0$,

$$\boldsymbol{j}_\alpha(x) = -\frac{e^2 N}{m} \boldsymbol{A}_\alpha(x) + \int P^R_{\alpha\beta}(x-y) \boldsymbol{A}_\beta(y) \, \mathrm{d}^4 y, \tag{37.20}$$

where we have introduced the notation

$$P^R_{\alpha\beta}(x-y) = \begin{cases} i \langle [\hat{\boldsymbol{j}}_\alpha(x), \hat{\boldsymbol{j}}_\beta(y)] \rangle & \text{for } t_x > t_y, \\ 0 & \text{for } t_x < t_y . \end{cases} \tag{37.21}$$

On Fourier transforming we get for the kernel of equation (37.3)

$$Q_{\alpha\beta}(\boldsymbol{k}, \omega) = \frac{e^2 N}{m} \delta_{\alpha\beta} - P^R_{\alpha\beta}(\boldsymbol{k}, \omega) .$$

We now consider the same problem in the technique at finite temperatures, $\boldsymbol{A}(\boldsymbol{r}, \tau)$ and $\boldsymbol{j}(\boldsymbol{r}, \tau)$ being formally regarded as functions of the "time" parameter τ. The following relation now occurs in place of (37.20):

$$\boldsymbol{j}_\alpha(\boldsymbol{r}, \tau) = -\frac{e^2 N}{m} \boldsymbol{A}_\alpha(\boldsymbol{r}, \tau) + \int \mathrm{d}^3 \boldsymbol{r} \int_0^\beta \mathscr{P}_{\alpha\beta}(x-y) \boldsymbol{A}_\beta(y) \, \mathrm{d}\tau_y ,$$

where

$$\mathscr{P}_{\alpha\beta}(\boldsymbol{r} - \boldsymbol{r}', \tau - \tau') = \langle T \left(\hat{\boldsymbol{j}}_{\alpha 1}(\boldsymbol{r}, \tau), \hat{\boldsymbol{j}}_{1\beta}(\boldsymbol{r}', \tau') \right) \rangle . \tag{37.22}$$

On introducing the Fourier components of the thermodynamic functions, we find that the role of the kernel $Q_{\alpha\beta}(\boldsymbol{k}, \omega_0)$ is played by

$$Q_{\alpha\beta}(\boldsymbol{k}, \omega_0) = \frac{e^2 N}{m} \delta_{\alpha\beta} - \mathscr{P}_{\alpha\beta}(\boldsymbol{k}, \omega_0) ,$$

where the frequencies ω_0 run over the discrete values $\omega_0 = 2n\pi T$.

We shall shortly prove in a general form that the Fourier components $P^R_{\alpha\beta}(\boldsymbol{k}, \omega)$ and $\mathscr{P}_{\alpha\beta}(\boldsymbol{k}, \omega_0)$ are values of the same function of a complex variable ω, analytic in the upper half-plane, taken in the first case on the real axis, and in the second at the points $\omega = i\omega_0$. The method of proof is precisely the same as in previous chapters. We expand (37.21) and (37.22) into sums over intermediate states. We now find for the Fourier components, $P^R_{\alpha\beta}$, $\mathscr{P}_{\alpha\beta}$, suitably defined in this case:

$$P^R_{\alpha\beta}(\boldsymbol{k}, \omega) = -\sum_{m,p} \varrho_{pm}(\boldsymbol{k}) \frac{1}{\omega - \omega_{pm} + i\delta} ,$$

$$\mathscr{P}_{\alpha\beta}(\boldsymbol{k}, \omega_0) = -\sum_{m,p} \varrho_{pm}(\boldsymbol{k}) \frac{1}{i\omega_0 - \omega_{pm}} , \tag{37.23}$$

where

$$\varrho_{pm}(\boldsymbol{k}) = \mathrm{e}^{(\Omega+\mu N_m - E_m)/T}\,(1-\mathrm{e}^{-\omega_{pm}/T})\,(\boldsymbol{j}_{\alpha 1})_{mp}(\boldsymbol{j}_{\beta 1})_{pm}(2\pi)^3\delta(\boldsymbol{k}-\boldsymbol{k}_{pm}).$$

It is clear from (37.23) that $P^R_{\alpha\beta}(\boldsymbol{k},\omega)$ is obtained from $\mathcal{P}_{\alpha\beta}(\boldsymbol{k},\omega_0)$ by replacing ω_0 by $-i\omega$, whilst the values of $P^R_{\alpha\beta}(\boldsymbol{k},\omega)$ on the real axis must be chosen as the limits when ω tends to the axis from above.

Thus, if we find $\mathcal{P}_{\alpha\beta}(k,\omega_0)$ in the Matsubara technique and continue it analytically onto real frequencies $P^R_{\alpha\beta}(\boldsymbol{k},\omega) = \mathcal{P}_{\alpha\beta}(\boldsymbol{k},-i\omega)$ in such a way that the function obtained has no singularities in the upper half-plane of ω, we can in principle find the kernel $Q_{\alpha\beta}(\boldsymbol{k},\omega)$ which determines the connection between $\boldsymbol{j}$ and $\boldsymbol{A}$ in an alternating electromagnetic field.

Whilst keeping this purpose in mind, let us consider formally the equations for the thermodynamic functions $\mathfrak{G}$ and $\mathfrak{F}^+$ in a field alternating in τ of the form $\boldsymbol{A}(\boldsymbol{r},\tau) = \boldsymbol{A}(\boldsymbol{k},\omega_0)e^{i\boldsymbol{kr}-i\omega_0\tau}$. Instead of (37.11), we now get for the Fourier components of the current $\boldsymbol{j}(\boldsymbol{k},\omega_0)$:

$$\boldsymbol{j}(\boldsymbol{k},\omega_0) = -\frac{2e^2}{(2\pi)^3 m^2}T\sum_{\omega}\int \boldsymbol{p}\,(\boldsymbol{p}\cdot\boldsymbol{A}(\boldsymbol{k},\omega_0))\,[\mathfrak{G}(p_+)\mathfrak{G}(p_-) + \mathfrak{F}(p_+)\mathfrak{F}^+(p_-)]\,\mathrm{d}^3\boldsymbol{p} - \frac{Ne^2}{m}\boldsymbol{A}(\boldsymbol{k},\omega_0)$$

(here $p_\pm = \{\boldsymbol{p}\pm 1/2\boldsymbol{k};\ \omega'\pm 1/2\omega_0\}$). On repeating the course of the arguments that led us, in the case of a constant field, to expression (37.15), we get after integration over ξ:

$$\overline{Q}(\boldsymbol{k},\omega_0) = \frac{3\pi T}{4}\sum_{\omega'}\int_{-1}^{+1}(1-\beta^2)\,\mathrm{d}\beta \times\left\{\frac{i(\omega_+ + \sqrt{\omega_+^2+\Delta^2})\,[i(\omega_- + \sqrt{\omega_+^2+\Delta^2}) - v|\boldsymbol{k}|\beta] + \Delta^2}{\sqrt{\omega_+^2+\Delta^2}\,[\omega_-^2+\Delta^2+(v|\boldsymbol{k}|\beta - i\sqrt{\omega_+^2+\Delta^2})^2]} + \frac{i(\omega_- + \sqrt{\omega_-^2+\Delta^2})\,[i(\omega_+ + \sqrt{\omega_-^2+\Delta^2}) + v|\boldsymbol{k}|\beta] + \Delta^2}{\sqrt{\omega_-^2+\Delta^2}\,[\omega_+^2+\Delta^2+(v|\boldsymbol{k}|\beta + i\sqrt{\omega_-^2+\Delta^2})^2]}\right\}. \tag{37.24}$$

As above, certain assumptions must be made in regard to $v|\boldsymbol{k}|$ in order further to simplify this expression. We shall confine ourselves below to the case of most practical interest, when $v|\boldsymbol{k}| \gg (T_c,\omega_0)$. In this case, as before, the main contribution to the expression for $\overline{Q}(\boldsymbol{k},\omega_0)$ comes from the range of angles where $\beta \sim T_c/v|\boldsymbol{k}|,\ \omega'/v|\boldsymbol{k}|$, so that we can neglect β^2 by comparison with unity in the numerator of the integrand. The remaining expression in curly brackets decreases when $\beta \gg T_c/v|\boldsymbol{k}|$, $\omega'/v|\boldsymbol{k}|$ rather more slowly than in the $T = 0$ case, namely, as $1/\beta$. It is therefore advisable to carry out a rearrangement of (37.24), separating out the more slowly decreasing terms:

$$\overline{Q}(\boldsymbol{k}, \omega_0)$$

$$= \frac{3\pi T}{4} \sum_{\omega'} \int_{-1}^{1} \mathrm{d}\beta \left\{ \frac{\Delta^2 - (\omega_+ + \sqrt{\omega_+^2 + \Delta^2})(\omega_- + \sqrt{\omega_-^2 + \Delta^2})}{\sqrt{\omega_+^2 + \Delta^2}\,[\omega_-^2 + \Delta^2 + (v\,|\boldsymbol{k}|\,\beta - i\sqrt{\omega_+^2 + \Delta^2})^2]} \right.$$

$$+ \frac{\Delta^2 - (\omega_- + \sqrt{\omega_-^2 + \Delta^2})(\omega_+ + \sqrt{\omega_+^2 + \Delta^2})}{\sqrt{\omega_-^2 + \Delta^2}\,[\omega_+^2 + \Delta^2 + (v\,|\boldsymbol{k}|\,\beta + i\sqrt{\omega_-^2 + \Delta^2})^2]}$$

$$- i \frac{\omega_+ + \sqrt{\omega_+^2 + \Delta^2}}{\sqrt{\omega_+^2 + \Delta^2}\,[v\,|\boldsymbol{k}|\,\beta - i(\sqrt{\omega_+^2 + \Delta^2} + \sqrt{\omega_-^2 + \Delta^2})]}$$

$$\left. + i \frac{\omega_- + \sqrt{\omega_-^2 + \Delta^2}}{\sqrt{\omega_-^2 + \Delta^2}\,[v\,|\boldsymbol{k}|\,\beta + i(\sqrt{\omega_+^2 + \Delta^2} + \sqrt{\omega_-^2 + \Delta^2})^2]} \right\}.$$

On carrying out the integration and passing to the limit as $v\,|\boldsymbol{k}| \to \infty$, we get

$$\overline{Q}(\boldsymbol{k}, \omega_0) = \frac{3\pi^2 T}{4v\,|\boldsymbol{k}|} \sum_{\omega_1} \left[1 + \frac{\Delta^2 - \omega'(\omega' - \omega_0)}{\sqrt{\omega'^2 + \Delta^2}\,\sqrt{(\omega' - \omega_0)^2 + \Delta^2}} \right]. \tag{37.25}$$

When $\omega_0 = 0$ this result tranforms to (37.18).

Since ω runs over the values $(2n+1)\pi T$ in the summation of (37.25), $\overline{Q}(\boldsymbol{k}, \omega_0)$ can be written as the contour integral:

$$\overline{Q}(\boldsymbol{k}, \omega_0) = \frac{3\pi i}{16 v\,|\boldsymbol{k}|} \int_C \tan \frac{\omega'}{2T} \left\{ 1 + \frac{\Delta^2 - \omega'(\omega' - \omega_0)}{\sqrt{\omega'^2 + \Delta^2}\,\sqrt{(\omega' - \omega_0)^2 + \Delta^2}} \right\} \mathrm{d}\omega, \tag{37.26}$$

where the contour C consists of two pieces C_+ and C_-, as illustrated in Fig. 100. The choice of the analytic branches of the functions $\sqrt{\omega'^2 + \Delta^2}$ and $\sqrt{(\omega' - \omega_0)^2 + \Delta^2}$ is clear from the same figure: the values of these functions on the cuts are purely imaginary, the imaginary part being positive to the right of the upper cut and to the left of the lower cut. We pass from an integration over the contours C_+ and C_- to an integration over the four contours $C_+^{(1,2)}$ and $C_-^{(1,2)}$ (Fig. 101). It is easily seen that the integrals over $C_+^{(2)}$ and $C_-^{(2)}$, considered formally as functions of ω_0, have singularities at $\omega_0 = (2n+1)\pi T$, since the contour of integration passes in this case through the point $\omega' = (2n+1)\pi T$, where $\tan \omega'/2T$ becomes infinite. Hence, in order to find the branch of the function, analytic in the upper half-plane of the variable $\omega = i\omega_0$, we have to transform expression (37.26) at particular points $\omega = 2n\pi T i$ in such a way that, on later extending this expression to arbitrary values of ω, the contour of integration does not pass through the singularities of the integrand. We observe for this, that if $\omega_0 = 2n\pi T$, by virtue of the

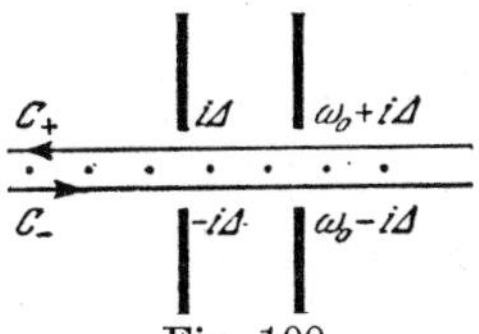

Fig. 100

periodicity of $\tan \omega'/2T$, the integral over the contour $C_-^{(2)}$ is equal to the integral over $C_+^{(1)}$ (and the same for contours $C_+^{(2)}$ and $C_-^{(1)}$). We can easily verify what has been said by replacing the variable of integration, say, in the integral over $C_-^{(2)}$ in accordance with $\omega' - \omega_0 = -u$. We can thus write (37.26) as

$$\bar{Q}(\boldsymbol{k}, \omega_0) = \frac{3\pi i}{8v|\boldsymbol{k}|}\left(\int\limits_{C_-^{(1)}} + \int\limits_{C_+^{(1)}}\right)\left\{\tan\frac{\omega'}{2T}\left[1 + \frac{\Delta^2 - \omega'(\omega' - \omega_0)}{\sqrt{\omega'^2 + \Delta^2}\sqrt{(\omega' - \omega_0)^2 + \Delta^2}}\right]\right\} d\omega.$$

The expression obtained, regarded formally as a function of $\omega = i\omega_0$, is analytic in the upper half-plane of ω, since the contour of integration, with $\operatorname{Im} \omega > 0$, never passes through a singularity of the integrand.

We can now write down directly the $\bar{Q}(\boldsymbol{k}, \omega)$ for $\omega > 0$ in which we are interested. Two cases arise: (a) $\omega < 2\Delta$ and (b) $\omega > 2\Delta$. The elementary calculation is performed in both cases in accordance with Fig. 102, where we have indicated in brackets the choice of sign of the imaginary part of the functions on the different sides of the cuts. Let us quote the results obtained [66, 67]:

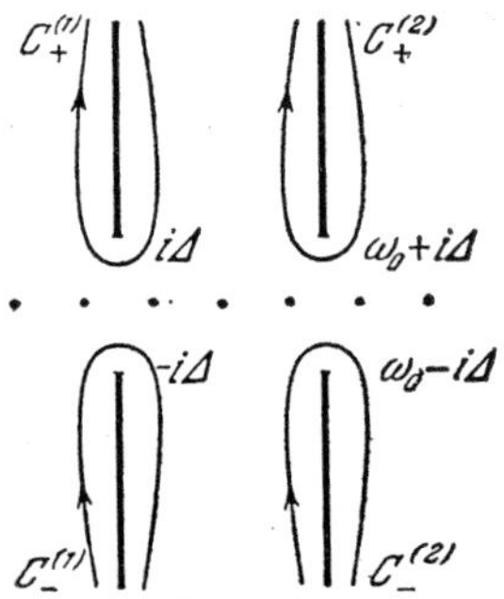

Fig. 101

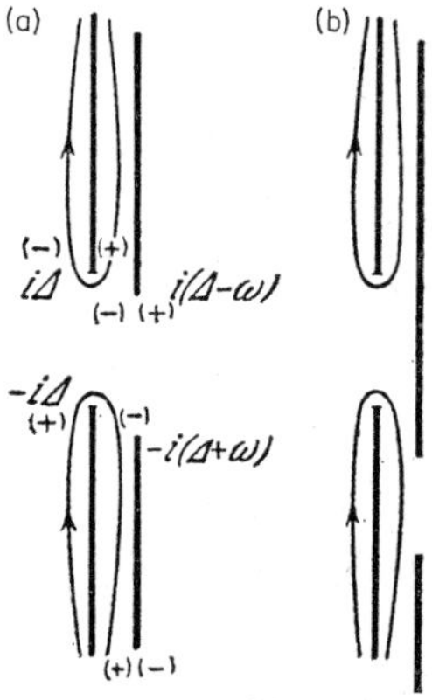

Fig. 102

(a) $\omega < 2\Delta$:

$$\bar{Q}(\boldsymbol{k}, \omega) = \frac{3\pi}{4v|\boldsymbol{k}|}\left\{\int\limits_{\Delta}^{\Delta+\omega} \tanh\frac{\omega'}{2T}\frac{\omega'(\omega' - \omega) + \Delta^2}{\sqrt{\omega'^2 - \Delta^2}\sqrt{\Delta^2 - (\omega' - \omega)^2}} d\omega' \right.$$

$$\left. + i\int\limits_{\Delta}^{\infty}\left[\tanh\frac{\omega'}{2T} - \tanh\frac{\omega' + \omega}{2T}\right]\frac{\omega'(\omega' + \omega) + \Delta^2}{\sqrt{\omega'^2 - \Delta^2}\sqrt{(\omega' + \omega)^2 - \Delta^2}} d\omega'\right\},$$

(b) $\omega > 2\Delta$:

$$\bar{Q}(\boldsymbol{k}, \omega) = \frac{3\pi}{4v|\boldsymbol{k}|}\left\{\int\limits_{\omega-\Delta}^{\omega+\Delta} \tanh\frac{\omega'}{2T}\frac{\omega'(\omega' - \omega) + \Delta^2}{\sqrt{\omega'^2 - \Delta^2}\sqrt{\Delta^2 - (\omega' - \omega)^2}} d\omega' \right.$$

$$+ i\int\limits_{\Delta}^{\omega-\Delta} \tanh\frac{\omega'}{2T}\frac{\omega'(\omega' - \omega) + \Delta^2}{\sqrt{\omega'^2 - \Delta^2}\sqrt{(\omega' - \omega)^2 - \Delta^2}} d\omega'$$

$$\left. + i\int\limits_{\Delta}^{\infty}\left[\tanh\frac{\omega'}{2T} - \tanh\frac{\omega' + \omega}{2T}\right]\frac{\omega'(\omega' + \omega) + \Delta^2}{\sqrt{\omega'^2 - \Delta^2}\sqrt{(\omega' + \omega)^2 - \Delta^2}} d\omega'\right\}.$$

§ 38. PROPERTIES OF SUPERCONDUCTORS CLOSE TO THE TRANSITION TEMPERATURE IN AN ARBITRARY MAGNETIC FIELD

The properties of superconductors close to the critical temperature provide a special case. Here the size of the gap is fairly small, as a result of which all the equations are greatly simplified. It is easily seen from the results of § 36 (expression (36.4)) that it is now possible to expand the equations in the quantity $1 - T/T_c \gg 1$. In addition, as we have already remarked in the previous section, close to T_c the depth of penetration of a weak magnetic field $\delta \ll \xi_0$, i.e. all the functions in the field, including the field itself, vary at distances which are much greater than the parameters $\xi_0 \sim v/T_c$ of the theory. This fact enables us to develop a theory in this temperature range (Gor'kov [68]), which describes the behaviour of superconductors in arbitrary magnetic fields (of the order of the critical field).

With this aim, we again rewrite equations (37.6):

$$\left\{i\omega + \frac{1}{2m}\left(\frac{\partial}{\partial \boldsymbol{r}} - ie\boldsymbol{A}(\boldsymbol{r})\right)^2 + \mu\right\} \mathfrak{G}_\omega(\boldsymbol{r}, \boldsymbol{r}') + \Delta(\boldsymbol{r}) \mathfrak{F}^+_\omega(\boldsymbol{r}, \boldsymbol{r}') = \delta(\boldsymbol{r} - \boldsymbol{r}'),$$

$$\left\{-i\omega + \frac{1}{2m}\left(\frac{\partial}{\partial \boldsymbol{r}} + ie\boldsymbol{A}(\boldsymbol{r})\right)^2 + \mu\right\} \mathfrak{F}^+_\omega(\boldsymbol{r}, \boldsymbol{r}') - \Delta^*(\boldsymbol{r}) \mathfrak{G}_\omega(\boldsymbol{r}, \boldsymbol{r}') = 0$$

together with the equation that determines the gap size:

$$\Delta^*(\boldsymbol{r}) = |\lambda|\, T \sum_\omega \mathfrak{F}^+_\omega(\boldsymbol{r}, \boldsymbol{r}). \tag{38.1}$$

Since $|\Delta|$ is small, we expand $\mathfrak{F}^+_\omega(\boldsymbol{r}, \boldsymbol{r}')$ in powers of $|\Delta|$ and, on substituting this expansion in (38.1), find an equation for $\Delta^*(\boldsymbol{r})$. It is useful here to introduce the Fourier components of the Green function $\tilde{\mathfrak{G}}^{(0)}_\omega(\boldsymbol{r}, \boldsymbol{r}')$ for the electrons in the normal metal (in the given field $\boldsymbol{A}(\boldsymbol{r})$). The equation which $\tilde{\mathfrak{G}}^{(0)}_\omega(\boldsymbol{r}, \boldsymbol{r}')$ satisfies can be written in two ways:

$$\left\{i\omega + \frac{1}{2m}\left(\frac{\partial}{\partial \boldsymbol{r}} - ie\boldsymbol{A}(\boldsymbol{r})\right)^2 + \mu\right\} \tilde{\mathfrak{G}}^{(0)}_\omega(\boldsymbol{r}, \boldsymbol{r}') = \delta(\boldsymbol{r} - \boldsymbol{r}') \tag{38.2}$$

or

$$\left\{i\omega + \frac{1}{2m}\left(\frac{\partial}{\partial \boldsymbol{r}'} + ie\boldsymbol{A}(\boldsymbol{r}')\right)^2 + \mu\right\} \tilde{\mathfrak{G}}^{(0)}_\omega(\boldsymbol{r}, \boldsymbol{r}') = \delta(\boldsymbol{r} - \boldsymbol{r}'). \tag{38.2'}$$

Taking the second form of the Green function, we can use it to reduce the system of equations for $\mathfrak{G}_\omega$ and $\mathfrak{F}^+_\omega$ to the integral form:

$$\begin{aligned} \mathfrak{G}_\omega(\boldsymbol{r}, \boldsymbol{r}') &= \tilde{\mathfrak{G}}^{(0)}_\omega(\boldsymbol{r}, \boldsymbol{r}') - \int \tilde{\mathfrak{G}}^{(0)}_\omega(\boldsymbol{r}, \boldsymbol{l})\, \Delta(\boldsymbol{l}) \mathfrak{F}^+_\omega(\boldsymbol{l}, \boldsymbol{r}')\, d^3\boldsymbol{l}, \\ \mathfrak{F}^+_\omega(\boldsymbol{r}, \boldsymbol{r}') &= \int \tilde{\mathfrak{G}}^{(0)}_{-\omega}(\boldsymbol{l}, \boldsymbol{r}) \Delta^*(\boldsymbol{l})\, \mathfrak{G}_\omega(\boldsymbol{l}, \boldsymbol{r}')\, d^3\boldsymbol{l}. \end{aligned} \tag{38.3}$$

Before proceeding further, we find the value of $\tilde{\mathfrak{G}}^{(0)}_\omega(\boldsymbol{r}, \boldsymbol{r}')$. In the absence of a magnetic field, $\tilde{\mathfrak{G}}^{(0)}_\omega(\boldsymbol{r} - \boldsymbol{r}')$ is equal to ($R = |\boldsymbol{r} - \boldsymbol{r}'|$):

$$\mathfrak{G}^{(0)}_\omega(R) = \begin{cases} -\dfrac{m}{2\pi R}\, e^{ip_0 R - |\omega| R/v} & \text{for } \omega > 0, \\[2ex] -\dfrac{m}{2\pi R}\, e^{-ip_0 R - |\omega| R/v} & \text{for } \omega < 0. \end{cases} \tag{38.4}$$

This can be verified either by direct substitution of (38.4) in (38.2) in the absence of field, or by using the now familiar expression for the Fourier component $\mathfrak{G}_\omega^{(0)}(\boldsymbol{p}) = [i\omega - \xi]^{-1}$:

$$\mathfrak{G}_\omega^{(0)}(R) = \frac{1}{(2\pi)^3}\int \mathfrak{G}_\omega^{(0)}(\boldsymbol{p})\mathrm{d}^3\boldsymbol{p}\, e^{i(\boldsymbol{p}\cdot\boldsymbol{R})}$$

$$= \frac{m}{(2\pi)^2 iR}\int \frac{e^{ip_0R+i\xi R/v} - e^{-ip_0R-i\xi R/v}}{i\omega - \xi}\mathrm{d}\xi$$

(we are of course interested in the form of $\mathfrak{G}_\omega^{(0)}(R)$ at distances which are larged compared with the atomic distances: $Rp_0 \gg 1$). On integrating over ξ, we obtain (38.4) immediately.

The function $\mathfrak{G}_\omega^{(0)}(R)$ oscillates rapidly. Since $p_0 R \gg 1$, this fact enables us to find $\tilde{\mathfrak{G}}_\omega^{(0)}(\boldsymbol{r}, \boldsymbol{r}')$ in a magnetic field by making use of the quasi-classical approximation. In fact, we look for $\tilde{\mathfrak{G}}_\omega^{(0)}(\boldsymbol{r}, \boldsymbol{r}')$ in the form

$$\tilde{\mathfrak{G}}_\omega^{(0)}(\boldsymbol{r}, \boldsymbol{r}') = e^{i\varphi(\boldsymbol{r},\boldsymbol{r}')}\mathfrak{G}_\omega^{(0)}(\boldsymbol{r} - \boldsymbol{r}'), \tag{38.5}$$

where $\varphi(\boldsymbol{r}, \boldsymbol{r}) = 0$. On substituting (38.5) into (38.2) and differentiating only the principal terms, we obtain an equation for the extra term $\varphi(\boldsymbol{r}, \boldsymbol{r}')$ in the action:

$$\left(\boldsymbol{n}\cdot\nabla_{\boldsymbol{r}}\varphi(\boldsymbol{r}, \boldsymbol{r}')\right) = e\left(\boldsymbol{n}, \boldsymbol{A}(\boldsymbol{r})\right); \quad \left(\boldsymbol{n} = \frac{\boldsymbol{R}}{|\boldsymbol{R}|}\right). \tag{38.6}$$

We have thrown away the terms quadratic in $\boldsymbol{A}$ in this equation, since the radius of curvature of the electrons ep_0/H in the fields of interest is extremely large by comparison with the depth of penetration: $p_0 \gg Ae \sim eH\delta$ (δ is the order of the depth of penetration).

We now return to equations (38.3) and carry out expansions in them in powers of $|\Delta(\boldsymbol{r})|$. As is clear from (38.4) and (38.6), it is sufficient to carry out this expansion in $\mathfrak{F}_\omega^+(\boldsymbol{r}, \boldsymbol{r}')$ up to the third power in $|\Delta|$. As regards the Green function $\mathfrak{G}_\omega(\boldsymbol{r}, \boldsymbol{r}')$, we only need to know it up to second order terms in $|\Delta|$:

$$\mathfrak{G}_\omega(\boldsymbol{r}, \boldsymbol{r}') = \tilde{\mathfrak{G}}_\omega^{(0)}(\boldsymbol{r}, \boldsymbol{r}') - \int \tilde{\mathfrak{G}}_\omega^{(0)}(\boldsymbol{r}, \boldsymbol{l})\Delta(\boldsymbol{l})\,\tilde{\mathfrak{G}}_\omega^{(0)}(\boldsymbol{m}, \boldsymbol{r}')\Delta^*(\boldsymbol{m})\tilde{\mathfrak{G}}_{-\omega}^{(0)}(\boldsymbol{m}, \boldsymbol{l})\,\mathrm{d}^3\boldsymbol{m}\,\mathrm{d}^3\boldsymbol{l}. \tag{38.7}$$

On substituting this expression in the second of equations (38.3), we obtain the expansion of $\mathfrak{F}_\omega^+(\boldsymbol{r}, \boldsymbol{r}')$, with the aid of which the following equation in $\Delta^*(\boldsymbol{r})$ is obtained from (38.1):

$$\Delta^*(\boldsymbol{r}) = |\lambda|\, T\sum_\omega \int \tilde{\mathfrak{G}}_\omega^{(0)}(\boldsymbol{l}, \boldsymbol{r})\Delta^*(\boldsymbol{l})\,\tilde{\mathfrak{G}}_{-\omega}^{(0)}(\boldsymbol{l}, \boldsymbol{r})\,\mathrm{d}^3\boldsymbol{l}$$
$$- |\lambda|\, T\sum_\omega \int\int\int \tilde{\mathfrak{G}}_\omega^{(0)}(\boldsymbol{l}, \boldsymbol{m})\Delta(\boldsymbol{m})\,\tilde{\mathfrak{G}}_\omega^{(0)}(\boldsymbol{s}, \boldsymbol{r})\Delta^*(\boldsymbol{s})\tilde{\mathfrak{G}}_{-\omega}^{(0)}(\boldsymbol{s}, \boldsymbol{m})$$
$$\times\Delta^*(\boldsymbol{l})\tilde{\mathfrak{G}}_{-\omega}^{(0)}(\boldsymbol{l}, \boldsymbol{r})\,\mathrm{d}^3\boldsymbol{l}\,\mathrm{d}^3\boldsymbol{m}\,\mathrm{d}^3\boldsymbol{s}. \tag{38.8}$$

The important distances in the integrals in this equation are of order ξ_0, since, as is clear from (38.4) and (38.5), the function $\tilde{\mathfrak{G}}_\omega^{(0)}(\boldsymbol{r}, \boldsymbol{r}')$ decreases exponentially for $|\boldsymbol{r}-\boldsymbol{r}'|>\xi_0$. The change in the gap $\Delta(\boldsymbol{r})$ and field $\boldsymbol{A}(\boldsymbol{r})$ occur at distances of the order of the penetration depth, which is much greater than ξ_0 close to the critical temperature. For the same reasons, the phase $\varphi(\boldsymbol{r}, \boldsymbol{r}')$ in (38.5) can be written as

$$\varphi(\boldsymbol{r}, \boldsymbol{r}') \simeq e(\boldsymbol{A}(\boldsymbol{r}) \cdot \boldsymbol{r}-\boldsymbol{r}').$$

Close to T_c, $\boldsymbol{A}(\boldsymbol{r}) \sim H\delta \sim \sqrt{1-(T/T_c)}$, so that the phase $\varphi(\boldsymbol{r}, \boldsymbol{r}')$ is small and the exponent can be expanded in powers of φ.

We start by considering the first term on the right-hand side. Let

$$K(\boldsymbol{l}, \boldsymbol{r}) = T\sum_\omega \mathfrak{G}^{(0)}_\omega(\boldsymbol{l}, \boldsymbol{r})\mathfrak{G}^{(0)}_{-\omega}(\boldsymbol{l}, \boldsymbol{r}) = K_0(\boldsymbol{l}-\boldsymbol{r})\exp[2ie(\boldsymbol{A}(\boldsymbol{r})\cdot\boldsymbol{l}-\boldsymbol{r})].$$

Using the representation (38.4) for the Green function in coordinate space and performing a summation over the frequencies, we get the following expression for $K_0(R)$:

$$K_0(R) = \frac{m^2 T}{(2\pi R)^2}\frac{1}{\sinh\dfrac{2\pi T R}{v}}. \tag{38.9}$$

As we have already remarked, all the functions change little over distances of the order of ξ_0, so that we can expand all the functions in the integral

$$\int K_0(\boldsymbol{l}-\boldsymbol{r})\exp[2ie(\boldsymbol{A}(\boldsymbol{r})\cdot\boldsymbol{l}-\boldsymbol{r})]\,\Delta^*(\boldsymbol{l})\,\mathrm{d}^3\boldsymbol{l}$$

into series in powers of $(\boldsymbol{l}-\boldsymbol{r})$ about the point $\boldsymbol{r}$. On confining ourselves to second order terms in $(\boldsymbol{l}-\boldsymbol{r})$, we get

$$\Delta^*(\boldsymbol{r})\int K_0(R)\,\mathrm{d}^3\boldsymbol{R} + \frac{1}{6}\left(\frac{\partial}{\partial\boldsymbol{r}} + 2ie\boldsymbol{A}(\boldsymbol{r})\right)^2 \Delta^*(\boldsymbol{r})\int K_0(R)R^2\,\mathrm{d}^3\boldsymbol{R}.$$

By (38.9), the function $K_0(R)$ tends to infinity as $1/R^3$ at $R=0$. The first of these two integrals is therefore formally divergent. It is more convenient to perform the cut-off of the divergent expression in momentum space. We obtain as a result the familiar expression:

$$\int K_0(R)\,\mathrm{d}^3\boldsymbol{R} = \frac{mp_0}{2\pi^2}\int_0^{\omega_D}\tanh\left(\frac{\xi}{2T}\right)\frac{\mathrm{d}\xi}{\xi}.$$

The second term contains no singularities on integrating close to $\boldsymbol{R}=0$. On evaluating this integral directly in the coordinate representation, we get:

$$\int R^2 K_0(R)\,\mathrm{d}^3\boldsymbol{R} = \frac{7\zeta(3)v^2}{8(\pi T)^2}\cdot\frac{mp_0}{2\pi^2}.$$

In the term of third order in $|\Delta|$ in equation (38.8), we can neglect the dependence of $\Delta(\boldsymbol{r})$ on the coordinates. This term is therefore equal to

$$\left(\frac{mp_0}{2\pi^2}\right)\Delta^*(\boldsymbol{r})\,|\Delta(\boldsymbol{r})|^2\, T\sum_\omega\int \mathrm{d}\xi\,\frac{1}{(\omega^2+\xi^2)^2} = \left(\frac{mp_0}{2\pi^2}\right)\frac{7\zeta(3)}{8(\pi T)^2}\Delta^*(\boldsymbol{r})\,|\Delta(\boldsymbol{r})|^2.$$

On collecting the results obtained, we find that $\Delta^*(\boldsymbol{r})$ satisfies the following equation close to the critical temperature:

$$\left\{\frac{1}{4m}\left(\frac{\partial}{\partial \boldsymbol{r}}+2ie\boldsymbol{A}(\boldsymbol{r})\right)^2+\frac{1}{\varrho}\left[\frac{T_c-T}{T_c}-\frac{7\zeta(3)}{8(\pi T_c)^2}|\Delta(\boldsymbol{r})|^2\right]\right\}\Delta^*(\boldsymbol{r})=0, \tag{38.10}$$

where

$$\varrho=\frac{7\zeta(3)}{6(\pi T_c)^2}\varepsilon_F.$$

In the absence of a field Δ is constant in space and equation (38.10) is the same as the first terms of expansion (36.4).

We now turn to evaluating the current density $\boldsymbol{j}(\boldsymbol{r})$. Obviously (37.5) of the previous section still holds in this case. If, however, we use the fact that Δ is small by comparison with T_c, we can also carry out here an expansion in $\boldsymbol{A}$ and Δ up to the first non-vanishing terms. Notice that the term in the current density (37.5)

$$-\frac{2e^2\boldsymbol{A}(\boldsymbol{r})}{m}\mathfrak{G}_{t'=t+0}(\boldsymbol{r},\boldsymbol{r})\equiv-\frac{e^2}{m}\boldsymbol{A}(\boldsymbol{r})N.$$

Here N is the density of the total number of electrons, which is equal to its value for a normal metal. Variation of N would contradict the condition for neutrality. (The solution of (38.7) satisfies this requirement, i.e. if

$$\mathfrak{G}_\omega(\boldsymbol{r},\boldsymbol{r}')=\tilde{\mathfrak{G}}_\omega^{(0)}(\boldsymbol{r},\boldsymbol{r}')+\delta\mathfrak{G}_\omega(\boldsymbol{r},\boldsymbol{r}'),\ \text{then}\ T\sum_\omega\delta\mathfrak{G}_\omega(\boldsymbol{r},\boldsymbol{r})\equiv 0.)$$

On substituting (38.7) in (37.5), we find that

$$T\sum_\omega\frac{ie}{m}(\nabla_{\boldsymbol{r}'}-\nabla_{\boldsymbol{r}})\,\tilde{\mathfrak{G}}_\omega^{(0)}(\boldsymbol{r},\boldsymbol{r}')_{\boldsymbol{r}'\to\boldsymbol{r}}-\frac{e^2}{m}N\boldsymbol{A}(\boldsymbol{r})\equiv 0,$$

since the current in a normal metal is zero in a constant magnetic field. Therefore

$$\boldsymbol{j}(\boldsymbol{r})=\frac{ie}{m}(\nabla_{\boldsymbol{r}'}-\nabla_{\boldsymbol{r}})_{\boldsymbol{r}'\to\boldsymbol{r}}\,T\sum_\omega\delta\mathfrak{G}_\omega(\boldsymbol{r},\boldsymbol{r}'),$$

where

$$\delta\mathfrak{G}_\omega(\boldsymbol{r},\boldsymbol{r}')=-\int\tilde{\mathfrak{G}}_\omega^{(0)}(\boldsymbol{r},\boldsymbol{l})\Delta(\boldsymbol{l})\tilde{\mathfrak{G}}_\omega^{(0)}(\boldsymbol{m},\boldsymbol{r}')\Delta^*(\boldsymbol{m})\tilde{\mathfrak{G}}_{-\omega}^{(0)}(\boldsymbol{m},\boldsymbol{l})\,d^3\boldsymbol{m}\,d^3\boldsymbol{l},$$

On substituting (38.5) in this, expanding all the functions about the point $\boldsymbol{r}$ up to first order terms and omitting terms which yield zero an integration over the angle, we get:

$$\boldsymbol{j}(\boldsymbol{r})=\left[\frac{ie}{m}\left(\Delta\frac{\partial\Delta^*}{\partial\boldsymbol{r}}-\Delta^*\frac{\partial\Delta}{\partial\boldsymbol{r}}\right)-\frac{4e^2|\Delta|^2}{m}\boldsymbol{A}(\boldsymbol{r})\right]C, \tag{38.11}$$

where

$$C=\frac{1}{3}T\sum_\omega\int\int\{[(\nabla_{\boldsymbol{r}}\mathfrak{G}_\omega^{(0)}(\boldsymbol{r}-\boldsymbol{l}))\,\mathfrak{G}_\omega^{(0)}(\boldsymbol{m}-\boldsymbol{r})-\mathfrak{G}_\omega^{(0)}(\boldsymbol{r}-\boldsymbol{l})$$

$$\times\left(\nabla_{\boldsymbol{r}}\mathfrak{G}_\omega^{0}(\boldsymbol{m}-\boldsymbol{r})\right)]\cdot\boldsymbol{m}\}\,\mathfrak{G}_{-\omega}^{(0)}(\boldsymbol{m}-\boldsymbol{l})\,d^3\boldsymbol{m}\,d^3\boldsymbol{l}.$$

The evaluation of C is best performed in Fourier components, after making the usual substitution $\boldsymbol{r} \to i(\mathrm{d}/\mathrm{d}\boldsymbol{p})$. We shall omit the details of this fairly simple calculation and give the final result:

$$C = \frac{7\zeta(3)N}{16(\pi T_c)^2}.$$

The system of equations (38.10) and (38.11) describes the properties of a superconductor in a constant magnetic field close to T_c. We introduce the wave function $\psi(\boldsymbol{r})$, proportional to $\Delta(\boldsymbol{r})$:

$$\psi(\boldsymbol{r}) = \sqrt{\frac{7\zeta(3)N}{8(\pi T_c)^2}}\,\Delta(\boldsymbol{r}). \tag{38.12}$$

On taking the complex conjugate of equation (38.10) and substituting (38.12) throughout, we can derive these equations as follows:

$$\begin{gathered} \left\{\frac{1}{4m}\left(\frac{\partial}{\partial \boldsymbol{r}} - 2ie\boldsymbol{A}(\boldsymbol{r})\right)^2 + \frac{1}{\varrho}\left[\frac{T_c - T}{T_c} - \frac{1}{N}|\psi|^2\right]\right\}\psi(\boldsymbol{r}) = 0, \\ \boldsymbol{j}(\boldsymbol{r}) = -\frac{2ie}{4m}(\psi^* \nabla\psi - \psi\nabla\psi^*) - \frac{(2e)^2}{2m}\boldsymbol{A}(\boldsymbol{r})\,|\psi|^2. \end{gathered} \tag{38.13}$$

The point in introducing the wave function $\psi(\boldsymbol{r})$ now becomes clear: the resulting equations are in a form similar to the quantum mechanical equations for particles with mass $2m$ and charge $2e$. This result is quite obvious physically, since $\Delta(\boldsymbol{r})$ has the significance of a function which is proportional to the wave function of a bound pair, or, more precisely, the wave function in the centre of mass coordinates. It is of interest that equations of a similar form where proposed in the phenomenological Ginzburg–Landau theory [69], where, however, a single elementary charge occurred. Apart from this essential difference, the new theory of superconductivity, which supports the correctness of the Ginzburg–Landau theory close to T_c, enables us to compute the constants featured in the latter theory.

Let us again mention in conclusion the fact that our derivation employed the smallness of the variation of all the functions over distances of the order ξ_0. Since, as may clearly be seen from (38.13) and (37.19), all the functions are in the general case variable at distances of the order of the London penetration depth close to T_c, whilst this depth becomes greater than ξ_0 only in the immediate neighbourhood of T_c in the case of Pippard metals, the domain of applicability of these equations for metals of the Pippard type is an extremly limited neighbourhood of temperatures close to the critical transition temperature. For metals of the London type, or intermediate metals, the equations are valid in a fairly wide temperature range close to T_c. This is a range of the greatest interest from the experimental point of view, and it must be remarked that equation (38.13) leads to a very good agreement between the theory and experimental results in this range.

§ 39. THEORY OF SUPERCONDUCTING ALLOYS

1. Statement of the problem

One of the interesting problems in the theory of superconductors concerns the properties of "alloys" (or dirty superconductors), i.e. a superconductor containing an admixture of atoms of other elements and other lattice disruptions (Abrikosov and Gor'kov [70]).

In the normal state these lattice defects determine the so-called residual resistance of the metal. In the superconducting state the impurities play a new role. As we have already indicated, in a superconductor the interaction between the electrons leads to the establishment of a definite spatial correlation between them. In particular, the dependence of certain Green functions in the coordinate representation on their spatial arguments at distances of the order ξ_0 (the effective dimension of a pair) changes substantially with the passage of the metal from the normal to the superconducting state. The presence of atoms of different elements or of other lattice defects leads to scattering of the electrons at the impurities. Since the scattering is random (over arbitrary angles), whilst the scattered electrons possess an extremely small wavelength, the correlations of the electrons will be very sensitive to the scattering effects. In other words, the scattering of electrons at impurity atoms must diminish the spatial correlation between them.

The role of the impurities is modest at extremely low concentrations.

An increase in the concentration of the impurities will evidently lead to a decrease in the radius of correlation of the electrons in the superconductor. In the case of a fairly concentrated alloy, the role of correlation parameter changes from ξ_0 to the mean free path of the electrons. At such concentrations we are justified in expecting the appearance of new characteristic properties of superconductors. Since we are not concerned in this book with a logical treatment of the theory of superconductivity, we shall only touch here on one aspect — the properties of the alloy in a constant weak magnetic field. This will nevertheless enable us to give a full demonstration of the special technique of field theory, which is extraordinarily useful in the investigation of this type of problem.

It has already been remarked in § 37 that the majority of actual superconductors belong, as regards their electromagnetic properties in a weak field, to the non-local (Pippard or intermediate) type. In other words, the current produced at a given point of the superconductor, when it is located in an electromagnetic field, is determined by the values of the field throughout some neighbourhood of this point. This nonlocal property can evidently be traced back to the theory based on

Cooper's idea of the formation of bound electron pairs. The pair dimensions result in the existence of electron correlation at distances of the order $\xi_0 \sim 10^{-4}$ cm, which makes itself felt as a non-local connection between the field and the current, provided the field is varying at distances considerably less than ξ_0 (these distances are of the order of the depth of penetration of the field). In the London case, on the contrary, the field is almost unchanged over distances of the order ξ_0, which are important in the integral equation (37.3), and can be taken outside the integral at the point $\boldsymbol{r}$.

The above discussion on the role of impurities in superconductors reveals that, given a sufficient impurity concentration, a superconducting alloy must belong to the second type. Since, when the impurities increase, the role of the correlation length starts to be played by the mean free path, an instant must arrive when this path becomes less than the field penetration depth, i.e. the London situation arises.

Before proceeding further, some explanation needs to be given of the following fact. In actual superconductors ξ_0 is of the order 10^{-4} cm. It follows from the foregoing, however, that the new properties make their appearance in the superconductor at concentrations where the mean free path becomes comparable with the field penetration depth, the order of which is $\sim 10^{-5}$ to 10^{-6} cm. It is of the greatest importance that these concentrations are still small (~ 1 per cent). The fact is that, at large impurity concentrations, we are in essence dealing with a new substance, the properties of which have nothing in common with the original superconductor. In particular, the properties of the electron-phonon interaction change, and hence the transition temperature also changes. These changes in the basic properties of the lattice can be neglected at fairly small concentrations.

But even small concentrations substantially change the behaviour of the superconductor in a magnetic field. Experiment supports the interesting point that its thermodynamic properties remain practically the same as for the pure superconductor.

The ordinary methods, based on the transport equation, and used for investigating, e.g. the residual resistance of a normal metal, prove to be unsuitable for solving the problem posed above. We shall therefore turn once again to the methods of quantum field theory.

2. *Residual resistance of normal metals*

In order to make the following treatment clearer, we shall first formulate our approach by taking the example of finding the residual resistance of a normal metal containing impurities at absolute zero (Abrikosov and Gor'kov [70], Edwards [71]). The results obtained in this case are of course completely equivalent to the well-known results obtained by the transport equation method.

It is well known that the presence of impurities in a normal metal leads to a finite conductivity σ, so that the current density $\boldsymbol{j}$ when a uniform electric field $\boldsymbol{E}$ of sufficiently low frequency is applied is given by

$$\boldsymbol{j} = \sigma \boldsymbol{E}.$$

On introducing the vector potential $\boldsymbol{A}(t)$ in the usual way, $\boldsymbol{E} = -\partial \boldsymbol{A}/\partial t$, we can write this equation as follows (for a monochromatic field component):

$$\boldsymbol{j}_\omega = i\omega\sigma \boldsymbol{A}_\omega.$$

In this form, the relation is the same as (37.3′). The kernel $Q(\boldsymbol{k}, \omega)$ is in this case simply

$$Q(\boldsymbol{k}, \omega) = -i\omega\sigma.$$

We shall determine $Q(\boldsymbol{k}, \omega)$ by the methods of quantum field theory below.

If we take into account the difference in the definition of the Green functions in the field technique at absolute zero and in the technique at $T \neq 0$, we obtain instead of (37.5):

$$\boldsymbol{j}(x) = \frac{e}{m}(\nabla_{r'} - \nabla_r) G(x, x')\Big|_{r' \to r, t' = t+0} - \frac{Ne^2}{m} \boldsymbol{A}(x).$$

On expanding $G(x, x')$ as usual up to terms linear in the field, we get

$$\boldsymbol{j}(x) = -\frac{ie^2}{2m^2}(\nabla_{r'} - \nabla_r)_{r' \to r} \int [\boldsymbol{A}(y), \; \nabla_{y'} - \nabla_y]_{y' \to y} G^{(0)}(x, y')$$
$$\times G^{(0)}(y, x')\, d^4 y - \frac{Ne^2}{m} \boldsymbol{A}(x). \qquad (39.1)$$

The $G^{(0)}(x, y)$ in this relationship are the Green functions of a normal metal in the absence of field. Notice that these functions no longer depend only on the difference of the arguments $x - y$, as has been the case throughout till now; we are assuming that the interaction of electrons with impurity atoms is taken into account in the $G^{(0)}(x, y)$. (In future, we shall use $G(x, y)$, etc. (without an index), to denote the Green functions of the metal containing impurities, and $G^{(0)}(x, y)$, etc., for the pure metal.) The Hamiltonian of the interaction of electrons with impurity atoms is

$$H_{int} = \sum_a H_a,$$

$$H_a = \int u(\boldsymbol{r} - \boldsymbol{r}_a) \psi^+(\boldsymbol{r}) \psi(\boldsymbol{r})\, d^3 \boldsymbol{r}.$$

We shall find the function $G(x, x')$ before carrying out further transformations of (39.1).

The Green function is not the same as (7.7) when impurities are present. We write it as

$$G(x, x') = (2\pi)^{-4} \int G(\boldsymbol{p}, \boldsymbol{p}'; \omega)\, e^{i(\boldsymbol{p}\cdot\boldsymbol{r})-i(\boldsymbol{p}'\cdot\boldsymbol{r}')-i\omega(t-t')}\, \mathrm{d}^3\boldsymbol{p}\, \mathrm{d}^3\boldsymbol{p}'\, \mathrm{d}^3\omega. \quad (39.2)$$

The function $G(\boldsymbol{p}, \boldsymbol{p}'; \omega)$ is expressible by the usual rules of field theory as the sum of the diagrams shown in Fig. 103. Each line corresponds to $G^{(0)}(p)$. We shall denote the impurity vertex by a cross. It corresponds

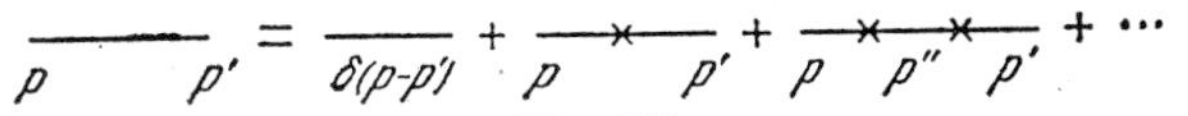

Fig. 103

to the factor $u(\boldsymbol{q}) \exp[i(\boldsymbol{q}\cdot\boldsymbol{r}_a)]\, \delta(\omega-\omega')$, where $u(\boldsymbol{q})$ is the Fourier component of the potential $u(\boldsymbol{r})$, and $\boldsymbol{q}$ the transferred momentum.

Summation of the diagrams leads to the integral equation for $G(\boldsymbol{p}, \boldsymbol{p}'; \omega)$:

$$G(\boldsymbol{p}, \boldsymbol{p}'; \omega) = \delta(\boldsymbol{p}-\boldsymbol{p}')G^{(0)}(p)$$

$$+ \frac{1}{(2\pi)^3} \sum_a G^{(0)}(p) \int u(\boldsymbol{p}-\boldsymbol{p}'')\, e^{i(\boldsymbol{p}-\boldsymbol{p}''\cdot\boldsymbol{r}_a)} G(\boldsymbol{p}'', \boldsymbol{p}'; \omega)\, \mathrm{d}^3\boldsymbol{p}''. \quad (39.3)$$

We are not interested in the exact solution of (39.3). Since the impurity atoms are distributed randomly over the metal, we have to average all the expressions over the position of each impurity atom. An important point here is that, by virtue of our assumption regarding the smallness of the atomic concentration, the mean distances between the impurity atoms are much greater than the atomic distances in the lattice of the metal, the result being that the averaging can be carried out in volumes with dimensions much greater than the interatomic distances. The Green function $G(\boldsymbol{p}, \boldsymbol{p}'; \omega)$ evidently becomes, after such averaging,

$$\overline{G(\boldsymbol{p}, \boldsymbol{p}'; \omega)} = G(p)\,\delta(\boldsymbol{p}-\boldsymbol{p}'). \quad (39.4)$$

The magnitudes of the momenta $\boldsymbol{p}$, $\boldsymbol{p}'$ of interest are of the order of the Fermi momentum p_0, which is in turn of the order of the reciprocal of the interatomic distance. This fact immediately simplifies the averaging process.

We carry out our calculations in the Born approximation, i.e. we assume $p_0^3 \int u(\boldsymbol{r})\,\mathrm{d}^3\boldsymbol{r} \ll \varepsilon_F$. It can be shown that the final results, expressed in terms of the collision time, will also hold in the general case.

The simplest diagram for $G(\boldsymbol{p}, \boldsymbol{p}'; \omega)$ contains just one cross. The value averaged over the positions of the impurity atoms is the constant $\overline{u(\boldsymbol{q})e^{i(\boldsymbol{q}\cdot\boldsymbol{r}_a)}} = u(0)$, which can be included in the ground state energy and is assumed zero in what follows. The diagram next in complexity contains two crosses (Fig. 104 *a*). If these crosses refer to different atoms, the matrix element contains the factor $u(\boldsymbol{p}''-\boldsymbol{p}')\, u(\boldsymbol{p}-\boldsymbol{p}'')$ $\exp[i(\boldsymbol{p}-\boldsymbol{p}''\cdot\boldsymbol{r}_a) + i(\boldsymbol{p}''-\boldsymbol{p}'\cdot\boldsymbol{r}_b)]$, the mean of which is zero. If the scattering is on the same atoms and $\boldsymbol{p} = \boldsymbol{p}'$ at both crosses, the mean value of this diagram (without external $G^{(0)}(p)$) is non-zero and equal to

$$\frac{1}{V}\int |u(\boldsymbol{p}-\boldsymbol{p}')|^2 G^{(0)}(p')\frac{d^3\boldsymbol{p}'}{(2\pi)^3}, \tag{39.5}$$

where V is the volume of the system. (In order to obtain this result, it is useful to change the integrals over the momenta to discrete sums in (39.2) and (39.3), then carry out the reverse process after performing the averaging.)

We shall be interested in future in values of $\boldsymbol{p}$ close to p_0 in absolute value. As in § 21, the integral in (39.5) can be split into two parts: over

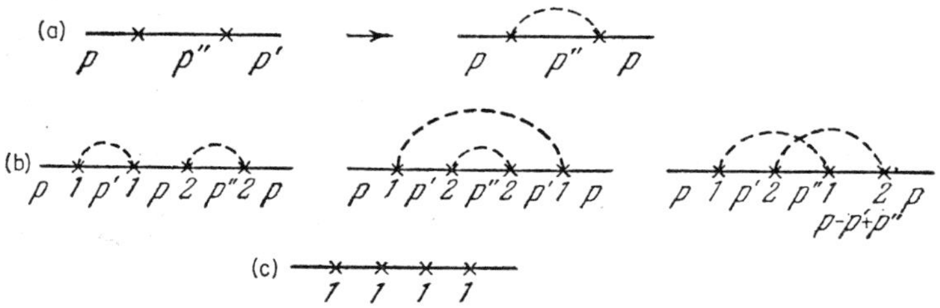

Fig. 104

the $\boldsymbol{p}'$ remote from the Fermi surface, and over the $\boldsymbol{p}'$ close to the Fermi surface (the limits of the second integral over $|\boldsymbol{p}'|$ can be taken as symmetrical with respect to $|\boldsymbol{p}'| = p_0$). The integral over the remote region yields a real constant, which, together with $u(0)$, is a renormalization of the chemical potential and can be disregarded. We can regard $u(\boldsymbol{p}-\boldsymbol{p}')$ as a slowly varying function in the second integral. On substituting (7.7) for $G^{(0)}(p)$ and summing over the impurity atoms (this means simply multiplying by the number of atoms), we get the essential contribution to the G-function:

$$\frac{i \operatorname{sign} \omega}{2\tau},$$

where

$$\frac{1}{\tau} = \frac{n m p_0}{(2\pi)^2}\int |u(\theta)|^2 d\Omega \tag{39.6}$$

(θ is the angle between the vectors $\boldsymbol{p}$, $\boldsymbol{p}'$. By (39.6), τ is the time between collisions in the Born approximation, and n is the number of impurity atoms per unit volume). It is clear from this that the main role in the integrals is played by the region close to the Fermi surface $(v(|\boldsymbol{p}|-p_0) \sim 1/\tau)$.

Not all the diagrams are equivalent from this point of view. Let us compare, for instance, the three diagrams illustrated in Fig. 104*b* (the dotted lines join crosses referring to one atom). It is easily seen that, in the first two expressions, the integration over $\boldsymbol{p}'$ and $\boldsymbol{p}''$ can be performed close to the Fermi surface when the momenta are at any angle to one another. Conversely, in the third integral, the requirement that all the arguments of the G-functions be close to the Fermi surface leads to a restriction of the angles. As a result, the contribution of this diagram is $1/vp_0G$ times less than in the case of the others. Since we shall require in future values of ω and $\xi \sim 1/\tau$, where τ is the time between collisions, the smallness of the "intersecting" diagrams can be estimated at $1/p_0l$, where $l = v\tau$ is the mean free path.

It is easily shown that a small contribution is also given by the diagrams containing more than two crosses from the same impurity atom. Let us compare, say, the total contribution (from all the impurity atoms) of the first type of diagram in Fig. 104*b* with the contribution of the diagrams of type 104*c*. The first diagrams yield a function of the order

$$\frac{1}{\tau^2} G^{(0)}(p) \sim \frac{1}{\tau},$$

and the second:

$$\frac{1}{\tau} \frac{u^2(\boldsymbol{q})}{v^2} p_0^4 \sim \frac{1}{\tau} \left[\frac{\int u \, \mathrm{d}^3\boldsymbol{r} \, p_0^3}{\varepsilon_F}\right]^2 \ll \frac{1}{\tau}$$

(this is a consequence of the Born approximation). It is clear from this that we should only consider those diagrams that contain two crosses per impurity atom.

On summing all the significant diagrams (i.e. only the "paired" type, and not those containing "intersections", like the third of the diagrams of Fig. 104*b*), we get the following equation for the G-function:

$$G(p) = G^{(0)}(p) + \frac{n}{(2\pi)^3} G^{(0)}(p) \int |u(\boldsymbol{p} - \boldsymbol{p}')|^2 G(p') \, \mathrm{d}^3\boldsymbol{p}' \, G(p). \quad (39.7)$$

If we depart from the Born approximation, we have to take into account the diagrams containing several crosses per impurity atom. It can be shown that the resulting change is simply to replace the Born amplitude $u(\theta)$ by the total scattering amplitude. This will also be the case for all our future calculations. In all the formulae, therefore, we can take $u(\theta)$ to mean the total scattering amplitude.

The solution of equation (39.7) is

$$G(p) = \frac{1}{\omega - \xi - \overline{G}_\omega},$$

where $\overline{G}_\omega$ satisfies the equation

$$\overline{G}_\omega = \frac{n}{(2\pi)^3} \int |u(\boldsymbol{p} - \boldsymbol{p}')|^2 \frac{1}{\omega - \xi' - \overline{G}_\omega} \mathrm{d}^3\boldsymbol{p}'.$$

On supposing $\overline{G}_\omega$ to be purely imaginary: $\overline{G}_\omega = -i\nu$, and finding the integral on the right-hand side by the same method as in (39.5), we get

$$\nu = \frac{\operatorname{sign} \nu}{2\tau},$$

where τ is defined by (39.6). On equating the $G(p)$ thus obtained with the result obtained when the amount of impurity is small ($G \to G^{(0)}$), we find that $\nu = \operatorname{sign} \omega/2\tau$ or

$$G(p) = \frac{1}{\omega - \xi + \dfrac{i\omega}{2|\omega|\tau}}. \tag{39.8}$$

It may easily be seen by changing to the x-representation that the change in G as compared with $G^{(0)}$ amounts to multiplying by the exponentially damped factor

$$G(x - x') = G^{(0)}(x - x')e^{-|\boldsymbol{r}-\boldsymbol{r}'|/2l}. \tag{39.9}$$

Indeed, we have, after integration over the angle

$$G(x - x') \sim \int p\,\mathrm{d}p\,\mathrm{d}\omega \frac{\sin pR\, e^{-i\omega(t-t')}}{R\left(\omega - \xi + \dfrac{i\omega}{2|\omega|\tau}\right)}$$

$$\sim m \int \mathrm{d}\xi\,\mathrm{d}\omega\, e^{-i\omega(t-t')} \frac{e^{i(p_0+\xi/v)R} - e^{-i(p_0+\xi/v)R}}{R\left(\omega - \xi + \dfrac{i\omega}{|\omega|2\tau}\right)}.$$

Integrating over ξ and taking the residue, we get (39.9).

Let us now find the kernel $Q(\boldsymbol{k}, \omega)$. Fourier transforming, we can write the expression for Q obtained from (39.1) conveniently as

$$Q_{\alpha\beta}(\boldsymbol{k}, \omega) = \frac{Ne^2}{m}\delta_{\alpha\beta} - \frac{2ie^2}{(2\pi)^4 m^2}\int \boldsymbol{p}'_\alpha \boldsymbol{\Pi}_\beta(p'_+, p'_-)\,\mathrm{d}^3\boldsymbol{p}'\,\mathrm{d}\omega', \tag{39.10}$$

where $p_\pm = (\boldsymbol{p}' \pm \boldsymbol{k}/2, \omega' \pm \omega/2)$. One of the photon vertices is distinguished in (39.10), the second vertex is contained in $\boldsymbol{\Pi}_\beta(p_+, p_-)$, which can be regarded, up to a coefficient, as the result of including the photon vertex $\boldsymbol{p}'_\beta$ in the electron line. On inserting this vertex in the electron Green function $G(\boldsymbol{p}, \boldsymbol{p}'; \omega)$, we get

$$\boldsymbol{\Pi}(p'_+, p'_-) = \frac{1}{(2\pi)^3}\int G\left(\boldsymbol{p}'_+, \boldsymbol{p}''_+; \omega' + \frac{\omega}{2}\right)\boldsymbol{p}'' G\left(\boldsymbol{p}''_-, \boldsymbol{p}'_-; \omega' - \frac{\omega}{2}\right)\mathrm{d}^3\boldsymbol{p}''. \tag{39.11}$$

The functions $G(\boldsymbol{p}, \boldsymbol{p}'; \omega)$ appearing in this correspond to the sum of the diagrams in Fig. 103 and satisfy equation (39.3). When averaging over the distribution of the impurity atoms, we have to remember that the average of the product of two Green functions is not equal to the product of their averages.

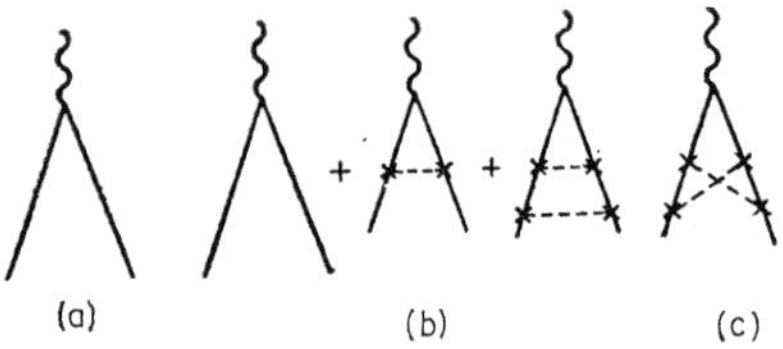

Fig. 105

In the case of a pure metal, equation (39.11) corresponds to the diagram illustrated in Fig. 105 *a*. After averaging over the position of the impurity atoms, apart from the simple diagrams corresponding to changing from the zero order Green functions $G^{(0)}(p)$ to the functions $G(p)$ (39.8), the diagrams shown in Fig. 105*b* prove to be important for the function (39.11). The large contribution from these corrections is connected with the fact that for the photon momentum at the vertex $k \ll p_0$, as a result of which the main contribution to the integral is provided by the region of momenta close to the Fermi surface. A diagram of another type, say that in Fig. 105*c*, contributes much less, since one of the integrations is over the region of momenta remote from the Fermi surface. The averaging of (39.11) therefore reduces to summation of the "ladders" of the diagrams in Fig. 105*b*.

The integral equation for $\Pi(p_+, p_-)$ reads as follows:

$$\boldsymbol{\Pi}(p_+, p_-) = G(p_+)G(p_-)\left[\boldsymbol{p} + \frac{n}{(2\pi)^3}\int |u(\boldsymbol{p}-\boldsymbol{p}')|^2 \boldsymbol{\Pi}(p'_+, p'_-)\,d^3\boldsymbol{p}'\right]. \tag{39.12}$$

Two limiting cases are possible:

(1) Anomalous skin effect ($|\boldsymbol{k}|\, v \gg 1/\tau$); it is easily shown that the integral on the right-hand side of (39.12) is negligibly small in this case, $\sim 1/|\boldsymbol{k}|\, v\tau \ll 1$;

(2) Normal skin effect ($|\boldsymbol{k}|\, v \ll 1/\tau$); it is this case that interests us. We can now assume $\boldsymbol{p}_+ = \boldsymbol{p}_-$ in equation (39.12).

The vector resulting from the integral on the right-hand side of (39.12) will obviously be directed along $\boldsymbol{p}$. We introduce the notation

$$\boldsymbol{p}\Lambda(\omega', \omega) = \frac{n}{(2\pi)^3}\int |u(\boldsymbol{p}-\boldsymbol{p}')|^2 \boldsymbol{\Pi}(p'_+, p'_-)\,d^3\boldsymbol{p}'. \tag{39.13}$$

Since $|\boldsymbol{p}| \simeq p_0$, $\Lambda(\omega', \omega)$ can be regarded as independent of $|\boldsymbol{p}|$. We multiply (39.12) by $[n/(2\pi)^3]\,|u(\boldsymbol{l}-\boldsymbol{p})|^2$ and integrate over $d^3\boldsymbol{p}$:

$$\boldsymbol{l}\Lambda(\omega', \omega) = \frac{n}{(2\pi)^3}\int |u(\boldsymbol{l}-\boldsymbol{p})|^2 \boldsymbol{p}G(p_+)G(p_-)[1+\Lambda(\omega', \omega)]\,d^3\boldsymbol{p}. \tag{39.14}$$

On substituting (39.8) for $G(p)$, we easily find that $\Lambda(\omega', \omega)$ is only non-zero for $|\omega'| < \omega/2$, since otherwise, by (7.7), both the poles in (39.14) will lie in the same half-plane when we integrate over ξ. In this interval $\Lambda(\omega', \omega)$ is independent of ω'. On integrating over ξ and using the relation

$$\cos\theta = \cos\theta' \cos\theta'' + \sin\theta' \sin\theta'' \cos(\varphi' - \varphi''), \tag{39.15}$$

we get

$$\Lambda(\omega', \omega) = \begin{cases} \dfrac{i}{\tau_1}\dfrac{1}{\omega + \dfrac{i}{\tau_{tr}}} & \text{for } \omega'^2 < \dfrac{\omega^2}{4}, \\[2ex] 0 & \text{for } \omega'^2 > \dfrac{\omega^2}{4}, \end{cases} \tag{39.16}$$

where $$\frac{1}{\tau_1} = \frac{1}{\tau} - \frac{1}{\tau_{tr}}, \quad \frac{1}{\tau_{tr}} = \frac{nmp_0}{(2\pi)^2} \int |u(\theta)|^2 (1 - \cos\theta) \mathrm{d}\Omega. \qquad (39.17)$$

After substituting (39.16) in (39.12) and (39.10) and integrating, on the assumption that $\omega\tau \ll 1$ we get

$$Q_{\alpha\beta}(\omega) = -i\omega\sigma\delta_{\alpha\beta}.$$

As must be the case, the conductivity $\sigma = Ne^2\tau_{tr}/m$ contains a "transport" time between the collisions.

We have thus seen that it is possible to assign a form of field technique of a special kind to the calculation of the various characteristics of a metal, averaged over the position of the impurity atoms. The averaging amounts in our case to a pair-wise averaging of the scattering on similar atoms, each of which can be associated on the diagram with a dotted line, joining two crosses. In the matrix elements a line carrying a momentum $\boldsymbol{q}$ corresponds to a factor $n|u(\boldsymbol{q})|^2$, playing the role of D-function for the dotted line. At the vertex from which the dotted line starts there is no change in the frequency of the electron line. It is extremely important that a small contribution is yielded by a dotted line that embraces a vertex at which a large change of momentum of the electron line occurs ($q \sim p_0$). In particular, it is for this reason that we can neglect the diagrams with intersecting dotted lines. The order of relative smallness of such diagrams is $1/p_0 l \ll 1$.

3. *Electromagnetic properties of superconducting alloys*

We now apply this method to an investigation of superconductors containing impurities. We shall consider the case of arbitrary temperatures right away. We write down the equations of the superconductor in the field of the impurities:

$$\left\{i\omega + \frac{\nabla^2}{2m} + \mu - \sum_{r_a} u(\boldsymbol{r} - \boldsymbol{r}_a)\right\} \mathfrak{G}_\omega(\boldsymbol{r}, \boldsymbol{r}') + \Delta(\boldsymbol{r})\, \mathfrak{F}^+_\omega(\boldsymbol{r}, \boldsymbol{r}') = \delta(\boldsymbol{r} - \boldsymbol{r}'),$$

$$\left\{-i\omega + \frac{\nabla^2}{2m} + \mu - \sum_{r_a} u(\boldsymbol{r} - \boldsymbol{r}_a)\right\} \mathfrak{F}^+_\omega(\boldsymbol{r}, \boldsymbol{r}') - \Delta^*(\boldsymbol{r})\, \mathfrak{G}_\omega(\boldsymbol{r}, \boldsymbol{r}') = 0.$$

As above, we are naturally only interested in $\mathfrak{G}$ and $\mathfrak{F}^*$ averaged over the impurities. In order to carry out the averaging, we need to expand all the Green functions in series in powers of the potential of the interaction with the impurities, similar to the equation (39.3) (see Fig. 103). It has to be borne in mind that the gap $\Delta(\boldsymbol{r})$ and $\Delta^*(\boldsymbol{r})$ also varies, generally speaking, on introducing the impurities. This could considerably complicate the diagram technique, since, by virtue of the condition $\Delta(\boldsymbol{r}) = |g|\, \mathfrak{F}(x, x)$, the corrections to Δ_0 must themselves be determined from some integral equation. Nevertheless, it turns out that, as a result of the averaging, $\Delta(\boldsymbol{r}) = \Delta_0$, and all these corrections vanish. It would be possible to prove this directly, by investigating the structure of the correction to $\Delta(\boldsymbol{r})$. We shall assume this fact in advance, however, and take $\Delta(\boldsymbol{r}) = \Delta_0$. Our assumption will be confirmed later by the final result, in accordance with which all the quantities of the type $\mathfrak{F}(x, x)$ remain unchanged on introducing non-magnetic impurities. Thus the only difference as compared with the above treatment lies in the fact that the superconductor is described by three Green functions — the functions $\mathfrak{G}$, $\mathfrak{F}$ and $\mathfrak{F}^*$. This circumstance leads to some modification of the diagram technique. This circumstance causes us to modify the diagram technique. It is easily seen that, in the present case, the modification is exactly similar to that of § 35 and amounts to the appearance of $\mathfrak{F}$- and $\mathfrak{F}^+$-lines in the diagrams for $\mathfrak{G}$ and, similarly, of $\mathfrak{G}$-lines in the diagrams for $\mathfrak{F}$ and $\mathfrak{F}^+$.

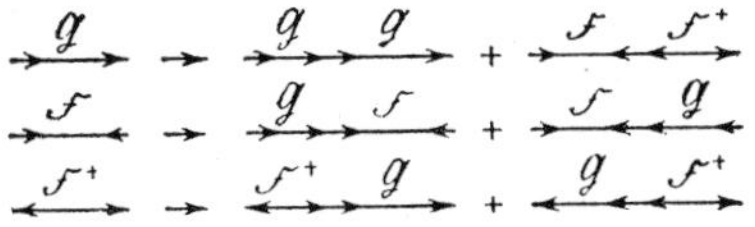
Fig. 106

The Hamiltonian of the interaction with impurities contains the operator product $\bar{\psi}\psi$. When an impurity vertex is included in an electron line, therefore, there are two possibilities each for the $\mathfrak{G}$, $\mathfrak{F}$ and $\mathfrak{F}^+$ lines. These possibilities are illustrated in Fig. 106. The result can be written as

$$\mathfrak{G}(x, x') \to \mathfrak{G}(x, y)\,\mathfrak{G}(y, x') - \mathfrak{F}(x, y)\,\mathfrak{F}^+(y, x'),$$

$$\mathfrak{F}^+(x, x') \to \mathfrak{F}^+(x, y)\,\mathfrak{G}(y, x') + \mathfrak{G}(y, x)\,\mathfrak{F}^+(y, x'),$$

$$\mathfrak{F}(x, x') \to \mathfrak{G}(x, y)\,\mathfrak{F}(y, x') + \mathfrak{F}(x, y)\,\mathfrak{G}(x', y).$$

Instead of (39.3), we have the following equations for the functions $\mathfrak{G}$ and $\mathfrak{F}^+$:

$$\begin{aligned}\mathfrak{G}(\boldsymbol{p}, \boldsymbol{p}'; \omega) &= \mathfrak{G}^{(0)}(p)\,\delta(\boldsymbol{p} - \boldsymbol{p}') \\ &+ \frac{1}{(2\pi)^3}\Big\{\mathfrak{G}^{(0)}(p) \int u(\boldsymbol{p} - \boldsymbol{p}'') \sum_a e^{i(\boldsymbol{p}-\boldsymbol{p}''\cdot\boldsymbol{r}_a)}\,\mathfrak{G}(\boldsymbol{p}'', \boldsymbol{p}', \omega)\,d^3\boldsymbol{p}'' \\ &\qquad - \mathfrak{F}^{(0)}(p) \int u(\boldsymbol{p} - \boldsymbol{p}'') \sum_a e^{i(\boldsymbol{p}-\boldsymbol{p}''\cdot\boldsymbol{r}_a)}\,\mathfrak{F}^+(\boldsymbol{p}'', \boldsymbol{p}'; \omega)\,d^3\boldsymbol{p}''\Big\},\end{aligned} \tag{39.18}$$

$$\begin{aligned}\mathfrak{F}^+(\boldsymbol{p}, \boldsymbol{p}'; \omega) &= \mathfrak{F}^{+(0)}(p)\,\delta(\boldsymbol{p} - \boldsymbol{p}') \\ &+ \frac{1}{(2\pi)^3}\Big\{\mathfrak{F}^{+(0)}(p) \int u(\boldsymbol{p} - \boldsymbol{p}'') \sum_a e^{i(\boldsymbol{p}-\boldsymbol{p}\cdot\boldsymbol{r}_a)}\,\mathfrak{G}(\boldsymbol{p}'', \boldsymbol{p}'; \omega)\,d^3\boldsymbol{p}'' \\ &\qquad + \mathfrak{G}^{(0)}(-p) \int u(\boldsymbol{p} - \boldsymbol{p}'') \sum_a e^{i(\boldsymbol{p}-\boldsymbol{p}''\cdot\boldsymbol{r}_a)}\,\mathfrak{F}^+(\boldsymbol{p}'', \boldsymbol{p}'; \omega)\,d^3\boldsymbol{p}''\Big\}.\end{aligned} \tag{39.19}$$

In principle, it is necessary to look for an equation for $\mathfrak{F}(\boldsymbol{p}, \boldsymbol{p}'; \omega)$ also. For a pure superconductor, $\mathfrak{F}^{(0)}(x, x') = \mathfrak{F}^{+(0)}(x, x')$ in the absence of field. We shall not dwell on the proof, but merely remark that the same will be true for alloys after the averaging of equations (39.18) and (39.19) over the position of the impurity atoms.

Elementary methods can be used for extending the averaging technique of the previous section to the case of finite temperatures in regard to superconductors. When an electron is scattered at a static impurity, only three components of its momentum vary. The dotted line is thus associated as before with a factor $n|u(\boldsymbol{q})|^2$, whilst the frequency of the electron line at the impurity vertex is retained. All the estimates remain in force, enabling us to neglect the intersection of dotted lines, as also diagrams in which a dotted line embraces a vertex with a momentum transfer of Fermi order. What are important for these estimates are the

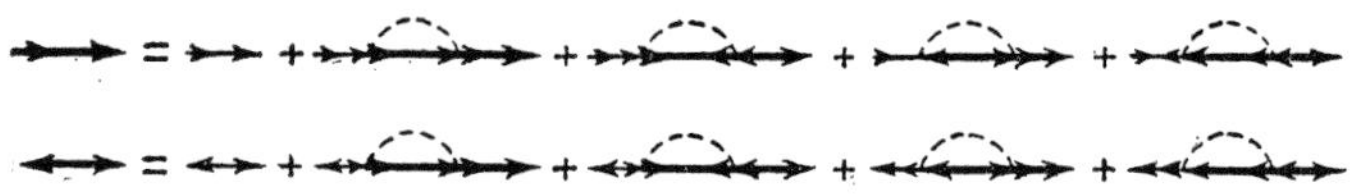

Fig. 107

properties of the Green functions of the normal metal, for which the superconducting transition temperatures are insignificantly small. The equations for the averaged functions $\mathfrak{G}(p)$ and $\mathfrak{F}^+(p)$ are illustrated schematically in Fig. 107. Their structure is clear without further explanation. Notice that, from the diagramatic point of view, the equations of Fig. 107 are similar to the equations of § 35 (see Fig. 96) for a system with electron-phonon interaction. The difference lies in the fact that, in the diagrams of Fig. 107, there are "zero" lines of all kinds:

$$\mathfrak{G}^{(0)}(p),\ \mathfrak{F}^{+(0)}(p) \text{ and } \mathfrak{F}^{(0)}(p).$$

The system of equations of Fig. 107 can be reduced to a very simple form by using the explicit expressions for the functions $\mathfrak{G}^{(0)}$ and $\mathfrak{F}^{+(0)}$ of a pure superconductor:

$$(i\omega - \xi - \overline{\mathfrak{G}}_\omega)\,\mathfrak{G}(p) + (\Delta + \overline{\mathfrak{F}_\omega^+})\,\mathfrak{F}^+(p) = 1,$$

$$(i\omega + \xi + \overline{\mathfrak{G}}_{-\omega})\,\mathfrak{F}^+(p) + (\Delta + \overline{\mathfrak{F}_\omega^+})\,\mathfrak{G}(p) = 0,$$

where

$$\overline{\mathfrak{G}}_\omega = \frac{n}{(2\pi)^3}\int |u(\boldsymbol{p}-\boldsymbol{p}')|^2\,\mathfrak{G}(p')\,\mathrm{d}^3\boldsymbol{p}',$$

$$\overline{\mathfrak{F}_\omega^+} = \frac{n}{(2\pi)^3}\int |u(\boldsymbol{p}-\boldsymbol{p}')|^2\,\mathfrak{F}^+(p')\,\mathrm{d}^3\boldsymbol{p}', \qquad (39.20)$$

$$\overline{\mathfrak{G}}_{-\omega}(p) = \overline{\mathfrak{G}}_\omega(-p)$$

(we have used the notation $p' = (\boldsymbol{p}'; \omega)$). The solution of the system is (it is shown below that $\overline{\mathfrak{G}}_\omega = -\overline{\mathfrak{G}}_{-\omega}$)

$$\mathfrak{G}(p) = -\frac{i\omega - \overline{\mathfrak{G}}_\omega + \xi}{(i\omega - \overline{\mathfrak{G}}_\omega)^2 + \xi^2 + (\Delta + \overline{\mathfrak{F}_\omega^+})^2},$$
$$\mathfrak{F}^+(p) = \frac{\Delta + \overline{\mathfrak{F}_\omega^+}}{(i\omega - \overline{\mathfrak{G}}_\omega)^2 + \xi^2 + (\Delta + \overline{\mathfrak{F}_\omega^+})^2}. \tag{39.21}$$

Substitution of these expressions in (39.20) yields two equations for $\overline{\mathfrak{G}}_\omega$ and $\overline{\mathfrak{F}_\omega^+}$. We see that, as before, $\overline{\mathfrak{G}}_\omega$ contains a constant term, denoting the additive correction to the chemical potential. This term does not depend on the temperature and is due to the integration over $d^3\boldsymbol{p}'$ remote from the Fermi surface. The term is therefore the same as for a normal metal:

$$\delta\mu \simeq \frac{n}{(2\pi)^3}\int |u(\boldsymbol{p} - \boldsymbol{p}')|^2 \frac{d\xi}{\xi}.$$

After subtracting this term, we find that the remaining part of $\overline{\mathfrak{G}}_\omega$ is determined to the same accuracy as $\overline{\mathfrak{F}_\omega^+}$, as is clear from equations (39.20) and (39.21). Hence

$$-\frac{\overline{\mathfrak{G}}_\omega}{i\omega} = \frac{\overline{\mathfrak{F}_\omega^+}}{\Delta}.$$

We introduce the notation

$$\tilde{\Delta} = \Delta + \overline{\mathfrak{F}_\omega^+} = \Delta\eta_\omega, \qquad i\tilde{\omega} = i\omega - \overline{\mathfrak{G}}_\omega = i\omega\eta_\omega.$$

We now obtain for the function η_ω the equation

$$\eta_\omega = 1 + \frac{\eta_\omega}{2\pi}\int \frac{d\xi}{\xi^2 + (\omega^2 + \Delta^2)\eta_\omega^2},$$

the solution of which is

$$\eta_\omega = 1 + \frac{1}{2\tau\sqrt{\omega^2 + \Delta^2}}. \tag{39.22}$$

Hence the functions $\mathfrak{G}(p)$ and $\mathfrak{F}^+(p)$, averaged over the positions of the impurity atoms, can be obtained from the functions for the pure superconductor by the substitution

$$\{\omega, \Delta\} \to \{\omega\eta_\omega, \Delta\eta_\omega\}. \tag{39.23}$$

It is easily shown that, as for the case of a normal metal, these functions imply, in the coordinate representation, multiplication of the zero-order functions by $e^{-R/2l}$. It follows from this, in particular, that

$$\Delta = |\lambda|\mathfrak{F}^+(x, x)$$

is the same in an alloy as the Δ for a pure superconductor. Since, as we saw in § 36, the thermodynamic functions depend only on Δ in a superconductor, we have now justified the assertion made above, that the thermodynamic properties of a superconductor do not change when an impurity is present, provided the concentration of the latter is sufficiently small(†).

We return to the question of the temperature dependence of the depth of penetration by a weak static magnetic field into a superconducting alloy. In accordance with (37.10), the expression for the current density $\boldsymbol{j}(\boldsymbol{r})$ in the approximation linear in the field is

$$\boldsymbol{j}(\boldsymbol{r}) = \frac{e^2}{m^2} T \sum_{\omega} (\nabla_{\boldsymbol{r}} - \nabla_{\boldsymbol{r}'})_{\boldsymbol{r}' \to \boldsymbol{r}} \int [\mathfrak{G}_\omega(\boldsymbol{r}, \boldsymbol{l}) \left(\boldsymbol{A}(\boldsymbol{l}), \nabla_{\boldsymbol{l}}\right) \mathfrak{G}_\omega(\boldsymbol{l}, \boldsymbol{r}') + \mathfrak{F}_\omega(\boldsymbol{l}, \boldsymbol{r}) \left(\boldsymbol{A}(\boldsymbol{l}), \nabla_{\boldsymbol{l}}\right) \mathfrak{F}^+_\omega(\boldsymbol{r}', \boldsymbol{l})] \mathrm{d}^3\boldsymbol{l} - \frac{Ne^2}{m} \boldsymbol{A}(\boldsymbol{r}),$$

where, however, the functions $\mathfrak{G}_\omega(\boldsymbol{r}, \boldsymbol{r}')$ and $\mathfrak{F}^+_\omega(\boldsymbol{r}, \boldsymbol{r}')$ include the interactions with impurity atoms. On averaging this equation over the positions of the impurities and Fourier transforming, we can write the kernel $Q_{\alpha\beta}(\boldsymbol{k})$ as follows:

$$Q_{\alpha\beta}(\boldsymbol{k}) = \frac{Ne^2}{m} \delta_{\alpha\beta} + \frac{2e^2 T}{(2\pi)^3 m^2} \sum_{\omega'} \int p'_\alpha \, \Pi^{(1)}_\beta (p'_+, p'_-) \mathrm{d}^3\boldsymbol{p}' \qquad (39.24)$$

($p'_\pm = p' \pm k/2$, $k = (\boldsymbol{k}, 0)$), where $\boldsymbol{\Pi}^{(1)}(p_+, p_-)$ is the Fourier component of

$$\boldsymbol{\Pi}^{(1)}(x - y, y - x') = -\frac{i}{2} (\nabla_y - \nabla_{y'})_{y' \to y} \times \left[\overline{\mathfrak{G}(x, y') \mathfrak{G}(y, x')} - \overline{\mathfrak{F}^+(y', x') \mathfrak{F}(x, y)}\right];$$

$$\boldsymbol{\Pi}^{(1)}(x - y, y - x') = \frac{T^2}{(2\pi)^6} \sum_{\omega_+, \omega_-} \int\int \boldsymbol{\Pi}^{(1)}(p_+, p_-) \times e^{i(\boldsymbol{p}_+ \cdot \boldsymbol{x} - \boldsymbol{y}) - i\omega_+(\tau_x - \tau_y)} e^{i(\boldsymbol{p}_- \cdot \boldsymbol{y} - \boldsymbol{x}') - i\omega_-(\tau_y - \tau_{x'})} \mathrm{d}^3\boldsymbol{p}_+ \mathrm{d}^3\boldsymbol{p}_-.$$

As in the case of a normal metal, the average of the product of two Green functions is not equal to the product of the averages. When carrying out the averaging over the positions of the impurity atoms, it is

(†) It can be shown that this deduction is only true up to terms of order $\sim 1/\omega_D \tau \sim 10^{-6}$ cm$/l$ (see (32.2), (34.37)). In the more realistic phonon model, a frequency cut-off occurs, and such terms no longer arise. The remaining results are the same in both models.

It should be noted that in an anisotropic superconductor the thermodynamic properties depend in the impurity concentration. The variation of T_c has, for instance, the form (see [78], [79])

$$1 - \frac{T_c}{T_{c0}} = \frac{\pi}{8 T_{c0} \tau} \left[\frac{\overline{\Delta^2}}{\bar{\Delta}^2} - 1\right], \quad \text{if } \frac{1}{T_{c0}\tau} \ll 1,$$

where $\bar{\Delta}$ and $\overline{\Delta^2}$ denote the averages of $\Delta(p)$ and $\Delta^2(p)$ over the directions of the momentum. In all known cases $\frac{\overline{\Delta^2}}{\bar{\Delta}^2} - 1$ is of the order of magnitude of 10^{-2}.

again necessary to sum a set of diagrams. Since superconductivity only distorts the Green functions close to the Fermi surface, the necessary diagrams will be of the "ladder" type, as in the previous section. The presence of three different Green functions in a superconductor implies,

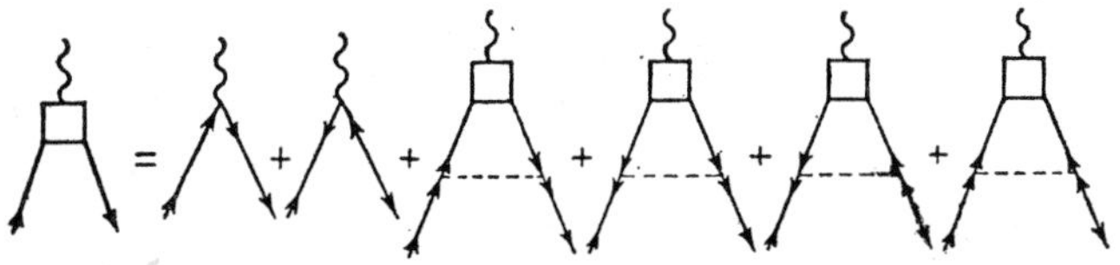

Fig. 108

however, that the equations are rather more complicated than (39.12), which describes the summation of the diagrams of Fig. 105 in a normal metal. It is clear from Fig. 108 that, to determine $\boldsymbol{\Pi}^{(1)}(p_+, p_-)$, we have to know three further quantities, that differ in the diagram from $\boldsymbol{\Pi}^{(1)}(p_+, p_-)$ by different directions of the arrows on the electron line. Each of these quantities corresponds to a special combination of $\mathfrak{G}$ and $\mathfrak{F}^+$:

$$\boldsymbol{\Pi}^{(2)}(x-y, y-x) = -\frac{i}{2}(\nabla_y - \nabla_{y'})_{y'\to y}\left[\overline{\mathfrak{F}^+(x, y')\,\mathfrak{G}(y, x')} + \overline{\mathfrak{G}(y, x)\,\mathfrak{F}^+(y', x')}\right],$$

$$\boldsymbol{\Pi}^{(3)}(x-y, y-x') = -\frac{i}{2}(\nabla_y - \nabla_{y'})_{y'\to y}\left[\overline{\mathfrak{G}(y, x)\,\mathfrak{G}(x', y')} - \overline{\mathfrak{F}^+(x, y')\,\mathfrak{F}(y, x')}\right],$$

$$\boldsymbol{\Pi}^{(4)}(x-y, y-x') = -\frac{i}{2}(\nabla_y - \nabla_{y'})_{y'\to y}\left[\overline{\mathfrak{G}(x, y')\,\mathfrak{F}(y, x')} + \overline{\mathfrak{F}(x, y)\,\mathfrak{G}(x', y')}\right].$$

Hence, instead of the simple equation (39.12), it becomes necessary in the present case to solve a system of four equations for the Fourier components $\boldsymbol{\Pi}^{(i)}(p_+, p_-)$ $(i = 1, \ldots, 4)$.

We introduce the notation

$$\boldsymbol{\Lambda}^{(i)}(\omega') = \frac{n}{(2\pi)^3}\int |u(\boldsymbol{p}-\boldsymbol{p}')|^2\,\boldsymbol{\Pi}^{(i)}(p'_+, p'_-)\,\mathrm{d}^3\boldsymbol{p}'. \tag{39.25}$$

The principle of construction of the equations in clear from Fig. 108. For instance, we have:

$$\boldsymbol{\Pi}^{(1)}(p_+, p_-) = \boldsymbol{p}\{\mathfrak{G}(p_+)\,\mathfrak{G}(p_-) + \mathfrak{F}(p_+)\,\mathfrak{F}^+(p_-)\} + \mathfrak{G}(p_+)\,\mathfrak{G}(p_-)\,\boldsymbol{\Lambda}^{(1)}(\omega) - \mathfrak{F}^+(p_+)\,\mathfrak{G}(p_-)\,\boldsymbol{\Lambda}^{(2)}(\omega) - \mathfrak{F}(p_+)\,\mathfrak{F}^+(p_-)\,\boldsymbol{\Lambda}^{(3)}(\omega) - \mathfrak{G}(p_+)\,\mathfrak{F}(p_-)\,\boldsymbol{\Lambda}^{(4)}(\omega).$$

The remaining four equations are similar. Substitution of them in (39.25) leads to a system of equations for the $\boldsymbol{\Lambda}^{(i)}(\omega)$. We shall not write it down in full, since it can be solved in the general form only in the case of spherically symmetric scattering. We are not interested in the solution for arbitrary relations between the functions. At small concentrations, the properties of a superconducting alloy are close to those of a pure metal. We have already pointed out that the majority of the latter belong to the Pippard or intermediate types. The introduction of impurities lessens the correlation radius and transforms the superconductor at sufficient impurity concentrations to an alloy of the London type, for which the electrodynamic properties are local. The criterion as to whether the properties are local or not is given by the ratio between the penetration depth δ and the mean free path for scattering of electrons at impurity atoms l (or, what amounts to the same thing, between the characteristic magnitudes $|k| \sim 1/\delta$ and $1/l$). We shall assume below that the impurity concentration is such that the alloy has become a London type ($|\boldsymbol{k}|\, l \ll 1$). In this situation the above-mentioned system of equations is greatly simplified. We can neglect $\boldsymbol{k}(\boldsymbol{p}_+ = \boldsymbol{p}_-)$ in $\boldsymbol{\Pi}^{(1)}(p_+, p_-)$; it turns out that now:

$$\boldsymbol{\Lambda}^{(1)}(\omega) = -\boldsymbol{\Lambda}^{(3)}(\omega);$$

$$\boldsymbol{\Lambda}^{(2)}(\omega) = \boldsymbol{\Lambda}^{(4)}(\omega)$$

and

$$\boldsymbol{\Lambda}^{(1)}(\omega) = \boldsymbol{p}\,\Lambda^{(1)}(\omega);$$

$$\boldsymbol{\Lambda}^{(2)}(\omega) = \boldsymbol{p}\,\Lambda^{(2)}(\omega).$$

Hence

$$\boldsymbol{\Pi}^{(1)}(p, p) = \boldsymbol{p}\{\mathfrak{G}^2(p) + \mathfrak{F}^{+2}(p)\}[1 + \Lambda^{(1)}(\omega)] - 2\,\boldsymbol{p}\,\mathfrak{G}(p)\,\mathfrak{F}^+(p)\,\Lambda^{(2)}(\omega), \tag{39.26}$$

and the system of equations for the $\Lambda^{(i)}(\omega)$ becomes

$$\Lambda^{(1)}(\omega) = [\overline{\mathfrak{G}_\omega^2} + \overline{\mathfrak{F}_\omega^{+2}}]\,[1 + \Lambda^{(1)}(\omega)] - 2\,\overline{\mathfrak{G}_\omega \mathfrak{F}_\omega^+}\,\Lambda^{(2)}(\omega),$$

$$\Lambda^{(2)}(\omega) = 2\,\overline{\mathfrak{G}_\omega \mathfrak{F}_\omega^+}\,[1 + \Lambda^{(1)}(\omega)] - [\overline{\mathfrak{G}_\omega^2} + \overline{\mathfrak{F}_\omega^{+2}}]\,\Lambda^{(2)}(\omega),$$

where

$$\boldsymbol{p}\,\overline{\mathfrak{G}_\omega^2} = \frac{n}{(2\pi)^3}\int |u(\boldsymbol{p}-\boldsymbol{p}')|^2\,\mathfrak{G}^2(p')\,\boldsymbol{p}'\,\mathrm{d}^3\boldsymbol{p}' = \frac{\Delta^2 \boldsymbol{p}}{4\tau_1\eta_\omega\left(\sqrt{\omega^2 + \Delta^2}\right)^3},$$

$$\boldsymbol{p}\,\overline{\mathfrak{F}_\omega^{+2}} = \frac{n}{(2\pi)^3}\int |u(\boldsymbol{p}-\boldsymbol{p}')|^2\,\mathfrak{F}^{+2}(p')\,\boldsymbol{p}'\,\mathrm{d}^3\boldsymbol{p}' = \frac{\Delta^2 \boldsymbol{p}}{4\tau_1\eta_\omega\left(\sqrt{\omega^2 + \Delta^2}\right)^3},$$

$$\boldsymbol{p}\,\overline{\mathfrak{F}_\omega^+ \mathfrak{G}_\omega} = \frac{n}{(2\pi)^3}\int |u(\boldsymbol{p}-\boldsymbol{p}')|^2\,\mathfrak{F}^{+}(p')\,\mathfrak{G}(p')\,\boldsymbol{p}'\,\mathrm{d}^3\boldsymbol{p}' = \frac{i\Delta\omega\boldsymbol{p}}{4\tau_1\eta_\omega\left(\sqrt{\omega^2 + \Delta^2}\right)^3}.$$

The solution leads to the following expression for $\Lambda^{(1)}(\omega)$ and $\Lambda^{(2)}(\omega)$:

$$\Lambda^{(1)}(\omega) = \frac{\Delta^2}{2\tau_1(\omega^2+\Delta^2)\left(\sqrt{\omega^2+\Delta^2}+\dfrac{1}{2\tau_{tr}}\right)},$$

$$\Lambda^{(2)}(\omega) = \frac{i\Delta\omega}{2\tau_1(\omega^2+\Delta^2)\left(\sqrt{\omega^2+\Delta^2}+\dfrac{1}{2\tau_{tr}}\right)}.$$

Substitution of these latter into (39.26) and substitution of $\Pi^{(1)}(\boldsymbol{p},\boldsymbol{p})$ into (39.24) yields the following expression for the kernel $Q(\boldsymbol{k})$:

$$Q \equiv Q(\boldsymbol{k}) = \frac{Ne^2}{m}\left\{1 + T\sum_{\omega}\int d\xi\left[(\xi^2-\omega^2\eta_\omega^2+\Delta^2\eta_\omega^2)\right.\right.$$

$$\left.\left.\times\left(1+\frac{\Delta^2}{2\tau_1(\omega^2+\Delta^2)^{3/2}\eta_{\omega tr}}\right)+\frac{2\Delta^2\omega^2\eta_\omega^2}{2\tau_1(\omega^2+\Delta^2)^{3/2}\eta_{\omega tr}}\right]\frac{1}{[\xi^2+(\omega^2+\Delta^2)\eta_\omega^2]^2}\right\} \tag{39.27}$$

(we have used the notation $\eta_{\omega tr} = 1 + (2\tau_{tr}\sqrt{\omega^2+\Delta^2})^{-1}$). Here, as in § 37, we again encounter a formally divergent integral. On the same basis, in the first place we must carry out the summation over the frequencies. After regrouping the terms in the curved brackets into the form $\xi^2+(\omega^2+\Delta^2)\eta_\omega^2-2\omega^2\eta_\omega^2$, we use Abel's transformation, which generalises for series the principle of integration by parts. In fact,

$$\sum_{n=1}^{k}(B_n-B_{n-1})u_n = B_k u_k - B_0 u_1 - \sum_{n=1}^{k-1}(u_{n+1}-u_n)B_n.$$

On applying this to the series

$$2\pi T\sum_{n=1}^{\infty}\frac{1}{\xi^2+(\Delta^2+\omega^2)\eta_\omega^2},$$

where

$$B_n = (2n+1)\pi T = \omega,$$

$$B_n - B_{n-1} = 2\pi T,$$

we can formally cancel the divergent terms in (39.27). We get after this:

$$Q = \frac{Ne^2}{m}2\pi T\Delta^2\sum_{n=1}^{\infty}\frac{1}{(\omega^2+\Delta^2)\left(\sqrt{\omega^2+\Delta^2}+\dfrac{1}{2\tau_{tr}}\right)}. \tag{39.28}$$

Only the "transport" time between collisions enters into this formula. We have for the penetration depth δ:

$$\delta = \frac{1}{\sqrt{4\pi Q}}.$$

When $1/\tau \to 0$, the formula transforms to the usual London formula:

$$\delta = \sqrt{\frac{m}{4\pi N_s e^2}},$$

where N_s is the "number of superconducting electrons".

In the opposite case $l \ll \xi_0$, we can neglect the radical in the bracket of (39.28). The remaining series is easily summed. We obtain as a result, for the penetration depth of dirty alloys:

$$\delta = \frac{1}{2\pi} \sqrt{\frac{1}{\Delta\sigma \tanh \dfrac{\Delta}{2T}}},$$

where σ is the conductivity of the normal metal.

CHAPTER VIII

TRANSPORT EQUATION FOR EXCITATIONS IN A FERMI LIQUID

§ 40. NON-EQUILIBRIUM PROPERTIES OF A FERMI LIQUID

1. *Introduction*

In the present chapter we shall show how one can apply the methods described in this book to transport problems. It is clear from the foregoing that a study of non-equilibrium processes, at least in the approximation which is linear in the external perturbation, requires in principle the analytical continuation with discrete Matsubara frequencies on the real axis. We gave a detailed example of this method in Chapter VII for the case of the electrodynamics of superconductors. At the same time, the procedure for the analytical continuation is not a well-defined one and in each separate case, when we are dealing with viscosity, heat conductivity, electrical conductivity, and so on, we must, generally speaking, start afresh. Moreover, the determination of these quantities may require numerical calculations which in a number of cases may be completely impossible, since one requires for the analytical continuation the knowledge of the exact analytical behaviour.

For the reasons given a moment ago it would be useful to obtain in those cases something like a transport equation for the excitations or particles, in which this step — the analytical continuation — already had been carried out. At the present time there exist rather many methods leading to a solution of this problem for different physical applications. It is not expedient, in our opinion, to give an exposition of all these methods in the present book since the problem of writing down a transport equation is, apparently, a very definite one. We shall consider here in detail the derivation of a transport equation for a Fermi liquid, given by Eliashberg [75], that is, the case of low temperatures and strongly interacting particles.

Let us now turn to the derivation of a transport equation for a Fermi liquid. We shall once again write down the transport equation (2.20) in the phenomenological theory:

$$\frac{\partial n}{\partial t} + \left(\frac{\partial n}{\partial \boldsymbol{r}} \cdot \frac{\partial \varepsilon}{\partial \boldsymbol{p}}\right) + \left(\frac{\partial n}{\partial \boldsymbol{p}} \cdot \frac{\partial \varepsilon}{\partial \boldsymbol{r}}\right) = I(n), \tag{40.1}$$

22*

where the collision integral $I(n)$ has the usual form

$$I(n) = -\operatorname{Tr}_{\sigma_2}\operatorname{Tr}_{\sigma_1'}\int \frac{d^3 p_2\, d^3 p_1'\, d^3 p_2'}{(2\pi)^6} |A(1, 2; 1', 2')|^2$$
$$\times [n_1 n_2 (1 - n_1')(1 - n_2') - (1 - n_1)(1 - n_2) n_1' n_2']$$
$$\times\ \delta(\varepsilon_1 + \varepsilon_2 - \varepsilon_1' - \varepsilon_2')\,\delta(\boldsymbol{p}_1 + \boldsymbol{p}_2 - \boldsymbol{p}_1' - \boldsymbol{p}_2'). \quad (40.2)$$

According to the basic assumption (2.7) in Landau's theory of a Fermi liquid, the energy of the excitations depends on the distribution function. It is therefore very important to bear in mind that the energy conservation law in (40.2) is only satisfied for the true excitation energies which correspond to the complete distribution function. This feature appears particularly distinctly when we write (40.1) in a linearised form. The linearisation of the total distribution function can be realised in two ways:

$$\text{(a)}\ \ n = n_0(\varepsilon_0) + \delta n; \qquad \text{(b)}\ \ n = n_0(\varepsilon) + \delta n'. \quad (40.3)$$

Here n_0 is the equilibrium distribution function in which we substitute in the one case the unperturbed energy ε_0, corresponding to total equilibrium, and in the other case the true energy value corresponding to the distribution function n. We applied the first method in § 2 when we considered the problem of the existence of zero sound; this was convenient as long as we neglected the term involving the collision integral. According to (2.7), the following relation holds between δn and $\delta n'$:

$$\delta n'_{\boldsymbol{p}\sigma} = \delta n_{\boldsymbol{p}\sigma} - \frac{\partial n_0}{\partial \varepsilon}\operatorname{Tr}_{\sigma'}\int \frac{d^3 \boldsymbol{p}'}{(2\pi)^3} f(\boldsymbol{p}\sigma; \boldsymbol{p}'\sigma')\,\delta n_{\boldsymbol{p}'\sigma'}.$$

We now linearise the transport equation itself. Let $V(\boldsymbol{r}, t)$ be the potential energy for the excitations in an external field. If we write

$$\delta n = \frac{\partial n_0}{\partial \varepsilon}\psi, \quad \delta n' = \frac{\partial n_0}{\partial \varepsilon}\varphi,$$

we obtain for the left-hand side of (40.1) in the Fourier representation

$$\text{(a)}\ \Big[-i\big(\omega - (\boldsymbol{k}\cdot\boldsymbol{v})\big)\psi_{\boldsymbol{p}}(\boldsymbol{k}, \omega) - i(\boldsymbol{k}\cdot\boldsymbol{v})\cdot 2\int \frac{d^3\boldsymbol{p}'}{(2\pi)^3} f(\boldsymbol{p}, \boldsymbol{p}')\frac{\partial n_0}{\partial \varepsilon'}\psi_{\boldsymbol{p}'}(\boldsymbol{k}, \omega)$$
$$- i(\boldsymbol{k}\cdot\boldsymbol{v})V(\boldsymbol{k}, \omega)\Big]\frac{\partial n_0}{\partial \varepsilon};$$

$$\text{(b)}\ \Big[-\big(i\omega - (\boldsymbol{k}\cdot\boldsymbol{v})\big)\varphi_{\boldsymbol{p}}(\boldsymbol{k}, \omega) - i\omega\cdot 2\int \frac{d^3\boldsymbol{p}'}{(2\pi)^3} g(\boldsymbol{p}, \boldsymbol{p}')\frac{\partial n_0}{\partial \varepsilon'}\varphi_{\boldsymbol{p}'}(\boldsymbol{k}, \omega)$$
$$- i\omega\, V(\boldsymbol{k}, \omega)\Big]\frac{\partial n_0}{\partial \varepsilon}. \quad (40.4)$$

The functions $g(\boldsymbol{p}, \boldsymbol{p}')$ and $f(\boldsymbol{p}, \boldsymbol{p}')$ are connected as follows:

$$g(\boldsymbol{p}, \boldsymbol{p}') = f(\boldsymbol{p}, \boldsymbol{p}') - \frac{p_0^2}{\pi^2 V}\int \frac{d\Omega''}{4\pi} f(\boldsymbol{p}, \boldsymbol{p}'')\, g(\boldsymbol{p}'' \boldsymbol{p}'). \quad (40.5)$$

The right-hand side of (40.1), the collision integral, has only its usual and convenient form, if it is expressed in terms of φ, as we mentioned earlier:

$$I(n) = \frac{1}{T} \mathrm{Tr}_{\sigma_2} \mathrm{Tr}_{\sigma_1'} \int \frac{d^3 p_2\, d^3 p_1'\, d^3 p_2'}{(2\pi)^6} |A(1, 2; 1', 2')|^2$$
$$\times n_{01} n_{02} (1 - n_{01}') (1 - n_{02}') (\varphi_1 + \varphi_2 - \varphi_1' - \varphi_2')$$
$$\times \delta(\varepsilon_1 + \varepsilon_2 - \varepsilon_1' - \varepsilon_2') \, \delta(\boldsymbol{p}_1 + \boldsymbol{p}_2 - \boldsymbol{p}_1' - \boldsymbol{p}_2'). \quad (40.6)$$

In what follows we shall derive the linearised transport equation (40.4b), (40.6) from a microscopic theory. As far as a derivation of the complete equation (40.1) is concerned, such a derivation is not yet known and it is not clear whether such a derivation exists at all.

2. *Statement of the problem*

Let the equilibrium be violated under the influence of an external field, corresponding to the following term in the Hamiltonian of the system:

$$H_{\mathrm{int}}(t) = \sum_{\boldsymbol{p},\boldsymbol{k},\sigma} V_{\boldsymbol{p}}(\boldsymbol{k}, t)\, a^{\dagger}_{\boldsymbol{p}-1/2\boldsymbol{k},\sigma}\, a_{\boldsymbol{p}+1/2\boldsymbol{k},\sigma}.$$

For the sake of simplicity we shall assume that the field does not act upon the spin. If the field varies slowly both in space and in time, the average of the quantity

$$\tilde{F}_{\boldsymbol{p}}(\boldsymbol{k}, t) = \tilde{a}^{\dagger}_{\boldsymbol{p}+1/2\boldsymbol{k}}(t)\, \tilde{a}_{\boldsymbol{p}-1/2\boldsymbol{k}}(t)$$

is the Fourier component of the quasi-classical momentum distribution function $F_{\boldsymbol{p}}(\boldsymbol{r}, t)$ of the particles.

According to (6.28), we can express the operator $\tilde{F}_{\boldsymbol{p}}(\boldsymbol{k}, t)$ in the interaction representation:

$$\tilde{F}_{\boldsymbol{p}}(\boldsymbol{k}, t) = S^{-1}(t)\, F_{\boldsymbol{p}}(\boldsymbol{k}, t)\, S(t),$$

where in our case

$$S(t) = T \exp\left\{ i \int_{-\infty}^{t} \sum_{\boldsymbol{p}',\boldsymbol{k},\sigma} V_{\boldsymbol{p}'}(\boldsymbol{k}, \tau)\, a^{\dagger}_{\boldsymbol{p}'+1/2\boldsymbol{k}}\, a_{\boldsymbol{p}'-1/2\boldsymbol{k}}\, d\tau \right\}.$$

In the approximation which is linear in V, we have

$$F_{\boldsymbol{p}}(\boldsymbol{k}, t) = \langle \tilde{F}_{\boldsymbol{p}}(\boldsymbol{k}, t) \rangle = - \int_{-\infty}^{+\infty} d\tau \sum_{\boldsymbol{p}'} K^{R}_{\boldsymbol{p},\boldsymbol{p}'}(\boldsymbol{k}, t - \tau)\, V_{\boldsymbol{p}'}(\boldsymbol{k}, \tau), \quad (40.7)$$

where

$$K^{R}_{\boldsymbol{p},\boldsymbol{p}'}(\boldsymbol{k}, t-\tau) = i \langle [a^{\dagger}_{\boldsymbol{p}'+1/2\boldsymbol{k}}(\tau), a_{\boldsymbol{p}'-1/2\boldsymbol{k}}(\tau)]_{-}, a^{\dagger}_{\boldsymbol{p}-1/2\boldsymbol{k}}(t)\, a_{\boldsymbol{p}+1/2\boldsymbol{k}}(t) \rangle\, \theta(t-\tau)$$

is a retarded function. The derivation is completely analogous to the derivation of the expression for the current in a superconductor given in § 37.2. We noted then that there is no direct method for calculating the components of $K^R_{\boldsymbol{p},\boldsymbol{p}'}(\boldsymbol{k}, \omega)$. The thermodynamic technique enables us to study the quantity $\mathcal{K}_{\boldsymbol{p},\boldsymbol{p}'}(\boldsymbol{k}, \omega_n)$ which through the relation

$$\mathcal{K}_{\boldsymbol{p},\boldsymbol{p}'}(\boldsymbol{k}, \omega_n) = \frac{1}{2} \int_{-1/T}^{1/T} d\tau \, e^{i\omega_n \tau} \mathcal{K}_{\boldsymbol{p},\boldsymbol{p}'}(\boldsymbol{k}, \tau)$$

is connected with the thermodynamic "causal" function

$$\mathcal{K}(\boldsymbol{k}, \tau_1, -\tau_2) = \langle T_\tau \{a^\dagger_{\boldsymbol{p}'+1/2\boldsymbol{k}}(\tau_2)\, a_{\boldsymbol{p}'-1/2\boldsymbol{k}}(\tau_2)\, a^\dagger_{\boldsymbol{p}-1/2\boldsymbol{k}}(\tau_1)\, a_{\boldsymbol{p}+1/2\boldsymbol{k}}(\tau_1)\}\rangle .$$

The relation between $K_{\boldsymbol{p},\boldsymbol{p}'}(\boldsymbol{k}, \omega)$ and $\mathcal{K}_{\boldsymbol{p},\boldsymbol{p}'}(\boldsymbol{k}, \omega_n)$ is established through the fact that they are both values of one and the same function which is analytical in the upper half-plane of ω; the first one in points on the real axis and the second one in the points $\omega = i\omega_n$.

The expression

$$\mathcal{K}_{\boldsymbol{p},\boldsymbol{p}'}(\boldsymbol{k}, \omega_n) = -T \sum_n \mathfrak{G}\left(\boldsymbol{p} + \frac{1}{2}\boldsymbol{k}\right) \mathfrak{G}\left(\boldsymbol{p} - \frac{1}{2}\boldsymbol{k}\right)$$
$$\times \left\{(2\pi)^3 \delta_{\boldsymbol{p}-\boldsymbol{p}'} + T \sum_{n'} \mathcal{T}(\boldsymbol{p}\boldsymbol{p}'; \boldsymbol{k})\, \mathfrak{G}\left(\boldsymbol{p}' + \frac{1}{2}\boldsymbol{k}\right) \mathfrak{G}\left(\boldsymbol{p}' - \frac{1}{2}\boldsymbol{k}\right)\right\} \tag{40.8}$$

corresponds to the diagram for $\mathcal{K}_{\boldsymbol{p},\boldsymbol{p}'}(\boldsymbol{k}, \omega_n)$. Fig. 109 shows the notation for the momenta in the vertex part. We need, according to (40.8), to know the analytical properties of the vertex part of the analytical continuation of $\mathcal{K}_{\boldsymbol{p},\boldsymbol{p}'}(\boldsymbol{k}, \omega)$. Let us now turn to that problem. First of all, we note merely that the function $F_{\boldsymbol{p}}(\boldsymbol{k}, t)$ is non-vanishing in a wide range of momenta, and not by any means just in the immediate vicinity of p_0. We cannot therefore interpret $F_{\boldsymbol{p}}(\boldsymbol{k}, t)$ as the distribution function of the excitations, and it is not possible to write down for it any equation similar to a transport equation. We shall, however, see that at a certain point we can introduce a function $\delta n_{\boldsymbol{p}}(\boldsymbol{k}, t)$ which is non-vanishing in the vicinity of the Fermi sphere and which is connected with $F_{\boldsymbol{p}}(\boldsymbol{k}, t)$ through an integral relation. The particle, momentum, and energy currents are connected with $\delta n_{\boldsymbol{p}}(\boldsymbol{k}, t)$ through the same relations as in the phenomenological theory. We can thus consider $\delta n_{\boldsymbol{p}}(\boldsymbol{k}, t)$ as the non-equilibrium part of the distribution function of the excitations.

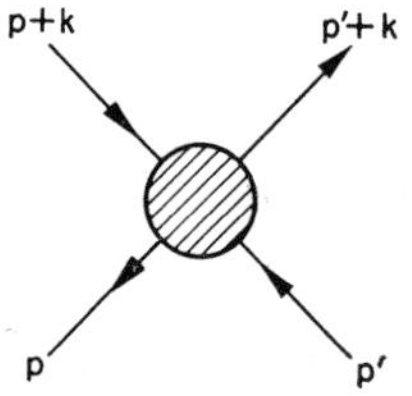

Fig. 109

§ 41. THE ANALYTICAL PROPERTIES OF THE VERTEX PART

The vertex part $\mathcal{T}(\boldsymbol{p}, \boldsymbol{p}'; \boldsymbol{k})$, defined in the temperature diagram technique, is given on a discrete set of its arguments ε_n, ε'_n, and ω_n. We shall consider $\mathcal{T}$ as a function of one of its arguments, say ε_n, while the values of the other arguments are fixed. The problem then arises how to define a set of functions of a complex argument z such that each of them is analytical in some region of the z-plane and that they are all equal to $\mathcal{T}(\varepsilon_n)$ in the points $z = i\varepsilon_n$ which lie in that region. It is known that for the Green functions such analytical functions are $G^R(z)$ and $G^A(z)$.

It is convenient for a study of the vertex part to use equation (16.5). which connects it with the Fourier components of the two-particle Green function $\mathfrak{G}^{\text{II}}_{\alpha\beta\gamma\delta}(p_1, p_2; p_3, p_4)$:

$$\mathfrak{G}(1, 2, 3, 4) = \langle T\{\psi(1)\,\psi(2)\,\psi^\dagger(3)\,\psi^\dagger(4)\}\rangle.$$

We use for it a spectral representation. To do this, we note that the chronological product splits the function in twenty-four parts corresponding to the different permutations of the ψ-operators. These permutations fall into six cycles with four permutations in each of them. The contribution from the cycle generated by the order (1 2 3 4) is equal to

$$\begin{aligned}\mathfrak{G}^{\text{II}}_1(x_1x_2x_3x_4) = \sum_{mnps} &\langle m\,|\psi(\boldsymbol{r}_1)|\,n\rangle\,\langle n\,|\psi(\boldsymbol{r}_2)|\,p\rangle\,\langle p\,|\psi^\dagger(\boldsymbol{r}_3)|\,s\rangle\\ &\times \langle s\,|\psi^\dagger(\boldsymbol{r}_4)|\,m\rangle\\ &\times \exp\{E_m(\tau_1-\tau_4)+E_n(\tau_2-\tau_1)+E_p(\tau_3-\tau_2)+E_s(\tau_4-\tau_3)\}\\ &\times [e^{-E_m/T}\,\theta(\tau_1-\tau_2)\,\theta(\tau_2-\tau_3)\,\theta(\tau_3-\tau_4)\\ &\quad - e^{-E_n/T}\,\theta(\tau_2-\tau_3)\,\theta(\tau_3-\tau_4)\,\theta(\tau_4-\tau_1)\\ &\quad + e^{-E_p/T}\,\theta(\tau_3-\tau_4)\,\theta(\tau_4-\tau_1)\,\theta(\tau_1-\tau_2)\\ &\quad - e^{-E_s/T}\,\theta(\tau_4-\tau_1)\,\theta(\tau_1-\tau_2)\,\theta(\tau_2-\tau_3)].\end{aligned} \tag{41.1}$$

All energies E_i are here reckoned from μN_i. Of the four differences $\tau_i - \tau_k$ only three are independent, for instance, $t_1 = \tau_1 - \tau_2$, $t_2 = \tau_2 - \tau_3$, and $t_3 = \tau_3 - \tau_4$. If we expand therefore in Fourier series in all four τ_i, we obtain the expression

$$\mathfrak{G}^{\text{II}}_1(\varepsilon_1, \varepsilon_2; \varepsilon_3, \varepsilon_4) = \frac{1}{T}\,\delta_{\varepsilon_1+\varepsilon_2-\varepsilon_3-\varepsilon_4}\,\mathfrak{G}^{\text{II}}_1(\varepsilon, \varepsilon', \omega),$$

where we have chosen $\varepsilon = \varepsilon_4$, $\varepsilon' = \varepsilon_2$, and $\omega = \varepsilon_1 - \varepsilon_4$ as independent variables. In terms of these variables we have

$$\begin{aligned}\mathfrak{G}^{\text{II}}_1(\varepsilon, \varepsilon', \omega) &= \sum_{mnps} A(mnps)\\ &\times\left\{\frac{e^{-E_m/T}}{(E_n-E_m-\varepsilon-\omega)\,(E_p-E_m-\varepsilon-\varepsilon'-\omega)\,(E_s-E_m-\varepsilon)}\right.\\ &-\frac{e^{-E_n/T}}{(E_m-E_n+\varepsilon+\omega)\,(E_p-E_n-\varepsilon')\,(E_s-E_n+\omega)}\\ &+\frac{e^{-E_p/T}}{(E_m-E_p+\varepsilon+\varepsilon'+\omega)\,(E_n-E_p+\varepsilon')\,(E_s-E_p+\varepsilon'+\omega)}\\ &\left.-\frac{e^{-E_s/T}}{(E_m-E_s+\varepsilon)\,(E_n-E_s-\omega)\,(E_p-E_s-\varepsilon')}\right\},\end{aligned} \tag{41.2}$$

where we have written for the sake of simplicity: $\varepsilon', \varepsilon = (2n+1)\pi i T$, $\omega = 2m\pi i T$. The quantity $A(mnps)$ is the product of the matrix elements in (41.1). We get similar expressions for the contributions from the other cycles. One easily obtains these expressions from (41.2) by simply permuting the indices. We merely note that apart from the denominators in (41.2) new denominators occur of the form $E_i - E_k - \varepsilon' + \varepsilon$.

Considering $\mathfrak{G}_1^{II}(\varepsilon, \varepsilon'; \omega)$ as a function of a complex argument $z \sim \varepsilon_n$, we see that the imaginary part of the energy denominators in (41.2) vanishes when Im z equals ε_n, 0, $-\omega_m$, or $-\varepsilon_{n'} - \omega_m$. We can divide the z-plane into parts (see Fig. 110) such that none of the denominators

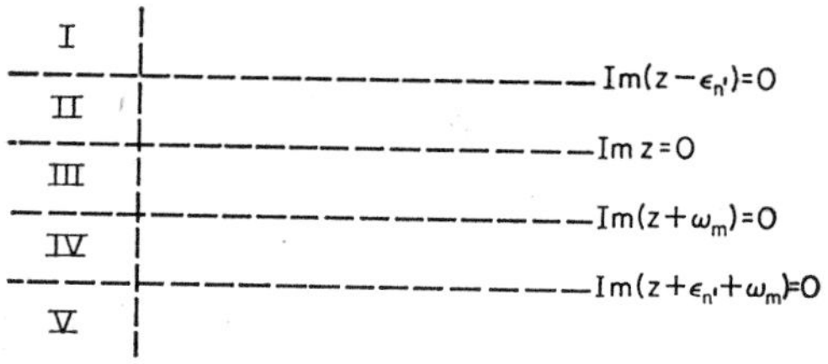

Fig. 110

vanishes within the limits of each of them when z is varied. We note that in I and V there is an infinite sequence of values $z = i\varepsilon_n$. The analytical continuation in those points is thus unique. We shall at the same time see in the following that the non-uniqueness of the analytical continuation in a finite number of points in the regions II, III, and IV does in no way affect the evaluation of physical quantities.

Turning now from the study of the study of the dependence of $\mathcal{T}$ on one of its variables to the general case, we can easily write down the conditions determining the boundaries of the regions of the analyticity of any of its variables:

$$\left.\begin{aligned} &\text{(a) Im}\, z = 0,\ \text{Im}\,(z+w) = 0,\ \text{Im}\, z' = 0,\ \text{Im}\,(z'+w) = 0;\\ &\text{(b) Im}\, w = 0,\ \text{Im}\,(z-z') = 0;\\ &\text{c) Im}\,(z+z'+w) = 0, \end{aligned}\right\} \qquad (41.3)$$

where z, z', and w correspond to the frequencies ε_n, $\varepsilon_{n'}$, and ω_m. Conditions (41.3a) determine at the same time the limits of the analyticity of the Green functions occurring in the definition (16.5). We shall return to this problem later on, but now we shall show only how one can by a direct analysis of perturbation theory diagrams verify that the vertex part possesses all the singularities of the two-particle Green function (41.2).

The information obtained about the analytical properties of the vertex part is sufficient to change the summations over n and n' in (40.8) to integrals. The ω_m remain here free arguments, and later we must look for the analytical continuation for them into the upper half-plane.

We can depict the regions of the analyticy of Γ in $z \sim \varepsilon_n$ and $z' \sim \varepsilon_{n'}$ for fixed values of $\omega_m = 2m\pi T$ $(m > 0)$ in the Im z, Im z'-plane, since

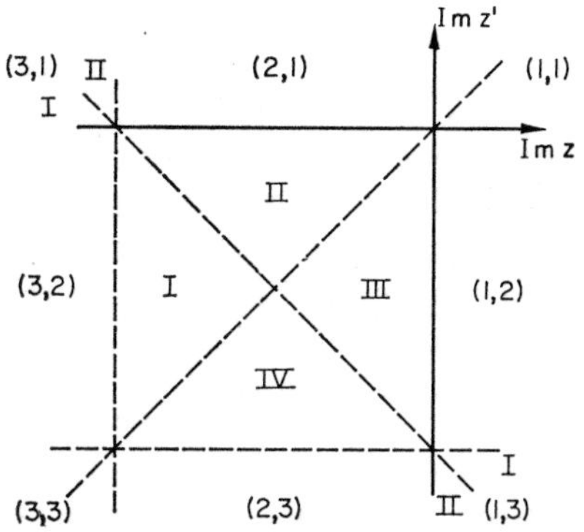

Fig. 111

the boundaries of these regions depend on the imaginary parts of the arguments. In all there are sixteen regions (Fig. 111) each of which corresponds to its own analytical function.

In (40.7) we consider $\mathcal{T}$ together with the Green functions corresponding to its limits:

$$\mathfrak{G}(\varepsilon_n + \omega_m)\,\mathfrak{G}(\varepsilon_n)\,\mathcal{T}(\varepsilon_n, \varepsilon_{n'}; \omega_m)\,\mathfrak{G}(\varepsilon_{n'} + \omega_m)\,\mathfrak{G}(\varepsilon_{n'}). \tag{41.4}$$

(Here and henceforth we shall omit the dependence on the three-dimensional momenta whenever this can not lead to confusion.) The vertical and horizontal lines in Fig. 111 are at the same time branch cuts for the Green functions of (41.4). Each of the nine regions formed by these lines corresponds thus, after the analytical continuation to its own combination of Green functions. As we assume that $m > 0$, there are three pairs of left-hand limits:

$$(1)\ G^R(\varepsilon+\omega)G^R(\varepsilon); \quad (2)\ G^R(\varepsilon+\omega)G^A(\varepsilon); \quad (3)\ G^A(\varepsilon+\omega)G^A(\varepsilon), \tag{41.5}$$

and the same pairs of right-hand limits. Numbering these pairs as in (41.5), we see that we can denote the above-mentioned nine analyticity regions and the functions Γ corresponding to them by (i, k), where $i, k = 1, 2, 3$. The diagonal lines in Fig. 111 determining the singularities in the transferred frequency $z - z'$ and the total frequency $z + z' + w$, divides some of the (i, k) regions into parts which are distinguished by Roman numerals. For instance, Γ_{11}^{I} corresponds to that method of analytical continuation where all external lines are associated with retarded Green functions and here $\mathrm{Im}\,(z - z') > 0$.

We shall now obtain expressions for the jumps in Γ on the diagonal cuts in Fig. 111. Let

$$\begin{aligned}&\varepsilon, \varepsilon', \omega \ll \mu; \quad |p-p_0|, |p'-p_0| \ll p_0; \quad |\mathbf{p}-\mathbf{p}'| \sim p_0;\\ &|\mathbf{p}+\mathbf{p}'| \sim p_0; \quad |\mathbf{k}| \ll p_0.\end{aligned} \tag{41.6}$$

In that case we can for the evaluation of the jumps in Γ use a method similar to the one used in § 19 to evaluate $\operatorname{Im}\Sigma(\varepsilon)$ for small ε. Indeed, the main contribution to the magnitude of the discontinuity will come from diagrams containing in the appropriate cuts the minimum number of lines. Thus, the jump when the sign of $\operatorname{Im}(z-z')$ changes is first and foremost connected with cuts such as those of Fig. 112. Let us consider one of such diagrams on Fig. 112. The vertices Γ_1 and Γ_2 can, in first approximation, since the range of frequencies important for the discontinuity is small, be assumed to be independent of the frequencies. Proceeding as in § 19, we can replace the summation over the frequencies in the matrix element of Fig. 112 by an integral over two Green functions:

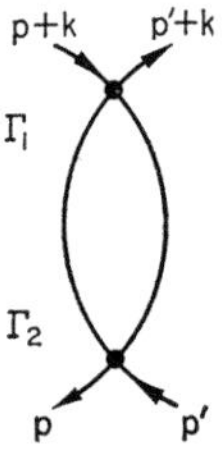

Fig. 112

$$\begin{aligned}\Gamma_1\Gamma_2 T \sum_{n''} \mathfrak{G}_1(\varepsilon_{n''})\, \mathfrak{G}_2(\varepsilon_{n''}+\varepsilon_n-\varepsilon_{n'})\\ = \frac{1}{2\pi}\int_{-\infty}^{+\infty} \mathrm{d}\varepsilon'' \Big[\tanh\frac{\varepsilon''}{2T} \operatorname{Im} G_1^R(\varepsilon'')\, G_2(\varepsilon''+\varepsilon_n-\varepsilon_{n'})\\ + \tanh\frac{\varepsilon''}{2T} G_1(\varepsilon''-\varepsilon_n+\varepsilon'_n) \operatorname{Im} G_2^R(\varepsilon'')\Big]\Gamma_1\Gamma_2.\end{aligned} \tag{41.7}$$

We see that expression (41.7) has a discontinuity when $\operatorname{Im}(z-z')$ passes through zero which equals

$$2\Gamma_1\Gamma_2 \int_{-\infty}^{+\infty} \frac{\mathrm{d}\varepsilon''}{2\pi}\left[\tanh\frac{\varepsilon''+\varepsilon'-\varepsilon}{2T} - \tanh\frac{\varepsilon''}{2T}\right] \times \operatorname{Im} G_1^R(\varepsilon'') \operatorname{Im} G_2^R(\varepsilon''+\varepsilon-\varepsilon').$$

Considering in the vertex part diagrams all cuts of Fig. 112 and summing, we can clearly replace Γ_1 and Γ_2 by the complete vertices. Finally, we have

$$\begin{aligned}\Delta_1(\varepsilon'-\varepsilon) &\equiv \Gamma(\varepsilon'-\varepsilon+i\delta) - \Gamma(\varepsilon'-\varepsilon-i\delta)\\ &= 2\int \frac{\mathrm{d}^4 p_1}{(2\pi)^4} \Gamma_{\alpha\nu\beta\mu}(\mathbf{p}, \mathbf{p}_1+\mathbf{p}'-\mathbf{p}; \mathbf{p}', \mathbf{p}_1)\, \Gamma_{\mu\beta\nu\alpha}(\mathbf{p}, \mathbf{p}'; \mathbf{p}_1+\mathbf{p}'-\mathbf{p}, \mathbf{p})\\ &\times \left[\tanh\frac{\varepsilon_1+\varepsilon'-\varepsilon}{2T} - \tanh\frac{\varepsilon_1}{2T}\right] \operatorname{Im} G^R(p_1) \operatorname{Im} G^R(p_1+p'-p).\end{aligned} \tag{41.8}$$

The discontinuity when $\text{Im}(z + z' + w)$ changes sign is connected with the diagrams of Fig. 113. Proceeding similarly as before we get

$$\Delta_2(\varepsilon + \varepsilon' + \omega) \equiv \Gamma(\varepsilon + \varepsilon' + \omega + i\delta) - \Gamma(\varepsilon + \varepsilon' + \omega - i\delta)$$
$$= -\int \frac{d^4p_1}{(2\pi)^4} \Gamma_{\alpha\beta\nu\mu}(\mathbf{p}, \mathbf{p}'; \mathbf{p} + \mathbf{p}' - \mathbf{p}_1, \mathbf{p}_1) \Gamma_{\mu\nu\beta\alpha}(\mathbf{p}_1, \mathbf{p} + \mathbf{p}' - \mathbf{p}_1; \mathbf{p}', \mathbf{p})$$
$$\times \left[\tanh\frac{\varepsilon_1}{2T} - \tanh\frac{\varepsilon_1 - \varepsilon' - \varepsilon - \omega}{2T}\right] \text{Im}\, G^R(p_1) \,\text{Im}\, G^R(p + p' + k - p_1). \quad (41.9)$$

The vertex parts occurring in these expressions must formally be taken at $\varepsilon = \varepsilon' = \varepsilon'' = 0$, $p = p' = p'' = p_0$, and $k = 0$. Strictly speaking, for Fermi quantities, the vertices are defined in the points $(2n + 1)\pi T i$. We implied this dependence only so that when $|\mathbf{p} - \mathbf{p}'|$ and $|\mathbf{p} + \mathbf{p}'|$ are not small compared to p_0 all vertices vary slowly when the frequency varies by an amount of order ω, $T \ll \mu$. Because of this, all quantities Γ depend only on the angles between the momenta and this enables us as in § 19 to integrate over $\xi_1 = v(p_1 - p_0)$. The result is

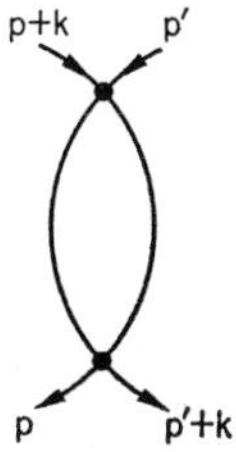

Fig. 113

$$\Delta_1 = \sinh\frac{\varepsilon' - \varepsilon}{2T} \frac{a^2 p_0}{2\pi^2 v^2} \int \frac{d\Omega'}{4\pi} |\Gamma_{\alpha\nu\beta\mu}|^2 \delta(|\mathbf{l}_1 + \mathbf{l}' - \mathbf{l}| - 1)$$
$$\times \int \frac{d\varepsilon_1}{\cosh(\varepsilon_1/2\text{T}) \cosh[(\varepsilon_1 + \varepsilon' - \varepsilon)/2T]}$$

and

$$\Delta_2 = -\sinh\frac{\varepsilon + \varepsilon' + \omega}{2T} \frac{a^2 p^0}{2\pi^2 v^2} \int \frac{d\Omega'}{4\pi} |\Gamma_{\alpha\beta\nu\mu}|^2 \delta(|\mathbf{l} + \mathbf{l}' - \mathbf{l}_1| - 1)$$
$$\times \int \frac{d\varepsilon_1}{\cosh(\varepsilon_1/2T) \cosh[(\varepsilon + \varepsilon' + \omega - \varepsilon_1)/2T]}.$$

We can easily integrate over ε_1:

$$\left.\begin{aligned} \Delta_1(\varepsilon' - \varepsilon) &= \frac{\pi^2}{a^2 p_0} (\varepsilon' - \varepsilon) A^{(1)}_{\alpha\beta\beta\alpha}(\mathbf{p}, \mathbf{p}'), \\ \Delta_2(\varepsilon + \varepsilon' + \omega) &= -\frac{\pi^2}{2a^2 p_0} (\varepsilon + \varepsilon' + \omega) A^{(2)}_{\alpha\beta\beta\alpha}(\mathbf{p}, \mathbf{p}'), \end{aligned}\right\} \quad (41.10)$$

where

$$A^{(1)}_{\alpha\beta\beta\alpha}(\mathbf{p}, \mathbf{p}'])$$
$$= \left[\frac{a^2 p_0}{4\pi v}\right]^2 \int \frac{d\Omega_1}{4\pi} |\Gamma_{\alpha\nu\beta\mu}(\mathbf{p}, \mathbf{p}_1 + \mathbf{p}' - \mathbf{p}, \mathbf{p}', \mathbf{p}_1)|^2 \delta(|\mathbf{l}_1 + \mathbf{l}' - \mathbf{l}| - 1),$$
$$A^{(2)}_{\alpha\beta\beta\alpha}(\mathbf{p}, \mathbf{p}') \quad (41.11)$$
$$= \left[\frac{a^2 p_0}{4\pi v}\right]^2 \int \frac{d\Omega_1}{4\pi} |\Gamma_{\alpha\beta\nu\mu}(\mathbf{p}, \mathbf{p}', \mathbf{p} + \mathbf{p}' - \mathbf{p}_1, \mathbf{p}_1)|^2 \delta(|\mathbf{l} + \mathbf{l}' - \mathbf{l}_1| - 1).$$

We now draw attention to the fact that one and the same diagram cannot contain simultaneously the two types of cuts depicted in Fig. 114. The quantities Δ_1 and Δ_2 are thus determined by diagrams which do not have cuts of the type 114a. In these diagrams we can split off two different

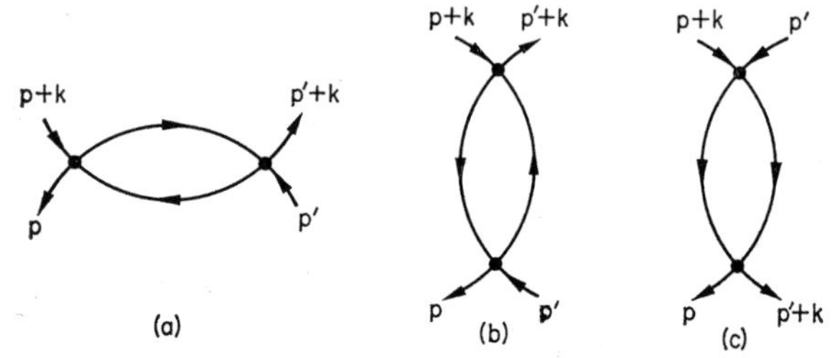

Fig. 114

groups possessing, respectively, the cuts 114b and 114c. The quantities Γ_{22} (Fig. 111) will play an important role in the following. What we have said so far enables us to write down the following expressions for these quantities:

$$\left.\begin{aligned}
\Gamma^{\mathrm{I}}_{22}(\varepsilon, \varepsilon'; \omega) &= \Gamma_{22}(\varepsilon, \varepsilon'; \omega) + \frac{i}{2}[\Delta_1(\varepsilon' - \varepsilon) - \Delta_2(\varepsilon + \varepsilon' + \omega)],\\
\Gamma^{\mathrm{II}}_{22}(\varepsilon, \varepsilon'; \omega) &= \Gamma_{22}(\varepsilon, \varepsilon'; \omega) + \frac{i}{2}[\Delta_1(\varepsilon' - \varepsilon) + \Delta_2(\varepsilon + \varepsilon' + \omega)],\\
\Gamma^{\mathrm{III}}_{22}(\varepsilon, \varepsilon'; \omega) &= \Gamma_{22}(\varepsilon, \varepsilon'; \omega) + \frac{i}{2}[-\Delta_1(\varepsilon' - \varepsilon) + \Delta_2(\varepsilon + \varepsilon' + \omega)],\\
\Gamma^{\mathrm{IV}}_{22}(\varepsilon, \varepsilon'; \omega) &= \Gamma_{22}(\varepsilon, \varepsilon'; \omega) + \frac{i}{2}[-\Delta_1(\varepsilon' - \varepsilon) - \Delta_2(\varepsilon + \varepsilon' + \omega)],
\end{aligned}\right\} \quad (41.12)$$

where $\Gamma_{22}(\varepsilon, \varepsilon'; \omega)$ is continuous for $\mathrm{Im}(z - z') = 0$ and $\mathrm{Im}(z + z' + \omega) = 0$.

§ 42. EQUATION FOR THE VERTEX PART; TRANSPORT EQUATION

We turn now to the solution of the problem of the analytical continuation of the function $\mathcal{K}(\omega_m)$ of (40.8). We shall write $\mathcal{K}(\omega_m)$ as follows:

$$\mathcal{K}(\omega_m) = -T \sum_n \mathcal{K}(\varepsilon_n, \omega_m), \quad (42.1)$$

where

$$\mathcal{K}(\varepsilon_n, \omega_m) = \mathfrak{G}(\varepsilon_n + \omega_m)\, \mathfrak{G}(\varepsilon_n) \Big[1 + T \sum_{n'} \mathcal{T}(\varepsilon_n, \varepsilon_{n'}; \omega_m)\, \mathfrak{G}_n(\varepsilon_{n'} + \omega_m)\, \mathfrak{G}(\varepsilon_{n'})\Big]. \quad (42.2)$$

First of all, we change in (42.1) the sum over n into an integral. To do this we use the fact that $\mathcal{K}(\varepsilon_n, \omega_m)$ considered as a function of the complex argument $z \sim \varepsilon_n$ has the regions of analyticity shown in Fig. 115. Indeed, the intermediate cuts in the diagrams for $\mathcal{K}(\varepsilon_n, \omega_m)$ depending on ε_n contain either simply ε_n, or $\varepsilon_n + \omega_m$. Let $K_i(z, \omega_m)$ $(i = 1, 2, 3)$ be

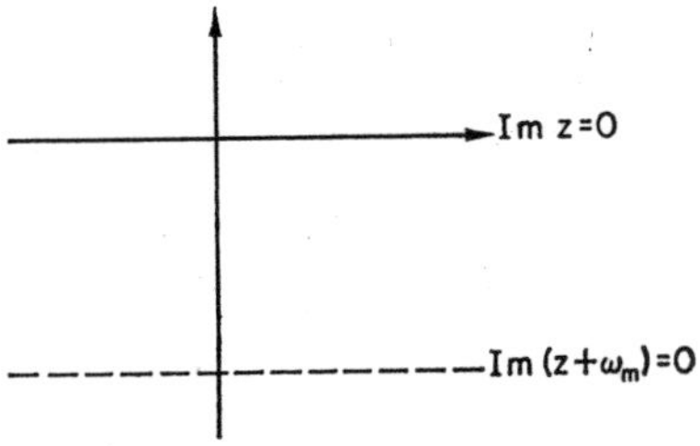

Fig. 115

functions, analytical in z for given ω_m in the appropriate regions of Fig. 115. Choosing the integration contour such that it goes round each of the three regions in the positive direction, we can write

$$T \sum_n \mathcal{K}(\varepsilon_n, \omega_m) = \frac{1}{4\pi i} \int_{-\infty}^{+\infty} d\varepsilon \left\{ \tanh \frac{\varepsilon}{2T} [K_1(\varepsilon, \omega_m) - K_2(\varepsilon, \omega_m)] + \tanh \frac{\varepsilon}{2T} [K_2(\varepsilon - \omega_m, \omega_m) - K_3(\varepsilon - \omega_m, \omega_m)] \right\}. \quad (42.3)$$

Although the analytical continuation in region 2 is not unique, the choice of the function $K_2(\varepsilon, \omega_m)$ does not affect the value of the integral (42.3).

We must analytically continue the functions K_i occurring in the integrand of (42.3) into the upper half-plane of the variable $w \sim \omega_m$. To do this we can write for the retarded function $K^R(\omega)$:

$$K^R(\omega) = -\frac{1}{4\pi i} \int_{-\infty}^{+\infty} d\varepsilon \left\{ \tanh \frac{\varepsilon}{2T} K_1(\varepsilon, \omega) + \left(\tanh \frac{\varepsilon + \omega}{2T} - \tanh \frac{\varepsilon}{2T} \right) K_2(\varepsilon\ \omega) - \tanh \frac{\varepsilon + \omega}{2T} K_3(\varepsilon, \omega) \right\}. \quad (42.4)$$

It is now necessary to express the K_i in terms of the Γ_{ik} of Fig. 111. To do this we change the sum over n' in (42.2) to an integral:

$$T \sum_{n'} \mathcal{T}(\varepsilon_n, \varepsilon_{n'}; \omega_m)\, \mathfrak{G}(\varepsilon_{n'} + \omega_m)\, \mathfrak{G}(\varepsilon_{n'})$$
$$= \frac{1}{4\pi i} \int_L dz' \tanh \frac{z'}{2T} \mathcal{T}(\varepsilon_n, z'; \omega_m)\, \mathfrak{G}(z' + \omega_m)\, \mathfrak{G}(z').$$

The contour L must go round all regions of Fig. 116 in the positive direction. Since the integrals over the arc of the large circle vanish, the relative position of the branch cuts does not play a role. Writing down the integrals along the edge of the cuts and putting indices on $\mathcal{T}$ in accordance with Fig. 111, we see that

$$K_i(\varepsilon, \omega) = g_i(\varepsilon, \omega)\left[1 + \frac{1}{4\pi i}\int_{-\infty}^{+\infty} d\varepsilon''\, \mathcal{L}_{ik}(\varepsilon, \varepsilon'; \omega)\, g_k(\varepsilon', \omega)\right]. \quad (42.5)$$

We have here denoted by g_i the pairs of Green functions of (41.5) ($i = 1, 2, 3$), while the $\mathcal{L}_{ik}$ are connected with the Γ_{ik} as follows:

$$\mathcal{L}_{11}(\varepsilon, \varepsilon'; \omega) = \tanh\frac{\varepsilon'}{2T}\Gamma_{11}^{\mathrm{I}} + \coth\frac{\varepsilon' - \varepsilon}{2T}[\Gamma_{11}^{\mathrm{II}} - \Gamma_{11}^{\mathrm{I}}], \quad (42.6)$$

$$\mathcal{L}_{13}(\varepsilon, \varepsilon'; \omega) = -\tanh\frac{\varepsilon' + \omega}{2T}\Gamma_{13}^{\mathrm{I}} - \coth\frac{\varepsilon + \varepsilon' + \omega}{2T}[\Gamma_{13}^{\mathrm{II}} - \Gamma_{13}^{\mathrm{I}}], \quad (42.7)$$

$$\mathcal{L}_{31}(\varepsilon, \varepsilon'; \omega) = \tanh\frac{\varepsilon'}{2T}\Gamma_{31}^{\mathrm{I}} + \coth\frac{\varepsilon + \varepsilon' + \omega}{2T}[\Gamma_{31}^{\mathrm{II}} - \Gamma_{31}^{\mathrm{I}}], \quad (42.8)$$

$$\mathcal{L}_{12}(\varepsilon, \varepsilon'; \omega) = \left[\tanh\frac{\varepsilon' + \omega}{2T} - \tanh\frac{\varepsilon'}{2T}\right]\Gamma_{12}, \quad (42.9)$$

$$\mathcal{L}_{32}(\varepsilon, \varepsilon'; \omega) = \left[\tanh\frac{\varepsilon' + \omega}{2T} - \tanh\frac{\varepsilon'}{2T}\right]\Gamma_{32}, \quad (42.10)$$

$$\mathcal{L}_{21}(\varepsilon, \varepsilon'; \omega) = \tanh\frac{2\varepsilon'}{2T}\Gamma_{21}, \quad (42.11)$$

$$\mathcal{L}_{23}(\varepsilon, \varepsilon'; \omega) = -\tanh\frac{\varepsilon' + \omega}{2T}\Gamma_{23}, \quad (42.12)$$

$$\begin{aligned}\mathcal{L}_{22}(\varepsilon, \varepsilon'; \omega) = &\left[\coth\frac{\varepsilon' - \varepsilon}{2T} - \tanh\frac{\varepsilon'}{2T}\right]\Gamma_{22}^{\mathrm{II}} \\ &+ \left[\coth\frac{\varepsilon + \varepsilon' + \omega}{2T} - \coth\frac{\varepsilon' - \varepsilon}{2T}\right]\Gamma_{22}^{\mathrm{III}} \\ &+ \left[\tanh\frac{\varepsilon' + \omega}{2T} - \coth\frac{\varepsilon + \varepsilon' + \omega}{2T}\right]\Gamma_{22}^{\mathrm{IV}}. \end{aligned} \quad (42.13)$$

Equations (42.5)—(42.13) solve the problem of the analytical continuation of the function $\mathcal{K}(\omega_m)$.

Earlier, in Chapter IV, we have already shown that at $T = 0$ the possibility to write down equation (18.9) for the zero-sound vibrations is connected with the singularity of the vertex part in the variables ω and k. We turn therefore again to a study of diagrams with intermediate cuts such as given by Fig. 114a:

$$\mathcal{T}' = T\sum_{n''}\Gamma^{(1)}(\varepsilon_n, \varepsilon_{n''}; \omega_m)\,\mathfrak{G}(\varepsilon_{n''} + \omega_m)\,\mathfrak{G}(\varepsilon_{n''})\,\Gamma^{(2)}(\varepsilon_{n''}, \varepsilon_{n'}; \omega_m).$$

Let $\mathscr{L}_{ik}^{(1)}$ and $\mathscr{L}_{ik}^{(2)}$ be quantities constructed from $\mathscr{T}$, $\Gamma^{(1)}$, and $\Gamma^{(2)}$ by means of equations (42.6)—(42.13). Changing the sum over n'' to an integral, we see then easily that the previous expression can be written as follows:

$$\mathscr{L}'_{ik}(\varepsilon, \varepsilon'; \omega_m) = \frac{1}{4\pi i} \int_{-\infty}^{+\infty} d\varepsilon'' \, \mathscr{L}_{ik}^{(1)}(\varepsilon, \varepsilon''; \omega_m) \, g_l(\varepsilon'', \omega_m) \, \mathscr{L}_{lk}^{(2)}(\varepsilon'', \varepsilon'; \omega_m), \tag{42.14}$$

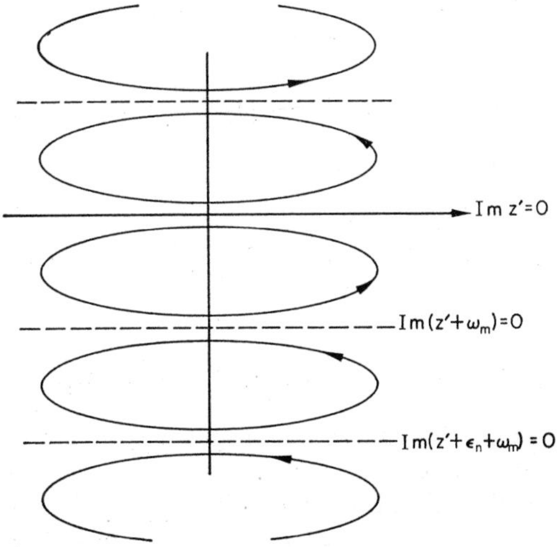

Fig. 116

where here and later on a summation over l is implied. Because of this property we can write for the complete quantity $\mathscr{L}_{ik}$ an equation which is the analogue of equation (18.3):

$$\mathscr{L}_{ik}(p, p'; k) = \mathscr{L}_{ik}^{(0)}(p, p'; k) + \frac{1}{2i} \int \frac{d^4 p''}{(2\pi)^4} \mathscr{L}_{ik}^{(0)}(p, p''; k) \, g_l(p'', k) \, \mathscr{L}_{lk}(p'', p'; k) \tag{42.15}$$

in which $\mathscr{L}^{(0)}$ corresponds to the analytical continuation of the diagrams which do not have the cuts of Fig. 114*a*. Let us consider in (42.15) the different cuts of (41.5). One sees easily that only the combination $g_2 = G^R G^A$ gives a contribution from the near regions when we integrate over d^4p''. The other combinations, g_1 and g_3, are only integrated over parts of momentum space far from the Fermi surface and over large frequencies. It is at the same time clear from equations (42.9), (42.10), and (42.13) that the structure of the quantities $\mathscr{L}_{i2}$ is such that they decrease exponentially when we go away from the Fermi surface. Only the near regions are therefore in fact important in the integrals containing g_2.

For small $\boldsymbol{k}$ and ω the dependence of the vertex parts on $\boldsymbol{k}$ and ω is mainly taken into account by considering the integrals over the near regions, and is thus determined by the cuts of g_2. It is therefore expedient to rearrange equation (42.15). Let us consider separately the quantity $\mathscr{L}_{22}$, the indices of which we shall drop in the following, and let us split off from it the part $\tilde{\mathscr{L}}$ which consists of diagrams containing no cuts of g_2. We can then write down the equation

$$\mathscr{L}(p, p'; k) = \tilde{\mathscr{L}}(p, p'; k) + \frac{1}{2i}\int \frac{d^4p''}{(2\pi)^4} \tilde{\mathscr{L}}(p, p''; k) g_2(p'', k)\, \mathscr{L}(p'', p'; k), \tag{42.16}$$

in which we can integrate over $\xi'' = v(p'' - p_0)$ on the basis of what we have just said:

$$\mathscr{L}(p\ p'; k) \tag{42.17}$$

$$= \tilde{\mathscr{L}}(p, p'; k) + \frac{a^2 p_0^2}{4\pi^2 v} \int\limits_{-\infty}^{+\infty} d\varepsilon'' \int \frac{d\Omega''}{4\pi} \tilde{\mathscr{L}}(p, p''; k) \frac{\mathscr{L}(p'', p'; k)}{\Omega(p'', k)}$$

where

$$\Omega(p'', k) = \omega - (\boldsymbol{v}\cdot\boldsymbol{k}) + i[\gamma(\varepsilon + \omega) + \gamma(\varepsilon)]. \tag{42.18}$$

The remaining quantities $\mathscr{L}_{ik}$ can be expressed in terms of $\mathscr{L}$ and the parts $\mathscr{L}_{ik}^{(0)}$ which do not contain the cuts of g_2. This connection is depicted graphically in Fig. 117. Substituting the appropriate expressions into (42.5) we can express the functions $K_i(\varepsilon, \omega)$ in terms of $\mathscr{L}\,(i \neq 2; l \neq 2)$:

$$\begin{aligned} K_i(\varepsilon, \omega) = K_i^{(0)} + 2g_i(\varepsilon, \omega)\Bigg\{&\frac{1}{i}\int \frac{d^4p'}{(2\pi)^4} \mathscr{L}_{i2}^{(0)}(p, p'; k)\, g_2(p', k) \\ &\times \left[1 + \frac{1}{2\pi i}\int d\varepsilon'' \mathscr{L}_{2l}^{(0)}(p', p''; k)\, g_l(p'', k)\right] \\ &- \int \frac{d^4p'}{(2\pi)^4}\int \frac{d^4p''}{(2\pi)^4} \mathscr{L}_{i2}^{(0)}(p, p'; k)\, g_2(p', k)\, \mathscr{L}(p', p''; k)\, g_2(p'', k) \\ &\times \left[1 + \frac{1}{2\pi i}\int d\varepsilon'''\, \mathscr{L}_{2l}^{(0)}(p'', p''', ; k)\, g_l(p''', k)\right]\Bigg\}, \end{aligned} \tag{42.19}$$

$$\begin{aligned} K_2(\varepsilon, \omega) = 2g_2(\varepsilon, \omega)\Bigg\{&1 + \frac{1}{2\pi i}\int d\varepsilon'\, \mathscr{L}_{2l}^{(0)}(p, p'; k)\, g_l(p', k) \\ &+ \frac{1}{i}\int \frac{d^4p'}{(2\pi)^4} \mathscr{L}(p, p'; k)\, g_2(p', k)\left[1 + \frac{1}{2\pi i}\int d\varepsilon''\, \mathscr{L}_{2l}^{(0)}(p', p''; k)\, g_l(p'', k)\right]\Bigg\}. \end{aligned} \tag{42.20}$$

Here $K_i^{(0)}$ does not contain the cuts of g_2 and is thus independent of ω and $\boldsymbol{k}$. By definition, the quantities $\mathscr{L}_{ik}^{(0)}$ do not contain the intermediate cuts of g_2.

Let us split off in an arbitrary diagram for $\mathscr{L}_{i2}^{(0)}$ the last cut on the right of g_l $(l \neq 2)$ (Fig. 118). The vertex $\mathscr{L}_{l2}^{(1)}$ separating this cut and the outer cut of g_2 contains according to (42.9) and (42.10) a factor $\tanh[(\varepsilon' + \omega)/2T] - \tanh(\varepsilon'/2T)$. It is convenient to introduce quantities $\Gamma_{i2}^{(0)}$ through the definition

$$\mathscr{L}_{l2}^{(0)} = \Gamma_{l2}^{(0)}\left[\tanh\frac{\varepsilon' + \omega}{2T} - \tanh\frac{\varepsilon'}{2T}\right]. \tag{42.21}$$

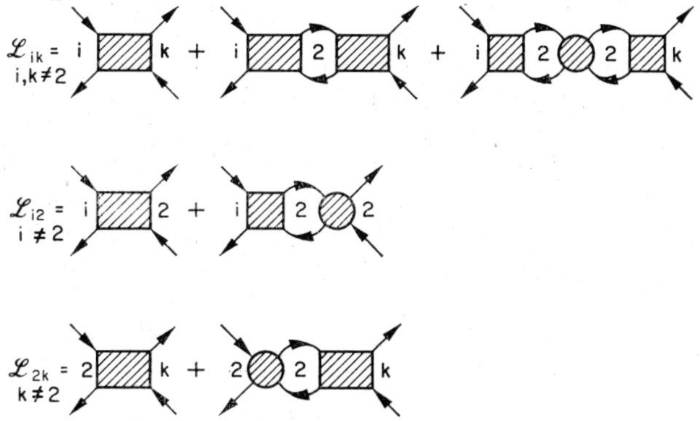

Fig. 117

Similarly, in correspondence with (42.11) and (42.12), we put for $\mathscr{L}_{2l}^{(0)}$:

$$\mathscr{L}_{21}^{(0)} = \Gamma_{21}^{(0)}\tanh\frac{\varepsilon'}{2T}; \qquad \mathscr{L}_{23}^{(0)} = -\Gamma_{23}^{(0)}\tanh\frac{\varepsilon' + \omega}{2T}. \tag{42.22}$$

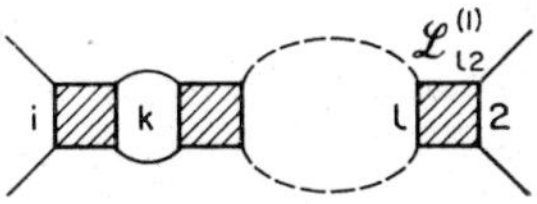

Fig. 118

Using these relations and substituting (42.19) and (42.20) into (42.4) we get the following expression for $K_{\boldsymbol{pp}'}^{R}(\omega, \boldsymbol{k})$:

$$K_{\boldsymbol{pp}'}^{R}(\omega, \boldsymbol{k}) = -\frac{1}{i}\int \frac{d^4p''}{(2\pi)^4} Q^{(1)}(\boldsymbol{p}, \boldsymbol{p}'')\left[\tanh\frac{\varepsilon'' + \omega}{2T} - \tanh\frac{\varepsilon''}{2T}\right]$$
$$\times g_2(p'', k)\, Q^{(2)}(\boldsymbol{p}'', \boldsymbol{p}') + \int \frac{d^4p''}{(2\pi)^4}\int \frac{d^4p'''}{(2\pi)^4} Q^{(1)}(\boldsymbol{p}, \boldsymbol{p}'')$$
$$\times\left[\tanh\frac{\varepsilon'' + \omega}{2T} - \tanh\frac{\varepsilon''}{2T}\right] \tag{42.23}$$
$$\times g_2(p'', k)\, \mathscr{L}(p'', p'''; k)\, g_2(p''', k)\, Q^{(2)}(\boldsymbol{p}''', \boldsymbol{p}') + K^{(0)},$$

where

$$\left.\begin{aligned} Q^{(1)}(\boldsymbol{p}, \boldsymbol{p}'') &= (2\pi)^3\,\delta(\boldsymbol{p}-\boldsymbol{p}'') + \frac{1}{2\pi i}\int \mathrm{d}\varepsilon \tanh\frac{\varepsilon}{2T} \\ &\quad\times [g_1(p)\,\Gamma^{(0)}_{12}(p, p'') - g_3(p)\,\Gamma^{(0)}_{32}(p, p'')], \\ Q^{(2)}(\boldsymbol{p}'', \boldsymbol{p}') &= (2\pi)^3\,\delta(\boldsymbol{p}''-\boldsymbol{p}') + \frac{1}{2\pi i}\int \mathrm{d}\varepsilon' \tanh\frac{\varepsilon'}{2T} \\ &\quad\times [g_1(p')\,\Gamma^{(0)}_{21}(p'', p') - g_3(p')\,\Gamma^{(0)}_{31}(p'', p')], \end{aligned}\right\} \quad (42.24)$$

while $K^{(0)}$ does not contain the cuts of g_2. Thanks to the factor $\tanh([\varepsilon''+\omega]/2T) - \tanh(\varepsilon''/2T)$ the integration over ε'' in (42.23) is limited to values $\varepsilon'' \approx \omega, T$. It is clear from (42.13) that also only small values of ε''' are important. It follows from this that also all integrals over momenta are restricted to the immediate vicinity of the Fermi sphere. In that region we may neglect the dependence of Q and $\mathscr{L}$ on the absolute magnitude of the momenta and integrate over ξ'' and ξ''':

$$\begin{aligned} K^R_{\boldsymbol{pp}'}(\omega, \boldsymbol{k}) = &-\frac{a^2 p_0^2}{2\pi^2 v}\int\frac{\mathrm{d}\Omega''}{4\pi}\int \mathrm{d}\varepsilon''\, Q(\boldsymbol{p}, \boldsymbol{p}'')\left[\tanh\frac{\varepsilon''+\omega}{2T} - \tanh\frac{\varepsilon''}{2T}\right] \\ &\times\frac{1}{\Omega(p'', k)}\,Q(\boldsymbol{p}'', \boldsymbol{p}') - \left[\frac{a^2 p_0^2}{2\pi^2 v}\right]^2\int\frac{\mathrm{d}\Omega''}{4\pi}\int\frac{\mathrm{d}\Omega'''}{4\pi} \\ &\times\iint \mathrm{d}\varepsilon''\,\mathrm{d}\varepsilon'''\, Q(\boldsymbol{p}, \boldsymbol{p}'')\left[\tanh\frac{\varepsilon''+\omega}{2T} - \tanh\frac{\varepsilon''}{2T}\right] \\ &\times\frac{1}{\Omega(p'', k)}\,\mathscr{L}(p'', p'''; k)\,\frac{1}{\Omega(p''', k)}\,Q(\boldsymbol{p}''', \boldsymbol{p}') + K^{(0)}. \end{aligned} \quad (42.25)$$

We have here taken into account that, as $Q^{(1)}$ and $Q^{(2)}$ are built up from diagrams which do not contain the cuts of g_2, we can substitute their values for $\omega = k = 0$. In that case

$$Q^{(1)} = Q^{(2)} = Q, \quad (42.26)$$

since one easily sees that $\Gamma^{(0)}_{2i}(p, p') = \Gamma^{(0)}_{i2}(p', p)$.

Let us now turn to equation (40.7). It is convenient to write the correction to the particle distribution function which is linear in the external field in the following form:

$$F'_{\boldsymbol{p}}(\boldsymbol{k}, \omega) = f^{(0)} + f^{(1)}_{\boldsymbol{p}}(\boldsymbol{k}, \omega),$$

where

$$f^{(0)}(\boldsymbol{k}, \omega) = -\int\frac{\mathrm{d}^3\boldsymbol{p}'}{(2\pi)^3}\,K^{(0)}_{pp'}\,V_{p'}(\boldsymbol{k}, \omega), \quad (42.27)$$

$$f^{(1)}_{\boldsymbol{p}}(\boldsymbol{k}, \omega) = -\int\frac{\mathrm{d}^3\boldsymbol{p}'}{(2\pi)^3}\,[K^R_{pp'} - K^{(0)}_{pp'}]\,V_{p'}(\boldsymbol{k}, \omega). \quad (42.28)$$

Let $\omega \ll T$. In that case we have for the difference between the hyperbolic tangents in (42.25):

$$\tanh\frac{\varepsilon+\omega}{2T}-\tanh\frac{\varepsilon}{2T}\approx-2\omega\frac{\partial n_0}{\partial\varepsilon},$$

and

$$K^R_{\boldsymbol{p}\boldsymbol{p}'}(\boldsymbol{k},\omega)-K(0)=\frac{a^2\,p_0}{\pi^2\,v}\int\frac{\mathrm{d}\Omega'}{4\pi}\,\mathrm{d}\varepsilon\,Q(\boldsymbol{p},\boldsymbol{p}')\,\frac{\partial\,n_0}{\partial\varepsilon}\,\frac{1}{\Omega(p',k)}$$
$$\times\left\{Q(\boldsymbol{p}',\boldsymbol{p}'')+\frac{a^2\,p_0}{2\pi^2\,v}\int\frac{\mathrm{d}\Omega'''}{4\pi}\,\mathrm{d}\varepsilon'''\,\frac{\mathscr{L}(p',p''';k)}{\Omega(p''',k)}\,Q(\boldsymbol{p}''',\boldsymbol{p}'')\right\}.$$

Substituting this expression into (42.28) and writing

$$\tilde{V}_{\boldsymbol{p}}(\boldsymbol{k},\omega)=a\int\frac{\mathrm{d}^3p'}{(2\pi)^3}\,Q(\boldsymbol{p},\boldsymbol{p}')\,V_{\boldsymbol{p}'}(\boldsymbol{k},\omega),$$

we determine the quantity $\delta n'_{\boldsymbol{p}}$ through the relation

$$\left.\begin{aligned} f'_{\boldsymbol{p}}(\boldsymbol{k},\omega)&=\frac{a\,p_0^2}{2\pi^2\,v}\int\frac{\mathrm{d}\Omega'}{4\pi}\int\mathrm{d}\varepsilon\,Q(\boldsymbol{p},\boldsymbol{p}')\,\delta n'_{\boldsymbol{p}'}(\boldsymbol{k},\omega),\\ \delta n'_{\boldsymbol{p}}&=\frac{\partial n_0}{\partial\varepsilon}\,\varphi_{\boldsymbol{p}}.\end{aligned}\right\}\qquad(42.29)$$

Here $\delta n'_{\boldsymbol{p}}$ is given by

$$\delta n'_{\boldsymbol{p}}=\frac{2\,\omega}{\Omega(p,k)}\left\{\tilde{V}_{\boldsymbol{p}}(\boldsymbol{k},\omega)+\frac{a^2\,p_0}{2\pi^2 v}\int\frac{\mathrm{d}\Omega'}{4\pi}\int\mathrm{d}\varepsilon'\,\frac{\mathscr{L}(p,p';k)}{\Omega(p',k)}\,\tilde{V}_{\boldsymbol{p}'}(\boldsymbol{k},\omega)\right\}.$$

Multiplying this relation by $\tilde{\mathscr{L}}(p'',p;k)$ and integrating over $\boldsymbol{p}$ and using (42.17), one sees easily that $\varphi_{\boldsymbol{p}}$ satisfies the equation

$$i[\omega-(\boldsymbol{k}\cdot\boldsymbol{v})]\,\varphi_{\boldsymbol{p}}=-i\omega\,\tilde{V}_{\boldsymbol{p}}(\boldsymbol{k},\omega)$$
$$+i\,\frac{a^2\,p_0^2}{2\pi^2 v}\int\frac{\mathrm{d}\Omega'}{4\pi}\int\mathrm{d}\varepsilon'\,\tilde{\mathscr{L}}(p,p';k)\,\varphi_{\boldsymbol{p}'}+2\gamma(\varepsilon)\,\varphi_{\boldsymbol{p}}.\qquad(42.30)$$

Let us study now the structure of the kernel $\tilde{\mathscr{L}}$ of this equation. This quantity consists of two parts (Fig. 119). The first part $\mathscr{L}^{(1)}$ is produced

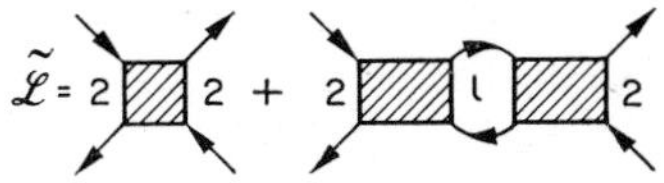

Fig. 119

by the diagrams which do not contain cuts with two lines g_l. Into the second part $\mathscr{L}^{(2)}$ enter those diagrams which although they do not contain cuts of g_2, may contain various numbers of cuts of g_l with $l \neq 2$. According to (42.13), $\mathscr{L}^{(1)}$ is constructed from the quantities $\Gamma^{(1)}$ which occur when we analytically continue the diagrams of the vertex part which do not contain cuts with two lines. We can then write for $\Gamma^{(1)}$ an expression such as (41.12) where the magnitudes of the jumps Δ_1 and Δ_2 are the same as for the complete vertex part. (Bear in mind that Δ_1 and Δ_2 are defined by cuts such as those of Figs. 112 and 113. All diagrams with such cuts occur in $\Gamma^{(1)}$.) Substituting (41.12) into (42.13) we get

$$\mathscr{L}^{(1)}(\varepsilon, \varepsilon'; \omega) = \Gamma^{(1)}\left[\tanh\frac{\varepsilon' + \omega}{2T} - \tanh\frac{\varepsilon'}{2T}\right] + i\,\mathscr{L}'(\varepsilon, \varepsilon'; \omega), \quad (42.31)$$

$$\mathscr{L}'(\varepsilon, \varepsilon'; \omega) = \frac{1}{2}\left[2\coth\frac{\varepsilon' - \varepsilon}{2T} - \tanh\frac{\varepsilon'}{2T} - \tanh\frac{\varepsilon' + \omega}{2T}\right]\Delta_1(\varepsilon' - \varepsilon)$$
$$+ \frac{1}{2}\left[2\coth\frac{\varepsilon + \varepsilon' + \omega}{2T} - \tanh\frac{\varepsilon'}{2T} - \tanh\frac{\varepsilon' + \omega}{2T}\right]\Delta_2(\varepsilon + \varepsilon' + \omega). \quad (42.32)$$

The second part $\mathscr{L}^{(2)}$ is proportional to $\tanh([\varepsilon' + \omega]/2T) - \tanh(\varepsilon'/2T)$ since according to Fig. 119 any diagram of $\mathscr{L}^{(2)}$ contains an element $\mathscr{L}^{(1)}_{i2}$ which contains such a factor according to (42.9) and (42.10).

We can thus write altogether for $\tilde{\mathscr{L}}$

$$\tilde{\mathscr{L}}(p, p'; k) = \Gamma^k(p, p')\left[\tanh\frac{\varepsilon' + \omega}{2T} - \tanh\frac{\varepsilon'}{2T}\right] + i\mathscr{L}'(p, p'; k). \quad (42.33)$$

It is clear that we cannot take into account the frequency dependence of $\Gamma^k(p, p')$. Moreover, we must assume Γ^k to be a real quantity since its imaginary part is connected with a different kind of real scattering processes and the decay of excitations. For small frequencies and low temperatures the imaginary part is small and is basically determined by the second term in (42.33).

We shall now show that $\Gamma^k(p, p')$ is the same as the k-limit of the vertex part, introduced by us earlier in § 18. To do this, we note that if we wished to obtain the analogue of that quantity in the temperature-diagram technique, we should consider $\mathscr{T}(p, p'; k)$ for $\omega_m = 0$. When we analytically continue this quantity only the Γ_{ik} with $i, k \neq 2$ appear, which are the same as the quantities defined in § 41, if we put $\omega = 0$. Let us now construct the combinations (42.6)—(42.8). The cuts of g_2 disappear in the equation (42.15) for $\mathscr{L}_{ik}(\omega = 0)$ $(i, k \neq 2)$: indeed, only the term $\mathscr{L}^{(0)}_{ik}$ is involved with the cuts with $g_2(\varepsilon')$ and this term is proportional to $\tanh([\varepsilon' + \omega]/2T) - \tanh(\varepsilon'/2T)$ which vanishes when $\omega = 0$. Finally, for small frequencies ε and ε' all quantities $\Gamma_{ik}(\varepsilon, \varepsilon')$ are the same, since the magnitude of the discontinuities on the cuts becomes small.

Writing equation (42.30) in the form

$$-i[\omega - (\boldsymbol{k}\cdot\boldsymbol{v})]\,\varphi_{\boldsymbol{p}} - i\,\omega\,\frac{a^2 p_0^2}{\pi^2 v}\int\frac{\mathrm{d}\Omega'}{4\pi}\,\Gamma^k(p, p')\int \mathrm{d}\varepsilon'\,\frac{\mathrm{d}n_0}{\mathrm{d}\varepsilon'} - i\,\omega\,\tilde{V}_{\boldsymbol{p}}$$

$$= \frac{a^2 p_0^2}{2\pi^2 v}\int\frac{\mathrm{d}\Omega'}{4\pi}\int \mathrm{d}\varepsilon'\;\mathscr{L}'(p, p')\,\varphi_{\boldsymbol{p}'} - 2\gamma(\varepsilon)\,\varphi_{\boldsymbol{p}},$$

we see that its left-hand side is the same as (40.4b), since the relations found in § 18, especially (18.8), show that the function $g(p, p')$ is the same as $a^2\Gamma^k(p, p')$. As far as the right-hand side is concerned, we must bear in mind that the quantity $\mathscr{L}'(p, p'; k)$ in (42.33) is determined by the magnitude of the discontinuities Δ_1 and Δ_2, while for $\tilde{\mathscr{L}}$ the discontinuities are the same as those of the complete vertex part, being connected with the singularities of diagrams such as those of Figs. 114*b* and *c*. For Δ_1 and Δ_2 we have equation (41.10). Comparing this with (40.6) we see that the function $\delta n'_{\boldsymbol{p}}$ in (42.29) has, indeed, the meaning of the linear correction to the excitation distribution function. There now remains only for us to show that we can use that function to write down expressions for the particle, momentum, and energy fluxes which are of the same form as in the phenomenological theory of a Fermi liquid.

Let us consider as an example the particle flux, which, clearly, because of the isotropy of the system is connected only with the part f' of the non-equilibrium particle distribution function which is given by (42.28):

$$\boldsymbol{j}(\boldsymbol{k}, \omega) = \int\frac{\mathrm{d}^3\boldsymbol{p}}{(2\pi)^3}\,\frac{\boldsymbol{p}}{m}\,f'_{\boldsymbol{p}}(\boldsymbol{k}, \omega).$$

Substituting here (42.29) we get

$$\boldsymbol{j}(\boldsymbol{k}, \omega) = \frac{a}{m}\int\frac{\mathrm{d}^3\boldsymbol{p}}{(2\pi)^3}\,(Q\boldsymbol{p})\;\delta n'_{\boldsymbol{p}}(\boldsymbol{k}, \omega),$$

where

$$(Q\boldsymbol{p}) = \int\frac{\mathrm{d}^3\boldsymbol{p}'}{(2\pi)^3}\,\boldsymbol{p}'\,Q(\boldsymbol{p}', \boldsymbol{p}) = \int\frac{\mathrm{d}^3\boldsymbol{p}'}{(2\pi)^3}\,Q(\boldsymbol{p}, \boldsymbol{p}')\,\boldsymbol{p}'$$

$$= \boldsymbol{p} + \frac{1}{i}\int\frac{\mathrm{d}^4\boldsymbol{p}'}{(2\pi)^4}\,\mathscr{L}_{2l}(\boldsymbol{p}, \boldsymbol{p}'')\,g_l(p')\,\boldsymbol{p}'. \tag{42.34}$$

For further transformations we use a relation for the retarded Green function which is similar to equation (19.2):

$$\frac{\partial}{\partial\boldsymbol{p}}\,[G^R_\varepsilon(\boldsymbol{p})] = -\frac{\boldsymbol{p}}{m} - \frac{1}{i}\int\frac{\mathrm{d}^4p'}{(2\pi)^4}\,\mathscr{L}_{1l}(p, p')\,g_l(p')\,\frac{\boldsymbol{p}'}{m}. \tag{42.35}$$

It is also a consequence of the gauge invariance of the theory and could have been obtained either by a direct study of the perturbation theory series or after the analytical continuation of the appropriate formulae for $\mathfrak{G}_{\boldsymbol{p}}(\varepsilon_n)$.

Near the Fermi surface it follows from (42.34) that

$$(Q\boldsymbol{p}) = -\, m \left[\frac{\partial G_{\boldsymbol{p}}^{-1}}{\partial \boldsymbol{p}}\right].$$

Using the definition of the effective mass,

$$\frac{p_0}{m^*} = -\, a\, \frac{\partial G^{-1}}{\partial p},$$

we find that

$$aQ = \frac{m}{m^*}.$$

The particle current density is thus connected with $\delta n'_{\boldsymbol{p}}$ by the usual relation:

$$\boldsymbol{j}(\boldsymbol{k}, \omega) = \int \frac{d^3\boldsymbol{p}}{(2\pi)^3} \frac{\boldsymbol{p}}{m^*} \delta n'_{\boldsymbol{p}}(\boldsymbol{k}, \omega).$$

Let us now turn to that part of the particle distribution function which is connected with $K^{(0)}$ and which is determined by diagrams which do not contain the cuts of g_2. As we can put ω and k equal to zero in $K^{(0)}$, we have

$$K^{(0)} \equiv K(0) = \frac{1}{4\pi i} \int d\varepsilon \,[K_1(\varepsilon) - K_3(\varepsilon)] \tanh \frac{\varepsilon}{2T}.$$

One verifies easily by changing the summation over discrete frequencies to an integration, as we have done so often before, that $K^{(0)}$ is the value of the Fourier component of the function $\mathcal{K}(k, \tau)$ of (42.2) for $\omega = 0$. Therefore, $f^{(0)}$ is the quasi-equilibrium correction to the particle distribution function which appears when we take a constant external field into account in the Gibbs ensemble.

We must in conclusion once more emphasise that the whole derivation given here refers only to a linearised transport equation for the excitations.

REFERENCES

1. L. D. Landau and E. M. Lifshitz, *Statistical Physics*, Pergamon Press, 1958.
2. R. E. Peierls, *Quantum Theory of Solids*, Oxford University Press, 1955.
3. L. D. Landau, *Zh. Exp. Teor. Fiz.* **11**, 592 (1941); *J. Phys. U.S.S.R.* **5**, 71 (1940); *Collected Papers*, Gordon & Breach and Pergamon Press, 1965, Ch. 46, p. 301.
4. L. D. Landau, *J. Phys. U.S.S.R.* **11**, 91 (1947); *Collected Papers*, Gordon & Breach and Pergamon Press, 1965, Ch. 63, p. 466.
5. L. P. Pitaevskii, *Zh. Exp. Teor. Fiz.* **31**, 536 (1956); *Soviet Phys. JETP* **4**, 439 (1957).
6. R. P. Feynman, *Phys. Rev.* **94**, 262 (1954).
7. J. L. Yarnell, G. P. Arnold, P. J. Bendt and E. C. Kerr, *Phys. Rev.* **113**, 1379 (1959).
8. I. M. Khalatnikov, *Usp. Fiz. Nauk* **59**, 673 (1956); **60**, 69 (1956). (German transl. in *Fortschr. Phys.* **5**, 211, 287 (1956)).
9. E. M. Lifshitz, Appendix to the Russian translation of the book by W. Keesom (see 10) [English translation published by Consultants Bureau (1959)].
10. W. H. Keesom, *Helium*, Elsevier, Amsterdam (1942).
11. L. D. Landau, *Zh. Exp. Teor. Fiz.* **30**, 1058 (1956); *Soviet Phys. JETP* **3**, 920 (1957); *Collected Papers*, Gordon & Breach and Pergamon Press, 1965, Ch. 90, p. 723.
12. L. D. Landau, *Zh. Exp. Teor. Fiz.* **32**, 59 (1957); *Soviet Phys. JETP* **5**, 101 (1957); *Collected Papers*, Gordon & Breach and Pergamon Press, 1965, Ch. 91, p. 731.
13. D. F. Brewer, J. G. Daunt and A. K. Sreedhar, *Phys. Rev.* **115**, 836 (1959).
14. E. C. Kerr, *Phys. Rev.* **96**, 551 (1954).
15. A. A. Abrikosov and I. E. Dzyaloshinskii, *Zh. Exp. Teor. Fiz.* **35**, 771 (1958); *Soviet Phys. JETP* **8**, 535 (1959).
16. L. D. Landau and E. M. Lifshitz, *Quantum Mechanics*, Pergamon Press, 1958.
17. N. N. Bogolyubov, *J. Phys. U.S.S.R.* **11**, 23 (1947).
18. K. Huang, C. N. Yang and J. M. Luttinger, *Phys. Rev.* **105**, 776 (1957).
19. K. Brueckner and K. Sawada, *Phys. Rev.* **106**, 1117 (1957).
20. K. Huang and C. N. Yang, *Phys. Rev.* **105**, 767 (1957).
21. T. D. Lee and C. N. Yang, *Phys. Rev.* **105**, 1119 (1957).
22. A. A. Abrikosov and I. M. Khalatnikov, *Zh. Exp. Teor. Fiz.* **33**, 1154 (1957); *Soviet Phys. JETP* **6**, 888 (1958).
23. V. M. Galitskii, *Zh. Exp. Teor. Fiz.* **34**, 151 (1958); *Soviet Phys. JETP* **7**, 104 (1958).
24. V. A. Belyakov, *Zh. Exp. Teor. Fiz.* **40**, 1210 (1961); *Soviet Phys. JETP* **13**, 850 (1961).
25. A. I. Akhiezer and V. B. Berestetskii, *Quantum Electrodynamics*, Fizmatgiz (1959). (English translation published by Interscience, 1965.)
26. M. Gell-Mann and F. Low, *Phys. Rev.* **84**, 350 (1951).
27. V. M. Galitskii and A. B. Migdal, *Zh. Exp. Teor. Fiz.* **34**, 139 (1958); *Soviet Phys. JETP* **7**, 96 (1958).
28. H. Lehmann, *Nuovo Cimento* **11**, 342 (1954).

29. A. B. Migdal, *Zh. Exp. Teor. Fiz.* **32**, 399 (1957); *Soviet Phys. JETP* **5**, 333 (1957).
30. T. Matsubara, *Progr. Theor. Phys.* **14**, 351 (1955).
31. A. A. Abrikosov, L. P. Gor'kov and I. E. Dzyaloshinskii, *Zh. Exp. Teor. Fiz.* **36**, 900 (1959); *Soviet Phys. JETP* **9**, 636 (1959).
32. E. S. Fradkin, *Zh. Exp. Teor. Fiz.* **36**, 1286 (1959); *Soviet Phys. JETP* **9**, 912 (1959).
33. L. D. Landau, *Zh. Exp. Teor. Fiz.* **34**, 262 (1958); *Soviet Phys. JETP* **7**, 182 (1958); *Collected Papers*, Gordon & Breach and Pergamon Press, 1965, Ch. 94, p. 749.
34. N. N. Bogolyubov and S. V. Tyablikov, *Dokl. Akad. Nauk* **126**, 53 (1959); *Soviet Phys. Dokl.* **4**, 604 (1959); see also D. N. Zubarev, *Usp. Fiz. Nauk.* **71**, 71 (1960); *Soviet Phys. Uspekhi* **3**, 320 (1961).
35. L. D. Landau, *Zh. Exp. Teor. Fiz.* **35**, 97 (1958); *Soviet Phys. JETP* **8**, 70 (1959); *Collected Papers*, Gordon & Breach and Pergamon Press, 1965, Ch. 95, p. 752.
36. L. P. Pitaevskii, *Zh. Exp. Teor. Fiz.* **37**, 1794 (1959); *Soviet Phys. JETP* **10**, 1267 (1960).
37. J. M. Luttinger and J. C. Ward, *Phys. Rev.* **118**, 1417 (1960), and J. M. Luttinger, *Phys. Rev.* **119**, 1153 (1960).
38. A. B. Migdal, *Zh. Exp. Teor. Fiz.* **34**, 1438 (1958); *Soviet Phys. JETP* **7**, 996 (1958).
39. M. Gell-Mann and K. A. Brueckner, *Phys. Rev.* **106**, 364 (1957).
40. A. A. Vedenov, *Zh. Exp. Teor. Fiz.* **36**, 641 (1959); *Soviet Phys. JETP* **9**, 446 (1959).
41. S. T. Belyaev, *Zh. Exp. Teor. Fiz.* **34**, 417 (1958); *Soviet Phys. JETP* **7**, 289 (1958).
42. S. T. Belyaev, *Zh. Exp. Teor. Fiz.* **34**, 433 (1958); *Soviet Phys. JETP* **7**, 299 (1958).
43. N. M. Hugenholtz and D. Pines, *Phys. Rev.* **116**, 489 (1959).
44. L. P. Pitaevskii, *Zh. Exp. Teor. Fiz.* **36**, 1168 (1959); *Soviet Phys. JETP*, **9**, 830 (1959).
45. L. P. Pitaevskii, *Zh. Exp. Teor. Fiz.* **39**, 216 (1960); *Soviet Phys. JETP* **12**, 155 (1961).
46. D. G. Henshaw, A. D. B. Woods and B. N. Brockhouse, *Bull. Amer. Phys. Soc.* **5**, No. 1, 12, C 3 (1960).
47. L. D. Landau and E. M. Lifshitz, *The Electrodynamics of Continuous Media*, Pergamon Press, 1960.
48. I. E. Dzyaloshinskii and L. P. Pitaevskii, *Zh. Exp. Teor. Fiz.* **36**, 1797 (1959); *Soviet Phys. JETP* **9**, 1282 (1959).
49. M. I. Ryazanov, *Zh. Exp. Teor. Fiz.* **32**, 1244 (1957); *Soviet Phys. JETP* **5**, 1013 (1957).
50. I. E. Dzyaloshinskii, E. M. Lifshitz and L. P. Pitaevskii, *Zh. Exp. Teor. Fiz.* **37**, 229 (1959); *Soviet Phys. JETP* **10**, 161 (1960).
51. E. M. Lifshitz, *Zh. Exp. Teor. Fiz.* **29**, 94 (1955); *Soviet Phys. JETP* **2**, 73 (1956).
52. R. Eisenschitz and F. London, *Zs. f. Physik* **60**, 491 (1930).
53. H. B. C. Casimir and D. Polder, *Phys. Rev.* **73**, 360 (1948).
54. L. P. Pitaevskii, *Zh. Exp. Teor. Fiz.* **37**, 577 (1959); *Soviet Phys. JETP* **10**, 408 (1960).
55. E. Maxwell, *Phys. Rev.* **78**, 447 (1959); **79**, 173 (1950); C. A. Reynolds B. Serin, W. H. Wright and L. B. Nesbitt, *Phys. Rev.* **78**, 487 (1950).
56. H. Fröhlich, *Phys. Rev.* **79**, 845 (1950).
57. L. N. Cooper, *Phys. Rev.* **104**, 1189 (1957).

58. J. Bardeen, L. N. Cooper and J. R. Schrieffer, *Phys. Rev.* **108**, 1175 (1957).
59. N. N. Bogolyubov, *Zh. Exp. Teor. Fiz.* **34**, 58, 73 (1958); *Soviet Phys. JETP* **7**, 41, 51 (1958).
60. N. N. Bogolyubov, V. V. Tolmachev and D. V. Shirkov, *A New Method in the Theory of Superconductivity*, Izd-vo Akad. Nauk SSSR (1958). (English transl. publ. by Consultants Bureau.)
61. L. P. Gor'kov, *Zh. Exp. Teor. Fiz.* **34**, 735 (1958); *Soviet Phys. JETP* **7**, 505 (1958).
62. G. M. Eliashberg, *Zh. Exp. Teor. Fiz.* **38**, 966 (1960); *Soviet Phys. JETP* **11**, 696 (1960).
63. A. A. Abrikosov and I. M. Khalatnikov, *Usp. Fiz. Nauk.* **65**, 551 (1958); *Adv. Phys.* **8**, 45 (1959).
64. H. London and F. London, *Proc. Roy. Soc.* A **149**, 71 (1935); *Physica* **2**, 341 (1935).
65. A. B. Pippard, *Proc. Roy. Soc.* A **216**, 547 (1953).
66. D. C. Mattis and J. Bardeen, *Phys. Rev.* **111**, 412 (1958).
67. A. A. Abrikosov, L. P. Gor'kov and I. M. Khalatnikov, *Zh. Exp. Teor. Fiz.* **35**, 265 (1958); **37**, 187 (1959); *Soviet Phys. JETP* 8, 182 (1959); **10**, 132 (1960).
68. L. P. Gor'kov, *Zh. Exp. Teor. Fiz.* **36**, 1918 (1959); *Soviet Phys. JETP* **9**, 1364 (1959).
69. V. L. Ginzburg and L. D. Landau, *Zh. Exp. Teor. Fiz.* **20**, 1064 (1958), (English transl. publ. in *Landau's Collected Papers*, Gordon & Breach and Pergamon Press, 1965, Ch. 73, p. 546.)
70. A. A. Abrikosov and L. P. Gor'kov, *Zh. Exp. Teor. Fiz.* **35**, 1558 (1958); **36**, 319 (1959); *Soviet Phys. JETP* 8, 1090 (1959); **9**, 220 (1959).
71. S. F. Edwards, *Phil. Mag.* **3**, 1020 (1958).
72. M. Cohen and R. P. Feynman, *Phys. Rev.* **107**, 13 (1957).
73. P. Nozières and D. Pines, *Phys. Rev.* **109**, 741 (1958).
74. G. M. Eliashberg, *Zh. Exp. Teor. Fiz.* **42**, 1658 (1962); *Soviet Phys. JETP* **15**, 1151 (1962).
75. G. M. Eliashberg, *Zh. Exp. Teor. Fiz.* **41**, 1241 (1961); *Soviet Phys. JETP* **14**, 886 (1962).
76. G. M. Eliashberg, *Zh. Exp. Teor. Fiz.* **43**, 1105 (1962); *Soviet Phys. JETP* **16**, 1000 (1963).
77. A. A. Abrikosov, *Zh. Exp. Teor. Fiz.* **41**, 569 (1961); *Soviet Phys. JETP* **14**, 408 (1962).
78. T. Tsuneto, Technical Report of the Institute of Solids State Physics, University of Tokyo, No. A 47 (1962).
79. P. C. Hohenberg, reprint.
80. A. S. Davydov, *Quantum Mechanics*, Pergamon Press, 1965.

INDEX

Made in Great Britain